Lecture Notes in Computer Science 14437

Founding Editors

Gerhard Goos
Juris Hartmanis

Editorial Board Members

The series Lecture Notes in Computer Science (LNCS), including its subseries Lecture Notes in Artificial Intelligence (LNAI) and Lecture Notes in Bioinformatics (LNBI), has established itself as a medium for the publication of new developments in computer science and information technology research, teaching, and education.

LNCS enjoys close cooperation with the computer science R & D community, the series counts many renowned academics among its volume editors and paper authors, and collaborates with prestigious societies. Its mission is to serve this international community by providing an invaluable service, mainly focused on the publication of conference and workshop proceedings and postproceedings. LNCS commenced publication in 1973.

Qingshan Liu · Hanzi Wang · Zhanyu Ma ·
Weishi Zheng · Hongbin Zha · Xilin Chen ·
Liang Wang · Rongrong Ji
Editors

Pattern Recognition and Computer Vision

6th Chinese Conference, PRCV 2023
Xiamen, China, October 13–15, 2023
Proceedings, Part XIII

Springer

Editors
Qingshan Liu (iD)
Nanjing University of Information Science
and Technology
Nanjing, China

Zhanyu Ma (iD)
Beijing University of Posts
and Telecommunications
Beijing, China

Hongbin Zha (iD)
Peking University
Beijing, China

Liang Wang
Chinese Academy of Sciences
Beijing, China

Hanzi Wang (iD)
Xiamen University
Xiamen, China

Weishi Zheng (iD)
Sun Yat-sen University
Guangzhou, China

Xilin Chen (iD)
Chinese Academy of Sciences
Beijing, China

Rongrong Ji (iD)
Xiamen University
Xiamen, China

ISSN 0302-9743   ISSN 1611-3349 (electronic)
Lecture Notes in Computer Science
ISBN 978-981-99-8557-9   ISBN 978-981-99-8558-6 (eBook)
https://doi.org/10.1007/978-981-99-8558-6

This Springer imprint is published by the registered company Springer Nature Singapore Pte Ltd.
The registered company address is: 152 Beach Road, #21-01/04 Gateway East, Singapore 189721, Singapore

Paper in this product is recyclable.

Preface

Welcome to the proceedings of the Sixth Chinese Conference on Pattern Recognition and Computer Vision (PRCV 2023), held in Xiamen, China.

PRCV is formed from the combination of two distinguished conferences: CCPR (Chinese Conference on Pattern Recognition) and CCCV (Chinese Conference on Computer Vision). Both have consistently been the top-tier conference in the fields of pattern recognition and computer vision within China's academic field. Recognizing the intertwined nature of these disciplines and their overlapping communities, the union into PRCV aims to reinforce the prominence of the Chinese academic sector in these foundational areas of artificial intelligence and enhance academic exchanges. Accordingly, PRCV is jointly sponsored by China's leading academic institutions: the Chinese Association for Artificial Intelligence (CAAI), the China Computer Federation (CCF), the Chinese Association of Automation (CAA), and the China Society of Image and Graphics (CSIG).

PRCV's mission is to serve as a comprehensive platform for dialogues among researchers from both academia and industry. While its primary focus is to encourage academic exchange, it also places emphasis on fostering ties between academia and industry. With the objective of keeping abreast of leading academic innovations and showcasing the most recent research breakthroughs, pioneering thoughts, and advanced techniques in pattern recognition and computer vision, esteemed international and domestic experts have been invited to present keynote speeches, introducing the most recent developments in these fields.

PRCV 2023 was hosted by Xiamen University. From our call for papers, we received 1420 full submissions. Each paper underwent rigorous reviews by at least three experts, either from our dedicated Program Committee or from other qualified researchers in the field. After thorough evaluations, 522 papers were selected for the conference, comprising 32 oral presentations and 490 posters, giving an acceptance rate of 37.46%. The proceedings of PRCV 2023 are proudly published by Springer.

Our heartfelt gratitude goes out to our keynote speakers: Zongben Xu from Xi'an Jiaotong University, Yanning Zhang of Northwestern Polytechnical University, Shutao Li of Hunan University, Shi-Min Hu of Tsinghua University, and Tiejun Huang from Peking University.

We give sincere appreciation to all the authors of submitted papers, the members of the Program Committee, the reviewers, and the Organizing Committee. Their combined efforts have been instrumental in the success of this conference. A special acknowledgment goes to our sponsors and the organizers of various special forums; their support made the conference a success. We also express our thanks to Springer for taking on the publication and to the staff of Springer Asia for their meticulous coordination efforts.

We hope these proceedings will be both enlightening and enjoyable for all readers.

October 2023

Qingshan Liu
Hanzi Wang
Zhanyu Ma
Weishi Zheng
Hongbin Zha
Xilin Chen
Liang Wang
Rongrong Ji

Organization

General Chairs

Hongbin Zha — Peking University, China

Xilin Chen — Institute of Computing Technology, Chinese Academy of Sciences, China

Liang Wang — Institute of Automation, Chinese Academy of Sciences, China

Rongrong Ji — Xiamen University, China

Program Chairs

Qingshan Liu — Nanjing University of Information Science and Technology, China

Hanzi Wang — Xiamen University, China

Zhanyu Ma — Beijing University of Posts and Telecommunications, China

Weishi Zheng — Sun Yat-sen University, China

Organizing Committee Chairs

Mingming Cheng — Nankai University, China

Cheng Wang — Xiamen University, China

Yue Gao — Tsinghua University, China

Mingliang Xu — Zhengzhou University, China

Liujuan Cao — Xiamen University, China

Publicity Chairs

Yanyun Qu — Xiamen University, China

Wei Jia — Hefei University of Technology, China

Local Arrangement Chairs

Xiaoshuai Sun   Xiamen University, China
Yan Yan   Xiamen University, China
Longbiao Chen   Xiamen University, China

International Liaison Chairs

Jingyi Yu   ShanghaiTech University, China
Jiwen Lu   Tsinghua University, China

Tutorial Chairs

Xi Li   Zhejiang University, China
Wangmeng Zuo   Harbin Institute of Technology, China
Jie Chen   Peking University, China

Thematic Forum Chairs

Xiaopeng Hong   Harbin Institute of Technology, China
Zhaoxiang Zhang   Institute of Automation, Chinese Academy of
                             Sciences, China
Xinghao Ding   Xiamen University, China

Doctoral Forum Chairs

Shengping Zhang   Harbin Institute of Technology, China
Zhou Zhao   Zhejiang University, China

Publication Chair

Chenglu Wen   Xiamen University, China

Sponsorship Chair

Yiyi Zhou   Xiamen University, China

Exhibition Chairs

Bineng Zhong   Guangxi Normal University, China
Rushi Lan   Guilin University of Electronic Technology, China
Zhiming Luo   Xiamen University, China

Program Committee

Baiying Lei   Shenzhen University, China
Changxin Gao   Huazhong University of Science and Technology,
                                China
Chen Gong   Nanjing University of Science and Technology,
                                China
Chuanxian Ren   Sun Yat-Sen University, China
Dong Liu   University of Science and Technology of China,
                                China
Dong Wang   Dalian University of Technology, China
Haimiao Hu   Beihang University, China
Hang Su   Tsinghua University, China
Hui Yuan   School of Control Science and Engineering,
                                Shandong University, China
Jie Qin   Nanjing University of Aeronautics and
                                Astronautics, China
Jufeng Yang   Nankai University, China
Lifang Wu   Beijing University of Technology, China
Linlin Shen   Shenzhen University, China
Nannan Wang   Xidian University, China
Qianqian Xu   Key Laboratory of Intelligent Information
                                Processing, Institute of Computing
                                Technology, Chinese Academy of Sciences,
                                China
Quan Zhou   Nanjing University of Posts and
                                Telecommunications, China
Si Liu   Beihang University, China
Xi Li   Zhejiang University, China
Xiaojun Wu   Jiangnan University, China
Zhenyu He   Harbin Institute of Technology (Shenzhen), China
Zhonghong Ou   Beijing University of Posts and
                                Telecommunications, China

Contents – Part XIII

Medical Image Processing and Analysis

Growth Simulation Network for Polyp Segmentation

Hongbin Wei[1], Xiaoqi Zhao[1], Long Lv[2,3], Lihe Zhang[1(✉)], Weibing Sun[2,3], and Huchuan Lu[1]

[1] Dalian University of Technology, Dalian, China
{weihongbin,zxq}@mail.dlut.edu.cn, {zhanglihe,lhchuan}@dlut.edu.cn
[2] Affiliated Zhongshan Hospital of Dalian University, Dalian, China
lvlong113@126.com, weibingsun_dyfemw@163.com
[3] Key Laboratory of Microenvironment Regulation and Immunotherapy of Urinary Tumors in Liaoning Province, Dalian, China

Abstract. Colonoscopy is a gold standard, while automated polyp segmentation can minimize missed rates and timely treatment of colon cancer at an early stage. But most existing polyp segmentation methods have borrowed techniques related to image semantic segmentation, and the main idea is to extract and fuse feature information of images more effectively. As we know, polyps naturally grow from small to large, thus they have strong rules. In view of this trait, we propose a Growth Simulation Network (GSNet) to segment polyps from colonoscopy images. First, the completeness map (i.e., ground-truth mask) is decoupled to generate Gaussian map and body map. Among them, Gaussian map is mainly used to locate polyps, while body map expresses the intermediate stages, which helps filter redundant information. GSNet has three forward branches, which are supervised by Gaussian map, body map and completeness map, respectively. What's more, we design a dynamic attention guidance (DAG) module to effectively fuse the information from different branches. Extensive experiments on five benchmark datasets demonstrate that our GSNet performs favorably against most state-of-the-art methods under different evaluation metrics. The source code will be publicly available at https://github.com/wei-hongbin/GSNet

Keywords: Colorectal cancer · Automatic polyp segmentation · Growth simulation · Dynamic attention guidance

1 Introduction

Colorectal cancer (CRC) is the third most common cancer worldwide, and it is considered to be the second most deadly cancer, accounting for 9.4% of all cancer deaths [18]. It usually develops from small, noncancerous clumps of cells called polyps in the colon. Hence, the best way to prevent CRC is to accurately identify and remove polyps in advance. Currently, colonoscopy is the primary method for prevention of colon cancer [19], but the rate of missed polyps is

Q. Liu et al. (Eds.): PRCV 2023, LNCS 14437, pp. 3–15, 2024.
https://doi.org/10.1007/978-981-99-8558-6_1

high because of the large variation (shape, size and texture) of polyps and the high dependence on professional doctors. Clinical studies have shown that the rate of missing colon polyps during endoscopy may range from 20% to 47% [1]. Therefore, an accurate and efficient approach of automatically identifying and segmenting polyps is highly demanded to improve the outcome of colonoscopy.

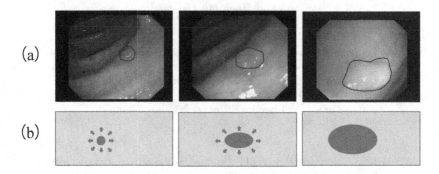

Fig. 1. Polyps in different stages. (a) From left to right, they indicate anterior, middle and late polyps, respectively. (b) Diagram of natural polyp growth.

Earlier automatic polyp segmentation models [22] mainly rely on hand-crafted features (e.g., color and texture), which are difficult to capture the global context information and have poor robustness. Recently, polyp segmentation algorithms based on deep learning have made good progress. For example, UNet [17], UNet++ [34] and ResUNet++ [9] have achieved higher accuracy than traditional methods. However, these methods only provide a better treatment of the polyp region but easily lead to a blurred boundary. To improve the boundary accuracy, PraNet [7] uses the reverse attention module to establish the relationship between region and boundary clues to obtain a more ideal boundary. To further improve model performance, many models consider targeted treatment of different levels of encoder features because there is a large semantic gap between high-level and low-level features. Polyp-PVT [3] obtains semantic and location information from high-level features and detail information from low-level features, and finally fuses them to improve the expression of features. Furthermore, MSNet [29,33] removes the redundant information in the every levels through multi-level subtraction, and effectively obtains the complementary information from lower order to higher order between different levels, so as to comprehensively enhance the perception ability.

Many above methods draw lessons from the general idea of semantic segmentation [13–15,30–32], that is, extracting and fusing feature information effectively. However, they are not targeted to the task of polyp segmentation. Polyps grow naturally from small to large, as shown in Fig. 1, so this scene has strong laws. Compared with other scenes (e.g., animals and plants) of nature, the regions of polyps are internally continuous, and their boundaries are blurred, without sharp edges.

In the Encoder-Decoder architecture, there are some similarities between the learning process of decoder and the natural growth process of polyps, which progress from simple to complex and from rough to fine. Inspired by this, we propose a novel growth simulation network (GSNet) to achieve polyp segmentation. First, the completeness map is decoupled to generate Gaussian map and body map, as shown in Fig. 2. Compared to the completeness map, the Gaussian map mainly characterizes the location cues and the body map describes more about the intermediate stages of polyp growth. Then, three feature learning branches are built, which are Gaussian branch, body branch and prediction branch. The Gaussian branch only uses high-level features to learn location information. The body branch uses more features and incorporates Gaussian branch features to learn main body representation. The prediction branch uses all-level features and body branch features to form the final prediction. The differentiated utilization of features in different branches can avoid introducing redundant information.

The information of the previous branch is used by the later branch. In order to effectively fuse it, we design a dynamic attention guidance (DAG) module, which forms the dynamic convolution kernel from the supervised feature of the previous branch to guide the feature learning of the later branch. Compared with conventional convolution, which takes the same parameters for all inputs, dynamic convolution can automatically change the convolution parameters and perform well when the model has strong correlation with input data. CondConv [28] and DyConv [4] are the earliest works about dynamic convolution, which are both implemented by generating different weights according to the input data, and then weighting and summing multiple convolution kernels. The above works all deal with one input. HDFNet [12] exploits dynamic convolution to fuse two strongly correlated features. It generates a convolution kernel based on one input and then performs convolution operation on the other input. Inspired by HDFNet, our DAG module makes substantial improvements. We first simplifies the generation process of dynamic convolution kernels, which greatly reduces the number of parameters and computational effort. Meanwhile, we borrow the non-local [25] idea and utilize dynamic convolution to generate the attention, thereby achieving information fusion of two input features.

In order to better restore its natural growth state, we further propose a Dynamic Simulation Loss (DSLoss), which makes the weight of each branch in the loss function change dynamically in different stages of training. Our main contributions can be summarized as follows:

- According to the characteristics of polyp images, we propose a novel growth simulation network, which is cascaded from Gaussian map to body map and finally forms a segmentation map. Meanwhile, the network avoids introducing redundant information and the training process is more smooth.
- To effectively fuse different branch features, we propose a dynamic attention guidance module, which can dynamically use features from the previous level to guide the feature generation of the next level.

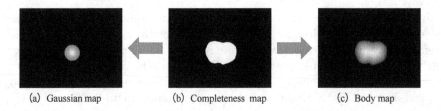

(a) Gaussian map   (b) Completeness map   (c) Body map

Fig. 2. Supervisions at different stages. Both Gaussian map and body map are obtained by decoupling completeness map.

- The GSNet can accurately segment polyps. Extensive experiments demonstrate that our GSNet achieves the state-of-the-art performance under different evaluation metrics on five challenging datasets.

2   The Proposed Method

2.1   Gaussian Map and Body Map

To eliminate redundant information and enable each branch to have a different focus, we decouple the completeness map to generate Gaussian map and body map, as shown in Fig. 2. Firstly, we introduce the generation of Gaussian map. Let I be a binary GT map, which can be divided into two parts: foreground I_{fg} and background I_{bg}. For each pixel p, $I(p)$ is its label value. If $p \in I_{fg}$, $I(p)$ equals 1, and 0 if $p \in I_{bg}$. All the foreground pixels are critically enclosed by a rectangular box. We define a transformation function $I'(p)$ to measure the distance between pixel p and center point p_c of the rectangle.

$$I'(p) = e^{\frac{-0.5 \times \|p - p_c\|_2}{R}}, \tag{1}$$

where $L2$-norm denotes the Euclidean distance between two pixels, and R denotes the minimum side length of the rectangle. We normalize the values I' of all pixels using a simple linear function:

$$\bar{I}' = \frac{I' - min(I')}{max(I') - min(I')}. \tag{2}$$

Thus, the Gaussian map is obtained by simple thresholding as follows:

$$O_{Gaussian} = \begin{cases} \bar{I}'(p), & \bar{I}'(p) > 0.5, \\ 0, & \bar{I}'(p) \leq 0.5. \end{cases} \tag{3}$$

For the body map, the transformation process refers to LDFNet [27]. We firstly compute the Euclidean distance $f(p, q)$ between pixels p and q. If pixel p belongs to the foreground, the function will first look up its nearest pixel q in the background and then use $f(p, q)$ to calculate the distance. If pixel p belongs to the background, their minimum distance is set to zero. The transformation can be formulated as:

$$I^*(p) = \begin{cases} \min_{q \in I_{bg}} f(p,q), & p \in I_{fg}, \\ 0, & p \in I_{bg}. \end{cases} \tag{4}$$

Then, the values I^* of all pixels are normalized by formula (2). To remove the interference of background, the body map is obtained as:

$$O_{body} = I \odot \bar{I}^*, \tag{5}$$

where $\odot$ denotes element-wise product.

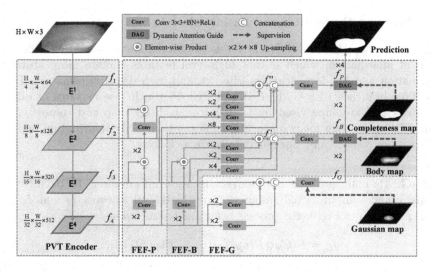

Fig. 3. The overall architecture of GSNet, which consists of a PVT encoder, features extraction and fusion (FEF) module including three branches FEF-G, FEF-B and FEF-P, and dynamic attention guidance (DAG) module for integrating and guiding features of different branches.

2.2 Overall Architecture

As shown in Fig. 3, the proposed GSNet consists of three key modules: pyramid vision transformer (PVT) encoder, feature extraction and fusion (FEF) module and dynamic attention guidance (DAG) module. Specifically, the PVT is used to extract multi-scale long-range dependencies features from the input image. The FEF is employed to collect the semantic cues through aggregating features in a progressive manner, which contains three different branches FEF-G, FEF-B and FEF-P. The DAG is adopted to fuse the features provided by the different FEF branches, effectively using simpler features to dynamically guide the generation of more complex features.

We adopt the PVTv2 [24] as the encoder to extract pyramid features and adjust their channel number to 32 through four convolutional units to obtain $f_i, i \in \{1, 2, 3, 4\}$, which are then fed to FEF. First, FEF-G fuses high-level features (f_3, f_4) and is supervised by Gaussian map, which mainly focuses on location. Second, FEF-B fuses more features (f_2, f_3, f_4) and is guided by the FEF-G to generate the body map, which already contains main body information. Finally, FEF-P fuses all features (f_1, f_2, f_3, f_4) and is guided by FEF-B to generate the final prediction. In particular, the DAG plays a role of guidance for full information fusion of multiple segmentation cues. During training, we optimize the model with a dynamic simulation loss, which makes the weight of each branch loss change adaptively along iteration step.

2.3   Features Extraction and Fusion Module

It is well known that the different levels of encoder features contain different types of information, which is that high-level features contain more abstract information and low-level features often contain rich detail information. FEF has three branches FEF-G, FEF-B and FEF-P, which fuse features of different levels and are supervised by Gaussian map, body map and completeness map. Therefore, they focus on location information, body information and all information, respectively. Specific details are as follows.

FEF-G. Its main purpose is to extract location information. To balance the accuracy and computational resources, only high-level features f_3 and f_4 are used. As shown in Fig. 3, the process is formulated as follows:

$$f_G = F(Cat(F(up(f_4)) \odot f_3, F(up(f_4)))), \tag{6}$$

where $\odot$ denotes the element-wise product, $up(\cdot)$ indicates up-sampling, and $Cat(\cdot)$ is the concatenation operation. $F(\cdot)$ is defined as a convolutional unit composed of a 3×3 convolutional layer with padding of 1, batch normalization and ReLU, and it is parameter independent.

FEF-B. Its main purpose is to extract body information. It combines high-level features $(f_3$ and $f_4)$ and mid-level feature f_2. First, these features are preliminarily aggregated to obtain a rich semantic representation as

$$f' = F(F(f_4) \odot f_3) \odot f_2. \tag{7}$$

Then, DAG utilizes f_G to guide further information combination,

$$f_B = DAG(f_G, F(Cat(F(up(f_4)), F(up(f_3)), f'))). \tag{8}$$

In this process, the features f_3 and f_4 are up-sampled and processed through convolution operation, which has the effect of residual learning.

FEF-P. It generates final prediction by using all level features (f_1, f_2, f_3 and f_4). Among them, low-level features are important to reconstruct details. Similar to FEF-B, the computation process can be written as

$$f'' = F(F(F(f_4) \odot f_3) \odot f_2) \odot f_1. \tag{9}$$

Then, DAG utilizes f_B to guide the combination of the above features,

$$f_P = DAG(f_B, F(Cat(F(f_4), F(f_3), F(f_2), f''))). \tag{10}$$

Finally, f_G, f_B and f_P are up-sampled to the same size as the input image and supervised by three different maps.

2.4   Dynamic Attention Guidance Module

The proposed network generates predictions in sequence from Gaussian map to body map and finally to prediction map. They are closely related. The former is the basis of the latter and the latter is an extension of the former. Since dynamic convolution can automatically change the convolution parameters according to inputs, we propose a dynamic attention guidance (DAG) module combining dynamic convolutions and attention mechanisms, which can leverage the simpler feature map to guide the blended features to produce a more complete feature.

Given feature map f_a with guidance information and f_b with rich semantic information, we fuse them through DAG. First, they pass a linear mapping process of 1×1 convolution operation to generate f_{a_1}, f_{b_1} and f_{b_2}, respectively. Second, we use kernel generation unit (KGU) [12] to yield dynamic kernels based on f_{a_1}. Then, f_{b_1} is adaptively filtered to obtain guidance feature f_{ab},

$$f_{ab} = KGU(f_{a_1}) \otimes f_{b_1}, \tag{11}$$

where $KGU(\cdot)$ denote the operations of the KGU module. $\otimes$ is an adaptive convolution operation. Finally, the feature combination is achieved as follows:

$$f_{out} = f_b \oplus (softmax(f_{ab}) \odot f_{b_2}), \tag{12}$$

where $\oplus$ and $\odot$ denote the element-wise addition and multiplication, respectively (Fig. 4).

2.5   Dynamic Simulation Loss

The total training loss function can be formulated as follows:

$$L_{total} = W_g * L_g + W_b * L_b + W_p * L_p, \tag{13}$$

where L_g, L_b and L_p separately represent loss of Gaussian branch, body branch and prediction branch. $L_g = L_b = L_{BCE}^w$ is standard binary cross entropy (BCE) loss, while $L_p = L_{IoU}^w + L_{BCE}^w$ is the weighted intersection over union (IoU) loss

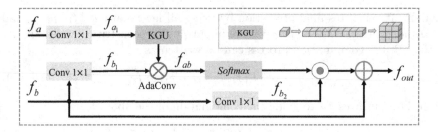

Fig. 4. Details of the DAG. KGU generates adaptive kernel tensors by adjusting convolutional channel, and then reshapes them to the regular form of convolution kernel.

and weighted BCE loss, which has been widely adopted in segmentation tasks [16,26]. W_g, W_b and W_p are the weighting coefficients of three branch losses, respectively. The expression is as follows:

$$W_g = e^{\frac{-x}{(n/2)}}, W_b = 1/(e^{\frac{-x}{(n/5)}} + 1), W_p = 1, \tag{14}$$

Table 1. Quantitative comparison. ↑ and ↓ indicate that the larger and smaller scores are better, respectively. The best results are shown in **bolded font**.

Datasets	Methods	$mDice$ ↑	$mIoU$ ↑	F_β^ω ↑	S_α ↑	E_ϕ^{max} ↑	MAE ↓
ColonDB	U-Net(MICCAI'15)	0.512	0.444	0.498	0.712	0.696	0.061
	U-Net++(DLMIA'18)	0.483	0.410	0.467	0.691	0.680	0.064
	PraNet(MICCAI'20)	0.712	0.640	0.699	0.820	0.847	0.043
	SANet(MICCAI'21)	0.753	0.670	0.726	0.837	0.869	0.043
	Polyp-PVT(CAAI AIR'23)	0.808	0.727	0.795	0.865	0.913	0.031
	GSNet(Ours)	**0.822**	**0.746**	**0.805**	**0.874**	**0.922**	**0.026**
ETIS	U-Net(MICCAI'15)	0.398	0.335	0.366	0.684	0.643	0.036
	U-Net++(DLMIA'18)	0.401	0.344	0.390	0.683	0.629	0.035
	PraNet(MICCAI'20)	0.628	0.567	0.600	0.794	0.808	0.031
	SANet(MICCAI'21)	0.750	0.654	0.685	0.849	0.881	0.015
	Polyp-PVT(CAAI AIR'23)	0.787	0.706	0.750	0.871	**0.906**	**0.013**
	GSNet(Ours)	**0.802**	**0.724**	**0.763**	**0.871**	0.906	**0.013**
Kvasir	U-Net(MICCAI'15)	0.818	0.746	0.794	0.858	0.881	0.055
	U-Net++(DLMIA'18)	0.821	0.743	0.808	0.862	0.886	0.048
	PraNet(MICCAI'20)	0.898	0.840	0.885	0.915	0.944	0.030
	SANet(MICCAI'21)	0.904	0.847	0.892	0.915	0.949	0.028
	Polyp-PVT(CAAI AIR'23)	0.917	0.864	0.911	0.925	0.956	0.023
	GSNet(Ours)	**0.930**	**0.883**	**0.923**	**0.934**	**0.968**	**0.020**
CVC-T	U-Net(MICCAI'15)	0.710	0.672	0.684	0.843	0.847	0.022
	U-Net++(DLMIA'18)	0.707	0.624	0.687	0.839	0.834	0.018
	PraNet(MICCAI'20)	0.871	0.797	0.843	0.925	0.950	0.010
	SANet(MICCAI'21)	0.888	0.815	0.859	0.928	0.962	0.008
	Polyp-PVT(CAAI AIR'23)	0.900	0.833	0.884	0.935	**0.973**	**0.007**
	GSNet(Ours)	**0.909**	**0.844**	**0.889**	**0.939**	0.972	**0.007**
ClinicDB	U-Net(MICCAI'15)	0.823	0.755	0.811	0.889	0.913	0.019
	U-Net++(DLMIA'18)	0.794	0.729	0.785	0.873	0.891	0.022
	PraNet(MICCAI'20)	0.899	0.849	0.896	0.936	0.963	0.009
	SANet(MICCAI'21)	0.916	0.859	0.909	0.939	0.971	0.012
	Polyp-PVT(CAAI AIR'23)	0.937	0.889	0.936	0.949	**0.985**	**0.006**
	GSNet(Ours)	**0.942**	**0.898**	**0.937**	**0.950**	0.984	**0.006**

where n represents the number of all epochs. The change is consistent with the growth law of polyp growth, that is, at the beginning, there is only the position information of small polyps. With the continuous growth of polyps, the position information remains basically stable, other information will be more important. Because the change of weight is very small when it is subdivided into each epoch, it will not affect the optimization of network back propagation.

3   Experiments

3.1   Settings

Datasets and Evaluation Metrics. Following the experimental setups in PraNet, we evaluate the proposed model on five benchmark datasets: ClinicDB [2], Kvasir [8], ColonDB [21], CVC-T [23] and ETIS [20]. To keep the fairness of the experiments, we adopt the same training set with PraNet, that is, 550 samples from the ClinicDB and 900 samples from the Kvasir are used for training. The remaining images and other three datasets are used for testing.

We adopt six widely-used metrics for quantitative evaluation: mean Dice, mean IoU, the weighted F-measure (F_β^ω) [11], mean absolute error (MAE), the recently released S-measure S_α [5] and E-measure(E_ϕ^{max}) [6] scores. The lower value is better for the MAE and the higher is better for others.

Implementation Details. Our model is implemented based on the PyTorch framework and trained on a single V100 GPU for 100 epochs with mini-batch size 16. We resize the inputs to 352 × 352 and employ a general multi-scale training strategy as the Polyp-PVT. For the optimizer, we adopt the AdamW [10], which is widely used in transformer networks. The learning rate and the weight decay all are adjusted to 1e−4. For testing, we resize the images to 352 × 352 without any post-processing optimization strategies.

3.2   Comparisons with State-of-the-art

To prove the effectiveness of the proposed GSNet, as shown in Table 1, five state-of-the-art models are used for comparison. On the five challenging datasets, our GSNet achieves the best performance against other methods. We also demonstrate qualitative results of different methods in Fig. 5.

3.3   Ablation Study

We take the PVTv2 as the baseline to analyze the contribution of each component. The results are shown in Table 2.

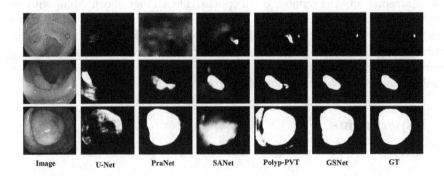

Image   U-Net   PraNet   SANet   Polyp-PVT   GSNet   GT

Fig. 5. Visual comparison of different methods.

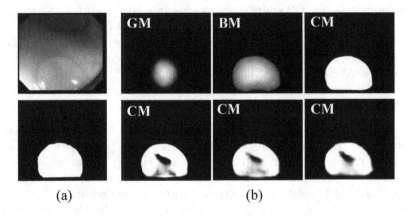

(a)   (b)

Fig. 6. Illustration of the benefits of using three decoupled maps. (a) Image and ground truth. (b) Visual comparison of different branches. The first row is the results of supervision by Gaussian, body and completeness maps, and the second row shows the result supervised by three completeness maps.

Table 2. Ablation study.

Bas.	FEF-P	FEF-B	FEF-G	DAG		ETIS				Kvasir		
					mDice	mIoU	F_β^ω	E_ϕ^{max}	mDice	mIoU	F_β^ω	E_ϕ^{max}
$\checkmark$					0.725	0.650	0.680	0.838	0.905	0.854	0.893	0.955
$\checkmark$	$\checkmark$				0.752	0.679	0.707	0.850	0.911	0.960	0.896	0.956
$\checkmark$	$\checkmark$	$\checkmark$		$\checkmark$	0.774	0.700	0.739	0.890	0.922	0.875	0.912	0.958
$\checkmark$	$\checkmark$	$\checkmark$	$\checkmark$		0.768	0.689	0.721	0.874	0.919	0.867	0.907	0.960
$\checkmark$	$\checkmark$	$\checkmark$	$\checkmark$	$\checkmark$	0.802	0.724	0.763	0.906	0.930	0.883	0.923	0.968
Only using CM as supervision					0.755	0.680	0.712	0.862	0.921	0.871	0.909	0.960

Effectiveness of FEF and DAG. By the comparisons of Row 1 vs. Row 2, Row 2 vs. Row 3, and Row 3 vs. Row 5, we can see that all three branches of FEF-G, FEF-B and FEF-G are effective. We remove the DAG module and use simple fusion (element-wise addition and convolution) to combine these branches. The results are shown in Row 4. By comparing Row 4 and Row 5, we can see that the DAG works. Since no data in the ETIS is used for training, the significant gains on this dataset indicate that the designs of FEF and DAG greatly enhance the generalization of the model. Some images in the Kvasir are used as training data, hence the baseline model performs much better, which results in that the performance improvement on this dataset is not as obvious as that on the ETIS.

Effectiveness of Decoupled Maps. We use the completeness map to supervise all three branches. The results are listed in Row 6 of Table 2. Comparing to Row 5, we find that the latter performs better than the former with a gain of 4.7% in terms of mDice on the ETIS.

The proposed GSNet structure is relatively simple and the different features are repeatedly introduced in the three branches. Gaussian branch and body branch mainly use high-level abstract features with small resolution, which does not learn the complete prediction perfectly. Enforcing the usage of the completeness map easily leads to the learning of broken features and introduces redundant information, as shown in Fig. 6.

4   Conclusion

In this work, we propose a novel growth simulation network (GSNet) for polyp segmentation. According to the characteristics of polyp images, we propose to decouple the completeness map into Gaussian map and body map and use different map parallel branches to learn, respectively. These labels contain partial information, which reduces the learning difficulty of the model. The Gaussian branch mainly learns the location, body branch focuses more on the main information, and then the prediction branch to refine the details to get the final prediction map. In this way, the entire learning process progresses from simple to complex. Simulating the polyp growth process, DSLoss dynamically adjusts the weight to further boost the performance during training.

Acknowledgements. This work was supported by the National Natural Science Foundation of China # 62276046 and the Liaoning Natural Science Foundation # 2021-KF-12-10.

References

1. Ahn, S.B., Han, D.S., Bae, J.H., Byun, T.J., Kim, J.P., Eun, C.S.: The miss rate for colorectal adenoma determined by quality-adjusted, back-to-back colonoscopies. Gut Liver **6**(1), 64 (2012)
2. Bernal, J., Sánchez, F.J., Fernández-Esparrach, G., Gil, D., Rodríguez, C., Vilariño, F.: WM-DOVA maps for accurate polyp highlighting in colonoscopy: validation vs. saliency maps from physicians. Comput. Med. Imaging Graph. **43**, 99–111 (2015)
3. Bo, D., Wenhai, W., Deng-Ping, F., Jinpeng, L., Huazhu, F., Ling, S.: Polyp-PVT: polyp segmentation with pyramidvision transformers (2023)
4. Chen, Y., Dai, X., Liu, M., Chen, D., Yuan, L., Liu, Z.: Dynamic convolution: attention over convolution kernels. In: Proceedings of the IEEE/CVF Conference on Computer Vision and Pattern Recognition, pp. 11030–11039 (2020)
5. Fan, D.P., Cheng, M.M., Liu, Y., Li, T., Borji, A.: Structure-measure: a new way to evaluate foreground maps. In: Proceedings of the IEEE International Conference on Computer Vision, pp. 4548–4557 (2017)
6. Fan, D.P., Gong, C., Cao, Y., Ren, B., Cheng, M.M., Borji, A.: Enhanced-alignment measure for binary foreground map evaluation. arXiv preprint arXiv:1805.10421 (2018)
7. Fan, D.-P., et al.: PraNet: parallel reverse attention network for polyp segmentation. In: Martel, A.L., et al. (eds.) MICCAI 2020. LNCS, vol. 12266, pp. 263–273. Springer, Cham (2020). https://doi.org/10.1007/978-3-030-59725-2_26
8. Jha, D., et al.: Kvasir-SEG: a segmented polyp dataset. In: Ro, Y.M., et al. (eds.) MMM 2020. LNCS, vol. 11962, pp. 451–462. Springer, Cham (2020). https://doi.org/10.1007/978-3-030-37734-2_37
9. Jha, D., et al.: ResUNet++: an advanced architecture for medical image segmentation. In: 2019 IEEE International Symposium on Multimedia (ISM), pp. 225–2255. IEEE (2019)
10. Loshchilov, I., Hutter, F.: Decoupled weight decay regularization. arXiv preprint arXiv:1711.05101 (2017)
11. Margolin, R., Zelnik-Manor, L., Tal, A.: How to evaluate foreground maps? In: Proceedings of the IEEE Conference on Computer Vision and Pattern Recognition, pp. 248–255 (2014)
12. Pang, Y., Zhang, L., Zhao, X., Lu, H.: Hierarchical dynamic filtering network for RGB-D salient object detection. In: Vedaldi, A., Bischof, H., Brox, T., Frahm, J.-M. (eds.) ECCV 2020. LNCS, vol. 12370, pp. 235–252. Springer, Cham (2020). https://doi.org/10.1007/978-3-030-58595-2_15
13. Pang, Y., Zhao, X., Xiang, T.Z., Zhang, L., Lu, H.: Zoom in and out: a mixed-scale triplet network for camouflaged object detection. In: CVPR, pp. 2160–2170 (2022)
14. Pang, Y., Zhao, X., Zhang, L., Lu, H.: Multi-scale interactive network for salient object detection. In: CVPR, pp. 9413–9422 (2020)
15. Pang, Y., Zhao, X., Zhang, L., Lu, H.: Caver: Cross-modal view-mixed transformer for bi-modal salient object detection. IEEE TIP **32**, 892–904 (2023)
16. Qin, X., Zhang, Z., Huang, C., Gao, C., Dehghan, M., Jagersand, M.: BASNet: boundary-aware salient object detection. In: Proceedings of the IEEE/CVF Conference on Computer Vision and Pattern Recognition, pp. 7479–7489 (2019)
17. Ronneberger, O., Fischer, P., Brox, T.: U-Net: convolutional networks for biomedical image segmentation. In: Navab, N., Hornegger, J., Wells, W.M., Frangi, A.F. (eds.) MICCAI 2015. LNCS, vol. 9351, pp. 234–241. Springer, Cham (2015). https://doi.org/10.1007/978-3-319-24574-4_28

18. Siegel, R.L., et al.: Colorectal cancer statistics, 2020. CA: a cancer journal for clinicians **70**(3), 145–164 (2020)
19. Siegel, R.L., et al.: Global patterns and trends in colorectal cancer incidence in young adults. Gut **68**(12), 2179–2185 (2019)
20. Silva, J., Histace, A., Romain, O., Dray, X., Granado, B.: Toward embedded detection of polyps in WCE images for early diagnosis of colorectal cancer. Int. J. Comput. Assist. Radiol. Surg. **9**(2), 283–293 (2014)
21. Tajbakhsh, N., Gurudu, S.R., Liang, J.: Automated polyp detection in colonoscopy videos using shape and context information. IEEE Trans. Med. Imaging **35**(2), 630–644 (2015)
22. Tajbakhsh, N., Gurudu, S.R., Liang, J.: Automatic polyp detection in colonoscopy videos using an ensemble of convolutional neural networks. In: 2015 IEEE 12th International Symposium on Biomedical Imaging (ISBI), pp. 79–83. IEEE (2015)
23. Vázquez, D., et al.: A benchmark for endoluminal scene segmentation of colonoscopy images. J. Healthcare Eng. **2017** (2017)
24. Wang, W., et al.: Pyramid vision transformer: a versatile backbone for dense prediction without convolutions. In: Proceedings of the IEEE/CVF International Conference on Computer Vision, pp. 568–578 (2021)
25. Wang, X., Girshick, R., Gupta, A., He, K.: Non-local neural networks. In: Proceedings of the IEEE Conference on Computer Vision and Pattern Recognition, pp. 7794–7803 (2018)
26. Wei, J., Wang, S., Huang, Q.: F^3net: fusion, feedback and focus for salient object detection. In: Proceedings of the AAAI Conference on Artificial Intelligence. vol. 34, pp. 12321–12328 (2020)
27. Wei, J., Wang, S., Wu, Z., Su, C., Huang, Q., Tian, Q.: Label decoupling framework for salient object detection. In: Proceedings of the IEEE/CVF Conference on Computer Vision and Pattern Recognition, pp. 13025–13034 (2020)
28. Yang, B., Bender, G., Le, Q.V., Ngiam, J.: CondConv: conditionally parameterized convolutions for efficient inference. In: Advances in Neural Information Processing Systems, vol. 32 (2019)
29. Zhao, X., et al.: M2SNet: multi-scale in multi-scale subtraction network for medical image segmentation. arXiv preprint arXiv:2303.10894 (2023)
30. Zhao, X., Pang, Y., Zhang, L., Lu, H.: Joint learning of salient object detection, depth estimation and contour extraction. IEEE TIP **31**, 7350–7362 (2022)
31. Zhao, X., Pang, Y., Zhang, L., Lu, H., Zhang, L.: Suppress and balance: a simple gated network for salient object detection. In: Vedaldi, A., Bischof, H., Brox, T., Frahm, J.-M. (eds.) ECCV 2020. LNCS, vol. 12347, pp. 35–51. Springer, Cham (2020). https://doi.org/10.1007/978-3-030-58536-5_3
32. Zhao, X., Pang, Y., Zhang, L., Lu, H., Zhang, L.: Towards diverse binary segmentation via a simple yet general gated network. arXiv preprint arXiv:2303.10396 (2023)
33. Zhao, X., Zhang, L., Lu, H.: Automatic polyp segmentation via multi-scale subtraction network. In: de Bruijne, M., et al. (eds.) MICCAI 2021. LNCS, vol. 12901, pp. 120–130. Springer, Cham (2021). https://doi.org/10.1007/978-3-030-87193-2_12
34. Zhou, Z., Rahman Siddiquee, M.M., Tajbakhsh, N., Liang, J.: UNet++: a nested U-Net architecture for medical image segmentation. In: Stoyanov, D., et al. (eds.) DLMIA/ML-CDS -2018. LNCS, vol. 11045, pp. 3–11. Springer, Cham (2018). https://doi.org/10.1007/978-3-030-00889-5_1

Brain Diffuser: An End-to-End Brain Image to Brain Network Pipeline

Xuhang Chen[1,2], Baiying Lei[3], Chi-Man Pun[2(✉)], and Shuqiang Wang[1(✉)]

[1] Shenzhen Institutes of Advanced Technology, Chinese Academy of Sciences, Beijing, China
sq.wang@siat.ac.cn
[2] University of Macau,Taipa, China
cmpun@umac.mo
[3] Shenzhen University, Shenzhen, China

Abstract. Brain network analysis is essential for diagnosing and intervention for Alzheimer's disease (AD). However, previous research relied primarily on specific time-consuming and subjective toolkits. Only few tools can obtain the structural brain networks from brain diffusion tensor images (DTI). In this paper, we propose a diffusion based end-to-end brain network generative model Brain Diffuser that directly shapes the structural brain networks from DTI. Compared to existing toolkits, Brain Diffuser exploits more structural connectivity features and disease-related information by analyzing disparities in structural brain networks across subjects. For the case of Alzheimer's disease, the proposed model performs better than the results from existing toolkits on the Alzheimer's Disease Neuroimaging Initiative (ADNI) database.

Keywords: Brain network generation · DTI to brain network · Latent diffusion model

1 Introduction

Medical image computation is an ubiquitous method to detect and diagnose neurodegenerative disease [1–4]. It is often related with computer vision to aid in the diagnosis of this form of brain disease [5–7], using CT or MRI to study the morphological feature. Deep learning has been extensively used in medical image processing. Due to its superior performance, several complicated medical imaging patterns may be discovered, allowing for the secondary diagnosis of illness [8,9].

Nonetheless, high-dimensional medical image patterns are challenging to discover and evaluate. Brain is a complex network, and cognition need precise coordination across regions-of-interest (ROIs). Hence, brain network is introduced to overcome the above problems. A brain network can be described as a collection of vertices and edges. ROIs of the brain are represented by the vertices of a brain network. The edges of a brain network indicate the interaction between

Q. Liu et al. (Eds.): PRCV 2023, LNCS 14437, pp. 16–26, 2024.
https://doi.org/10.1007/978-981-99-8558-6_2

brain ROIs. Brain network as a pattern of interconnected networks may more accurately and efficiently represent latent brain information. There are two fundamental forms of connection in brain networks: functional connectivity (FC) and structural connectivity (SC). FC is the dependency between the blood oxygen level-dependent BOLD signals of two ROIs. SC is the strength of neural fiber connections between ROIs. Many studies have indicated that by employing FC or SC to collect AD-related features, more information may be obtained than conventional neuroimaging techniques [10–12].

Nevertheless, the aforementioned brain networks are produced by the use of particular preprocessing toolkits, such as PANDA [13]. These toolkits are sophisticated and need a large amount of manual processing steps to go through the pipeline, which might lead to inconsistent findings if researchers make parametric errors or omit critical processing stages. Therefore, we construct an end-to-end brain network generation network to simplify the uniform generation of brain networks and accurate AD development prediction. We summarizes our contribution as follows:

1. A diffusion-based model, Brain Diffuser, is proposed for brain network generation. Compared to existing toolkits, it is an end-to-end image-to-graph model that can investigate potential biomarkers in view of brain network.
2. A novel diffusive generative loss is designed to characterize the differences between the generated and template-based brain networks, therefore minimizing the difference in intrinsic topological distribution. It can considerably increase the proposed model's accuracy and resilience.
3. To generate brain networks embedded with disease-related information, the latent cross-attention mechanism is infused in Brain Diffuser so that the model focuses on learning brain network topology and features.

2   Related Work

Recent years have seen the rise of graph-structured learning, which covers the whole topology of brain networks and has considerable benefits in expressing the high-order interactions between ROIs. Graph convolutional networks (GCN) are widely used in graph-structured data such as brain networks. Abnormal brain connections have been effectively analyzed using GCN approaches [14, 15]. Pan *et al.* introduced a novel decoupling generative adversarial network (DecGAN) to detect abnormal neural circuits for AD [16]. CT-GAN [17] is a cross-modal transformer generative adversarial network to combine functional information from resting-state functional magnetic resonance imaging (fMRI) with structural information from DTI. Kong *et al.* introduced a structural brain-network generative model (SBGM) based on adversarial learning to generate the structural connections from neuroimaging [18].

Diffusion Model [19] is a probabilistic generative models designed to learn a data distribution by gradually denoising a normally distributed variable, which corresponds to learning the reverse process of a Markov Chain. Diffusion models have been used in medical imaging for anomaly detection [20,21] and reconstruction [22]. And its strong capability in image synthesis also can be used to generated graph data [23]. Therefore we adopt diffusion model as our generation module to synthesize brain networks.

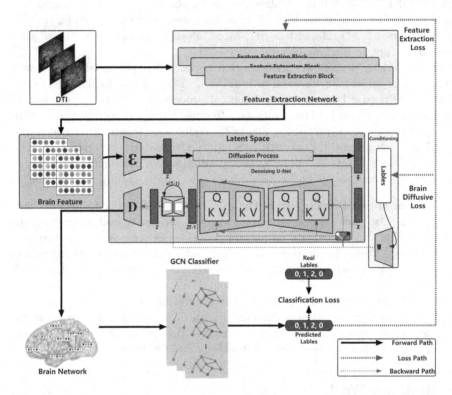

Fig. 1. The overall architecture of our Brain Diffuser. Feature extraction network is applied to DTI images to extract brain network topology features. After passing the features to the diffusion model, a complete brain network is synthesized. The GCN classifier will incorporate classification knowledge to improve feature extraction and diffusion model further.

3   Methodology

3.1   Feature Extraction Module

To construct the brain network's topology, vertex and edge properties must be used. We propose the Feature Extraction Net (FENet) to extract these structural attributes from DTI.

Combining standard convolutional layer with depthwise separable convolutional blocks, with each block including a depth-wise convolutional layer and a pointwise convolutional layer, FENet retrieves features from DTI.

DTI is a single-channel, three-dimensional image with the shape $109 \times 91 \times 91$ after preprocessing. Each feature map serves as a structural feature vector of the relevant brain area, allowing us to derive the structural feature matrix $P \in R^{90 \times 80}$ of the brain region.

3.2   Brain Diffuser

As show in Fig. 1, the proposed model is based on the latent diffusion model [24]. It begins with an encoding module $\mathcal{E}$ and a decoding module $\mathcal{D}$. $\mathcal{E}$ encodes our structural feature matrix to latent space processed by the diffusion module. $\mathcal{D}$ decodes the latent diffusion model output into brain network $B \in R^{90 \times 90}$. These two modules concentrate on the crucial, semantic portion of the data and save a substantial amount of processing resources.

The primary component of this model is the capacity to construct the underlying U-Net [25], and to further concentrate the goal on the brain network topology using the reweighed bound as shown in Eq. 1:

$$L_{LDM} = E_{\mathcal{E}(x), \epsilon \sim \mathcal{N}(0,1), t}\left[\|\epsilon - \epsilon_\theta(z_t, t)\|_2^2\right] \tag{1}$$

The network backbone $\epsilon_\theta(\circ, t)$ is a time-conditional U-Net. Since the forward process is fixed, z_t can be efficiently obtained from $\mathcal{E}$ during training, and samples from $p(z)$ can be decoded to image space with a single pass through $\mathcal{D}$.

To provide more adaptable results, we combine U-Net with the cross-attention technique [26]. We introduce a domain specific encoder τ_θ hat projects input to an intermediate representation $\tau_\theta(y) \in R^{M \times d_\tau}$, which is then mapped to the intermediate layers of the U-Net via a cross-attention layer implementing $Attention(Q, K, V) = softmax(\frac{QK^T}{\sqrt{d}}) \cdot V$, with

$$Q = W_Q^{(i)} \cdot \varphi_i(z_t), \ K = W_K^{(i)} \cdot \tau_\theta(y), \ V = W_V^{(i)} \cdot \tau_\theta(y)$$

Here, $\varphi_i(z_t) \in R^{N \times d_\epsilon^i}$ denotes a (flattened) intermediate representation of the U-Net implementing ϵ_θ and $W_V^{(i)} \in R^{d \times d_\epsilon^i}$, $W_Q^{(i)} \in R^{d \times d_\tau}$ & $W_K^{(i)} \in R^{d \times d_\tau}$ are learnable projection matrices [27].

Based on input pairs, we then learn the conditional LDM via

$$L_{LDM} = E_{\mathcal{E}(x), y, \epsilon \sim \mathcal{N}(0,1), t}\left[\|\epsilon - \epsilon_\theta(z_t, t, \tau_\theta(y))\|_2^2\right], \tag{2}$$

both τ_θ and ϵ_θ are jointly optimized via Eq. 2. This conditioning mechanism is flexible as τ_θ can be parameterized with domain-specific experts e.g. the GCN classifier.

3.3   GCN Classifier

Our GCN classifier comprises of a GCN layer [28] and a fully connected layer. The GCN layer is used to fuse the structural characteristics of each vertex's surrounding vertices, while the fully connected layer classifies the whole graph based on the fused feature matrix. The classifier is used to categorize EMCI, LMCI, and NC, but its primary function is to guide the generation workflow utilizing classification information.

3.4 Loss Function

Classification Loss L_C. Taking into account the structural properties of each vertex individually, we add an identity matrix I to brain network $\hat{A}$. As demonstrated in Eq. 3, we utilize the output results of the classifier and the class labels to compute the cross-entropy loss as the loss of the classifier.

$$L_C = -\frac{1}{N} \sum_{i=1}^{N} p(y_i|x_i) log[q(\hat{y}_i|x_i)] \qquad (3)$$

Where N is the number of input samples, $p(y_i \mid x_i)$ denotes the actual distribution of sample labels, and $q(\hat{y}_i \mid x_i)$ denotes the distribution of labels at the output of the classifier.

Brain Diffusive Loss L_B. As previously mentioned, the outcome of the diffusion model is contingent on the input and label-guided conditioning mechanism. FENet extracts the DTI feature as its input. And GCN classifier is applied to categorize brain networks and generate biomarkers that serve as labels. Consequently, Brain Diffuse Loss consists of Feature Extraction Loss L_{FE}, LDM Loss L_{LDM}, Classification Loss L_C. We use L_1 loss as L_{FE} to retrieve disease related information according to the brain atlas. So our Brain Diffusive Loss L_B can be described as Eq. 4

$$L_B = L_{FENet} + L_{LDM} + L_C \qquad (4)$$

4 Experiments

4.1 Dataset and Preprocessing

Our preprocessing workflow began with employing the PANDA toolkit [13] that used the AAL template. DTI data were initially transformed from DICOM format to NIFTI format, followed by applying head motion correction and cranial stripping. The 3D DTI voxel shape was resampled to $109 \times 91 \times 91$. Using the Anatomical Automatic Labeling (AAL) brain atlas, the whole brain was mapped into 90 regions of interest (ROIs), and the structural connectivity matrix was generated by calculating the number of DTI-derived fibers linking brain regions. Since the quantity of fiber tracts significantly varies across brain regions, we normalized the structural connectivity matrix between 0 and 1 based on the input's empirical distribution.

Table 1. ADNI dataset information in this study.

Group	NC	EMCI	LMCI
Male/Female	42M/45F	69M/66F	35M/41F
Age (mean±SD)	74.1 ± 7.4	75.8 ± 6.4	75.9 ± 7.5

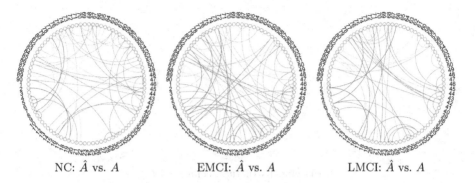

NC: $\hat{A}$ vs. A EMCI: $\hat{A}$ vs. A LMCI: $\hat{A}$ vs. A

Fig. 2. Variations in structural brain connectivity at various developmental stages.

4.2 Experiment Configuration

We employed DTI, structural connectivity matrix A and each subject's matching label as model inputs. We divided 298 samples into groups of 268 training samples and 30 testing samples (9 for NC, 14 for EMCI, 7 for LMCI). The specific data information is presented in Table 1.

Table 2. Prediction performance of the PANDA-generated brain networks and our reconstructed brain networks.

	Accuracy	Sensitivity	Specificity	F1-score
PANDA	66.14%	66.14%	83.51%	67.15%
Ours	87.83%	87.83%	92.66%	87.83%

Table 3. Ablation study on prediction performance of different classifiers.

	Accuracy	Sensitivity	Specificity	F1-score
GAT [29]	25.66%	25.66%	63.15%	21.44%
GIN [30]	75.93%	75.93%	87.72%	73.31%
Graphormer [31]	80.69%	80.69%	88.31%	77.38%
GraphSAGE [32]	74.87%	74.87%	86.72%	72.89%
Ours	87.83%	87.83%	92.66%	87.83%

The performance of classification is defined by detection accuracy, sensitivity, specificity, and F1-score. We used PyTorch to implement our Brain Diffuser model. Our experiments were conducted using NVIDIA GeForce GTX 2080 Ti. The Adam optimizer was used with a batch size of 2, and we set the learning rate empirically to be 10^{-4}.

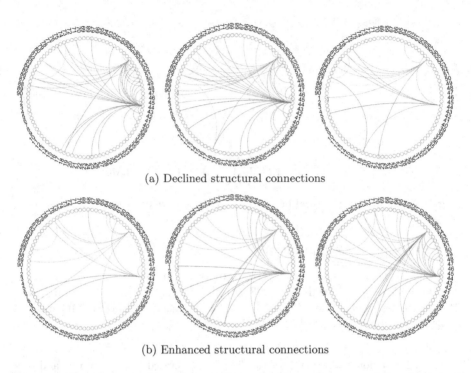

(a) Declined structural connections

(b) Enhanced structural connections

Fig. 3. Chord diagrams of connectivity change based on the empirical distribution of our synthesized connection matrix. From left to right: NC versus EMCI, NC versus LMCI, EMCI versus LMCI.

4.3   Results and Discussion

Classification Performance. The results presented in Table 2 and Fig. 4 demonstrate that our reconstructed results outperform those generated by the PANDA toolkit. This suggests that our reconstructed pipeline has more advantages for classifiers detecting the potential biomarkers of MCI. Our generator's performance is significantly superior to that of PANDA, which has great potential for future research in AD detection and intervention. The ablation study in Table 3 and Fig. 4 demonstrate that our classifier outperforms others, suggesting that our classifier can effectively utilize a greater number of disease-related features or prospective disease-related biomarkers.

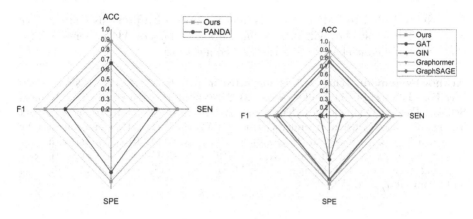

Fig. 4. Radar charts of classification performance.

Brain Structural Network Connectivity Analysis. To determine whether the structural connectivity matrix $\hat{A}$ generated by our proposed model differed significantly from the PANDA outcome A, we conducted a paired-samples T-test at a threshold value of 0.05 between $\hat{A}$ and A following Kong *et al.*'s setting [18]. The chord diagram in Fig. 2 indicates connections with significant changes. The structural connectivity matrix $\hat{A}$ derived by our model has altered connectivity between numerous brain regions compared to the static software generation.

To establish whether there was a significant difference between the structural connectivity matrix $\hat{A}$ generated by our proposed model and the PANDA result A, we performed a paired-samples T-test at a threshold value of 0.05 between $\hat{A}$ and A using Kong *et al.*'s setting [18]. The connections that exhibited substantial changes are illustrated in the chord diagram in Fig. 2. Our model generated structural connectivity matrix $\hat{A}$ demonstrated significant variations in connectivity between numerous brain regions compared to the static software generation.

Figure 3 compares the changes in brain connectivity in different stages. The alteration in connectivity between brain regions was more pronounced in LMCI stage than in NC. Similarly, there was a significant decline in connectivity between brain regions in LMCI patients compared to EMCI patients. These alterations and tendencies reveal a sequential progression toward AD pathology in NC subjects: a progressive decrease in structural brain network connectivity.

5 Conclusion

This paper presents Brain Diffuser, a novel approach for directly generating structural brain networks from DTI. Brain Diffuser provides an entire pipeline for generating structural brain networks from DTI that is free of the constraints inherent in existing software toolkits. Our method enables us to study structural brain network alterations in MCI patients. Using Brain Diffuser, we discovered that the structural connectivity of the subjects' brains progressively decreased from NC to EMCI to LMCI, which is in accordance with previous neuroscience research. Future studies will locate which brain regions of AD patients exhibit the most significant changes in structural connectivity.

Acknowledgement. This work is supported in part by the National Natural Science Foundations of China under Grant 62172403, in part by the Distinguished Young Scholars Fund of Guangdong under Grant 2021B1515020019, in part by the Excellent Young Scholars of Shenzhen under Grant RCYX20200714114641211 and in part by Shenzhen Key Basic Research Project under Grant JCYJ20200109115641762, in part by the University of Macau under Grant MYRG2022-00190-FST, in part by the Science and Technology Development Fund, Macau SAR, under Grant 0034/2019/AMJ and Grant 0087/2020/A2.

References

1. You, S., et al.: Fine perceptive GANs for brain MR image super-resolution in wavelet domain. IEEE Trans. Neural Netw. Learn. Syst. **34**, 8802–8814 (2022)
2. Hu, B., Zhan, C., Tang, B., Wang, B., Lei, B., Wang, S.Q.: 3-D brain reconstruction by hierarchical shape-perception network from a single incomplete image. IEEE Trans. Neural Netw. Learn. Syst. (2023)
3. Hu, S., Lei, B., Wang, S., Wang, Y., Feng, Z., Shen, Y.: Bidirectional mapping generative adversarial networks for brain MR to PET synthesis. IEEE Trans. Med. Imaging **41**(1), 145–157 (2021)
4. Yu, W., et al.: Morphological feature visualization of Alzheimer's disease via multi-directional perception GAN. IEEE Trans. Neural Netw. Learn. Syst. **34**, 4401–4415 (2022)
5. Wang, H., et al.: Ensemble of 3D densely connected convolutional network for diagnosis of mild cognitive impairment and Alzheimer's disease. Neurocomputing **333**, 145–156 (2019)
6. Hu, S., Yu, W., Chen, Z., Wang, S.: Medical image reconstruction using generative adversarial network for Alzheimer disease assessment with class-imbalance problem. In: 2020 IEEE 6th International Conference on Computer and Communications (ICCC), pp. 1323–1327. IEEE (2020)
7. Lei, B., et al.: Predicting clinical scores for Alzheimer's disease based on joint and deep learning. Exp. Syst. Appl. **187**, 115966 (2022)
8. Wang, S., et al.: An ensemble-based densely-connected deep learning system for assessment of skeletal maturity. IEEE Trans. Syst. Man Cybern. Syst. **52**(1), 426–437 (2020)
9. Jing, C., Gong, C., Chen, Z., Lei, B., Wang, S.: TA-GAN: transformer-driven addiction-perception generative adversarial network. Neural Comput. Appl. **35**(13), 9579–9591 (2023). https://doi.org/10.1007/s00521-022-08187-0
10. Wang, S., Hu, Y., Shen, Y., Li, H.: Classification of diffusion tensor metrics for the diagnosis of a myelopathic cord using machine learning. Int. J. Neural Syst. **28**(02), 1750036 (2018)
11. Jeon, E., Kang, E., Lee, J., Lee, J., Kam, T.-E., Suk, H.-I.: Enriched representation learning in resting-state fMRI for early MCI diagnosis. In: Martel, A.L., et al. (eds.) MICCAI 2020. LNCS, vol. 12267, pp. 397–406. Springer, Cham (2020). https://doi.org/10.1007/978-3-030-59728-3_39

12. Wang, S., Shen, Y., Chen, W., Xiao, T., Hu, J.: Automatic recognition of mild cognitive impairment from MRI images using expedited convolutional neural networks. In: Lintas, A., Rovetta, S., Verschure, P.F.M.J., Villa, A.E.P. (eds.) ICANN 2017. LNCS, vol. 10613, pp. 373–380. Springer, Cham (2017). https://doi.org/10.1007/978-3-319-68600-4_43
13. Cui, Z., Zhong, S., Xu, P., He, Y., Gong, G.: PANDA: a pipeline toolbox for analyzing brain diffusion images. Front. Hum. Neurosci. **7**, 42 (2013)
14. Parisot, S., et al.: Disease prediction using graph convolutional networks: application to autism spectrum disorder and Alzheimer's disease. Med. Image Anal. **48**, 117–130 (2018)
15. Ktena, S.I., et al.: Distance metric learning using graph convolutional networks: application to functional brain networks. In: Descoteaux, M., Maier-Hein, L., Franz, A., Jannin, P., Collins, D.L., Duchesne, S. (eds.) MICCAI 2017. LNCS, vol. 10433, pp. 469–477. Springer, Cham (2017). https://doi.org/10.1007/978-3-319-66182-7_54
16. Pan, J., Lei, B., Wang, S., Wang, B., Liu, Y., Shen, Y.: DecGAN: decoupling generative adversarial network detecting abnormal neural circuits for Alzheimer's disease. arXiv preprint arXiv:2110.05712 (2021)
17. Pan, J., Wang, S.: Cross-modal transformer GAN: a brain structure-function deep fusing framework for Alzheimer's disease. arXiv preprint arXiv:2206.13393 (2022)
18. Kong, H., Pan, J., Shen, Y., Wang, S.: Adversarial learning based structural brain-network generative model for analyzing mild cognitive impairment. In: Yu, S., et al. (ed.) Pattern Recognition and Computer Vision, PRCV 2022. LNCS, vol. 13535, pp. 361–375. Springer, Cham (2022). https://doi.org/10.1007/978-3-031-18910-4_30
19. Sohl-Dickstein, J., Weiss, E., Maheswaranathan, N., Ganguli, S.: Deep unsupervised learning using nonequilibrium thermodynamics. In: ICML, pp. 2256–2265. PMLR (2015)
20. Wolleb, J., Bieder, F., Sandkühler, R., Cattin, P.C.: Diffusion models for medical anomaly detection. In: Wang, L., Dou, Q., Fletcher, P.T., Speidel, S., Li, S. (eds.) Medical Image Computing and Computer Assisted Intervention, MICCAI 2022. LNCS, vol. 13438, pp. 35–45. Springer, Cham (2022). https://doi.org/10.1007/978-3-031-16452-1_4
21. Pinaya, W.H.L., et al.: Fast unsupervised brain anomaly detection and segmentation with diffusion models. In: Wang, L., Dou, Q., Fletcher, P.T., Speidel, S., Li, S. (eds.) Medical Image Computing and Computer Assisted Intervention, MICCAI 2022. LNCS, vol. 13438, pp. 705–714. Springer, Cham (2022). https://doi.org/10.1007/978-3-031-16452-1_67
22. Chung, H., Ye, J.C.: Score-based diffusion models for accelerated MRI. Med. Image Anal. **80**, 102479 (2021)
23. Jo, J., Lee, S., Hwang, S.J.: Score-based generative modeling of graphs via the system of stochastic differential equations. In: ICML, pp. 10362–10383. PMLR (2022)
24. Rombach, R., Blattmann, A., Lorenz, D., Esser, P., Ommer, B.: High-resolution image synthesis with latent diffusion models. In: CVPR, pp. 10684–10695 (2022)
25. Ronneberger, O., Fischer, P., Brox, T.: U-Net: convolutional networks for biomedical image segmentation. In: Navab, N., Hornegger, J., Wells, W.M., Frangi, A.F. (eds.) MICCAI 2015. LNCS, vol. 9351, pp. 234–241. Springer, Cham (2015). https://doi.org/10.1007/978-3-319-24574-4_28
26. Vaswani, A., et al.: Attention is all you need. In: NIPS, vol. 30 (2017)

27. Jaegle, A., Gimeno, F., Brock, A., Vinyals, O., Zisserman, A., Carreira, J.: Perceiver: general perception with iterative attention. In: ICML, pp. 4651–4664. PMLR (2021)
28. Kipf, T.N., Welling, M.: Semi-supervised classification with graph convolutional networks. In: ICLR (2017)
29. Veličković, P., Cucurull, G., Casanova, A., Romero, A., Liò, P., Bengio, Y.: Graph attention networks. In: ICLR (2018)
30. Xu, K., Hu, W., Leskovec, J., Jegelka, S.: How powerful are graph neural networks? In: ICLR (2019)
31. Ying, C., et al.: Do transformers really perform badly for graph representation? In: NIPS, vol. 34, pp. 28877–28888 (2021)
32. Hamilton, W.L., Ying, Z., Leskovec, J.: Inductive representation learning on large graphs. In: NIPS (2017)

CCJ-SLC: A Skin Lesion Image Classification Method Based on Contrastive Clustering and Jigsaw Puzzle

Yuwei Zhang, Guoyan Xu$^{(\boxtimes)}$, and Chunyan Wu

College of Computer and Information, Hohai University, Nanjing 211100, China
{zhangyuwei,gy_xu,wuchunyan}@hhu.edu.cn

Abstract. Self-supervised learning has been widely used in natural image classification, but it has been less applied in skin lesion image classification. The difficulty of existing self-supervised learning is the design of pretext tasks, intra-image pretext tasks and inter-image pretext tasks often focus on different features, so there is still space to improve the effect of single pretext tasks. We propose a skin lesion image classification method based on contrastive clustering and jigsaw puzzles from the perspective of designing a hybrid pretext task that combines intra-image pretext tasks with inter-image pretext tasks. The method can take advantage of the complementarity of high-level semantic features and low-level texture information in self-supervised learning to effectively solve the problems of class imbalance in skin lesion image datasets and large intra-class differences in skin lesion images with small inter-class differences, thus improving the performance of the model on downstream classification tasks. Experimental results show that our method can achieve supervised learning classification results and outperform other self-supervised learning classification models on the ISIC-2018 and ISIC-2019 datasets, moreover, it performs well on the evaluation metrics of unbalanced datasets.

Keywords: Self-supervised Learning · Skin Lesion Classification · Contrastive Learning · Clustering · Jigsaw Puzzle

1 Introduction

Skin cancer is one of the most dangerous diseases in the world, among which melanoma is one of the more common skin cancers. Although the survival rate of melanoma patients in the first 5 years is close to 95%, the late survival rate is only about 15%, a very significant difference that highlights the importance of timely identification and detection of melanoma in the course of melanoma treatment [1]. Therefore, for patients with skin disease, early detection and treatment is the only way to effectively cure the skin disease.

Unlike the intuitively obvious features of natural image datasets, skin lesion image datasets have relatively unique features of medical image datasets. There are three main characteristics of the skin lesion image dataset: (1) The cost of skin lesion image labels is

Q. Liu et al. (Eds.): PRCV 2023, LNCS 14437, pp. 27–39, 2024.
https://doi.org/10.1007/978-981-99-8558-6_3

high. Medical image labels require specialized experts, which leads to hard work in producing medical image datasets. (2) The distribution of skin lesion images in the dataset is unbalanced. In most skin lesion datasets, benign lesion images usually account for the majority of the images. Meanwhile, many skin lesion image datasets have significant quantitative inequalities in the number of samples from different skin lesion categories. (3) The phenomenon of small inter-class variation and large intra-class variation exists for skin lesion images, as shown in Fig. 1. Small inter-class variation means that there are many similarities between skin lesion images of different categories; large intra-class variation means that skin lesion images of the same category differ in color, features, structure, size, location and other characteristics, like melanoma.

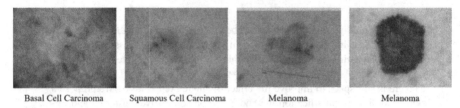

| Basal Cell Carcinoma | Squamous Cell Carcinoma | Melanoma | Melanoma |

Fig. 1. The phenomenon of small inter-class variation and large intra-class variation.

To alleviate the pressure on the cost of producing skin lesion image datasets, the researchers tried to use self-supervised learning methods. Existing self-supervised learning methods usually use a single pretext task, where the intra-image pretext task focuses on the feature information of a single image and the inter-image pretext task focuses on the similarities and differences of multiple pairs of images. Therefore, we propose a model that effectively combines the two types of pretext tasks and applied them in the classification of skin lesion images. To address the problem of class imbalance in the skin lesion image dataset, we adopt contrastive learning to alleviate the model from overlearning features of the majority class. Contrastive learning could increase the focus of the model on the minority class without negatively affecting the majority class by getting rid of the influence of real labels on the instance discrimination. However, each anchor image will be considered as a single class in contrastive learning, and defaulting the same class of objects to different classes will result in the model not being able to fully learn the image features. This negative impact will be exacerbated because of the increase in positive and negative sample pairs after cut of the image and the potentially large image differences in the same class of skin lesion images. To mitigate this effect, we add a pretext task based on clustering to the contrastive learning. The clustering task clusters the feature vectors extracted from the same image to mitigate the effect of each anchor image being a separate class caused by instance discrimination in contrastive learning, and by integrating clustering with contrastive learning, we could optimize the learning ability of the model in unbalanced datasets. To address the problem of small class-to-class differences and large class-internal differences in skin lesion image datasets, we propose a jigsaw pretext task that restores image patches cut out from the images to their relative positions in the original image. The jigsaw pretext task could focus on comparing the differences between image patches and fully utilizing the features within the images, causing the neural network model to learn global and local features to distinguish different locations of image patches in the original image, and restore the original image from multiple input images. Solving the jigsaw puzzle tells the model what parts of the same class of disease are composed of and what those parts are. Additionally, the jigsaw pretext task requires image patches, similar to the contrastive clustering task, to reduce model complexity.

Compared to other applications of self-supervised learning in the classification of skin lesion images, the main innovations of our proposed model are summarized as follows:

- We propose an inter-image pretext task combining contrastive learning with clustering task and propose an intra-image pretext task based on a hybrid image jigsaw task.
- We propose a skin lesion image classification model based on contrastive clustering and jigsaw puzzle (CCJ-SLC), which combines an inter-image pretext task (contrastive clustering) and an intra-image pretext task (jigsaw puzzle).
- The model is able to overcome the limitations of existing self-supervised learning by combining high-level semantic features with low-level texture information, further enhancing the model's ability to learn unique features of skin lesion images.

The structure of the paper is as follows. In the next section we introduce research related to the classification of skin lesion images. Section 3 describes the proposed method. Section 4 gives experiments and results. We conclude in Sect. 5.

2 Related Work

Self-Supervised Learning. Self-supervised learning mainly utilizes pretext tasks to mine its own supervised information from large-scale unsupervised data, and trains the network with this constructed supervisory information. Self-supervised learning can be categorized into three main directions: intra-image pretext task methods [2, 3], inter-image pretext task methods [4, 5], and hybrid task methods [6].

Skin Lesion Image Classification. At present, most skin lesion classification studies use methods based on convolutional neural network [7, 8]. Self-supervised learning is still in its emerging stage in medical image classification. Azizi et al. [9] studied the effectiveness of self-supervised learning as a pre-training strategy for medical image classification, and proposed a multi-instance contrast learning (MICLe) method. Wang et al. [10] proposed an unsupervised skin cancer image classification method based on self-supervised topological clustering network. Kwasigroch et al. [11] used rotation transformation and jigsaw puzzle as pretext tasks to achieve skin lesion image classification by designing twin neural network model.

3 Methodology

3.1 Overview of Our Method

Our proposed model specifically consists of three components, which are data sampling pre-processing module, backbone network module and pretext task module, as shown in Fig. 2. Firstly, the model splits the unlabeled skin lesion images, the multiple image parts are recombined into different images and then data augmentation is performed to highlight the image features, after that, the hybrid images are fed into the backbone network, which extracts the image features. Then, the model shuffles the feature maps extracted by the backbone network and feeds them into the task branches, which combine the inter-image pretext task with the intra-image pretext task. Finally, the model computes the loss functions of different pretext tasks.

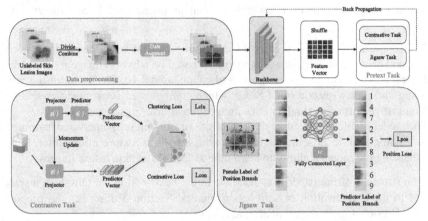

Fig. 2. The overview framework of our CCJ-SLC model. Each input image is split into $m \times m$ images, shuffled and recombined into a new image for input to the backbone network. ResNet network architecture is used in the backbone network. The orange part is the diagram of the Contrastive Clustering task and the green part is the diagram of the Jigsaw Puzzle task.

3.2 Contrastive Clustering

Contrastive learning can learn better, label-independent information about initialized features from unbalanced datasets. In the instance-based contrastive learning, each sample is considered as a distinct class in the dataset, however, this makes it appear that the same class of images is defaulted to a different class of images when comparing the input samples with other samples of the class to which the original samples belong. Therefore, we add the clustering task to the contrastive learning to reduce the bias caused by the contrastive learning, based on which a hybrid image of different image patches is used as input to help the model learn the features between images.

Assume that the backbone network will generate vectors of $n \times m \times m \times c$ dimensions per batch and feed them into the contrastive clustering branch, which will consider m^2 feature vectors from the same image as the same class. In order to project the feature vectors to the same vector space dimension to measure the distance, we combine the methods used in previous contrastive learning models [5, 12–14] by setting a projection head and a prediction head for mapping in the positive sample branch and only one projection head for mapping in the negative sample branch. The specific network structure of the projection head and the prediction head is shown in Fig. 3. The vector after passing through the projection head and the prediction head can be expressed as:

$$z_i = g(h_i) = projector(h_i) = projector(f(x_i)) \qquad (1)$$

$$y_i = q(z_i) = q(g(h_i)) = predictor(projector(f(x_i))) \qquad (2)$$

where $f(\odot)$ represents the encoding layer network.

We use the cosine similarity measure to measure the distance between two feature vectors, and the cosine similarity is calculated as follows:

$$sim(u, v) = \frac{u^T \times v}{||u|| \times ||v||} = \frac{\sum\limits_{i=1}^{n} u_i \times v_i}{\sqrt{\sum\limits_{i=1}^{n}(u_i)^2} \times \sqrt{\sum\limits_{i=1}^{n}(v_i)^2}} \qquad (3)$$

where u, v are n-dimensional vectors, respectively.

Fig. 3. Projector network structure diagram and Predictor network structure diagram.

In contrastive tasks, researchers usually use the NCE loss function [15] and its variant InfoNCE loss function [12] to optimize the model. Since we split and recombine the skin lesion images, so that there will be multiple positive sample instances per image in contrastive learning. To enhance the learning ability of the model under multiple pairs of instances, we use SupCon Loss [16] as the loss function for contrastive learning. The SupCon Loss is defined as follows.

$$\mathcal{L}_{con} = \sum_{i \in I} \frac{-1}{|P(i)|} \sum_{p \in P(i)} \log \frac{\exp(f_q \cdot f_p / \tau)}{\sum_{s \in S(i)} \exp(f_q \cdot f_s / \tau)} \qquad (4)$$

where I is the set of input vectors, f_q is the anchor in this training, which is used to distinguish positive and negative samples. f_p denotes positive sample, f_s denotes negative sample. τ is the temperature hyperparameter used to smooth the probability distribution. $P(i) \equiv \{p \in I : \tilde{y}_p = \tilde{y}_i\}$ is the set of all positive samples in the batch, i.e., the image feature vector sampled from a single image sample. $|P(i)|$ is the number of samples in the set. $S(i)$ is the set of negative samples in this batch relative to the anchor vector.

The clustering task is to divide the image patches into different clusters, each cluster contains only patches from the same image. The loss function of the clustering task is similar to the contrastive learning loss function. For each pair of image patches in the same cluster, the loss function is as follows:

$$\mathcal{L}_{i,j} = -\log \frac{\exp(sim(z_i, z_j)/\tau)}{\sum_{k=1}^{nmm} \mathbb{I}_{k \neq i} \exp(sim(z_i, z_k)/\tau)} \qquad (5)$$

where $\mathbb{I}$ denotes the indicator function and holds when $k \neq i$. $sim(u, v)$ is the cosine similarity of the vector u and v. (i, j) is the same image sample pair, and (i, k) is different image sample pair.

Since $m \times m$ image patches will be sampled in a single image, when a vector is selected as a positive sample for comparison in the clustering task, $mm - 1$ patches of image patches belonging to the same image remain to be computed, the loss function for a particular chunk of an image is defined as follows:

$$\mathcal{L}_i = \frac{1}{mm - 1} \Sigma_{j \in \mathbb{C}_i} \mathcal{L}_{i,j} \tag{6}$$

where $\mathbb{C}_i$ is the set of image patch indexes in the same class cluster of i.

Ultimately, the clustering loss function needs to compute all pairs from different clusters, as follows:

$$\mathcal{L}_{clu} = \frac{1}{nmm} \Sigma_i \left(\frac{1}{mm - 1} \Sigma_{j \in \mathbb{C}_i} \mathcal{L}_{i,j} \right) \tag{7}$$

3.3 Jigsaw Puzzle

To address the problem of small inter-class variation and large intra-class variation in the skin lesion image dataset, and that the inter-image pretext task will ignore the intra-image feature relationships to some extent, we propose a jigsaw puzzle pretext task to focus on the intra-image local feature relationships in order to make full use of the feature vectors after image splitting. The goal of the jigsaw puzzle pretext task is to force the network to learn as representative and discriminative features as possible for each image part region in order to determine their relative positions.

We use images made by mixing image patches from different images instead of image patches from a single image as input to the network, which will greatly increase the difficulty of pretext task. At the same time, combining intra-image pretext tasks with inter-image pretext tasks enables learning another branch that the original traditional model cannot accommodate at the same memory overhead. By establishing the complementary relationship between local features and contextual information through the jigsaw puzzle task, the model will observe the relationship between individual image patches and the complete image, removing some ambiguity from feature edges with the increase of image patches. The jigsaw puzzle task will be designed as a classification problem in which each image patch will generate a pseudo-label when it is divided to indicate its position in the original image. The jigsaw puzzle task is seen as predicting the label of each image patch, and a fully connected layer is attached to the input vector as a classifier, so the cross-entropy loss function is adopted as follows:

$$\mathcal{L}_{jigsaw} = CrossEntropy(Y, \tilde{Y})$$
$$= -\frac{1}{nmm} \sum_{i=1}^{nmm} \left[y^{(i)} \log \tilde{y}^{(i)} + \left(1 - y^{(i)} \right) \log \left(1 - \tilde{y}^{(i)} \right) \right] \tag{8}$$

where Y is the truth location label, $\tilde{Y}$ is the predicted location label.

3.4 Loss Function

The final loss function proposed in collaboration with the above loss function is shown as follows:

$$\mathcal{L} = \alpha(\mathcal{L}_{clu} + \mathcal{L}_{con}) + \mathcal{L}_{Jigsaw} \tag{9}$$

where α is hyperparameter to balance these three tasks. In our experiments, $\alpha = 0.75$.

4 Experiments

4.1 Dataset and Evaluation Metrics

Datasets. To fairly compare with other methods, we select publicly available datasets for training and evaluation, which are ISIC-2018 [17, 18], ISIC-2019 [17, 19, 20] respectively. ISIC-2018 includes a total of 10,015 skin lesion images, which are divided into 7 categories of skin lesions. These are actinic keratosis (AK), basal cell carcinoma (BCC), benign keratosis (BKL), dermatofibroma (DF), melanoma (MEL), melanocytic nevus (NV) and vascular skin lesions (VASC). ISIC 2019 is composed of three dermatology data image datasets: HAM dataset, MSK dataset, and BCN_20000 dataset. There are 25,331 skin lesion images in this dataset, which are divided into 8 categories of skin lesions, these are AK, BCC, BKL, DF, MEL, NV, VASC and squamous cell carcinoma (SCC).

Training Details. Our implementation is based on Pytorch on two NVIDIA Tesla V100 cards. In this paper, the momentum gradient descent method is used to optimize the network model, the momentum is set to 0.9, the initial learning rate is set to 0.03 and is attenuated by cosine annealing strategy, and the weight attenuation is set to 0.0001. The temperature parameter of the loss function is set to 0.3. The Batch Size is set to 256 and all models will be trained for 400 epochs. The backbone network uses the ResNet-50 network model. We uniformly resize the images 224×224 to the input model. The data augmentation setting follows [5] such as color enhancement and geometric broadening. We divide ISIC-2019 into 80% training set and 20% test set, ISIC-2018 was the test set.

Evaluation Metrics. We compute accuracy (ACC), balanced accuracy (BACC), F1-Score to evaluate the model and adopts the Macro-average method to optimize the index. The accuracy is the proportion of correct samples predicted by the classifier to all samples. BACC is used to measure the detection ability of the model for unbalanced datasets. F1-Score is an index that considers both precision and recall.

4.2 Baseline Performance

We conducted multiple sets of comparison experiments with supervised and self-supervised learning methods on ISIC-2018 and ISIC-2019 datasets respectively and analyzed the different experimental results.

Table 1. Results on ISIC-2018 and ISIC-2019 datasets.

Dataset	Method	Mode	ACC (%)	BACC (%)	F1-Score (%)
ISIC-2018	Densenet-161 [21]	Supervised	87.33	86.71	88.41
	CSLNet [22]	Supervised	88.75	85.35	86.11
	EfficieneNetB4-CLF [23]	Supervised	89.97	87.45	88.64
	STCN [10]	Self-Supervised	80.06	79.54	81.28
	BYOL [4]	Self-supervised	80.58	78.63	75.99
	AdUni [24]	Self-Supervised	85.16	80.34	81.81
	CL-ISLD [25]	Self-supervised	86.77	85.75	82.11
	CCJ-SLC (ours)	Self-Supervised	87.62	88.53	87.96
ISIC-2019	EW-FCM [26]	Supervised	84.56	80.03	79.65
	DenseNet-161 [21]	Supervised	85.42	79.64	71.20
	EfficientNet-B4 [27]	Supervised	86.12	79.30	84.33
	Moco-v2 [13]	Self-Supervised	74.51	71.81	72.15
	BYOL [4]	Self-Supervised	74.66	70.32	71.17
	SimCLR [5]	Self-Supervised	76.16	71.35	73.20
	Fusion [28]	Self-Supervised	78.43	75.78	74.99
	CCJ-SLC (ours)	Self-Supervised	83.13	80.45	80.56

The classification results of the ISIC-2018 are shown in Table 1. From Table 1, we can find that CCJ-SLC improves the classification accuracy by 7.04%, the balanced accuracy by 9.90%, and the F1-Score by 11.97% compared to BYOL. Compared with other self-supervised learning methods, the performance of CCJ-SLC is better, with all its indicators higher than other self-supervised learning methods. Compared with supervised learning methods, it is closer in classification accuracy and F1-Score values. The balance accuracy of the supervised learning method is exceeded in all detection metrics for the imbalanced dataset. These indicators show that the CCJ-SLC as a self-supervised learning method has achieved a similar effect to supervised learning methods in the ISIC-2018 data set classification, and has a better performance in facing imbalanced classification tasks.

The classification results of the ISIC-2019 model are shown in Table 1. From Table 1, it can be found that among the self-supervised learning methods, CCJ-SLC has the best performance with its classification accuracy of 83.13%, balanced accuracy of 80.45%, and F1-Score of 80.56%. CCJ-SLC improves 6.97% relative to the SimCLR method with the highest classification accuracy and 8.64% relative to the Moco-v2 method with the highest balance accuracy, which indicates that CCJ-SLC has better classification ability in the skin lesion image classification task relative to the single contrast learning method. The Fusion method is similar to the ideological approach of this paper, which combines the SimCLR method with the rotation pretext task. The metrics of CCJ-SLC are all better than the Fusion, which improves the classification accuracy, balance accuracy, and F1-Score by 4.70%, 4.67%, and 5.57%, respectively. The increase in the amount of ISIC-2019 data makes the classification accuracy of self-supervised learning and supervised learning have a certain gap, compared with EW-FCM, Densenet-161, EfficientNet-B4, which are supervised learning methods, although the classification accuracy decreases by 1.43%, 2.29%, and 2.99%, but the balance accuracy improved by 0.42%, 0.81%, and 1.15%, respectively. This indicates that under supervised learning, the true label of the main class with a large amount of data in the imbalanced data set will affect the model to learn the features of the few samples, and the model will be biased to the feature learning of the large samples, while CCJ-SLC can alleviate the impact of imbalanced datasets to a certain extent.

4.3 Ablation Experiment

In order to prove the effectiveness of each pretext task of CCJ-SLC, we will conduct ablation experiments on ISIC-2019, mainly including:

– CCJ-SLC removes jigsaw task (-w/o Jigsaw);
– CCJ-SLC removes the contrastive learning task from the contrastive clustering task (-w/o Contrastive Learning);
– CCJ-SLC removes the clustering task in the contrast clustering task (-w/o Clustering);

Table 2. CCJ-SLC ablation experiments

Method	Jigsaw	Contraction	Cluster	Accuracy (%)
-w/o Jig		✓	✓	76.24
-w/o Con	✓		✓	78.12
-w/o Clu	✓	✓		79.65
-w/o Jig,Con			✓	73.49
-w/o Jig,Clu		✓		75.28
-w/o Con,Clu	✓			55.42
CCJ-SLC	✓	✓	✓	83.13

The experimental results are shown in Table 2. By comparing the results, it can be found that: (1) the classification accuracy of the model only using the jigsaw puzzle task is low, which indicates that the jigsaw puzzle task cannot learn the position information in the complex mixed image to restore the image. (2) The complete CCJ-SLC method proposed in this paper can achieve the best performance, indicating the effectiveness and necessity of each pretext task in each branch task of CCJ-SLC.

4.4 Analysis

Visualization. Figure 4 presents the visualization results of feature maps using class activation mapping (CAM), which clearly shows that our proposed method has a stronger ability to extract detailed features.

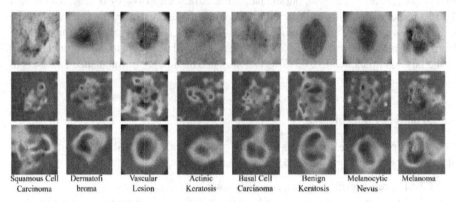

| Squamous Cell Carcinoma | Dermatofi broma | Vascular Lesion | Actinic Keratosis | Basal Cell Carcinoma | Benign Keratosis | Melanocytic Nevus | Melanoma |

Fig. 4. Visualization of skin lesion images. The first row is the original image, the second row is layer 3 of the backbone network, and the third row is layer 4 of the backbone network.

The Impact of Data Augmentation. Since our method requires image division and recombination to enhance the learning ability of the model for skin lesion image features, the sequence of data augmentation and division and recombination may affect the learning ability of the model. The experimental results are shown in Table 3. We could find that if data augmentation is put before division and recombination, the classification accuracy will be greatly reduced. We guess that the reason for the lower classification accuracy is that if the data augmentation is performed before the division and recombination, the image patches may be located at any position of the original image after the horizontal flip, rotation and other operations. On this basis, the image position considered by the model is the position after data augmentation transformation rather than the original position (Table 4).

Table 3. Comparison of classification results of different data Augmentation sequence

Augmentation Position	Accuracy(%)
Augmentation before split	40.15
Augmentation after split (ours)	83.13

Influence of Overlapping Area Size. The experimental results show that when the ratio of overlapping area is 0, it means that each image patches is independent and fragmented, which will make it difficult for the model to learn the relationship and

Table 4. Experimental results on the effect of the ratio of overlapping regions

Ratio	Accuracy(%)	Ratio	Accuracy(%)
0.0	75.68	0.3(ours)	83.13
0.1	78.13	0.4	81.94
0.2	82.25	0.5	79.27

difference between image patches. When the ratio of overlapping regions is more than 0.3, the classification performance of the model decreases. We guess that this may be because too many overlapping regions lead to fuzzy learning features of the model, and the image patches obtained under a large ratio will not form effective positive and negative sample pairs. Therefore, in the experiment of this paper, the ratio of image overlapping area will be set to 0.3.

5 Conclusion

In this paper, we propose a skin lesion image classification model that combines the contrastive clustering and jigsaw tasks, CCL-SLC. Experimental results demonstrate that our method can achieve the effect of supervised learning methods and outperforms other self-supervised learning methods. Our method only considers medical image features, and the learned features are relatively formulaic, while medical images are often heavily influenced by the patient's own condition. In future work we will consider the application of patient information combined with lesion images in self-supervised learning.

References

1. Barbaric, J., Laversanne, M., Znaor, A.: Malignant melanoma incidence trends in a Mediterranean population following socioeconomic transition and war: results of age-period-cohort analysis in Croatia, 1989–2013. Melanoma Res. **27**, 498–502 (2017)
2. Zhang, R., Isola, P., Efros, A.A.: Colorful image colorization. In: 14th European Conference on Computer Vision (ECCV), pp. 649–666 (2016)
3. Larsson, G., Maire, M., et al.: Learning representations for automatic colorization. In: 14th European Conference on Computer Vision (ECCV), pp. 577–593 (2016)
4. Grill, J.-B., Strub, F., et al.: Bootstrap your own latent-a new approach to self-supervised learning. Adv. Neural. Inf. Process. Syst. **33**, 21271–21284 (2020)
5. Chen, T., Kornblith, S., Norouzi, M., Hinton, G., Assoc Informat, S.: A simple framework for contrastive learning of visual representations. In: 25th Americas Conference on Information Systems of the Association-for-Information-Systems(AMCIS) (2019)
6. Chen, S., Xue, J.-H., Chang, J., Zhang, J., Yang, J., Tian, Q.: SSL++: improving self-supervised learning by mitigating the proxy task-specificity problem. IEEE Trans. Image Process. **31**, 1134–1148 (2022)
7. Barata, C., Celebi, M.E., Marques, J.S.: Explainable skin lesion diagnosis using taxonomies. Pattern Recogn. **110** (2021)

8. Datta, S.K., Shaikh, M.A., Srihari, S.N., Gao, M.C.: Soft attention improves skin cancer classification performance. In: MICCAI, pp. 13–23 (2021)
9. Azizi, S., et al.: IEEE: big self-supervised models advance medical image classification. In: 18th IEEE/CVF International Conference on Computer Vision (ICCV), pp. 3458–3468 (2021)
10. Wang, D., Pang, N., Wang, Y.Y., Zhao, H.W.: Unlabeled skin lesion classification by self-supervised topology clustering network. Biomed. Signal Process. Control **66** (2021)
11. Kwasigroch, A., Grochowski, M., Mikolajczyk, A.: Self-supervised learning to increase the performance of skin lesion classification. Electronics **9** (2020)
12. He, K., Fan, H., Wu, Y., Xie, S., Girshick, R.: Momentum contrast for unsupervised visual representation learning. In: Proceedings of the IEEE/CVF Conference on Computer Vision and Pattern Recognition, pp. 9729–9738 (2020)
13. Chen, X., Fan, H., Girshick, R., He, K.: Improved baselines with momentum contrastive learning. arXiv preprint arXiv:2003.04297 (2020)
14. Chen, X., Xie, S., He, K.: An empirical study of training self-supervised vision transformers. In: ICCV, pp. 9640–9649 (2021)
15. Wu, Z.R., Xiong, Y.J., Yu, S.X., Lin, D.H., IEEE: unsupervised feature learning via non-parametric instance discrimination. In: 31st IEEE/CVF Conference on Computer Vision and Pattern Recognition (CVPR), pp. 3733–3742 (2018)
16. Khosla, P., Teterwak, P., Wang, C., Sarna, A., Tian, Y., Isola, P., et al.: Supervised contrastive learning. Adv. Neural. Inf. Process. Syst. **33**, 18661–18673 (2020)
17. Tschandl, P., Rosendahl, C., Kittler, H.: The HAM10000 dataset, a large collection of multi-source dermatoscopic images of common pigmented skin lesions. Sci. data **5**, 1–9 (2018)
18. Codella, N., Rotemberg, V., Tschandl, P., et al.: Skin lesion analysis toward melanoma detection 2018: a challenge hosted by the international skin imaging collaboration (isic). arXiv preprint arXiv:1902.03368 (2019)
19. Combalia, M., et al.: Bcn20000: Dermoscopic lesions in the wild. arXiv preprint arXiv:1908. 02288 (2019)
20. Codella, N.C., Gutman, D., et al.: Skin lesion analysis toward melanoma detection: a challenge at the 2017 international symposium on biomedical imaging (isbi). In: 2018 IEEE 15th International Symposium on Biomedical Imaging (ISBI 2018), pp. 168–172. IEEE (2018)
21. Huang, G., Liu, Z., Van Der Maaten, L., Weinberger, K.Q.: Densely connected convolutional networks. In: Proceedings of the IEEE Conference on Computer Vision and Pattern Recognition, pp. 4700–4708 (2017)
22. Iqbal, I., Younus, M., Walayat, K., Kakar, M.U., Ma, J.: Automated multi-class classification of skin lesions through deep convolutional neural network with dermoscopic images. Comput. Med. Imaging Graph. **88** (2021)
23. Pham, T.-C., Doucet, A., Luong, C.-M., Tran, C.-T., Hoang, V.-D.: Improving skin-disease classification based on customized loss function combined with balanced mini-batch logic and real-time image augmentation. IEEE Access **8**, 150725–150737 (2020)
24. Cong, C., et al.: Adaptive unified contrastive learning for imbalanced classification. In: 13th International Workshop on Machine Learning in Medical Imaging (MLMI), pp. 348–357 (2022)
25. Shi, Y., Duan, C., Chen, S.: IEEE: contrastive learning based intelligent skin lesion diagnosis in edge computing networks. In: IEEE Global Communications Conference (GLOBECOM) (2021)
26. Hoang, L., Lee, S.-H., et al.: Multiclass skin lesion classification using a novel lightweight deep learning framework for smart healthcare. Appl. Sci. Basel **12** (2022)

27. Jaisakthi, S.M., et al.: Classification of skin cancer from dermoscopic images using deep neural network architectures. Multimedia Tools Appl. **82**, 15763–15778 (2023)
28. Verdelho, M.R., Barata, C., IEEE: on the impact of self-supervised learning in skin cancer diagnosis. In: 19th IEEE International Symposium on Biomedical Imaging (IEEE ISBI) (2022)

A Real-Time Network for Fast Breast Lesion Detection in Ultrasound Videos

Qian Dai[1], Junhao Lin[1], Weibin Li[2], and Liansheng Wang[1(✉)]

[1] School of Informatics, Xiamen University, Xiamen, China
lswang@xmu.edu.cn
[2] School of Medicine, Xiamen University, Xiamen, China

Abstract. Breast cancer stands as the foremost cause of cancer-related deaths among women worldwide. The prompt and accurate detection of breast lesions through ultrasound videos plays a crucial role in early diagnosis. However, existing ultrasound video lesion detectors often rely on multiple adjacent frames or non-local temporal fusion strategies to enhance performance, consequently compromising their detection speed. This study presents a simple yet effective network called the Space Time Feature Aggregation Network (STA-Net). Its main purpose is to efficiently identify lesions in ultrasound videos. By leveraging a temporally shift-based space-time aggregation module, STA-Net achieves impressive real-time processing speeds of 54 frames per second on a single GeForce RTX 3090 GPU. Furthermore, it maintains a remarkable accuracy level of 38.7 mean average precision (mAP). Through extensive experimentation on the BUV dataset, our network surpasses existing state-of-the-art methods both quantitatively and qualitatively. These promising results solidify the effectiveness and superiority of our proposed STA-Net in ultrasound video lesion detection.

Keywords: Ultrasound Video · Breast Lesion · Detection

1 Introduction

Breast cancer is a prevalent malignant tumor among women, and its incidence has been increasing in recent years. Early detection plays a crucial role in improving the prognosis of this disease. In clinical practice, breast ultrasound (US) serves as an effective diagnostic tool for identifying breast abnormalities and cancer. It provides precise information about the nature and extent of the disease, enabling clinicians to deliver targeted and effective treatment. Nonetheless, accurately identifying breast lesions in US videos poses a complex and challenging task. The presence of artifacts leads to blurred and unevenly distributed breast lesions, and the accuracy of detection relies heavily on the operator's skill level and experience. Therefore, the objective of our study is to propose an efficient and accurate automatic detection system for breast lesions, which aims to assist doctors in diagnosing and treating breast cancer more effectively in a clinical setting.

Q. Liu et al. (Eds.): PRCV 2023, LNCS 14437, pp. 40–50, 2024.
https://doi.org/10.1007/978-981-99-8558-6_4

The progress of deep learning, specifically Convolutional Neural Networks (CNNs), has made significant strides in the realm of computer-aided diagnosis [4]. However, in the context of ultrasound-based breast cancer diagnosis using deep learning, the emphasis has predominantly been on classifying benign and malignant tumors [7,25]. This approach falls short of meeting the clinical requirements. In clinical breast diagnosis, the accurate detection and identification of lesions, specifically suspicious regions, in US images are of utmost importance.

Several methods have been proposed for identifying lesions in US images [21, 23]. While [9,14,20] aim to achieve automatic segmentation of breast lesions, which requires accurate frame-wise annotation at pixel level and is high-cost than lesion detection. Furthermore, existing methods primarily focus on detecting lesions on 2D US images [5,22]. However, compared to image detection, detecting lesions in US videos is more in line with clinical practice as it provides additional temporal information for the target objects.

While the existing benchmark method CVA-Net [12] has demonstrated promising results in breast lesion detection in ultrasound videos, it heavily relies on a transformer-based attention mechanism to learn temporal features. However, this approach is time-consuming, with a frame rate of 11 frames per second (FPS), making it impractical for real-time clinical applications. Furthermore, ultrasound images are often affected by speckle noise, weak boundaries, and overall low image quality. Therefore, there is significant potential for improvement in breast lesion detection in ultrasound videos. To address these challenges, we present a real-time network called the Space Time Feature Aggregation Network (STA-Net) aimed at enhancing breast lesion detection in ultrasound videos. The proposed STA-Net efficiently learns temporal features by incorporating a temporally shift-based global channel attention module. This module effectively fuses temporal features from adjacent frames into the key feature of the current frame, thereby enhancing the accuracy of lesion detection. Remarkably, our approach achieves a high inference speed of 54 FPS while maintaining superior performance. Experimental results unequivocally demonstrate that our network surpasses state-of-the-art methods in terms of both breast lesion detection in ultrasound videos and video polyp detection benchmark datasets.

2   Method

Figure 1 illustrates a comprehensive visualization of our proposed Space Time Feature Aggregation Network (STA-Net). We begin by considering an ultrasound frame I_t, accompanied by two adjacent video frames (I_{t-1} and I_{t-2}). These frames are initially fed into an Encoder (e.g., the backbone of YOLO-V7 [18]), enabling the acquisition of three distinct features denoted as $f_t \in \mathbb{R}^{C \times WH}$, $f_{t-1} \in \mathbb{R}^{C \times WH}$, and $f_{t-2} \in \mathbb{R}^{C \times WH}$. Here, C, W, and H denote the channels, width, and height of the feature maps, respectively. Subsequently, we introduce a space time feature aggregation (STA) module to facilitate the integration of spatial-temporal features from each video frame. This module plays a pivotal role in combining and leveraging the inherent relationships between the frames. After

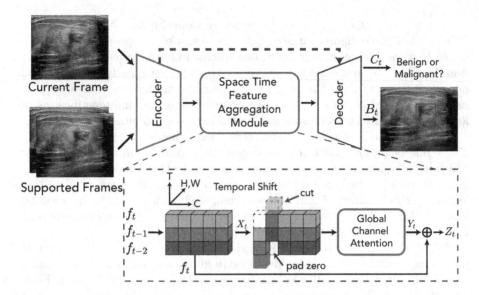

Fig. 1. The schematic illustration showcases our STA-Net, designed for breast lesion detection in ultrasound videos. Our STA-Net incorporates a space-time feature aggregation module, which effectively merges temporal features extracted from neighboring frames, ultimately generating key features for the current frame.

the STA module, the resulting output features Y_t are passed on to the decoder of YOLO-V7. This decoder plays a crucial role in detecting and classifying breast lesions. Leveraging the capabilities of this robust decoder, we achieve enhanced accuracy in identifying and precisely localizing breast lesions.

Loss Function. To effectively address the scale variance of breast lesions in ultrasound video, we employ a fusion approach that combines the binary cross-entropy (BCE) loss function with the mean squared error (MSE) loss function. This fusion enables us to compute the overall lesion detection loss ($\mathcal{L}_{total}$) for the current input frame I_t:

$$\mathcal{L}_{total} = \mathcal{L}_{BCE}(C_t, G_t^{cls}) + \lambda\mathcal{L}_{MSE}(B_t, G_t^{box}) \tag{1}$$

Here, $\mathcal{L}_{BCE}(\cdot)$ represents the BCE loss, and $\mathcal{L}_{MSE}(\cdot)$ represents the MSE loss. C_t refers to the predicted classification of the breast lesion, while G_t^{cls} corresponds to the ground-truth label for classification. Similarly, B_t represents the predicted bounding box of the breast lesion, and G_t^{box} corresponds to the ground-truth label for the bounding box. λ is to control the weight of classification loss, and we empirically set $\lambda = 0.5$ in our experiments.

2.1   Space Time Feature Aggregation (STA) Module

In contrast to the state-of-the-art breast lesion detection method, CVA-Net [12], which relies on self-attention to combine intra/inter video-level features, our proposed approach introduces the Space Time Feature Aggregation (STA) module based on temporal shift. Notably, our module offers a significant advantage by enabling real-time application without the time-consuming overhead associated with the self-attention approach. As depicted in Fig. 1, our STA module utilizes three input features, namely f_t, f_{t-1}, and f_{t-2}. Initially, we concatenate these three features along the temporal dimension, resulting in a new tensor $X_t \in \mathbb{R}^{T \times C \times WH}$, where T represents the number of frames. In Fig. 1, different colors are used to represent features at distinct time stamps in each row. Along the temporal dimension, a portion of the channels is shifted by -1, another portion by $+1$, while the remaining channels remain un-shifted. Afterward, we pass the temporally shifted tensor X_t into our designed global channel attention block to obtain a refined feature map Y_t. We first use L2-norm to aggregate the feature map X_t into a vector $Gx \in \mathbb{R}^{C \times T}$, then we pass it through a linear layer, a ReLU activation function, a linear layer, and a sigmoid function to compute an attention map. Lastly, we perform an element-wise multiplication between the temporally shifted tensor X_t and the obtained attention map. We then compute the average value along the temporal dimension to obtain the refined tensor Y_t. Algorithm 1 shows the pseudo code of our global channel attention module.

Algorithm 1: Pseudo Code of Global Channel Attention Block

Data:
x: input feature, B: batch size, C: dimension of channel,
T: number of frames, H, W: spatial size of input feature

1 Gx = torch.norm(inputs, $p = 1$, dim = (2, 3), keepdim = True) # (1, $C \times T$, 1, 1)
2 x = nn.Linear(Gx)
3 x = F.relu(x)
4 x = nn.Linear(x)
5 x = torch.sigmoid(x).view(1, $C \times T$, 1, 1)
6 inputs = inputs * x
7 y = inputs.view(B, T, C, H, W).mean(dim = 1)
8 return y

---

3 Experiments and Results

Implementation Details. We initial the backbone of our network by pretraining on the COCO dataset, and the rest of our network is trained from scratch. All the training video frames are resized to 512×512 before feeding into the network. Our network is implemented on Pytorch [15] and uses SGD optimizer [13] with 100 epochs, an initial learning rate of 5×10^{-5}, and a batch-size of 20. The whole architecture is trained on four GeForce RTX 3090 GPUs.

Dataset. In line with CVA-Net, we conducted our experiments using the publicly available BUV dataset [12]. Our proposed method was evaluated on this dataset, which comprises 113 malignant videos and 75 benign videos. Furthermore, we followed the official data split for our evaluation.

3.1 Comparisons with State-of-the-Arts

We compare our network against eleven state-of-the-art methods, including five image-based methods and six video-based methods. Five image-based methods are GFL [10], Faster R-CNN [16], VFNet [24], RetinaNet [11], and YOLO-V7 [18], while six video-based methods are DFF [27], FGFA [26], SELSA [19], Temporal ROI Align [6], CVA-Net [12], and FAQ [3]. We employ the same evaluation metrics, namely mean average precision, as utilized in the CVA-Net [12].

Table 1. Quantitative comparisons of our network and compared 11 state-of-the-art methods.

Method	Type	AP	AP$_{50}$	AP$_{75}$	Size	Params (M)	FLOPs(G)	FPS
GFL [10]	image	23.4	46.3	22.2	512	32	53	53
Faster R-CNN [16]	image	25.2	49.2	22.3	512	42	64	59
VFNet [24]	image	28.0	47.1	31.0	512	33	49	35
RetinaNet [11]	image	29.5	50.4	32.4	512	38	61	37
YOLO-V7 [18]	image	31.6	53.0	35.9	512	36	103	**63**
DFF [27]	video	25.8	48.5	25.1	512	101	71	60
FGFA [26]	video	26.1	49.7	27.0	512	103	709	25
SELSA [19]	video	26.4	45.6	30	512	70	810	14
Temporal ROI Align [6]	video	29.0	49.9	33.1	512	78	237	7
CVA-Net [12]	video	36.1	65.1	38.5	512	41	330	11
FAQ [3]	video	32.2	52.3	37.4	512	47	503	9
Ours	video	**38.7**	**65.5**	**41.5**	512	39	103	54

Quantitative Comparisons. Table 1 presents a summary of the quantitative results obtained from our network and eleven other breast lesion detection methods compared in this study. Upon analyzing the quantitative results depicted in Table 1, it becomes apparent that video-based approaches generally outperform image-based methods in terms of the evaluated metrics. Among the eleven methods compared, CVA-Net [12] achieves the highest mean average precision (AP), while YOLO-V7 [18] exhibits the fastest inference speed. Significantly, our STA-Net surpasses CVA-Net [12] in both mean AP and FPS. More specifically, STA-Net improves the AP score from 36.1 to 38.7 and increases the FPS score from 11 to 54. It can be seen that YOLO-V7 [18] has the fastest inference speed, which reflects the advanced nature of its network architecture, however, our improved STA-Net based on it strikes a better balance between performance and speed, offering superior trade-offs in these aspects.

Qualitative Comparisons. Figure 2 displays visual comparisons of breast lesion detection results between our network and the state-of-the-art method CVA-Net [12] across different input video frames. Notably, our method exhibits precise detection of breast lesions in the input ultrasound video frames, effectively capturing target lesions of various sizes and diverse shapes.

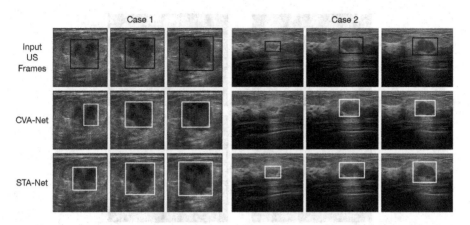

Fig. 2. Visual comparisons of video breast lesion detection results obtained by our network and the state-of-the-art method CVA-Net are showcased for multiple frames of two ultrasound videos. The ground truth is represented by a red bounding box, while the prediction result is denoted by a green bounding box. (Color figure online)

3.2   Ablation Study

To assess the effectiveness of the major components in our network, we developed three baseline networks. The first one, referred to as "Basic", excluded our core-designed STA module, which corresponds to YOLO-V7 [18]. The second and third baseline networks, named "Basic+TS" and "Basic+GCA", respectively, integrated the temporal shift (TS) operation and the global channel attention (GCA) module into the basic network. In Table 2, we present the quantitative results of our method and the three baseline networks. It is evident that "Basic+TS" and "Basic+GCA" outperform "Basic" in terms of metric performance. This finding demonstrates that our TS operation and GCA module effectively enhance the detection performance of breast lesions in ultrasound videos. Furthermore, the superior performance of "Ours" compared to "Basic+TS" and "Basic+GCA" illustrates that the combination of our TS operation and GCA module produces more accurate detection results.

Table 2. Quantitative comparisons results of ablation study experiments.

Method	TS	GCA	AP	AP_{50}	AP_{75}
Basic	×	×	31.6	53.0	35.9
Basic+TS	✓	×	33.4	55.6	38.3
Basic+GCA	×	✓	34.6	57.1	39.0
Ours	✓	✓	**38.7**	**65.5**	**41.5**

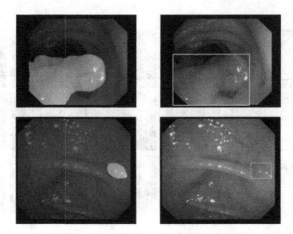

Fig. 3. Shadow in left column images is the pixel-wise notation, bounding box with yellow rectangle is the notation used during our training for detection task. (Color figure online)

Table 3. Quantitative comparisons of our network and compared 11 state-of-the-art methods in video polyp detection dataset.

Method	Type	AP	AP_{50}	AP_{75}
GFL [10]	image	44.9	78.2	47.4
Faster R-CNN [16]	image	43.9	75.2	45.4
VFNet [24]	image	48.4	83.5	48.9
RetinaNet [11]	image	44.3	78.9	46.2
YOLO-V7 [18]	image	50.6	85.2	49.4
DFF [27]	video	12.2	28.8	8.23
FGFA [26]	video	40.7	81.5	37.0
SELSA [19]	video	49.3	84.1	51.1
Temporal ROI Align [6]	video	46.9	81.2	49.5
CVA-Net [12]	video	40.6	82.6	35.0
FAQ [3]	video	39.2	81.2	33.5
Ours	video	**58.5**	**90.7**	**66.9**

3.3 Generalizability of Our Network

We further evaluate the effectiveness of our STA-Net by testing it on the task of video polyp detection. We select the three most common video-based polyp datasets (*i.e.*, CVC-300 [2], CVC-612 [1] and ASU-Mayo [17]) as our video polyp detection benchmark datasets. All the annotations provided by these datasets are pixel-wise masks, whereas our task requires the bounding boxes of ground truth. So we converted all the masks to rectangle boxes and got the bounding box information, as shown in Fig. 3. We only adopt 10 positive samples from

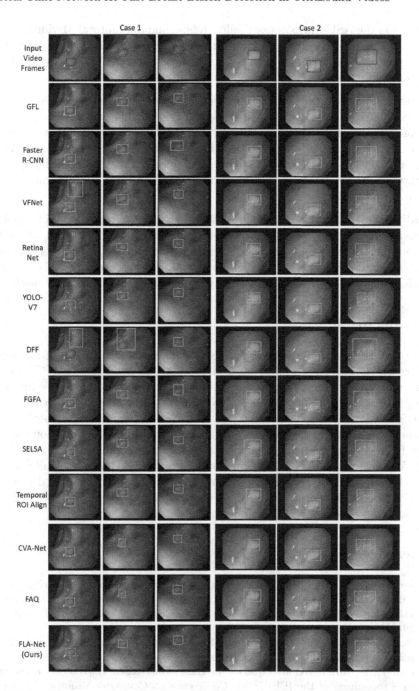

Fig. 4. Visual comparisons of video polyp detection results obtained by our network and 11 state-of-the-art methods are showcased for multiple frames of two videos. The ground truth is represented by a red bounding box, while the prediction result is denoted by a green bounding box. (Color figure online)

patients in the ASU-Mayo dataset following [8], and obtain total 51 videos along with CVC-300 (12 videos) and CVC-612 (29 videos). All videos belong to the same category - polyp, and then we split the videos into 70% (35 videos) for training and 30% (16 videos) for validation.

Quantitative Comparisons. Table 3 reports the results of our STA-Net and the remaining 11 state-of-the-art methods on the video polyp detection dataset. Apparently, our method clearly outperforms the state-of-the-art methods on all three metrics. Specifically, our approach increased the mean AP to 58.5, the AP_{50} over 90, and the AP_{75} over 65. Therefore, our superior experimental results in the extra video polyp detection dataset show the effectiveness and generalizability of our proposed STA-Net.

Qualitative Comparisons. Figure 4 shows visual comparisons of video polyp detection results between our STA-Net and the 11 state-of-the-art methods across different input video frames. Clearly, our method also exhibits superior performance on the video polyp detection task. In the absence of missed detection and false detection, the detection of polyps is more precise than other methods.

4 Conclusion

In this study, we have presented STA-Net, a temporally shift-based framework for accurately detecting lesions from ultrasound videos in real-time (54 FPS). Our framework integrates a basic space time feature aggregation module that can easily be incorporated into existing CNN-based architectures. Through extensive experiments, we have demonstrated that STA-Net outperforms other methods on two publicly available medical video object detection datasets. Ablation studies have further confirmed the effectiveness of the core components in our STA-Net. We believe that STA-Net can not only advance the field of ultrasound video lesion detection but also contribute to other related video-based medical detection tasks. Our future work will involve evaluating the performance of STA-Net on larger video-based medical detection datasets.

References

1. Bernal, J., Sánchez, F.J., Fernández-Esparrach, G., Gil, D., Rodríguez, C., Vilariño, F.: WM-DOVA maps for accurate polyp highlighting in colonoscopy: validation vs. saliency maps from physicians. Comput. Med. Imaging Graph. **43**, 99–111 (2015)
2. Bernal, J., Sánchez, J., Vilarino, F.: Towards automatic polyp detection with a polyp appearance model. Pattern Recogn. **45**(9), 3166–3182 (2012)
3. Cui, Y.: Feature aggregated queries for transformer-based video object detectors. In: Proceedings of the IEEE/CVF Conference on Computer Vision and Pattern Recognition (CVPR), pp. 6365–6376, June 2023
4. Doi, K.: Computer-aided diagnosis in medical imaging: historical review, current status and future potential. Comput. Med. Imaging Graph. **31**(4–5), 198–211 (2007)

5. Drukker, K., Giger, M.L., Horsch, K., Kupinski, M.A., Vyborny, C.J., Mendelson, E.B.: Computerized lesion detection on breast ultrasound. Med. Phys. **29**(7), 1438–1446 (2002)
6. Gong, T., et al.: Temporal ROI align for video object recognition. In: Proceedings of the AAAI Conference on Artificial Intelligence, vol. 35, pp. 1442–1450 (2021)
7. Huang, X., Lin, Z., Huang, S., Wang, F.L., Chan, M.T., Wang, L.: Contrastive learning-guided multi-meta attention network for breast ultrasound video diagnosis. Front. Oncol. **12**, 952457 (2022)
8. Ji, G.P., et al.: Progressively normalized self-attention network for video polyp segmentation. In: Cattin, P.C., et al. (eds.) MICCAI 2021. LNCS, vol. 12901, pp. 142–152. Springer, Cham (2021). https://doi.org/10.1007/978-3-030-87193-2_14
9. Li, J., et al.: Rethinking breast lesion segmentation in ultrasound: a new video dataset and a baseline network. In: Wang, L., Dou, Q., Fletcher, P.T., Speidel, S., Li, S. (eds.) Proceedings of the 25th International Conference on Medical Image Computing and Computer Assisted Intervention, MICCAI 2022, Part IV, 18–22 September 2022, Singapore, pp. 391–400. Springer, Cham (2022). https://doi.org/10.1007/978-3-031-16440-8_38
10. Li, X., et al.: Generalized focal loss: Learning qualified and distributed bounding boxes for dense object detection. arXiv preprint arXiv:2006.04388 (2020)
11. Lin, T.Y., Goyal, P., Girshick, R., He, K., Dollár, P.: Focal loss for dense object detection. In: Proceedings of the IEEE International Conference on Computer Vision, pp. 2980–2988 (2017)
12. Lin, Z., Lin, J., Zhu, L., Fu, H., Qin, J., Wang, L.: A new dataset and a baseline model for breast lesion detection in ultrasound videos. In: Wang, L., Dou, Q., Fletcher, P.T., Speidel, S., Li, S. (eds.) Medical Image Computing and Computer Assisted Intervention, MICCAI 2022, , vol. 13433, pp. 614–623. Springer, Cham (2022). https://doi.org/10.1007/978-3-031-16437-8_59
13. Montavon, G., Orr, G., Müller, K.R.: Neural Networks: Tricks of the Trade, 2nd edn., January 2012. https://doi.org/10.1007/978-3-642-35289-8
14. Ning, Z., Zhong, S., Feng, Q., Chen, W., Zhang, Y.: SMU-Net: saliency-guided morphology-aware u-net for breast lesion segmentation in ultrasound image. IEEE Trans. Med. Imaging **41**(2), 476–490 (2021)
15. Paszke, A., et al.: PyTorch: an imperative style, high-performance deep learning library. In: Wallach, H., Larochelle, H., Beygelzimer, A., d' Alché-Buc, F., Fox, E., Garnett, R. (eds.) Advances in Neural Information Processing Systems, vol. 32, pp. 8024–8035. Curran Associates, Inc. (2019). https://papers.neurips.cc/paper/9015-pytorch-an-imperative-style-high-performance-deep-learning-library.pdf
16. Ren, S., He, K., Girshick, R., Sun, J.: Faster R-CNN: towards real-time object detection with region proposal networks. arXiv preprint arXiv:1506.01497 (2015)
17. Tajbakhsh, N., Gurudu, S.R., Liang, J.: Automated polyp detection in colonoscopy videos using shape and context information. IEEE Trans. Med. Imaging **35**(2), 630–644 (2015)
18. Wang, C.Y., Bochkovskiy, A., Liao, H.Y.M.: YOLOv7: trainable bag-of-freebies sets new state-of-the-art for real-time object detectors. In: Proceedings of the IEEE/CVF Conference on Computer Vision and Pattern Recognition, pp. 7464–7475 (2023)
19. Wu, H., Chen, Y., Wang, N., Zhang, Z.: Sequence level semantics aggregation for video object detection. In: 2019 IEEE/CVF International Conference on Computer Vision (ICCV), pp. 9216–9224 (2019)
20. Xue, C., et al.: Global guidance network for breast lesion segmentation in ultrasound images. Med. Image Anal. **70**, 101989 (2021)

21. Yang, Z., Gong, X., Guo, Y., Liu, W.: A temporal sequence dual-branch network for classifying hybrid ultrasound data of breast cancer. IEEE Access **8**, 82688–82699 (2020)
22. Yap, M.H., et al.: Automated breast ultrasound lesions detection using convolutional neural networks. IEEE J. Biomed. Health Inform. **22**(4), 1218–1226 (2017)
23. Zhang, E., Seiler, S., Chen, M., Lu, W., Gu, X.: BIRADS features-oriented semi-supervised deep learning for breast ultrasound computer-aided diagnosis. Phys. Med. Biol. **65**(12), 125005 (2020)
24. Zhang, H., Wang, Y., Dayoub, F., Sunderhauf, N.: VarifocalNet: an IoU-aware dense object detector. In: IEEE/CVF Conference on Computer Vision and Pattern Recognition (CVPR), pp. 8510–8519 (2021)
25. Zhao, G., Kong, D., Xu, X., Hu, S., Li, Z., Tian, J.: Deep learning-based classification of breast lesions using dynamic ultrasound video. Eur. J. Radiol. **165**, 110885 (2023)
26. Zhu, X., Wang, Y., Dai, J., Yuan, L., Wei, Y.: Flow-guided feature aggregation for video object detection. In: IEEE International Conference on Computer Vision (ICCV), pp. 408–417 (2017)
27. Zhu, X., Xiong, Y., Dai, J., Yuan, L., Wei, Y.: Deep feature flow for video recognition. In: Proceedings of the IEEE Conference on Computer Vision and Pattern Recognition, pp. 2349–2358 (2017)

CBAV-Loss: Crossover and Branch Losses for Artery-Vein Segmentation in OCTA Images

Zetian Zhang[1], Xiao Ma[1], Zexuan Ji[1], Na Su[2], Songtao Yuan[2], and Qiang Chen[1](✉)

[1] School of Computer Science and Engineering, Nanjing University of Science and Technology,
Nanjing 210094, China
chen2qiang@njust.edu.cn
[2] Department of Ophthalmology, The First Affiliated Hospital of Nanjing Medical University,
Nanjing 210029, China

Abstract. Obvious errors still exist in the segmentation of artery-vein (AV) in retinal optical coherence tomography angiography (OCTA) images, especially near crossover and branch points. It is believed that these errors occur because the existed segmentation method cannot effectively identify the crossover and branch points of AV. In this study, we proposed a Crossover Loss and a Branch Loss (CBAV-Loss), which are two novel structure-preserving loss functions. By restricting the crossover and branch points of arteries and veins, the segmentation accuracy can be improved by correcting the segmentation errors near the crossover and branch points. The experimental results on a manually annotated AV dataset with 400 OCT and OCTA cubes demonstrate that the crossover and branch losses can effectively reduce errors for AV segmentation and preserve vascular connectivity to a certain extent.

Keywords: Artery-vein Segmentation · Crossover Loss · Branch Loss · Optical Coherence Tomography Angiography

1 Introduction

The retina is a complex neurovascular network that can often reveal eye diseases [20]. Early diagnosis of disease and effective treatment assessment are essential to prevent vision loss. Artery-vein (AV) segmentation can provide a reliable basis for the diagnosis of retinal diseases, such as diabetes [2, 3]. The width of Retinal vessel is highly related to the incidence or risks of many diseases, and the retinal arteriolar-to-venular ratio has become an important indicator of diseases [4, 5]. Therefore, AV segmentation provides a stable basis for the quantitative analysis of retinal arteries and veins [12].

At present, automatic AV segmentation works are mainly based on color fundus photographs [13–22], which provide limited resolution and sensitivity to identity microvascular [11]. Optical coherence tomography angiography (OCTA) is a non-invasive fundus imaging technique that can identify retinal blood signals with high resolution. Currently, OCTA has been widely used to quantify vessels [1], including artery and vein. Automatic AV segmentation has also been applied to OCTA images. AV-Net provides a platform for

Q. Liu et al. (Eds.): PRCV 2023, LNCS 14437, pp. 51–60, 2024.
https://doi.org/10.1007/978-981-99-8558-6_5

deep learning based AV segmentation in OCTA, an FCN based on a modified U-shaped CNN architecture [3]. MF-AV-Net deeply analyzes how OCT-OCTA multimodal fusion strategy affects AV segmentation performance [6].

The existing methods of AV segmentation in OCTA still have the following limitations: (1) Segmentation errors near AV crossover points [7]. (2) Segmentation errors near AV branch points. (3) Inadequate connectivity in vessels, such as small segments of mis-segmented vein in an artery. (4) Segmentation errors at the end of the vessel. Figure 1 shows (1) and (2) in the yellow box and (3) in the green box. In these limitations, (3) & (4) lead to incomplete segmentation details, but (1) & (2) lead to obvious segmentation errors. Therefore, compared with the limitations (3) & (4), it is more important to solve the limitations (1) & (2). In this paper, we proposed Crossover Loss and Branch Loss (CBAV-Loss) for limitations (1) & (2).

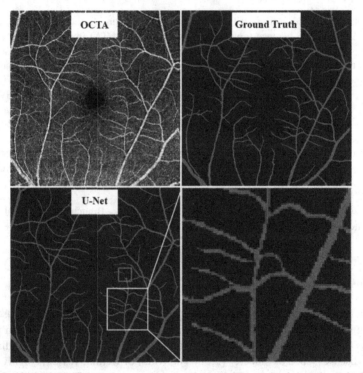

Fig. 1. Motivation: The figure shows a 6 mm × 6 mm OCTA projection at the retinal macular region, where a typical example of segmentation errors is shown and segmented by U-Net model, red and blue represent arteries and veins, respectively. In the yellow box, there are many crossover points of arteries and veins, while there are many branches on the arteries and veins themselves. In this complex area of AV intersection, the model is most prone to segmentation errors, and we find that this characteristic error is also the most common type of error in AV segmentation. In the green box, there is a segmentation discontinuity problem, which is also a common problem in retinal vessel segmentation. In this work, we mainly address the specific AV segmentation errors that exist in the yellow box. (Color figure online)

In order to solve the problems in AV segmentation, we proposed new loss functions, crossover-loss and branch-loss, which are used to constrain the segmentation results near the crossover and branch points of arteries and veins, respectively. The experiment results on a dataset containing 400 OCT and OCTA cubes show that our proposed crossover-loss and branch-loss can effectively improve the accuracy of AV segmentation.

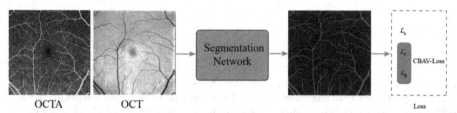

Fig. 2. Overview of the AV segmentation network with our proposed CBAV-Loss. The input of the model is OCT and OCTA images, and the output is pixel-level AV segmentation map. In addition to the basic loss $\mathcal{L}_b$, which is the liner combination of cross-entropy loss and dice loss, we also use crossover loss $\mathcal{L}_C$ and branch loss $\mathcal{L}_B$, which are collectively referred to as CBAV-Loss. (Color figure online)

2   Methods

2.1   Overview

In this paper, we proposed two losses to maintain the accuracy of the local structure of the AV segmentation, namely Crossover Loss and Branch Loss, which are collectively referred to as CBAV-Loss. As shown in Fig. 2, the same as the conventional AV segmentation, the input is OCT and OCTA images, and the output is pixel-level AV segmentation map, where, red, blue and black represent arteries, veins and back-ground, respectively.

2.2   Crossover Loss and Branch Loss

In AV segmentation, the most common error occurs near the crossover and branch points, because the segmentation network cannot accurately identify the crossover and branch points. Near the crossover point, two kinds of blood vessels intersect, while near the branch point, one blood vessel produces two or more branches. Therefore, the number of artery and vein pixels near the crossover point should be similar, while the number of artery and vein pixels near the branch points should have a large difference. As shown in Fig. 3, if the segmentation network divides the crossover points into branch points by mistake, the area where the artery and vein intersect will be divided into the same blood vessel; if the segmentation network divides the branch points into crossover points by mistake, the area of the same blood vessel will be divided into artery and vein intersection areas. To constrain the segmentation results near the AV crossover points and branch points, we proposed two novel structure-preserving losses, namely Crossover Loss and Branch Loss.

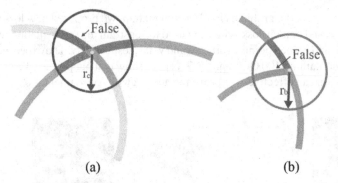

Fig. 3. Common errors for AV segmentation in OCTA images. (a) Common errors near the crossover points. (b) Common errors near the branch points. In these two images, the arrows point to the wrong segments. r_c and r_b represent the operating radius of $\mathcal{L}_{cross}$ and $\mathcal{L}_{branch}$ respectively. Red and blue curves represent arteries and veins respectively. (Color figure online)

Therefore, we restrict the pixels near the crossover and branch points to reduce the error rate in those areas. From the point of view of the aim, both the Crossover Loss and the Branch Loss are the statistics of the pixel error rate in the target area, so we will uniformly define these two losses.

We consider four binary masks: the artery ground truth mask (G_A), the vein ground truth mask (G_V), the artery predicted segmentation mask (P_A) and the vein predicted segmentation mask (P_V). First, the artery skeleton S_A and the vein skeleton S_V are extracted from G_A and G_V respectively, meanwhile, the artery branch $M_{branchA}$ and the vein branch $M_{branchV}$ are extracted from S_A and S_V. To get $M_{branchA}$ and $M_{branchV}$, for S_A and S_V, we believe that each pixel with more than or equal to 3 neighbors is a branch point. Subsequently, we compute the crossover-loss $\mathcal{L}_{cross}$, and the branch-loss $\mathcal{L}_{branch}$ as defined bellow:

$$\mathcal{L}_{cross} = \frac{\|M_{cross}*D_{A,c}\|}{r_c^2*\|M_{cross}\|} + \frac{\|M_{cross}*D_{V,c}\|}{r_c^2*\|M_{cross}\|} \tag{1}$$

$$\mathcal{L}_{branch} = \frac{\|M_{branchA}*D_{A,b}\|}{r_b^2*\|M_{branchA}\|} + \frac{\|M_{branchV}*D_{V,b}\|}{r_b^2*\|M_{branchV}\|} \tag{2}$$

$$D_{\alpha,\beta} = Cov(P_\alpha - G_\alpha, R_\beta) \tag{3}$$

where, $M_{cross} = S_A \cap S_V$, $Cov(A, R)$ represents the convolution of matrix A and the all-one matrix R with the size of $(1 + 2r) * (1 + 2r)$. R_c and R_b represent the all-one matrix with radius r_c and r_b, respectively, meanwhile, r_c and r_b represent the $\mathcal{L}_{cross}$ and $\mathcal{L}_{branch}$ operation ranges, respectively. In order to cover the widest vessel, the value of r_c should be large enough but not too big. $\mathcal{L}_{branch}$ is a consistent constraint for the same class of vessels, so r_b should be small. In this paper, r_c and r_b are set to 8 and 5, respectively.

2.3 Loss Function

In this paper, all segmentation networks were trained using a joint loss, including cross-entropy loss, dice loss [23], our proposed crossover-loss and branch-loss, and it was defined as

$$\mathcal{L}_{total} = \lambda_1 \mathcal{L}_{CE} + \lambda_2 \mathcal{L}_{dice} + \lambda_3 \mathcal{L}_{cross} + \lambda_4 \mathcal{L}_{branch} \tag{4}$$

where $\mathcal{L}_{CE}$ is the cross-entropy loss, $\mathcal{L}_{dice}$ is the dice loss, $\mathcal{L}_{cross}$ and $\mathcal{L}_{branch}$ are defined in Sect. 2.2. λ_1, λ_2, λ_3 and λ_4 are the weights to balance different loss terms. In this paper, according to experience, λ_1, λ_2, λ_3 and λ_4 are set to 1.0, 0.6, 0.005 and 0.005, respectively.

For convenience, $\lambda_1 \mathcal{L}_{CE} + \lambda_2 \mathcal{L}_{dice}$ is called $\mathcal{L}_b$, $\mathcal{L}_{cross}$ and $\mathcal{L}_{branch}$ are abbreviated as $\mathcal{L}_C$ and $\mathcal{L}_B$ respectively, and $\lambda_3 \mathcal{L}_{cross} + \lambda_4 \mathcal{L}_{branch}$ is collectively called CBAV-Loss.

3 Experiments

3.1 Data

The proposed crossover-loss and branch-loss were validated on a database that includes 400 OCTA volumes and the corresponding OCT volumes. Of these, 100 eyes were provided by Jiangsu Province Hospital and another 300 eyes were from the OCTA-500 dataset [25]. Among 300 eyes from the OCTA-500 dataset, the proportion of subjects with ophthalmic diseases is 69.7%. The diseases include age-related macular degeneration (AMD), diabetic retinopathy (DR), choroidal neovascularization (CNV), central serous chorioretinopathy (CSC), retinal vein occlusion (RVO), and other. The other 100 eyes are from people with myopia of different diopter. The research was approved by the institutional human subjects committee and followed the tenets of the Declaration of Helsinki. For each eye, OCTA volumes and OCT volumes were obtained together from AngioVue (Version 2017.1.0.151; Optovue, Fremont, CA). The size of the OCT volume is $640 \times 400 \times 400$ voxels corresponding to a 2 mm $\times$ 6 mm $\times$ 6 mm volume centered at the retinal macular region. The size of OCTA volume is $160 \times 400 \times 400$ voxels.

The ground truth of AV segmentation was drawn on the OCTA maximum projection maps between the Internal limiting membrane (ILM) layer and the outer plexiform layer (OPL). The ILM layer and OPL layer were generated by software (OCTExplorer3.8). Two experts participated in ground truth annotation and one senior specialist checked. Since arteries and veins may overlap at the same position, and the axial position of arteries and veins cannot be accurately determined, the arterial and venous ground truth were annotated separately.

3.2 Experimental Settings

In this paper, all experiments were implemented with Python (v3.9) using torch (v1.12.0) on GeForce RTX 3090 GPU and Intel(R) Xeon(R) Gold 6226R CPU. The size of all input and output images is 400×400. The segmentation network was trained and optimized with Adam optimizer with a batch size of 4 and an initial learning rate of 0.00005. The maximum number of epochs is set to 300, and the frequency of model saving is per epoch. In all experiments, we followed a 360/40 train/test split procedure.

3.3 Evaluation Metrics

To evaluate the network performance, we introduced three metrics to measure the segmentation results quantitatively, including Dice similarity coefficient (Dice), Intersection over union (IoU) and vascular connectivity [26]. Among them, Dice and IoU are common evaluation metrics for image segmentation, and vascular connectivity is used to evaluate the performance of the vascular connectivity, which is defined as the radio of the number of the connected pixels to the total number of pixels on the skeleton map [9]. The connected pixels are defined as any connected domain with at a length of at least 10 (including diagonal connections) [10]. Different from the general studies that only consider vascular connectivity, the connectivity of arteries and veins need to be evaluated. Therefore, the results of vascular connectivity are the mean of arteries and veins in this paper.

3.4 Ablation Study on CBAV-Loss

In order to verify the effectiveness of our proposed CBAV-Loss, we conducted an ablation study on the combination of different loss functions using U-Net [8] as the backbone segmentation network. We set up four experiments: 1) only the basic segmentation loss $\mathcal{L}_b$, i.e. linear combinations of cross-entropy loss and Dice loss. 2) the basic segmentation loss $\mathcal{L}_b$ and the Crossover Loss $\mathcal{L}_C$. 3) the basic segmentation loss $\mathcal{L}_b$ and the Branch Loss $\mathcal{L}_B$. 4) the basic segmentation loss $\mathcal{L}_b$ and two proposed losses $\mathcal{L}_C$ and $\mathcal{L}_B$ (CBAV-Loss). The quantitative experimental results are shown in Table 1. It shows that the IoU, Dice and vascular connectivity of experiments 2) and 3) are higher than those of experiment 1), which indicates that both $\mathcal{L}_C$ and $\mathcal{L}_B$ can improve the performances of the segmentation network and vascular connectivity. The IoU, Dice and vascular connectivity of experiment 4) achieve the highest and much higher than those of experiment 1), which indicates that the segmentation network with the CBAV-Loss is the most effective network for AV segmentation, and better than only with $\mathcal{L}_b$.

Figure 4 shows one example with different combinations of loss functions. It can be seen that the result with $\mathcal{L}_C$ loss is slightly better than that with $\mathcal{L}_B$ loss, and they are much better than that with only $\mathcal{L}_b$ loss, which shows that $\mathcal{L}_C$ and $\mathcal{L}_B$ play a positive role in AV segmentation. The result of $\mathcal{L}_C$ and $\mathcal{L}_B$ losses used at the same time is better than that of using one of them alone. As shown in the yellow box in Fig. 4, this is an area with many crossover and branch points. In the results with the loss $\mathcal{L}_b$ (Fig. 4c), due to the complex vascular structure in this area, there are many errors near the crossover and branch points. In the results with the losses $\mathcal{L}_b$ and $\mathcal{L}_B$ (Fig. 4e), the errors near the branch points have been corrected, but the errors near the crossover points still exist. In the results with the losses $\mathcal{L}_b$, $\mathcal{L}_C$ and $\mathcal{L}_B$ (Fig. 4f), the errors near both the crossover and branch points have been corrected well. Therefore, we can find that using $\mathcal{L}_C$ and $\mathcal{L}_B$ can effectively solve the problems of some obvious segmentation errors. It demonstrates that the advantage of our proposed CBAV-Loss is to enhance the network attention to crossover and branch points.

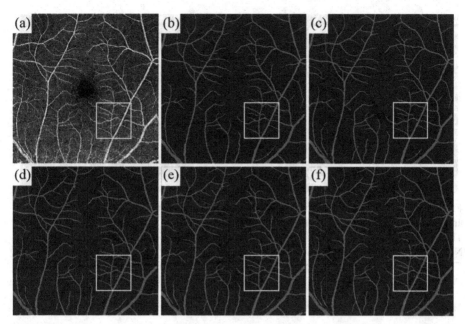

Fig. 4. Qualitative comparison of different combinations of loss functions. This is a sample with ametropia from U-Net model. (a) The OCTA projection map. (b) The ground truth. (c) The result with the loss $\mathcal{L}_b$ only. (d) The result with the losses $\mathcal{L}_b$ and $\mathcal{L}_C$. (e) The result with the losses $\mathcal{L}_b$ and $\mathcal{L}_B$. (f) The result with the losses $\mathcal{L}_b$, $\mathcal{L}_C$ and $\mathcal{L}_B$. In the images, red, blue and black represent arteries, veins and background, respectively. (Color figure online)

Table 1. Quantitative results under the supervision of different combination of loss functions.

Loss	IoU(%)		Dice(%)		Connectivity
	Artery/Vein	Mean	Artery/Vein	Mean	
$\mathcal{L}_b$	82.62/80.59	81.61	90.44/89.17	89.81	98.24
$\mathcal{L}_b, \mathcal{L}_C$	83.44/81.54	82.49	90.93/89.75	90.34	98.50
$\mathcal{L}_b, \mathcal{L}_B$	83.05/81.22	82.14	90.69/89.55	90.12	98.33
$\mathcal{L}_b, \mathcal{L}_C, \mathcal{L}_B$	**84.10/83.06**	**83.58**	**91.32/90.67**	**91.00**	**98.59**

3.5 Influence of the Proposed Loss on Different Segmentation Networks

To evaluate the performance of the proposed CBAV-Loss, we selected 4 segmentation models as the baselines, which are AV-Net [11], MF-AV-Net [6], Attention U-Net [24] and U-Net [8]. For each baseline network, we conducted two sets of experiments, namely (1) $\mathcal{L}_b$ and (2) the liner combination of $\mathcal{L}_b$ and CBAV-Loss. The comparative results are summarized in Table 2. For each baseline method, whether IoU, Dice or vascular connectivity, CBAV-Loss is used to achieve the best segmentation results.

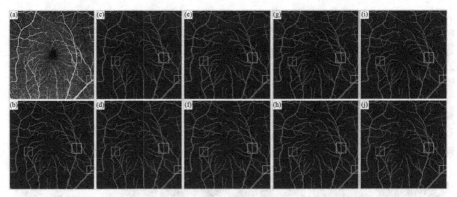

Fig. 5. Qualitative comparison of four baseline segmentation models. This is a sample with ametropia. (a) The OCTA projection map. (b) The ground truth. (c) The result of AV-Net. (d) The result of AV-Net with CBAV-Loss. (e) The result of MF-AV-Net. (f) The result of MF-AV-Net with CBAV-Loss. (g) The result of Attention U-Net. (h) The result of Attention U-Net with CBAV-Loss. (i) The result of U-Net. (j) The result of U-Net with CBAV-Loss. (Color figure online)

Table 2. Quantitative comparison for the AV segmentation of different methods with CBAV-Loss.

Method	Loss	IoU (%)		Dice (%)		Connectivity
		Artery/Vein	Mean	Artery/Vein	Mean	
AV-Net	$\mathcal{L}_b$	78.70/76.31	77.51	87.99/86.42	87.21	96.82
	$\mathcal{L}_{total}$	79.19/77.42	78.31	88.28/87.15	87.72	97.22
MF-AV-Net	$\mathcal{L}_b$	77.41/75.93	76.67	87.16/86.20	86.68	96.65
	$\mathcal{L}_{total}$	78.65/76.96	77.81	87.93/86.81	87.37	97.22
Attention U-Net	$\mathcal{L}_b$	82.23/80.67	81.45	90.14/89.18	89.66	97.72
	$\mathcal{L}_{total}$	82.62/80.74	81.68	90.41/89.24	89.83	97.87
U-Net	$\mathcal{L}_b$	82.62/80.59	81.61	90.44/89.17	89.81	98.24
	$\mathcal{L}_{total}$	84.10/83.06	83.58	91.32/90.67	91.00	98.59

Figure 5 shows the qualitative comparison for a sample with ametropia. In the orange, yellow and green boxes, there are some obvious segmentation errors in the results of AV-Net (Fig. 5c) and MF-AV-Net (Fig. 5e), which occur near the crossover and branch points. These errors have been resolved under the supervision of CBAV-Loss. It can be seen from the results of other methods that, under the supervision of CBAV-Loss, better results have also been achieved in the areas that are difficult to be segmented by the existing networks.

Figure 6 shows another example of one DR case, where there are many mis-segmentations in the result of U-Net (Fig. 6c). For example, in the yellow, blue and green boxes, there are various errors near the crossover and branch points; In the purple box, arteries are discontinuous; The mis-segmentation of the end of vessels appears in

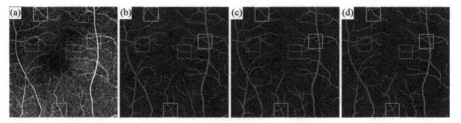

Fig. 6. This is a sample in the case of DR. (a) The OCTA projection map. (b) The ground truth. (c) The result of U-Net. (d) The result of U-Net with CBAV-Loss. (Color figure online)

the orange box. In contrast, the U-Net with CBAV-Loss (Fig. 6d) achieves the better result near the crossover and branch points, and the vessel structure is more complete.

4 Conclusion

In this paper, we proposed Crossover Loss and Branch Loss for the existing problems of AV segmentation, which are two novel structure-preserving loss functions and can make some common segmentation errors correct. Experimental results show that CBAV-Loss can effectively improve the accuracy of AV segmentation. Finally, the best method with CBAV-Loss achieved the mean IoU of 83.58%, the mean Dice of 91.00% and vascular connectivity of 98.59% for AV segmentation on the test data.

Acknowledgments. This work was supported by National Science Foundation of China under Grants (62172223, 62072241), and the Fundamental Research Funds for the Central Universities (30921013105).

References

1. Kashani, A.H., et al.: Optical coherence tomography angiography: a comprehensive review of current methods and clinical applications. Prog. Retin. Eye Res. **60**, 66–100 (2017)
2. Alam, M., Lim, J.I., Toslak, D., Yao, X.: Differential artery-vein analysis improves the performance of OCTA staging of sickle cell retinopathy. Transl. Vis. Sci. Technol. **8**(2), 3 (2019)
3. Alam, M., Tosklak, D., Lim, J.I., Yao, X.: Color fundus image guided artery-vein differentiation in optical coherence tomography angiography. Invest. Ophthalmol. Visual Sci. **59**(12), 4953–4962 (2018)
4. Ikram, M.K., et al.: Retinal vessel diameters and risk of impaired fasting glucose or diabetes: the Rotterdam study. Diabetes **55**(2), 506–510 (2006)
5. Wong, T.Y., et al.: Retinal arteriolar narrowing and risk of coronary heart disease in men and women: the Atherosclerosis Risk in Communities Study. JAMA **287**(9), 1153–1159 (2002)
6. Abtahi, M., Le, D., Lim, J.I., Yao, X.: MF-AV-Net: an open-source deep learning network with multimodal fusion options for artery-vein segmentation in OCT angiography. Biomed. Opt. Express **13**(9), 4870–4888 (2022)
7. Le, D., Alam, M., Son, T., Lim, J.I., Yao, X.: Deep learning artery-vein classification in OCT angiography. Ophthal. Technol. XXXI. SPIE **11623**, 54–60 (2021)

8. Ronneberger, O., Fischer, P., Brox, T.: U-net: Convolutional networks for biomedical image segmentation. In: Navab, N., Hornegger, J., Wells, W.M., Frangi, A.F. (eds.) Medical Image Computing and Computer-Assisted Intervention – MICCAI 2015: 18th International Conference, Munich, Germany, October 5-9, 2015, Proceedings, Part III, pp. 234–241. Springer International Publishing, Cham (2015). https://doi.org/10.1007/978-3-319-24574-4_28

9. Wang, J., et al.: Reflectance-based projection-resolved optical coherence tomography angiography. Biomed. Opt. Express 8(3), 1536–1548 (2017)

10. Gao, M., Guo, Y., Hormel, T.T., Sun, J., Hwang, T.S., Jia, Y.: Reconstruction of high-resolution 6 × 6-mm OCT angiograms using deep learning. Biomed. Opt. Express 11(7), 3585–3600 (2020)

11. Alam, M., et al.: AV-Net: deep learning for fully automated artery-vein classification in optical coherence tomography angiography. Biomed. Opt. Express 11(9), 5249–5257 (2020)

12. Alam, M.N., Le, D., Yao, X.: Differential artery-vein analysis in quantitative retinal imaging: a review. Quant. Imaging Med. Surg. 11(3), 1102 (2021)

13. Wendy Aguilar, M., et al.: Graph-based methods for retinal mosaicing and vascular characterization. In: Escolano, F., Vento, M. (eds.) Graph-Based Representations in Pattern Recognition, pp. 25–36. Springer Berlin Heidelberg, Berlin, Heidelberg (2007). https://doi.org/10.1007/978-3-540-72903-7_3

14. Chrástek, R., et al.: Automated Calculation of Retinal Arteriovenous Ratio for Detection and Monitoring of Cerebrovascular Disease Based on Assessment of Morphological Changes of Retinal Vascular System. MVA, pp. 240–243 (2002)

15. Grisan, E., Ruggeri, A.: A divide et impera strategy for automatic classification of retinal vessels into arteries and veins. In: Proceedings of the 25th Annual International Conference of the IEEE Engineering in Medicine and Biology Society (IEEE Cat. No. 03CH37439), vol. 1, pp. 890–893. IEEE (2003)

16. Jelinek, H.F., et al.: Towards vessel characterization in the vicinity of the optic disc in digital retinal images. Image Vis. Comput. Conf. 2(7) (2005)

17. Li, H., Hsu, W., Lee, M.L., Wang, H.: A piecewise Gaussian model for profiling and differentiating retinal vessels. In: Proceedings 2003 International Conference on Image Processing (Cat. No. 03CH37429), vol. 1, p. I-1069. IEEE (2003)

18. Niemeijer, M., van Ginneken, B., Abràmoff, M.D.: Automatic classification of retinal vessels into arteries and veins. Medical Imaging 2009: Computer-Aided Diagnosis. SPIE vol. 7260, pp. 422–429 (2009)

19. Rothaus, K., Jiang, X., Rhiem, P.: Separation of the retinal vascular graph in arteries and veins based upon structural knowledge. Image Vis. Comput. 27(7), 864–875 (2009)

20. Simó, A., de Ves, E.: Segmentation of macular fluorescein angiographies A statistical approach. Pattern Recogn. 34(4), 795–809 (2001)

21. Vázquez, S., Barreira, N., Penedo, M., Penas, M., PoseReino, A.: Automatic classification of retinal vessels into arteries and veins. In: 7th International Conference Biomedical Engineering (BioMED 2010), pp. 230–236 (2010)

22. Vázquez, S.G., et al.: Improving retinal artery and vein classification by means of a minimal path approach. Mach. Vis. Appl. 24(5), 919–930 (2013)

23. Milletari, F., Navab, N., Ahmadi, S.A.: V-net: fully convolutional neural networks for volumetric medical image segmentation. In: 2016 Fourth International Conference on 3D vision (3DV), pp. 565–571. IEEE (2016)

24. Ozan, O., et al.: Attention U-Net: Learning Where to Look for the Pancreas. arXiv preprint arXiv:1804.03999 (2018)

25. Ming, L., et al.: Ipn-v2 and octa-500: methodology and dataset for retinal image segmentation. arXiv preprint arXiv:2012.07261 (2020)

26. Yali, J., et al.: Split-spectrum amplitude-decorrelation angiography with optical coherence tomography. Opt. Express 20(4), 4710–4725 (2012)

Leveraging Data Correlations for Skin Lesion Classification

Junzhao Hao, Chao Tan[✉], Qinkai Yang, Jing Cheng, and Genlin Ji

School of Computer and Electronic Information/School of Artificial Intelligence, Nanjing Normal University, Nanjing, China
tutu_tanchao@163.com

Abstract. Over the past decade, deep convolutional neural networks (DCNN) have been widely adopted for medical image classification and some of them have achieved the diagnostic outcomes comparable or even superior to those of dermatologists. However, broad implementation of DCNN does not fully utilized valuable medical prior knowledge for semantic reasoning to guide the network, which leads to a lack of understanding of the correlation between lesion classes across the medical dataset. In this paper, based on the smooth assumption that similar lesion features tend to be grouped into close lesion categories, we propose a Graph-based pathological feature Enhancement structure and integrate Variance regularization into Label Distribution Learning to construct GEVLDL framework. This framework makes use of multiple images in a dataset to classify a single image, allowing the data in the dataset to be utilized multiple times. Furthermore, learning from the different degrees of contribution of all labels enables the full exploitation of the disease lesion relationship. The framework is verified on the benchmark skin diseases dataset ACNE04. Experiments demonstrate that our proposed method performs favorably against state-of-the-art methods.

Keywords: Medical image classification · Transformer · Graph convolutional network · Label distribution learning

1 Introduction

Computer-aided diagnosis based on medical image content analysis is a valuable tool for assisting professionals in decision making and patient screening. Automatic classification of skin disease is of great importance in the medical field. Skin disease is one of the most common malignancies in the world with significantly increased incidence over the past decade, among which about 80% of adolescents suffer from acne [4]. Manual observation of skin lesion classification is costly as it relies on expert knowledge and experience. And there are still relatively few practitioners in the field of dermatology. Reliable computer-aided diagnosis is therefore required for effective and efficient treatment.

Earlier approaches for computer-aided dermoscopic image analysis rely on extraction of hand-crafted features to be fed into classifiers [20]. Then, the rapid

Q. Liu et al. (Eds.): PRCV 2023, LNCS 14437, pp. 61–72, 2024.
https://doi.org/10.1007/978-981-99-8558-6_6

progress of CNN application has provided a new technical direction for solving the skin lesion classification task [23]. Recently, transformer has been adopted and shows promise in the computer vision(CV) field. Medical image analysis, as a critical branch of CV, also greatly benefits from this state-of-the-art technique [8]. Despite these promising research progresses, further improvement of the diagnostic accuracy is hindered by several limitations, not the least of which is that most of publicy accessible datasets for skin lesions do not have a sufficient sample size. It is difficult for CNN and transformer model to perform without the support of a large dataset. These existing deep learning models learn one image at a time, ignoring the correlation information between skin lesion features.

Based on the perception that skin lesion of similar class share similar pathological features and similar pathological features usually belong to a close class, we propose a new structure that can leverage both the feature space and label space correlations hidden in the adjacency to classify the skin lesion. First, we build a graph structure in feature space. Specificly, the feature of the lesion area in each face image is treated as a node of the graph. We connect all nodes based on the similarity of lesion appearance to get a fully-connected graph. Then the fully-connected graph is separated into several sub-graphs by cutting the low-weight edges. Retained sub-graphs contain instances with similar pathological features. Next, fuse the node features in each sub-graph to get the lesion-representative feature. The lesion-representative features make downstream grading tasks easier. In label space, we follow [21] to utilize the label distribution learning to define an increase and decrease of skin lesion as a process and introduce variance regularization [14] to constrain this process to focus on the dominant labels. Our contributions can be summarized as follows:

- *A state-of-the-art framework for skin lesion classification.* GEVLDL takes advantage of existing data and alleviate the scarcity of skin lesion dataset.
- *A graph-based pathological feature enhancement structure.* After the enhancement, the features of individuals converge to the lesion-representative features.

We conduct extensive experiments on benchmark dataset ACNE04 [21]. Our method significantly outperforms SOTA methods by a large margin up to ~5%.

2   Related Work

2.1   Skin Lesion Classification

In recent years, skin lesion analysis has achieved substantial advances. Wu et al. [21] address the problem of acne classification via Label Distribution Learning (LDL) [5] which considers the ambiguous information among acne severity. The key information of acne grading is face skin. Lin et al. [11] develop an adaptive image preprocessing method to filter meaningless information before the image is fed into the deep learning model. Alzahrani et al. [1] utilize the bounding boxes on the image to generate the ground-truth kernel density maps and then train a U-net model to learn this mapping method. The model treats each point in the predicted mapping outcome as an acne. By counting the number of acne to get the final grading severity. A growing body of work is devoted to the use of computer science as an aid to dermatological diagnosis.

2.2   Correlation Mining

Skin lesion is a progressive and moderate process. The various levels and symptoms are not independent of each other. Traditional single-label learning ignores the relevant information involved. Geng et al. [5] first propose a new machine learning paradigm, i.e., label distribution learning. It covers a certain number of labels, representing the degree to which each label describes the instance. Then there are already many works using LDL to take advantage of label correlation to better solve label ambiguity problems(e.g., emotion detection [24] and visual sentiment analysis [22]). Graph has been proved to be another effective label structure modeling. Graph Convolutional Network(GCN) was introduced in [9] and it has been successfully utilized for non-grid structured data modeling. The most current mainstream usage of GCN is to propagate information among labels and merge label information with CNN features at the final classification stage(e.g., [2] for multi-label classification). Drawing on this thought, we treat each image as a node, construct graphs in feature space, and then utilize GCN to propagate information among pathological features. In this way, correlation in feature space can be fully exploited.

3   Methodology

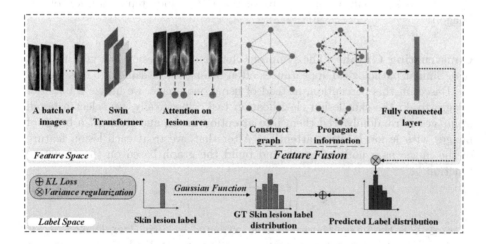

Fig. 1. The overview of GEVLDL framework.

In this section, we introduce a Graph-based pathological feature Enhancement structure and integrate Variance regularization into Label Distribution Learning to construct GEVLDL framework for skin lesion classification. The goal of our framework is to output the grading result of skin lesion(including *Mild, Moderate, Severe* and *Very severe*) for an input image following Hayashi criterion [6]. The main flowchart of the proposed skin lesion image classification frame is illustrated in Fig. 1. GEVLDL consists of two main stages: feature enhancement stage and label distribution learning stage. The details of each step are presented in the following subsections.

3.1   Feature Enhancement Stage

The feature enhancement stage consists of three main parts: To ensure that the parts we enhanced are pathological features, we first need to focus the model attention on lesion features. Next, aggregate images with similar pathological features into a graph. Finally, fuse the features in each graph.

Attention on Lesion Features. Pathological appearance of skin lesion varies from person to person. When dermatologists make the diagnosis of lesion severity, they focus on the lesion area and compare pathological changes between normal skin and diseased skin. So instead of utilizing CNN to extract the common features of the lesion area, Transformer which focuses on the comparison between features is better suited to this type of task. Given that skin lesion is a kind of localized lesion, Transformer can see all pixels in the image, which is not necessary. For studying the lesion area in the image, the local mechanism of the Swin Transformer [12] is preferable. The structure of Swin Transformer is not described here, please refer to [12] for details. After Swin Transformer completes all blocks computation, average pooling is conducted on each small window of the last block to output the feature vector as the node representation of the image.

Constructing Graphs. The symptom change of skin lesion is a gradual process. Similar pathological appearances often belong to the same or a close severity. Based on this perception, instead of using one image, we utilize a batch of images to do the skin lesion classification task. Specifically, we select a batch of images X randomly. Fed them into attention model and then get a batch of images with lesion location attention. After that, we treat each lesion feature of the image as a node of graph and build the graph based on the Gaussian function:

$$A_{ij} = \begin{cases} exp(-\frac{\|x_i - x_j\|^2}{2}) & i \neq j \\ 0 & i = j \end{cases} \tag{1}$$

$x_i, x_j \in X$ is the image feature matrix extracted by Swin Transformer, A_{ij} is the Gaussian distance between image x_i and image x_j. With this step, we get an affinity matrix A which represents a fully-connected graph in feature space.

Each element of A represents a pathological appearance similarity between two images. While what we need is the cluster with a similar pathological appearance. In the next step, a threshold τ is set to cut the edge with low weight and binary the weight of the edge:

$$A_{ij} = \begin{cases} 0 & A_{ij} < \tau \\ 1 & A_{ij} \geq \tau \end{cases} \tag{2}$$

After this, the fully-connected graph is separated into several sub-graphs. Images within the same sub-graph have a similar pathological appearance.

Pathological Feature Enhancement. For each sub-graph, we perform fusion operations on them with the thoughts of unsupervised learning. In this work, we utilize graph convolution network [9] to fuse the pathological features in each sub-graph. The essential idea of GCN is to update the node representations by propagating information between nodes. We follow a common practice as was done in [2,9] to apply graph convolution. The difference is that our graph is constructed in feature space. Each layer of GCN can be described as follows:

$$H^{(l+1)} = \Phi(\tilde{D}^{-\frac{1}{2}} \tilde{A} \tilde{D}^{-\frac{1}{2}} H^{(l)} W^{(l)}) \tag{3}$$

Among them, $H^{(l+1)}$ is the enhanced feature matrix after the l-th layer, and $H^{(0)}$ is the original feature matrix extracted by Swin Transformer. $\Phi(\cdot)$ is an activation function. $\tilde{A} = A + I$ is the adjacency matrix with added self-connections where I is the identity matrix and A is calculated by Eq. (1) and (2). $\tilde{D}$ denotes the diagonal matrix and $\tilde{D}_{ii} = \sum_j \tilde{A}_{ij}$. $W^{(l)}$ is a layer-specific matrix to be learned in model training. Using multiple images during inference actually introduce increased randomness. While this randomness is individual feature randomness. Instead, for clusters, the aggregation of more similar individuals enhances the robustness of cluster-representative features. During the propagation, the correlated pathological information is used to provide the positive feedback of each connected node in the same cluster to enhance the acne symptom features. By fusing pathological features in each sub-graph, the lesion features within each sub-graph converge towards the representative features of that sub-graph. The feature demarcation between each node within the same sub-graph tends to blur. Inter-cluster differentiability is improved and intra-cluster features converge leading to the result that images in the same severity are closer in feature space. Enhanced features make downstream classification tasks easier.

3.2   Label Distribution Learning Stage

Traditional machine learning models use only one true label to guide model training. In the label space, there is some label semantic information that can be utilized based on prior knowledge. Label distribution learning [5] covers a certain number of labels, representing the degree to which each label describes the instance. This can leverage the label correlations to provide more detailed guidance for the model learning. The skin lesion process is an ordered sequence, so we can use label distribution learning to allow the model to learn not only the ground-true label, but also the labels near the ground-true at the same time. In this study, we adopt Gaussian function for skin lesion encoding which defines an increase and decrease of lesion severity as a process that is more consistent with the development of skin diseases.

Specifically, given an image x_i labeled with the acne level z_i, the single label is transformed to a discrete probability distribution over the whole range of acne label which follows a Gaussian distribution centered at z_i:

$$d_{x_i}^j = \frac{1}{\sqrt{2\pi}\sigma M} exp(-\frac{(j - z_i)^2}{2\sigma^2}), \quad M = \frac{1}{\sqrt{2\pi}\sigma} \sum_{j=1}^{Z} exp(-\frac{(j - z_i)^2}{2\sigma^2}) \quad (4)$$

Where $j \in [1, \cdots, Z]$ is the scope of lesion labels, $d_{x_i}^j$ denotes the degree of description of the image x_i by the j-th label. σ is the standard deviation of the Gaussian distribution which is set as 3 in this paper. The normalization factor M ensures that $d_{x_i}^j \in [0, 1]$ and $\sum_{j=1}^{Z} d_{x_i}^j = 1$. In this way, we get ground-truth acne label distribution. For the input instance x_i, its predicted acne label distribution can be calculated as follows:

$$p_{x_i}^j = \frac{exp(\theta_j)}{\sum_{m=1}^{Z} exp(\theta_m)} \quad (5)$$

where θ_j is the predicted score corresponding to the j-th class output by the last fully connected layer. Here, we introduce variance regularization [14] so that the predicted label distribution is concentrated on the main labels around the ground truth label.

After the softmax of the last fully connected layer, the mean (m_{x_i}) and the variance (v_{x_i}) of the acne label distribution for the image x_i can be calculated as:

$$m_{x_i} = \sum_{j=1}^{Z} j * p_{x_i}^j, \quad v_{x_i} = \sum_{j=1}^{Z} p_{x_i}^j * (j - m_{x_i})^2 \quad (6)$$

Based on Eqs. (6), we compute the variance loss as follow:

$$L_v = v_{x_i} = \sum_{j=1}^{Z} p_{x_i}^j * (j - \sum_{\hat{j}=1}^{Z} \hat{j} * p_{x_i}^{\hat{j}})^2 \quad (7)$$

We apply Kullback-Leibler (KL) divergence which is used to measure the amount of information lost when using a distribution to approximate another distribution to minimize the deviation between ground-truth severity distribution and predicted severity distribution. At the training procedure, the learning loss is defined as:

$$L = \sum_{j=1}^{Z} (d_{x_i}^j ln p_{x_i}^j + p_{x_i}^j * (j - \sum_{\hat{j}=1}^{Z} \hat{j} * p_{x_i}^{\hat{j}})^2) \quad (8)$$

In the testing stage, the model follows the testing approach used in single-label classification. The lesion severity distribution of the given images can be predicted, where the label with the highest probability is considered as the predicted severity.

4 Experiments

4.1 Experiment Settings

Our experiments are applied on the benchmark dataset ACNE04. The ACNE04 dataset is released by Wu et al. [21] and consists of 1,457 images. Following the Hayashi criterion requirements [6], all images are taken at a 70-degree elevation from the front of patients. Each image is labeled with acne count and one of the acne severities, such as Mild, Moderate, Severe, and Very severe. We randomly split the dataset into 5 folds, using one fold as the test set and the rest as training set, repeated for 5 times.

Considering that the work of the acne severity grading is related to medical image processing, not only the commonly used precision and accuracy are applied but also utilize several metrics from the medical field, such as sensitivity to represent the recall rate, specificity to reflect the ability to rule out the disease and Youden Index to represent the comprehensive ability of diagnosis.

The proposed model is implemented on the open-source platform Pytorch. The experiments are performed on a computer with an NVIDIA GeForce GTX TITAN V GPU and Intel i7 CPU. The Stochastic Gradient Descent (SGD) method is adopted with the momentum 0.9 and the weight decay 5e−4 as the model optimizer. All models are trained for 120 epochs with a batch size of 32 ensuring that the loss on the training set is stable. The learning rate is started at 0.001 and decay it by 0.5 every 30 epochs. The input images are resized into $224 \times 224 \times 3$ pixels and normalized to the range of [0,1] in RGB channels.

4.2 Hyper Parameters Setting

The Impact of Batch Size. Batch size is to control the scale of graph in feature space. If the graph is too small, the features of the nodes may be over-smooth after the information is propagated. As illustrated in Fig. 2, with increasing batch size, the model performs better within a certain range. The model obtains the good performance in the accuracy evaluation index when the batch size is 32.

The Impact of GCN Layers. It is not the case that the deeper the layer of the GCN, the better the performance. When more GCN layers are used, the propagation information between nodes will be accumulated, which may lead to over-smoothing of nodes. It can be seen from Fig. 2 that when L is 2, the model has the relatively good performance.

The Impact of τ. We vary the values of the threshold τ in Eq. (2) for correlation matrix binarization, and show the results in Fig. 2. Note that if we do not filter any edges, the model will not converge. However, when too many edges are filtered out, the accuracy drops since correlated neighbors will be ignored as well. The optimal value of τ is 1e−5.

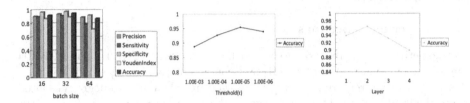

Fig. 2. Performance of GEVLDL framework with different *batch size, Threshold* and *GCN Layer* hyperparameters.

4.3   Comparison with State-of-the-Art Methods

To demonstrate the efficiency of the proposed model, state-of-the-art methods of regression-based deep learning approaches, detection-based approaches, skin lesion specific classification approaches are chosen as peer competitors on ACNE04 dataset. Results are summarised in Table 1. The regression-based machine learning approaches like SIFT [13] and HOG [3] send the hand-crafted features to Support Vector Machine model classifier. For the regression-based deep learning approaches such as Inception [18], VGGNet [17], and ResNet [7], we extract the features and utilize a fully connected layer to classify the severity into four levels. Considering the size of the acne lesions is small and overlapped, detection-based methods including YOLO [15] and F-RCNN [16] are hard to detect because these detection methods perform well in a sparse object condition. Recently, more specific works have been proposed to solve the acne severity grading task. LDL [21] method utilizes the ambiguous information among levels of acne severity. Image preprocessing approach [11] filters meaningless information and enhances key information. For the attention guided regressor approach [1], the attention mechanism guides the regressor model to look for the acne lesions by locating the most salient features related to the understudied acne lesions. Compared with state-of-the-art methods, our model first makes full use of the information both in the feature space and label space and achieves state-of-the-art performance that surpasses existing methods in all evaluation metrics.

4.4   Ablation Studies

Considering that our framework is based on the label distribution learning paradigm, we set the LDL [21] as the baseline model to analyze the efficiency of each component in our proposed method.

Table 1. Comparison with the regression-based deep learning approaches(HOG, SIFT, VGGNet, Inceptionv3, ResNet), detection-based approaches(YOLOv3, RCNN, Efficient), and existed specific acne grading SOTA methodes(LDL, KIEGLFN, AGR).

Method/Criteria	Precision	Sensitivity	Specificity	Youden Index	Accuracy
HOG [3]	39.10	38.10	77.91	16.10	41.30
SIFT [13]	42.59	39.09	78.44	17.53	45.89
VGGNet [17]	72.65	72.71	90.60	63.31	75.17
Inception [18]	74.26	72.77	90.95	63.72	76.44
ResNet [7]	75.81	75.36	91.85	67.21	78.42
YOLOv3 [15]	67.01	51.68	85.96	37.63	63.70
F-RCNN [16]	56.91	61.01	90.32	51.34	73.97
Efficient [19]	70.34	69.32	90.38	59.70	74.52
LDL [21]	84.37	81.52	93.80	75.32	84.11
KIEGLFN [11]	83.58	81.95	94.11	76.06	84.52
AGR [1]	88.25	83.00	93.00	76.00	91.50
GEVLDL	**95.34**	**94.94**	**98.63**	**93.57**	**96.36**

Attention on Lesion Area. The attention mechanism of Swin Transformer could locate the lesion area which provides more accurate features for the downstream works. Illustrated in the Table 2. After locating the lesion area with attention mechanism, all the metrics scores are improved.

Feature Fusion. Referring to the ambiguous information among acne severity in label space, there are also pathological appearance correlations in feature space. We leverage these correlations by propagating the information in the feature space. As shown in Table 2, we introduce the feature fusion into the baseline model and get better scores than the single image model.

Lesion Feature Enhancement. Combining the lesion attention model with feature fusion model, we get the lesion feature enhancement model. Observing the Table 2, we get a huge breakthrough in scores of all metrics after feature enhancement. This can prove that we create a framework that achieves "1+1>2" effect. The rationale is that when only feature fusion is performed, useless background information is also mixed in; after attention localization, only the lesion features are enhanced. We visualize all the image representation vectors after the feature enhancement by T-SNE [10]. As shown in Fig. 3, it shows that after the feature enhancement, the images related to the same or close severity are clustered together.

Variance Regularization. The above works are all implemented in feature space. Label distribution learning tries to learn all the labels contribution, while

Table 2. Ablation experiments demonstrating the effectiveness of different modules on the ANCE04 dataset.

LDL (base)	Attention	FeatureFusion	VarianceLoss	Criteria				
				Pre	Sen	Sp	YI	Acc
✓				84.37	81.52	93.80	75.32	84.11
✓	✓			84.45	83.72	94.15	77.88	85.36
✓		✓		86.82	83.57	94.40	77.97	85.96
✓	✓	✓		94.31	93.51	97.48	90.99	94.81
✓	✓	✓	✓	**95.34**	**94.94**	**98.63**	**93.57**	**96.36**

tail labels have little contribution on the task. It can be seen in the Table 2 that after reducing the contribution of tail labels, scores under all the metrics are improved.

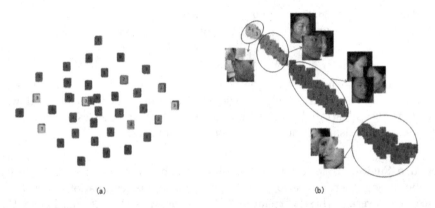

(a)   (b)

Fig. 3. Visualization of GEVLDL using T-SNE on ACNE04. Each point denotes an image-level vector learned via GEVLDL. Different colors indicate different acne severity labels. (a) denotes the untrained feature space, and (b) denotes the feature space processed by feature enhancement.

5 Conclusion

In this paper, we propose an skin lesion classification framework by leveraging the correlations both in feature space and label space. In feature space, we utilize Swin Transformer whose shifted window attention mechanism works like a dermatologist to locate the lesion area. Then we pick a batch of images with attention on the lesion area to construct a graph based on the similarity of pathological appearance and cut the edge with low similarity to get several sub-graphs. After completing the preliminary clustering work, we utilize GCN to fuse pathological features in each sub-graph to make the features more distinguishable. In label space, based on the label distribution learning work which leverages the ambiguous information among acne severity, we introduce variance regularization to guide the model lower the influence of tail labels and focus on the dominated labels. Experiments proved that our framework makes a great breakthrough in all metrics on the benchmark dataset ACNE04, thus it can provide more accurate diagnostic reference than state-of-the-art acne grading methods.

References

1. Alzahrani, S., Al-Bander, B., Al-Nuaimy, W.: Attention mechanism guided deep regression model for acne severity grading. Comput. **11**(3), 31 (2022)
2. Chen, Z., Wei, X., Wang, P., Guo, Y.: Multi-label image recognition with graph convolutional networks. In: IEEE Conference on Computer Vision and Pattern Recognition, CVPR 2019, Long Beach, CA, USA, 16–20 June 2019, pp. 5177–5186. Computer Vision Foundation/IEEE (2019)
3. Dalal, N., Triggs, B.: Histograms of oriented gradients for human detection. In: 2005 IEEE Computer Society Conference on Computer Vision and Pattern Recognition (CVPR 2005), 20–26 June 2005, San Diego, CA, USA, pp. 886–893. IEEE Computer Society (2005)
4. Krowchuk, D.P.: Managing acne in adolescents. Pediatr. Clin. North Am. **47**(4), 841–854 (2000)
5. Geng, X., Smith-Miles, K., Zhou, Z.: Facial age estimation by learning from label distributions. In: Fox, M., Poole, D. (eds.) Proceedings of the Twenty-Fourth AAAI Conference on Artificial Intelligence, AAAI 2010, Atlanta, Georgia, USA, 11–15, July 2010. AAAI Press (2010)
6. Hayashi, N., Akamatsu, H., Kawashima, M.: Establishment of grading criteria for acne severity. J. Dermatol. **35**(5), 255–260 (2008)
7. He, K., Zhang, X., Ren, S., Sun, J.: Deep residual learning for image recognition. In: 2016 IEEE Conference on Computer Vision and Pattern Recognition, CVPR 2016, Las Vegas, NV, USA, 27–30 June 2016, pp. 770–778. IEEE Computer Society (2016)
8. He, X., Tan, E., Bi, H., Zhang, X., Zhao, S., Lei, B.: Fully transformer network for skin lesion analysis. Med. Image Anal. **77**, 102357 (2022)
9. Kipf, T.N., Welling, M.: Semi-supervised classification with graph convolutional networks. In: 5th International Conference on Learning Representations, ICLR 2017, Toulon, France, 24–26 April 2017, Conference Track Proceedings. OpenReview.net (2017)
10. Laurens, V.D.M., Hinton, G.: Visualizing data using t-SNE. J. Mach. Learn. Res. **9**(2605), 2579–2605 (2008)
11. Lin, Y., et al.: KIEGLFN: a unified acne grading framework on face images. Comput. Meth. Program. Biomed. **221**, 106911 (2022)
12. Liu, Z., et al.: Swin transformer: hierarchical vision transformer using shifted windows. In: 2021 IEEE/CVF International Conference on Computer Vision, ICCV 2021, Montreal, QC, Canada, 10–17 October 2021, pp. 9992–10002. IEEE (2021)
13. Lowe, D.G.: Distinctive image features from scale-invariant keypoints. Int. J. Comput. Vis. **60**(2), 91–110 (2004)
14. Pan, H., Han, H., Shan, S., Chen, X.: Mean-variance loss for deep age estimation from a face. In: 2018 IEEE Conference on Computer Vision and Pattern Recognition, CVPR 2018, Salt Lake City, UT, USA, 18–22 June 2018, pp. 5285–5294. Computer Vision Foundation/IEEE Computer Society (2018)

15. Redmon, J., Farhadi, A.: YOLOv3: an incremental improvement. CoRR abs/1804.02767 (2018)
16. Ren, S., He, K., Girshick, R.B., Sun, J.: Faster R-CNN: towards real-time object detection with region proposal networks. In: Cortes, C., Lawrence, N.D., Lee, D.D., Sugiyama, M., Garnett, R. (eds.) Advances in Neural Information Processing Systems 28: Annual Conference on Neural Information Processing Systems 2015, 7–12 December 2015, Montreal, Quebec, Canada, pp. 91–99 (2015)
17. Simonyan, K., Zisserman, A.: Very deep convolutional networks for large-scale image recognition. In: Bengio, Y., LeCun, Y. (eds.) 3rd International Conference on Learning Representations, ICLR 2015, San Diego, CA, USA, 7–9 May 2015, Conference Track Proceedings (2015)
18. Szegedy, C., Vanhoucke, V., Ioffe, S., Shlens, J., Wojna, Z.: Rethinking the inception architecture for computer vision. In: 2016 IEEE Conference on Computer Vision and Pattern Recognition, CVPR 2016, Las Vegas, NV, USA, 27–30 June 2016, pp. 2818–2826. IEEE Computer Society (2016)
19. Tan, M., Le, Q.V.: EfficientNet: rethinking model scaling for convolutional neural networks. CoRR abs/1905.11946 (2019)
20. Tourassi, G.D., Armato, S.G., Abas, F.S., Kaffenberger, B., Bikowski, J., Gurcan, M.N.: Acne image analysis: lesion localization and classification. In: Medical Imaging: Computer-aided Diagnosis, p. 97850B (2016)
21. Wu, X., : Joint acne image grading and counting via label distribution learning. In: 2019 IEEE/CVF International Conference on Computer Vision, ICCV 2019, Seoul, Korea (South), October 27 - November 2, 2019, pp. 10641–10650. IEEE (2019)
22. Yang, J., Sun, M., Sun, X.: Learning visual sentiment distributions via augmented conditional probability neural network. In: Singh, S., Markovitch, S. (eds.) Proceedings of the Thirty-First AAAI Conference on Artificial Intelligence, 4–9 February 2017, San Francisco, California, USA, pp. 224–230. AAAI Press (2017)
23. Yuan, Y., Chao, M., Lo, Y.: Automatic skin lesion segmentation using deep fully convolutional networks with Jaccard distance. IEEE Trans. Med. Imaging **36**(9), 1876–1886 (2017)
24. Zhang, Z., Lai, C., Liu, H., Li, Y.: Infrared facial expression recognition via gaussian-based label distribution learning in the dark illumination environment for human emotion detection. Neurocomputing **409**, 341–350 (2020)

CheXNet: Combing Transformer and CNN for Thorax Disease Diagnosis from Chest X-ray Images

Xin Wu[1], Yue Feng[1(✉)], Hong Xu[1,2], Zhuosheng Lin[1], Shengke Li[1], Shihan Qiu[1], QiChao Liu[1], and Yuangang Ma[1]

[1] Faculty of Intelligent Manufacturing, Wuyi University, Jiangmen 529020, Guangdong, China
`J002443@wyu.edu.cn`
[2] Institute for Sustainable Industries and Liveable Cities, Victoria University, Melbourne 8001, Australia

Abstract. Multi-label chest X-ray (CXR) image classification aims to perform multiple disease label prediction tasks. This concept is more challenging than single-label classification problems. For instance, convolutional neural networks (CNNs) often struggle to capture the statistical dependencies between labels. Furthermore, the drawback of concatenating CNN and Transformer is the lack of direct interaction and information exchange between the two models. To address these issues, we propose a hybrid deep learning network named CheXNet. It consists of three main parts in the CNN and Transformer branches: Label Embedding and Multi-Scale Pooling module (MEMSP), Inner Branch module (IB), and Information Interaction module (IIM). Firstly, we employ label embedding to automatically capture label dependencies. Secondly, we utilize Multi-Scale Pooling (MSP) to fuse features from different scales and an IB to incorporate local detailed features. Additionally, we introduce a parallel structure that allows interaction between the CNN and the Transformer through the IIM. CNN can provide richer inputs to the Transformer through bottom-up feature extraction, whilst the Transformer can guide feature extraction in the CNN using top-down attention mechanisms. The effectiveness of the proposed method has been validated through qualitative and quantitative experiments on two large-scale multi-label CXR datasets with average AUCs of 82.56% and 76.80% for CXR11 and CXR14, respectively.

Keywords: Hybird network · Multi-label · Chest X-ray image

This work is supported by the Basic Research and Applied Basic Research Key Project in General Colleges and Universities of Guangdong Province, China (2021ZDZX1032); the Special Project of Guangdong Province, China (2020A1313030021); and the Scientific Research Project of Wuyi University (2018TP023, 2018GR003).

Supplementary Information The online version contains supplementary material available at https://doi.org/10.1007/978-981-99-8558-6_7.

1   Introduction

Chest X-ray (CXR), as a painless examination method, plays an important role in auxiliary clinical diagnosis. Meanwhile, it is one of the most common radiology tests used to screen for and diagnose a variety of lung conditions. However, achieving highly reliable diagnostic results for thoracic diseases using CXRs remains challenging due to the dependence on the expertise of radiologists.

In the past few years, CNNs in particular have shown remarkable performance in the diagnosis of various thoracic diseases [11]. Pesce et al. [12] utilized a CNN to extract features and input them into a classifier and a locator for detecting lung lesions. Sahlol et al. [13] employed a pre-trained MobileNet to extract features from CXR images. Baltruschat et al. [2] evaluated the performance of various methods for classifying 14 disease labels using an extended ResNet50 architecture and text data.

The great success of Transformers [16] has inspired researchers [3,8,15] to try to introduce Transformers into the field of computer vision. Furthermore, some studies have utilized Transformers to capture multi-label information in images and improve classification performance. Taslimi et al. [14] introduced a Swin Transformer backbone, which predicts each label by sharing components across models. Xiao et al. [17] utilized masked auto encoders to pre-train Vision Transformers (ViT), reconstructing missing pixel images from a small portion of separate X-ray image.

However, there are still some challenges that need to be addressed in the classification of multi-label CXR images. Firstly, there may exist interdependencies between different labels in multi-label CXR images, such as certain lung abnormalities being related to cardiac abnormalities. The second is the imbalance of labels. Thirdly, there can be prominent local lesion features and scattered global features in the images.

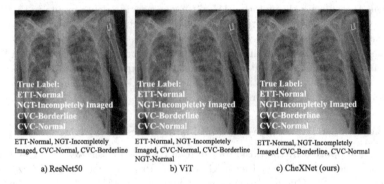

Fig. 1. Examples of recognition results of the CNN (ResNet50), the Transformer (ViT) and our proposed CheXNet. The true labels are in red font, the incorrectly identified labels are in green font, and the correctly predicted labels with small probability are in blue font (Color figure online).

In this paper, we propose CheXNet to address the aforementioned challenges. Firstly, we introduce self-attention operations on label embedding. This approach adaptively captures the correlations among labels without relying on manually predefined label relationships. Secondly, we employ cross-attention between image features and label features, allowing the model to weigh the image features based on the importance of each label. Additionally, we introduce the MSP block in the Transformer branch to extract features at different levels and then fuse them. Finally, we utilize a parallel structure that allows interaction between the CNN and the Transformer. To demonstrate the superiority of our approach, we visualize ResNet50, ViT and the proposed CheXNet, as shown in Fig. 1.

The main contributions of this study are summarized as follows:

(1) We propose a CheXNet model for multi-label CXR image classification, which captures both short local features and global representations.
(2) For the Transformer branch, we introduce the MSP block to perform multiscale pooling, aiming to enhance the richness and diversity of feature representations. Additionally, the label embedding, using self-attention, adaptively captures the correlations between labels. For the CNN branch, an embedded residual structure is employed to learn more detailed information. The IIM supports cross-branch communication and helps to explore implicit correlations between labels.
(3) We evaluate the CheXNet on two publicly available datasets, CXR11 and CXR14. The experimental results demonstrate that the CheXNet outperforms existing models on both datasets in terms of performance.

2 Related Work

In this section, we discuss label dependency and balance issues observed in multi-label classification methods, as well as multi-label CXR image classification methods for a wide range of lesion locations.

2.1 Label Dependency and Imbalance

In multi-label CXR image classification, the challenges of label dependency and label imbalance are common. These issues significantly impact accurate classification and model performance evaluation. To address these challenges, researchers have employed various methods, including weighted loss functions, hierarchical classification, and transfer learning. Allaouzi et al. [1] proposed a method that combines a CNN model with convolutional filters capable of detecting local patterns in images. By learning the dependencies between features and labels in multi-label classification tasks, they enhanced the accuracy and reliability of disease diagnosis. Lee et al. [7] introduced a hybrid deep learning model. The model consists of two main modules: image representation learning and graph representation learning. Yang et al. [19] utilized a triple network ensemble learning framework consisting of three CNNs and a classifier. The framework aimed to learn combined features and address issues such as class imbalance and network ensemble.

2.2   Extensive Lesion Location

In CXR images, there can be multiple lesion locations, with each lesion corresponding to a different pathology. Some researchers have adopted different methods, such as region localization, and attention mechanism. Ma et al. [10] proposed a cross attention network approach for the automated classification of thoracic diseases in CXR images. This method efficiently extracts more meaningful representations from the data and improves performance through cross-attention, requiring only image-level annotations. Guan et al. [4] introduced category residual attention learning to address the problem of pathologies interfering with the recognition of target-related pathologies. This approach enables the prediction of multiple pathologies' presence in the attention view of specific categories. The goal of this method is to suppress interference from irrelevant categories by assigning higher weights to relevant features, thereby achieving automatic classification of thoracic diseases.

3   Approaches

The multi-label CXR image classification method of CheXNet consists of three main stages, as shown in Fig. 2 . The first stage involves extracting initial features, using the Stem module. These features are then split into two branches, one sent to the Transformer branch and the other to the CNN branch. In the second stage, the Transformer branch utilizes MEMSP, while the CNN branch utilizes a nested IB. The stacking of MEMSP modules and IB modules corresponds to the number of layers in a vanilla Transformer, which is set to 12. Additionally, the IIM consists of the CNN branch to the Transformer branch (C2T) and the Transformer branch to the CNN branch (T2C) components, which progressively interactively fuse feature maps. Finally, after obtaining the features T and features C of the two branches separately, we directly sum up the branch fusions.

3.1   Label Embedding and MSP Block

For an input image x, the Stem module first extracts feature. Then, in the Transformer branch, the label embedding is used to query the MEMSP. However, most existing works primarily focus on regression from inputs to binary labels, while overlooking the relationship between visual features and the semantic vectors of labels. Specifically, we obtain the features extracted by the convolution as the K and V inputs and the label embeddings as the query $Q_i \in \mathbb{R}^{C \times d}$, with d denoting dimensionality, cross-noting the desired K, V and Q. We use a Transformer-like architecture, which includes a self-attention module, a cross-attention block, an MSP block, and an FFN block. When using the self-attention block, label embedding is the conversion of labels into vector representations so that the computer can better understand and process them. By incorporating label embeddings into the MEMSP module, it can effectively and automatically capture the semantics

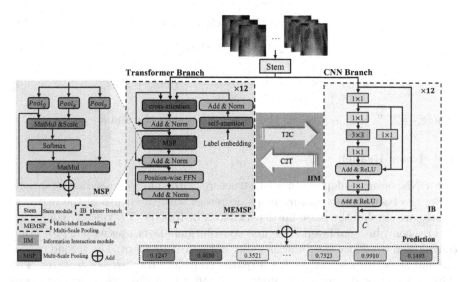

Fig. 2. An overview diagram of the proposed CheXNet framework, where the Transformer branch uses the MEMSP block (green part), the CNN branch uses the IB block (blue part), and the IIM uses C2T and T2C (orange part). T and C denote the final output characteristics of MEMSP and IB, respectively (Color figure online).

of the labels and make more accurate predictions for multiple labels associated with an input sample. K, V, and Q are label features, all denoted as Q_{i-1}. The specific formula is as follows:

$$\text{self-attention} : Q_i^{(1)} = \text{MultiHead}\left(Q_{i-1}, Q_{i-1}, Q_{i-1}\right) \tag{1}$$

$$\text{coss-attention} : Q_i^{(2)} = \text{MultiHead}\left(Q_i^{(1)}, K, V\right) \tag{2}$$

$\text{MultiHead}(Q, K, V)$ and $Q_i = \text{FNN}(x)$ have the same decoder definition as the standard Transformer [16]. We did not use masked multi-head attention, but instead used self-attention, as autoregressive prediction is not required in multi-label image classification. In the MSP block, multiple pooling operations $(2 \times 2, 3 \times 3, 4 \times 4)$ are applied separately to Q, K, and V, allowing each pooling size to extract features at different levels. These pooled features are then fused together. Specifically, given the input feature F, three copies of F are created to obtain pools, denoted as $Pool_i$, where i represents q, k, v. The $Pool_q$ and $Pool_k$ of different scales are multiplied element-wise, resulting in $Pool_q'$ and $Pool_k'$. The $Pool_v$ of different scales are summed element-wise, resulting in $Pool_v'$. Then, the Softmax function is applied to normalize the values, ensuring that the sum of all elements is equal to 1. The normalized values are used as weights to linearly combine the value vectors in $Pool_v'$, and the residual of $Pool_q'$ is added to the weighted sum. The final output is the sum of these two terms.

$$Pool_q' = \text{SUM}(\text{MatMul}(Pool_q^i)) \tag{3}$$

$$Pool_k' = \text{SUM}(\text{MatMul}(Pool_k^i)) \tag{4}$$

$$Pool_v' = \text{SUM}(Pool_v^i) \tag{5}$$

$$MSP = \text{Softmax}\left(\frac{Pool_q' Pool_k'}{\sqrt{d_k}}\right) Pool_v' + Pool_q' \tag{6}$$

Where $i = 2, 3, 4$, d_k represents the vector dimension of q and k.

3.2   Inner Branch

The CNN branch adopts a nested structure with multiple residual branches, where the resolution of feature maps decreases with the depth of the network. Firstly, we employ the basic bottleneck block of ResNet [5], typically consisting of three convolutional layers. The first convolutional layer uses a 1×1 projection convolution to reduce the dimensionality of the feature maps. The second convolutional layer utilizes a larger 3×3 spatial convolution to extract features. The third convolutional layer again employs a smaller 1×1 projection convolution to further reduce the dimensionality of the feature maps. It also includes a residual connection between the input and output. The IB modifies the residual block by introducing an additional nested branch, replacing the main 3×3 convolution with a residual block. The CNN branch can continuously contribute localized feature details to the Transformer branch through the C2T module within the IIM module. The IB module significantly enhances the model's feature representation capacity, particularly in the context of multi-label CXR classification tasks.

3.3   C2T and T2C in IIM

For the CNN branch, mapping features to the Transformer branch is a crucial problem. Similarly, for the Transformer branch, embedding patch representations into the CNN branch is also important. CNN features are represented as $[B, C, H, W]$, where B denotes the batch size, C denotes the number of channels, H denotes the image height, and W denotes the image width. On the other hand, Transformer features are represented as $[B, \_, C]$, where '$\_$' represents the sum of the number of image patches and the number of class tokens, usually equal to $H \times W + 1$. To address this issue, we propose the C2T and T2C to progressively integrate the feature maps in an interactive manner, as shown in Fig. 3.

The C2T method involves dimensionality transformation of the feature maps using 1×1 convolutions. Additionally, we combine features from different channels to enhance the expressive power of the features. We utilize average pooling to downsample the feature maps, reducing their spatial dimensions while preserving the essential information. The GELU activation function is employed for fast convergence and reduced training time, thereby improving the efficiency of the model training. Layer normalization is used for feature regularization.

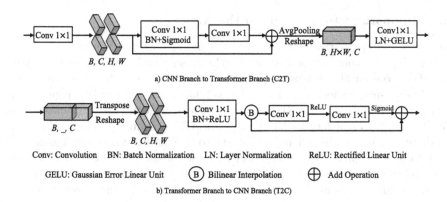

Fig. 3. Structure of Information Interaction Module (IIM), which includes the C2T and T2C.

The T2C process involves aligning the spatial scale by employing appropriate up-sampling techniques. Batch normalization is used to regularize the features. We use the commonly used ReLU activation function in the convolutional operations. Bilinear interpolation is applied to upsample the feature maps, enhancing the spatial resolution and capturing finer details. Similar to C2T, multiple 1×1 convolutions are utilized for optimizing feature information exchange. After the 1×1 convolutions, we incorporate both ReLU and Sigmoid activations in a cross-interaction manner to improve the nonlinear fitting capability. Finally, a residual connection is introduced between the output of the bilinear interpolation and the output after a series of operations to preserve important information and enhance the model's representational capacity.

4   Experiments

4.1   Dataset

The Catheter and Line Position Challenge on Kaggle[1] is a competition that involves classifying 40000 images to detect misplaced catheters. In this study, 30083 CXR image training data were used as multi-label sample classification, which was named CXR11.

The NIH ChestX-ray14 dataset[2], which was named CXR14, includes 112120 frontal X-ray images from 30805 unique patients with annotations for 14 common diseases. Limited by computer equipment, we only use part of the data. A detailed description of CXR11 and CXR14 is provided in the supplementary material.

---

[1] CXR11: kaggle.com/competitions/ranzcr-clip-catheter-line-classification/data
[2] CXR14: nihcc.app.box.com/v/ChestXray-NIHCC

4.2   Comparison to the State-of-the-Arts

To ensure fair comparisons, we utilized the aforementioned parameter settings and classical methods to calculate the AUC scores for each category and the average AUC score for all diseases, as presented in Tables 1 and 2 for the CXR11 and CXR14 datasets, respectively. To further validate the feasibility of our proposed approach, we compared it with eight other state-of-the-art medical image classification networks, achieving the best performance in multi-label classification. Furthermore, we conducted paired t-tests to assess the statistical significance of performance differences between our proposed model and those proposed by other authors. Based on the p-values, we can conclude that there are statistically significant differences in the performance of each model for this specific task.

Table 1. Comparison of the classification performance of different models on CXR11 datasets, where Swan Ganz denotes Swan Ganz Catheter Present. The best results are shown in bold.

AUC score (%)		ResNet34 [5]	ResNet50 [5]	ResNeXt50 [18]	SEReNet50 [6]	ViT [3]	Swin Transformer [8]	ConvNeXt [9]	DeiT [15]	CheXNet (ours)
ETT	Abnormal	78.26	82.97	74.16	85.95	77.68	67.23	74.96	79.36	**92.49**
	Borderline	88.51	89.33	85.96	**89.96**	83.02	78.49	81.99	83.56	88.04
	Normal	96.96	97.37	96.95	97.29	88.18	83.84	94.96	87.95	**97.92**
NGT	Abnormal	78.15	80.69	79.63	77.26	77.18	74.09	72.54	76.48	**81.38**
	Borderline	78.10	79.35	81.14	78.58	73.41	69.74	69.40	74.51	**81.72**
	Incompletely Imaged	92.94	93.09	93.20	92.37	87.88	82.97	88.66	87.25	**94.19**
	Normal	91.19	**92.72**	91.57	91.55	85.09	81.86	88.74	85.93	92.72
CVC	Abnormal	59.64	61.26	62.11	61.66	60.60	59.29	56.05	61.44	**62.88**
	Borderline	58.80	58.89	59.26	58.99	56.36	56.18	58.57	57.44	**59.36**
	Normal	57.20	59.14	**61.32**	57.89	55.80	53.38	59.19	56.23	60.70
Swan Ganz		92.95	95.72	**97.11**	93.44	87.48	84.99	90.51	89.31	96.72
Mean		79.34	80.96	80.22	80.45	75.70	72.01	75.97	76.32	**82.56**
p-value		.0108	.0435	.0896	.0054	.0001	.0001	.0006	.0001	-

Table 1 presents the results of different models for CXR11 classification. From Table 1, it can be observed that our proposed method achieved the highest average AUC score (82.56%). Among the compared models, ResNet50 attained the highest average AUC score of 80.96%, while Swin Transformer achieved the lowest average AUC score of 72.01%, which is 1.60% and 10.56% lower than our proposed method, respectively. Regarding ETT-Abnormal, our method outperformed the other top-performing model, ResNet50, by 9.52%.

Table 2 provides the results of different models for CXR14 classification. From Table 2 , it can be seen that our proposed method obtained the highest average AUC score (76.80%). Among the compared models, ConvNeXt achieved the highest average AUC score of 76.73%, while ViT obtained the lowest average AUC score of 54.00%, which are only 0.07% and 22.80% lower than our proposed method, respectively. Although ConvNeXt performed similarly to CheXNet, our method exhibited superior performance in multiple categories. For example, for Fibrosis, CheXNet outperformed ConvNeXt by 4.33%.

Table 2. Comparison of the classification performance of our different models on CXR14 datasets. The best results are shown in bold.

AUC score (%)	ResNet34 [5]	ResNet50 [5]	ResNeXt50 [18]	SEReNet50 [6]	ViT [3]	Swin Transformer [8]	ConvNeXt [9]	DeiT [15]	CheXNet (ours)
Atelectasis	73.35	73.87	73.51	73.81	53.53	62.48	73.31	70.19	**74.24**
Cardiomegaly	91.00	89.18	89.03	89.72	50.18	56.67	91.42	80.81	**91.76**
Consolidation	70.19	71.72	**72.98**	71.73	52.43	64.33	71.19	68.06	71.69
Edema	85.40	85.08	84.32	**86.02**	68.78	76.52	84.89	83.85	83.76
Effusion	81.02	81.79	82.09	81.40	44.74	64.95	**82.77**	77.17	81.96
Emphysema	76.39	**83.29**	80.63	80.88	53.19	55.94	80.79	75.45	81.09
Fibrosis	76.66	76.37	78.26	77.18	60.02	69.27	73.94	73.84	**78.27**
Hernia	69.00	79.62	80.96	78.88	67.29	69.35	88.84	69.99	**89.04**
Infiltration	69.12	68.85	69.24	**69.78**	56.07	61.89	68.87	67.16	68.41
Mass	74.96	**77.04**	76.07	74.92	47.56	57.07	73.97	66.77	76.29
Nodule	69.65	68.94	71.36	69.34	50.83	61.15	**72.73**	65.79	69.71
Pleural Thickening	67.83	66.99	69.23	67.43	46.46	57.09	**70.51**	63.99	67.39
Pneumonia	59.02	59.21	58.32	**60.39**	48.97	54.07	55.68	50.50	59.61
Pneumothorax	82.39	83.32	81.81	**85.75**	55.90	61.65	85.34	74.86	81.96
Mean	74.71	76.09	76.27	76.23	54.00	62.32	76.73	70.60	**76.80**
p-value	.0706	.1638	.2041	.2400	.0000	.0000	.4559	.0001	-

For the CXR11 and CXR14 datasets, CheXNet is compared with other algorithms and the overall performance of the network is demonstrated. In the supplementary material, a qualitative analysis of the classification performance of each compared method and the AUC for each disease is presented.

4.3   Ablation Study

To assess the effectiveness of the proposed CheXNet network and the contribution of each module within the overall network, we conducted a series of step-wise ablation experiments on the CXR11 dataset. The following models were compared to evaluate their performance:

Baseline: The Transformer branch is a standard Transformer encoder, and the CNN branch consists of residual blocks from the ResNet network, without any gated mechanisms for information exchange and communication.

Model 1: Based on the baseline, the standard Transformer encoder is replaced with the MEMSP module, which includes label embedding but not the MSP block.

Model 2: Based on the baseline, the residual blocks of the ResNet network are replaced with IB, introducing internal nesting.

Model 3: Built upon Model 1, the residual blocks are replaced with IB.

Model 4: Built upon Model 2, the IIM modules are added.

Model 5: Built upon Model 3, the IIM modules are added. In this case, the MEMSP module does not include the MSP block.

CheXNet (ours): Built upon Model 5, the MSP module is added.

Table 3. Classification performance of different models in our system on the CXR11 dataset, where Swan Ganz denotes Swan Ganz Catheter Present. The best results are in bold.

AUC socre (%)		Baseline	Model 1	Model 2	Model 3	Model 4	Model 5	CheXNet (ours)
ETT	Abnormal	80.63	93.70	83.74	90.87	88.75	90.08	**92.49**
	Borderline	**88.18**	85.00	87.08	87.93	84.93	85.46	88.04
	Normal	97.57	97.05	97.85	**97.88**	97.84	97.84	97.92
NGT	Abnormal	78.39	74.76	81.02	79.39	81.22	80.92	**81.38**
	Borderline	78.64	78.97	80.40	81.53	81.86	**82.05**	81.72
	Incompletely Imaged	92.86	92.17	93.39	93.60	94.26	**94.33**	94.19
	Normal	91.96	89.93	91.73	92.57	92.18	92.18	**92.72**
CVC	Abnormal	60.14	63.90	64.05	**64.56**	63.11	61.83	62.88
	Borderline	57.66	58.63	58.74	56.39	59.99	**60.00**	59.36
	Normal	58.04	58.32	60.02	57.32	58.33	58.70	**60.70**
Swan Ganz		**98.11**	94.68	97.73	97.50	95.65	96.36	96.72
Mean		80.20	80.65	81.50	81.78	81.65	81.79	**82.56**

Table 3 provides a quantitative analysis of the experimental results on the CXR11 dataset for different modules. Compared to the Baseline and Models 1 to 5, CheXNet demonstrates improvements in AUC of 2.36%, 1.91%, 1.06%, 0.78%, 0.91%, and 0.77%, respectively. The proposed CheXNet achieves the best classification results through a combination of several modules. From the AUC values for each category in Table 3 , it can be observed that each module plays a role, confirming the effectiveness of these modules. Figure 4 provides a visual representation of our classification results on the CXR11 dataset.

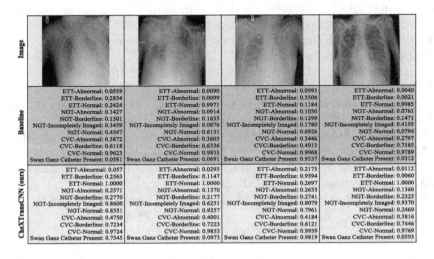

Fig. 4. Comparison of the number of cases for each disease in the CXR11 dataset.

5 Conclusion

In this paper, we propose a hybrid deep learning network named CheXNet. The label embedding automatically captures label dependencies, effectively alleviating label dependency issues. We incorporate multi-scale pooling to fuse features from different scales and an inner branch to capture more locally detailed features. Moreover, we employ the IIM module to facilitate information interaction between the CNN and Transformer, enabling the network to effectively utilize both local and global lesion features. We conducted one ablation experiment and two comparative experiments to analyze our method. The extensive experimental results on the two CXR datasets demonstrate the effectiveness and generalization ability of our approach in the field of multi-label medical classification.

References

1. Allaouzi, I., Ben Ahmed, M.: A novel approach for multi-label chest x-ray classification of common thorax diseases. IEEE Access **7**, 64279–64288 (2019)
2. Baltruschat, I.M., Nickisch, H., Grass, M., Knopp, T., Saalbach, A.: Comparison of deep learning approaches for multi-label chest x-ray classification. Sci. Rep. **9**(1), 1–10 (2019)
3. Dosovitskiy, A., et al.: An image is worth 16 ×16 words: Transformers for image recognition at scale. arXiv preprint arXiv:2010.11929 (2020)
4. Guan, Q., Huang, Y.: Multi-label chest x-ray image classification via category-wise residual attention learning. Pattern Recogn. Lett. **130**(SI), 259–266 (2020)
5. He, K., Zhang, X., Ren, S., Sun, J.: Deep residual learning for image recognition. In: Proceedings of the IEEE Conference on Computer Vision and Pattern Recognition (CVPR) 2016, Seattle, WA, June 27–30, (2016)
6. Hu, J., Shen, L., Albanie, S., Sun, G., Wu, E.: Squeeze-and-excitation networks. IEEE Comput. Soc. **42**, 2011–2023 (2020)
7. Lee, Y.W., Huang, S.K., Chang, R.F.: CheXGAT: a disease correlation-aware network for thorax disease diagnosis from chest x-ray images. Artif. Intell. Med. **132**, 102382 (2022)
8. Liu, Z., et al.: Swin transformer: Hierarchical vision transformer using shifted windows. In: 2021 IEEE/CVF International Conference on Computer Vision (ICCV 2021). pp. 9992–10002 2021, eLECTR Network, Oct 11–17 (2021)
9. Liu, Z., Mao, H., Wu, C.Y., Feichtenhofer, C., Darrell, T., Xie, S.: A convnet for the 2020s. In: Proceedings of the IEEE/CVF Conference on Computer Vision and Pattern Recognition (CVPR), pp. 11976–11986 June 2022, new Orleans, LA, JUN 18–24 (2022)
10. Ma, C., Wang, H., Hoi, S.C.H.: Multi-label thoracic disease image classification with cross-attention networks. In: Shen, D., et al. (eds.) MICCAI 2019. LNCS, vol. 11769, pp. 730–738. Springer, Cham (2019). https://doi.org/10.1007/978-3-030-32226-7_81
11. Majkowska, A., et al.: Chest radiograph interpretation with deep learning models: assessment with radiologist-adjudicated reference standards and population-adjusted evaluation. Radiology **294**(2), 421–431 (2020)
12. Pesce, E., et al.: Learning to detect chest radiographs containing pulmonary lesions using visual attention networks. Med. Image Anal. **53**, 26–38 (2019)

13. Sahlol, A.T., Abd Elaziz, M., Tariq Jamal, A., Damaševičius, R., Farouk Hassan, O.: A novel method for detection of tuberculosis in chest radiographs using artificial ecosystem-based optimisation of deep neural network features. Symmetry **12**(7), 1146 (2020)

14. Taslimi, S., Taslimi, S., Fathi, N., Salehi, M., Rohban, M.H.: Swinchex: multi-label classification on chest x-ray images with transformers. arXiv preprint arXiv:2206.04246 (2022)

15. Touvron, H., Cord, M., Douze, M., Massa, F., Sablayrolles, A., Jegou, H.: Training data-efficient image transformers & distillation through attention. In: International Conference On Machine Learning, vol. 139, pp. 7358–7367. ELECTR NETWORK, JUL 18–24 (2021)

16. Vaswani, A., et al.: Attention is all you need. In: Advances in Neural Information Processing Systems. vol. 30. Curran Associates, Inc. (2017)

17. Xiao, J., Bai, Y., Yuille, A., Zhou, Z.: Delving into masked autoencoders for multi-label thorax disease classification. In: Proceedings of the IEEE/CVF Winter Conference on Applications of Computer Vision (WACV), pp. 3588–3600 (January), Los Angeles, CA (2023)

18. Xie, S., Girshick, R., Dollar, P., Tu, Z., He, K.: Aggregated residual transformations for deep neural networks. In: Proceedings of the IEEE Conference on Computer Vision and Pattern Recognition (CVPR) (July), Honolulu, HI (2017)

19. Yang, M., Tanaka, H., Ishida, T.: Performance improvement in multi-label thoracic abnormality classification of chest x-rays with noisy labels. Int. J. Comput. Assist. Radiol. Surg. **18**(1, SI), 181–189 (2023)

Cross Attention Multi Scale CNN-Transformer Hybrid Encoder Is General Medical Image Learner

Rongzhou Zhou[1], Junfeng Yao[1,2,3(✉)], Qingqi Hong[4(✉)], Xingxin Li[1], and Xianpeng Cao[1]

[1] Center for Digital Media Computing, School of Film, School of Informatics, Xiamen University, Xiamen, China
[2] Key Laboratory of Digital Protection and Intelligent Processing of Intangible Cultural Heritage of Fujian and Taiwan, Ministry of Culture and Tourism, Xiamen, China
[3] Institute of Artificial Intelligence, Xiamen University, Xiamen 361005, China
yao0010@xmu.edu.cn
[4] Xiamen University, Xiamen 361005, China
hongqq@xmu.edu.cn

Abstract. Medical image segmentation plays a crucial role in medical artificial intelligence. Recent advancements in computer vision have introduced multiscale ViT (Vision Transformer), revealing its robustness and superior feature extraction capabilities. However, the independent processing of data patches by ViT often leads to insufficient attention to fine details. In medical image segmentation tasks like organ and tumor segmentation, precise boundary delineation is of utmost importance. To address this challenge, this study proposes two novel CNN-Transformer feature fusion modules: SFM (Shallow Fusion Module) and DFM (Deep Fusion Module). These modules effectively integrate high-level and low-level semantic information from the feature pyramid while maintaining network efficiency. To expedite network convergence, the Deep Supervise method is introduced during the training phase. Additionally, extensive ablation experiments and comparative studies are conducted on well-known public datasets, namely Synapse and ACDC, to evaluate the effectiveness of the proposed approach. The experimental results not only demonstrate the efficacy of the proposed modules and training method but also showcase the superiority of our architecture compared to previous methods. The code and trained models will be available soon.

Keywords: Medical image segmentation · Feature Fusion · Transformer

Q. Liu et al. (Eds.): PRCV 2023, LNCS 14437, pp. 85–97, 2024.
https://doi.org/10.1007/978-981-99-8558-6_8

1 Introduction

Medical image segmentation is considered to be one of the most fundamental problems in medical imaging.[1] Medical images often undergo significant pixel distortions during different image acquisitions, making their accurate segmentation challenging. Additionally, the scarcity of expensive pixel-level annotations is another issue that leads to performance gaps. To achieve efficient and effective segmentation, models need to better understand not only their local semantic features to capture more subtle organ structures but also their global feature dependencies to capture the relationships between multiple organs [1].

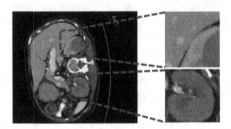

Fig. 1. There are typically two challenges in medical image segmentation, particularly in the context of multi-object segmentation. The first challenge is how to capture the interrelationships between different objects, similar to the compressive interactions between organs (Top). The second challenge pertains to achieving finer delineation of the boundaries of enclosed objects (Bottom).

Due to the strong feature extraction capabilities of convolutional operations and their inductive bias, image segmentation algorithms based on Convolutional Neural Networks (CNNs), such as UNet [2] and its variants [3–6], have achieved significant success in medical image segmentation. However, due to the limited receptive field of convolutional operations, they are unable to capture long-range relationships in the semantic information of images. As a result, CNN-based methods have certain limitations in extracting global features.

To address the inherent limitations of convolutional neural networks in the field of medical image segmentation, an increasing number of studies have opted for ViT [7] as the backbone for feature extraction. ViT incorporates a multi-head self-attentive mechanism (MHSA) and feedforward neural network (FFN), enabling the transformation of input image patches into tokens within the network to extract their global semantic features. Unlike computer vision, where the concept of scale is not heavily emphasized, ViT emerged primarily from the field of natural language processing. For instance, TransUNet [8] serves as a pioneering work that combines the Transformer module with a CNN encoder in a

[1] This work is supported by the Natural Science Foundation of China (No. 62072388), the industry guidance project foundation of science technology bureau of Fujian province in 2020 (No. 2020H0047), and Fujian Sunshine Charity Foundation.

cascaded fashion, forming a hybrid encoder and integrating a U-shaped network to concatenate low-level features into the decoder. By employing CNN for scale transformation and stacking Transformer modules for global feature extraction, TransUNet achieved high-accuracy image segmentation. However, the computational intensity associated with the utilization of a single-scale Transformer module imposes certain limitations. Consequently, recent research has shifted towards utilizing ViT as the backbone network for encoders, particularly with the introduction of multi-scale ViT [9,10]. These proposals have the capacity to extract multi-scale features, making them more suitable for intensive prediction tasks. SwinUNet [11] employs the multi-scale Swin Transformer Block as the encoding-decoding structure of the model, employing a pure Transformer structure for high-performance semantic segmentation and surpassing TransUNet comprehensively in terms of results. Polyp-PVT [12] utilizes the Pyramid Vision Transformer (PVT) [10] as the encoder and a meticulously designed decoder for polyp segmentation, yielding favorable outcomes. Collectively, these works demonstrate the tremendous potential of multi-scale ViT in the field of medical image segmentation.

However, both the multi-scale ViT and the convolutional neural networks with feature pyramid structures fail to address the two challenges presented in Fig. 1. Therefore, it is a natural idea to combine the strengths of Transformer and CNN by fusing them together. However, most existing studies employ a fixed-scale ViT fused with CNN, which incurs high computational costs and can only apply global attention to the high-level semantic features encoded by CNN, lacking attention to details. Hence, we propose a dual-encoder architecture that combines multi-scale ViT and convolutional neural networks, enabling the integration of full-scale feature maps from both Transformer and CNN. Specifically, we design the Shallow Fusion Module (SFM) and Deep Fusion Module (DFM) for the fusion of low-level and high-level semantic information in the feature pyramid, respectively. The design of SFM is mainly based on convolutional neural networks since low-level feature maps are typically large in size, making it impractical to use self-attention-based fusion methods due to the high computational overhead. Additionally, low-level feature maps are often concatenated with the decoder to restore low-order semantic information of the image, making the use of convolutional neural networks an efficient and elegant choice for this module. For DFM, we introduce a Dual Path cross-attention mechanism that takes inputs from both the CNN encoder and ViT encoder. Through cross-attention, the feature maps encoded by one branch interact with those encoded by the other branch, achieving high-quality feature fusion. To facilitate training, we adopt the deep supervision training approach, which involves upsampling and concatenating the features at each scale of the decoder, and finally obtaining the predicted segmentation map through a double convolution operation with the Ground Truth computational loss. We have conducted comprehensive ablation experiments to demonstrate the effectiveness of all our proposed modules and training methods.

Our contributions can be summarized in three aspects: (1) We propose a dual-encoder architecture that combines multi-scale ViT and convolutional neural networks for general medical image segmentation. This design enables the model to simultaneously capture both local and global features. (2) We introduce the SFM and DFM modules to fuse high and low-level feature maps, ensuring both performance improvement and computational efficiency. We also incorporate deep supervision training to accelerate the convergence speed of our model. (3) We conduct extensive experiments on two well-known publicly available datasets, ACDC and Synapse, achieving average DSC scores of 91.38 and 82.15, respectively. These results demonstrate the competitiveness of our approach compared to previous methods.

2   Methods

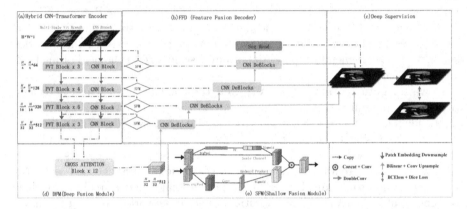

Fig. 2. The pipeline of our proposal's framework. H, W represents the height and width of the picture, respectively, and the following number represents the number of channels of the picture. Classes indicate the number of classes present in the segmentation result.

This section introduces the overall architecture of our proposal. The pipeline of our method is shown in Fig. 2, which includes the multi-scale ViT and CNN encoders. Specifically, for the multi-scale encoders used in our experiments, we employ the Pyramid Vision Transformer [10] and the encoder of the UNet [2]. These encoders are widely used in dense prediction tasks such as semantic segmentation and have demonstrated good performance. They also provide sufficient baselines for comparison with our method. The figure also illustrates our two feature fusion modules, namely SFM and DFM. These modules are designed to fuse low-level and high-level semantic information, respectively. SFM is based on the attention mechanism in convolutional neural networks, while DFM leverages cross multi-head self-attention mechanisms. These designs balance the computational cost of our method while providing high-quality fusion results. To facilitate the training of our network, we incorporate the deep supervision method. This ensures that each scale's feature maps in our model are supervised by the ground truth, leading to faster convergence. In this section, we will provide a detailed explanation of all the modules in our method.

2.1   Dual Encoder

The dual-encoder design has been proven effective in many scenarios [13], where the combination of two encoders with distinct structures, along with carefully designed feature fusion modules, often achieves complementary effects. In medical image segmentation, we need to consider the compression and relationships between different organs or lesions, as well as the precise extraction of individual lesion boundaries. Inspired by this, we simultaneously adopt the PVT and UNet encoders as the feature encoding backbone networks in our model. PVT, unlike methods such as TransUNet and TranFuse that use fixed-scale encoding or sliding window approaches, utilizes a pyramid structure that retains multiscale features and employs spatial reduction methods to reduce computational burden. This is crucial for dense prediction tasks like semantic segmentation. As for the CNN branch, we directly adopt the encoder part of UNet, which is widely used in semantic segmentation networks. Its encoder part has a very small parameter count, resulting in acceptable computational overhead. In fact, compared to deeper backbones like ResNet50 [14], our composite encoder formed by the combination of two encoders has lower floating-point calculations and parameter counts.

2.2   Shallow Fusion Module

When it comes to low-level feature fusion, there are often challenges related to large feature maps and scattered semantic information. Directly fusing features at the pixel level incurs significant computational overhead, which is not feasible. Therefore, we incorporate the concepts of spatial attention from the Convolutional Neural Network's CBAM [15] module and channel attention from the squeeze-and-Excitation [16] module to design a comprehensive feature fusion module called the shallow fusion module, as depicted in part (e) of Fig. 2. It consists of dual-path inputs and a single-path output. Firstly, let us define the computation process for channel attention as Eqs. 1, 2 and spatial attention mechanisms as Eq. 3.

$$Channel_{weight} = \mathcal{C}\left(P_{\max}(T_i)\right) + \mathcal{C}\left(P_{\text{avg}}\left(T_i\right)\right) \qquad (1)$$

$$\text{Attn}_c(T_i) = Softmax\left(Channel_{weight}\right) \odot T_i \qquad (2)$$

$$\text{Attn}_s(T_i) = Softmax\left(\mathcal{H}\left[R_{\max}(T_i), R_{\text{avg}}\left(T_i\right)\right]\right) \odot T_i \qquad (3)$$

where $\mathcal{C}$ consists of a convolutional layer with 1×1 kernel size to reduce the channel dimension 16 times, followed by a ReLU layer and another 1×1 convolutional layer to recover the original channel dimension. $P_{max}(\cdot)$ and $P_{avg}(\cdot)$ denote adaptive maximum pooling and adaptive average pooling functions, respectively. In spatial attention, where $R_{max}(\cdot)$ and $R_{avg}(\cdot)$ represent the maximum and average values obtained along the channel dimension, respectively. $\mathcal{H}$ represents a 1×1 convolutional layer to reduce the channel dimension of 1.

If we use $Input_{cnn}$ and $Input_{transformer}$ to represent the features obtained from the CNN branch and Transformer branch, respectively, the entire SFM process can be described as Eq. 4.

$$SFM(Input_t, Input_c) = C_{proj}([Attn_c(Input_t), Attn_s(Input_c)]) \qquad (4)$$

Where $Attn_c(Input_t)$ and $Attn_s(Input_c)$ represent the features encoded by PVT and CNN, respectively, after being weighted by channel attention and spatial attention. C_{proj} denotes the channel projection operation, which reduces the feature dimensionality of the concatenated features in the channel dimension to match the dimensionality before concatenation, resulting in the fused key feature.

2.3   Deep Fusion Module

Low-level features are crucial for capturing fine details in image restoration tasks. Additionally, in medical image segmentation, where multi-object segmentation is more common, the extraction of high-level semantic information is equally important. However, the fusion of high-level features is more complex compared to low-level fusion. Modeling high-level features requires more precise representation. Therefore, we employ pixel-level matching based on cross-attention to fuse the features extracted from the CNN and Transformer encoders. The specific process is illustrated as Fig. 3.

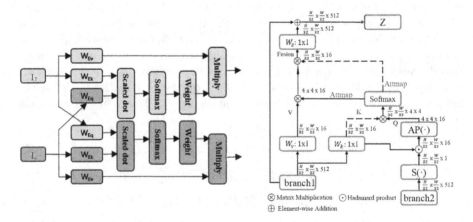

Fig. 3. The pipeline of deep feature fusion is shown. The left diagram illustrates how the features from the two encoders are matched at a pixel level, achieved by exchanging the positions of the query matrices for feature querying. The right diagram demonstrates the specific computational process involved in a matching operation.

From the left diagram in Fig. 3, we can observe that our feature fusion consists of two symmetric branches. Therefore, we will provide a detailed explanation of the operation process for one branch. Firstly, our feature fusion relies on the matrices K and V computed by branch1, as well as the matrix Q obtained from branch2. The specific operation process is shown in the Eqs. 5 and 6.

$$K = W_K * \text{branch1}, \quad V = W_v * \text{branch1}, \tag{5}$$

$$Q = \text{AP}\,(K \odot \mathbf{S}\,(\text{branch2})) \tag{6}$$

where the symbol * and $\odot$ represent the convolution operation and Hadamard product, respectively. The correlation between K and V is then established, and the specific process can be described as Eq. 7:

$$\text{Attmap} = \text{Softmax}\,(Q \otimes K^\top) \tag{7}$$

where K^T is the transpose of K and $\otimes$ is the matrix multiplication. Then we can perform the fusion of the Attention map and original features by matrix multiplication and graph neural network, and the process is shown by the following Eq. 8:

$$\text{Fusion} = \text{Attmap}^\text{T} \otimes (\text{Attmap} \otimes V) \tag{8}$$

After the features are fused and added element by element with the initial branch2, the dimensionality of the Fusion feature matrix needs to be transformed to be the same as the branch2 before that, again using a 1×1 size convolution to transform the dimensionality. The process can be described as Eq. 9:

$$Z = \text{branch1} + W_z * \text{Fusion} \tag{9}$$

where $Z \in \mathbb{R}^{\frac{H}{32} \times \frac{W}{32} \times 512}$ is the fusion result of CNN branch and Transformer branch.

We refer to all the individual branch computations mentioned above as Cross-Attention, denoted as CA(.). By stacking 12 such DFM modules, we achieve the fusion of high-level features from the dual branches. Afterwards, we reshape and concatenate the fused features, followed by dimension reduction along the channel dimension. This process of DFM can be described as Eq. 10.

$$DFM(I_F, I_C) = C_{proj}([CA_{12}...(CA_1(I_F, I_C), CA(I_C, I_F))]) \tag{10}$$

Where I_T and I_C represent the inputs of the Transformer and CNN, respectively. CA_{12} represents the 12th stacked CA module. Through this approach, we accomplish the fusion of high-level features. Subsequent comparative experiments and ablation studies have demonstrated the effectiveness of this approach in learning fused information with high-level semantics.

2.4   Deep Supervision

To train our model, we adopt a different strategy compared to most segmentation networks. Instead of using the traditional approach of obtaining a binarized image and Ground Truth to calculate the loss in the final segmentation result, we employ the method of deep supervision. This training method involves generating an additional branch for each layer of the model during the upsampling process. These branches are then convolved using bilinear interpolation to obtain segmentation results of the same size as the labels, as depicted in part (c) of Fig. 2. For these four feature maps, we concatenate them and apply a 1×1 convolution to transform their size to $H \times W \times Classes$, where H and W represent the height and width of the feature maps, respectively, and $Classes$ denotes the number of classes. The final segmentation result is then used to calculate the loss and train the model. This training approach has demonstrated several benefits. It enhances the stability of the training process and reduces the challenges associated with model fusion in networks with divergent coding and decoding structures.

3   Experiments and Results

3.1   Dateset

Synapse Multi-organ Segmentation Dataset: The Synapse dataset [17] includes 30 cases with 3779 axial abdominal clinical CT images. Following, 18 samples are divided into the training set and 12 samples into the testing set. And the average Dice-Similarity Coefficient (DSC) and average 95% Hausdorff Distance (HD95) are used as evaluation metrics to evaluate our method on 8 abdominal organs (aorta, gallbladder, spleen, left kidney, right kidney, liver, pancreas, spleen, stomach).

Automated Cardiac Diagnosis Challenge Dataset (ACDC): The ACDC dataset [18] is collected from different patients using MRI scanners. For each patient's MR image, the left ventricle (LV), right ventricle (RV), and myocardium (MYO) are labelled. The dataset is split into 70 training samples, 10 validation samples, and 20 testing samples.

Both datasets are partitioned strictly according to the index in TransUNet [8].

3.2   Implementation Details

Our method was trained on an Nvidia RTX 3090 using the PyTorch framework. The PVT encoder of the model initializes weights obtained from training on the ImageNet dataset, while the encoder of UNet is trained from scratch. During training, we applied data augmentation techniques such as random flipping, random rotation, and random center cropping. We set the batch size to 16, the base learning rate to 5e−4, and the epoch size to 200. The learning rate decay strategy is determined by Eq. 11, where $iter_{num}$ represents the current iteration number and $iter_{max}$ denotes the maximum number of iterations. We used the AdamW optimizer and a hybrid loss function given by $\mathcal{L}_{\text{total}} = \frac{1}{2}\mathcal{L}_{\text{BCE}} + \frac{1}{2}\mathcal{L}_{\text{Dice}}$, where $\mathcal{L}_{\text{Dice}}$ is the Dice-Similarity Coefficient loss and $\mathcal{L}_{\text{BCE}}$ is the Binary Cross-Entropy loss.

$$lr = baselr \times \left(1 - \frac{iter_{num}}{iter_{max}}\right)^{0.9} \tag{11}$$

3.3   Comparison with Other Methods

The experimental results of our method on two datasets are presented in Table 1, Table 2, and Fig. 4. Specifically, Table 1 displays the performance of our method on the Synapse dataset.

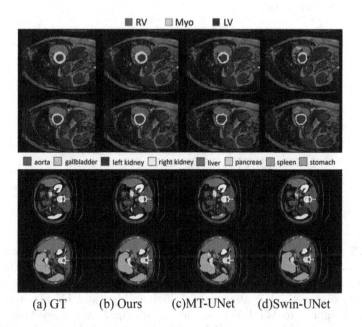

(a) GT   (b) Ours   (c)MT-UNet   (d)Swin-UNet

Fig. 4. Visualization of segmentation results on two datasets with the current better-performing model. ACDC dataset at the top, Synapse dataset at the bottom.

Our method demonstrates strong competitiveness compared to previous top-performing models in terms of the DSC metric. Specifically, our method outperforms SwinUNet by nearly 2.5% in DSC, and it achieves a 1.6 mm lower HD95

Table 1. Segmentation accuracy of different methods on the Synapse multi-organ CT dataset. The first two columns indicate the mean DSC and HD95 of all tested samples, and the next eight columns indicate the DSC of different organ segmentations.

Method	DSC↑	HD95↓	Aorta	Gallbladder	Kidney(L)	Kidney(R)	Liver	Pancreas	Spleen	Stomach
DARR [19]	74.68	-	74.74	53.77	72.31	73.24	94.08	54.18	89.90	45.96
R50 UNet [8]	74.68	36.87	84.18	62.84	79.19	71.29	93.35	48.23	84.41	73.92
UNet [2]	76.85	39.70	89.07	69.72	77.77	68.60	93.43	53.98	86.67	75.58
AttnUNet [8]	77.77	36.02	**89.55**	68.88	77.98	71.11	93.57	58.04	87.30	75.75
R50 ViT [8]	71.29	32.87	73.73	55.13	75.80	72.20	91.51	45.99	81.99	73.95
TransUNet [8]	77.48	31.69	87.23	63.13	81.87	77.02	94.08	55.86	85.08	75.62
MT-UNet [20]	78.59	26.59	87.92	64.99	81.47	77.29	93.06	59.46	87.87	76.81
Swin-UNet [11]	79.13	21.55	85.47	**66.53**	83.28	**79.61**	94.29	56.58	**90.66**	76.60
Polyp-PVT [12]	79.52	21.37	85.92	56.72	81.09	80.38	94.85	62.59	87.12	83.37
Ours	**82.15**	**19.77**	88.54	65.34	**87.67**	79.34	**95.87**	**67.02**	89.38	**84.06**

compared to Polyp-PVT. When it comes to specific organ segmentation, our method performs well for Kidney(L), Liver, Pancreas, and Stomach. Particularly for Pancreas, our method outperforms the previous best method by nearly 7%. The visualization in the bottom of Fig. 4 supports this conclusion, showing that our method is capable of achieving accurate segmentation even for small fractions of the Pancreas.

Table 2. Experimental results of the ACDC Dataset.

Method	DSC↑	RV	Myo	LV
R50 UNet [8]	87.60	84.62	84.52	93.68
ViT-CUP [8]	83.41	80.93	78.12	91.17
R50 ViT [8]	86.19	82.51	83.01	93.05
TransUNet [8]	89.71	86.67	87.27	95.18
Swin-UNet [11]	88.07	85.77	84.42	94.03
MT-UNet [20]	90.43	86.64	**89.04**	**95.62**
Ours	**91.38**	**89.98**	88.92	95.23

Table 2 shows the results of our method on the ACDC dataset, where our method was 0.95% higher than the previous best method in the mean DSC. Also in the segmentation results of RV, compared to the previous best-performing TransUNet, our method is almost 3.5% higher. Myo and LV are also almost the same compared to MT-UNet. At the top of Fig. 4 we can see that the segmentation accuracy of our method for RV is significantly higher than the other methods. The model and hyperparameters we use are identical in both datasets, demonstrating the generalization and robustness of our approach.

3.4   Ablation Studies

To verify the effectiveness of the modules in our proposed method and the supervised approach, we conducted ablation experiments on the Synapse dataset, and the results of the ablation experiments can be found in Table 3.

Table 3. Ablation experiment for the proposed module on Synapse dataset.

Module Name			Mean Metric	
DFM	SFM	DS	DSC↑	HD95↓
✗	✗	✗	79.12	35.86
✓	✗	✗	80.05	30.23
✗	✓	✗	80.43	28.35
✗	✗	✓	79.51	33.57
✓	✓	✗	81.72	21.56
✓	✓	✓	**82.15**	**19.77**

The abbreviation DS in the table refers to Deep Supervision. DFM and SFM remain consistent with the previously mentioned modules. When DFM, SFM, and DS are added to the model, the DSC coefficients for the test results increased by 0.93%, 1.31%, and 0.39% respectively. It can be observed that the introduction of DFM and SFM significantly improves the results, indicating the importance of the feature fusion module designed to integrate Transformer and CNN. In our study, we explored the integration of DFM and SFM into the network architecture. This integration improved performance by 2.64% compared to the baseline model. It is observed that fusing high-level semantic information with low-level semantic information allows the model to extract better representations. Additionally, we conducted ablation experiments on the Deep Supervision (DS) module. The introduction of DS had a positive impact on the model's test results. Specifically, adding the DS module to the baseline model and the complete model improved performance by 0.39% and 0.42% respectively. These findings suggest that deep supervision can help the model find optimal solutions and converge to better states. Overall, the modules we proposed, whether used individually or combined in our network, have a positive effect on our task, demonstrating the effectiveness of the architecture we proposed.

To ensure dimension consistency, when SFM and DFM are ablated, a concatenation operation is performed along the channel dimension, followed by a 1×1 convolutional operation to reduce the dimensionality. In other words, the ablated modules are replaced by a 1×1 convolutional operation.

4   Conclusion

In this paper, we propose a highly effective dual-encoder architecture for medical image segmentation. The dual-encoder consists of a multi-scale Vision Transformer and a convolutional neural network, specifically the PVT and UNet encoders, respectively. To enable the fusion of features from the two encoders with significantly different structures, we introduce the Spatial Feature Module and the Depth Feature Module for the fusion of low-level and high-level features. Additionally, to ensure a stable training process, we employ the method of deep supervision. The resulting trained model achieves competitive results on two publicly available datasets. In the future, we plan to further improve our approach to accommodate 3D image computations, enabling its application in a broader range of medical image computing scenarios.

References

1. Liu, Q., Kaul, C., Anagnostopoulos, C., Murray-Smith, R., Deligianni, F.: Optimizing vision transformers for medical image segmentation and few-shot domain adaptation. arXiv preprint arXiv:2210.08066 (2022)
2. Ronneberger, O., Fischer, P., Brox, T.: U-Net: convolutional networks for biomedical image segmentation. In: Navab, N., Hornegger, J., Wells, W.M., Frangi, A.F. (eds.) MICCAI 2015. LNCS, vol. 9351, pp. 234–241. Springer, Cham (2015). https://doi.org/10.1007/978-3-319-24574-4_28

3. Soucy, N., Sekeh, S.Y.: CEU-Net: ensemble semantic segmentation of hyperspectral images using clustering. arXiv preprint arXiv:2203.04873 (2022)

4. Diakogiannis, F.I., Waldner, F., Caccetta, P., Wu, C.: ResUNet-a: a deep learning framework for semantic segmentation of remotely sensed data. ISPRS J. Photogrammetry Remote Sens. **162**, 94–114 (2020)

5. Huang, H., Tong, R., Hu, H., Zhang, Q.: UNet 3+: a full-scale connected UNet for medical image segmentation. In: International Conference on Acoustics, Speech and Signal Processing (2020)

6. Zhou, Z., Rahman Siddiquee, M.M., Tajbakhsh, N., Liang, J.: UNet++: a nested U-Net architecture for medical image segmentation. In: Stoyanov, D., et al. (eds.) DLMIA/ML-CDS -2018. LNCS, vol. 11045, pp. 3–11. Springer, Cham (2018). https://doi.org/10.1007/978-3-030-00889-5_1

7. Dosovitskiy, A., et al.: An image is worth 16 × 16 words: transformers for image recognition at scale. In: ICLR 2021 (2021)

8. Chen, J., et al.: TransUNet: transformers make strong encoders for medical image segmentation. arXiv Computer Vision and Pattern Recognition (2021)

9. Liu, Z., et al.: Swin transformer: hierarchical vision transformer using shifted windows. In: International Conference on Computer Vision (2021)

10. Wang, W., et al.: Pyramid vision transformer: a versatile backbone for dense prediction without convolutions. In: International Conference on Computer Vision (2021)

11. Cao, H., et al.: Swin-Unet: Unet-like pure transformer for medical image segmentation. arXiv Image and Video Processing (2021)

12. Dong, B., Wang, W., Fan, D.-P., Li, J., Fu, H., Shao, L.: Polyp-PVT: polyp segmentation with pyramid vision transformers. arXiv Computer Vision and Pattern Recognition (2021)

13. Li, W., Yang, H.: Collaborative transformer-CNN learning for semi-supervised medical image segmentation. In: IEEE International Conference on Bioinformatics and Biomedicine, BIBM 2022, Las Vegas, NV, USA, 6–8 December 2022, pp. 1058–1065. IEEE (2022)

14. Verma, A., Qassim, H., Feinzimer, D.: Residual squeeze CNDS deep learning CNN model for very large scale places image recognition. In: 8th IEEE Annual Ubiquitous Computing, Electronics and Mobile Communication Conference, UEMCON, New York City, NY, USA, 19–21 October 2017, pp. 463–469. IEEE (2017)

15. Woo, S., Park, J., Lee, J.-Y., Kweon, I.S.: CBAM: convolutional block attention module. In: Ferrari, V., Hebert, M., Sminchisescu, C., Weiss, Y. (eds.) ECCV 2018. LNCS, vol. 11211, pp. 3–19. Springer, Cham (2018). https://doi.org/10.1007/978-3-030-01234-2_1

16. Hu, J., Shen, L., Albanie, S., Sun, G., Wu, E.: Squeeze-and-excitation networks. IEEE Trans. Pattern Anal. Mach. Intell. **42**(8), 2011–2023 (2020). https://doi.org/10.1109/TPAMI.2019.2913372

17. Landman, B., Xu, Z., Igelsias, J., Styner, M., Langerak, T., Klein, A.: MICCAI multi-atlas labeling beyond the cranial vault-workshop and challenge. In: Proceedings of the MICCAI Multi-Atlas Labeling Beyond Cranial Vault-Workshop Challenge, vol. 5, p. 12 (2015)

18. Bernard, O., Lalande, A., et al.: Deep learning techniques for automatic MRI cardiac multi-structures segmentation and diagnosis: is the problem solved? IEEE Trans. Med. Imaging **37**(11), 2514–2525 (2018)

19. Fu, S., et al.: Domain adaptive relational reasoning for 3D multi-organ segmentation. In: Martel, A.L., et al. (eds.) MICCAI 2020. LNCS, vol. 12261, pp. 656–666. Springer, Cham (2020). https://doi.org/10.1007/978-3-030-59710-8_64
20. Wang, H., et al.: Mixed transformer U-Net for medical image segmentation. arXiv preprint arXiv:2111.04734 (2022)

Weakly/Semi-supervised Left Ventricle Segmentation in 2D Echocardiography with Uncertain Region-Aware Contrastive Learning

Yanda Meng[1], Yuchen Zhang[3], Jianyang Xie[4], Jinming Duan[5], Yitian Zhao[6], and Yalin Zheng[1,2(✉)]

[1] Department of Eye and Vision Sciences, University of Liverpool, Liverpool, UK
[2] Liverpool Centre for Cardiovascular Science, University of Liverpool and Liverpool Heart and Chest Hospital, Liverpool, UK
yalin.zheng@liverpool.ac.uk
[3] Center for Bioinformatics, Peking University, Beijing, China
[4] Department of Electrical Engineering and Electronics and Computer Science, University of Liverpool, Liverpool, UK
[5] School of Computer Science, University of Birmingham, Birmingham, UK
[6] Cixi Institute of Biomedical Engineering, Ningbo Institute of Industrial Technology, Chinese Academy of Sciences, Ningbo, China

Abstract. Segmentation of the left ventricle in 2D echocardiography is essential for cardiac function measures, such as ejection fraction. Fully-supervised algorithms have been used in the past to segment the left ventricle and then offline estimate ejection fraction, but it is costly and time-consuming to obtain annotated segmentation ground truths for training. To solve this issue, we propose a weakly/semi-supervised framework with multi-level geometric regularization of vertex (boundary points), boundary and region predictions. The framework benefits from learning discriminative features and neglecting boundary uncertainty via the proposed contrastive learning. Firstly, we propose a multi-level regularized semi-supervised paradigm, where the regional and boundary regularization is favourable owing to the intrinsic geometric coherence of vertex, boundary and region predictions. Secondly, we propose an uncertain region-aware contrastive learning mechanism along the boundary via a hard negative sampling strategy for labeled and unlabeled data at the pixel level. The uncertain region along the boundary is omitted, enabling generalized semi-supervised learning with reliable boundary prediction. Thirdly, for the first time, we proposed a differentiable ejection fraction estimation module along with 2D echocardiographic left ventricle segmentation for end-to-end training without

---

Y. Meng and Y. Zhang—co-first authors.

Supplementary Information The online version contains supplementary material available at https://doi.org/10.1007/978-981-99-8558-6_9.

offline post-processing. Supervision on ejection fraction can also serve as weak supervision for the left ventricular vertex, region and boundary predictions without further annotations. Experiments on the EchoNet-Dynamic datasets demonstrate that our method outperforms state-of-the-art semi-supervised approaches for segmenting left ventricle and estimating ejection fraction.

Keywords: Weakly/Semi-supervised Left Ventricle Segmentation · Contrastive Learning · End-to-End Ejection Fraction Estimation

1   Introduction

On the basis of echocardiography, the precise diagnosis of cardiovascular disease relies on the accurate segmentation of important structures and the assessment of critical metrics. For example, the relative size of the 2D echocardiographic left ventricle at end-diastolic (LV_{ED}) and end-systolic (LV_{ES}) time points can be used to formulate the ejection fraction (LV_{EF}), which is essential for diagnosing cardiomyopathy, atrial fibrillation and stroke, *etc.* Thus, accurate LV_{ES} and LV_{ED} segmentation is critical for the central metric of cardiac function of echocardiographic imaging.

In the past few years, there have been many new deep learning models used for the diagnosis and segmentation of medical images [17,18,21–24,27,31,32], including left ventricle (LV) in echocardiography [3,8,12]. They have greatly improved the ability to segment LV_{ED} and LV_{ES} in 2D echocardiographic images. These supervised segmentation models required a fully supervised mechanism that relied on extensive annotations. In addition, additional offline post-processing steps were needed to calculate LV_{EF}. By directly learning from a limited collection of labeled data and a huge amount of unlabeled data, semi-supervised learning frameworks could provide high-quality segmentation results [7,13,14,16,20,25,26,35]. Many investigations of unsupervised consistency regularization have been conducted. For example, researchers [7,35] introduced data-level disturbances into unlabeled inputs and demanded consistency between original and disturbed data predictions. Other studies, such as [13,16], perturbed the model's features at the output level to ensure consistency across different output branches. However, task-level consistency regularization in semi-supervised learning has received less attention until recently [14,20]. For instance, if we can map the predictions of multiple tasks into the same preset space and then assess them using the same criteria, the results would surely be suboptimal due to the perturbations in predictions across tasks [14].

To this end, we proposed a multi-level task-based geometric regularization of vertices K_p, boundaries B_p and regions R_p as shown in Fig. 1. Firstly, we derived the boundary B_r from the region prediction R_p and applied the unsupervised losses to enforce the boundary regularization between B_r and B_p with a large amount of unlabeled data. Furthermore, given the potential uncertainty of object boundary in medical image segmentation tasks [34,37], an uncertain

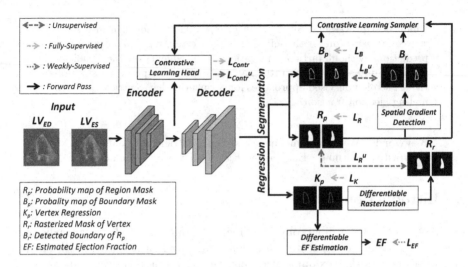

Fig. 1. A simplified overview of the proposed learning pipeline. L_R (*Dice* loss), L_B (*Dice* loss) and L_K (*MSE* loss) denote supervised losses with ground truth on labeled data. Please refer to Fig. 2 for the detailed structure of the contrastive learning module and the differentiable LV_{EF} estimation process.

region-aware contrastive learning mechanism for both labeled and unlabeled data was proposed. Recent prototype-level of contrastive segmentation methods [7] applied contrastive loss on high-level feature embeddings. Differently, ours enabled the mining of positive and hard negative samples at the pixel level. Our proposed novel hard negative sampling strategy omitted the uncertain background regions, resulting in a more accurate B_p and thus benefited boundary regularization in the proposed semi-supervised pipeline. Secondly, we formed a polygon by connecting each vertex of K_p sequentially and produced a rasterized region mask R_r using a differentiable rasterization process. Then the unsupervised loss was applied to learn the underlying regional association for more accurate K_p predictions, thus leading to a precise LV_{EF} estimation. On the other hand, the previous weakly supervised segmentation method [15] used scribbles as supervision, to reduce the workload of annotations. Differently, we exploited the clinical function assessment of LV_{EF} metric as weak supervision for other tasks in this framework. Specifically, given the K_p and R_r of LV_{ES} and LV_{ED}, an end-to-end LV_{EF} estimation mechanism was proposed based on a novel differentiable process to estimate the LV long axis and area. As a result, the information gained from LV_{EF} ground truth can weakly supervise R_p, B_p and K_p of LV_{ES} and LV_{ED} simultaneously.

This article demonstrates how to rationally leverage geometric associations between vertices, regions and boundaries on semi-supervised regularization and differentiable end-to-end weakly-supervised LV_{EF} estimation.

2   Methods

2.1   Multi-level Regularization of Semi-supervision

The regularization between regions R_p and vertices K_p is achieved by producing rasterized region masks R_r from K_p by a differentiable rasterizer [11]. Firstly, a polygon is designed by sequentially connecting each vertex of K_p. Given K_p of polygon and a desired rasterization pixel size of $H \times W$, for pixel x at location (h, w), where $0 \leq h \leq H$, $0 \leq w \leq W$, the PnP algorithm [1] is used to determine whether x falls inside the polygon, such as $I(K_p, x)$, where $I = 1$ if it falls within and -1 otherwise. Then, x is projected onto the nearest segment of the polygon's boundary, and the resulting distance is recorded as $D(K_p, x)$. Finally, the rasterized region mask R_r can be defined as:

$$R_r(x) = Sigmoid(\frac{I(K_p, x) * D(K_p, x)}{\epsilon}), \tag{1}$$

where $Sigmoid$ is a non-linear function; ϵ is a hyper-parameter that controls the sharpness/smoothness of R_r and is empirically set as 0.1. With Eq. 1, for all of the unlabeled input, we apply a $Dice$ loss (L_{R^u}) between R_p and R_r to enforce the unsupervised region-level regularization.

The regularization between regions R_p and boundaries B_p is achieved by producing derived boundary masks B_r from R_p via a derived spatial gradient. Specifically, inspired by the previous method [2], we extract a certain range of wide boundaries instead of narrow contours due to the limited effectiveness of regularization in a semi-supervised manner. Given the spatial gradient detection function F_{sg}, the derived boundary masks B_r can be defined as:

$$F_{sg}(R_p) = B_r = R_p + Maxpooling2D(-R_p). \tag{2}$$

A simple yet efficient $Maxpooling2D$ operation conducts the same feature map size as its input. Notably, the boundary width of B_r can be chosen by adjusting the kernel size, stride, and padding value of the $Maxpooling2D$ function. The width is empirically set as 4 in this work. $Dice$ loss (L_{B^u}) between B_p and B_r is applied to enforce the unsupervised boundary-level regularization of the unlabeled data.

2.2   Uncertain Region-Aware Contrastive Learning

An uncertain region-aware contrastive learning (shown in Fig. 2) is proposed for discriminative boundary feature representations, leading to an improved boundary regularization mechanism in a semi-supervised manner. Generally, contrastive learning consists of an encoded query q, a positive key k^+, and other negative keys k^-. A typical contrastive loss function InfoNCE (L_q^{NCE}) [29] in image-level tasks is to pull q close to k^+ while pushing it away from negative keys k^-:

$$L_q^{NCE} = -\log \frac{\exp(q \cdot k^+/\tau)}{\exp(q \cdot k^+/\tau) + \sum_{k^-} \exp(q \cdot k^-/\tau)}. \tag{3}$$

However, it is hard to adopt the negative sampling strategy in classical contrastive learning, from image-level to pixel-level tasks. Because it is very likely to select false negative samples given the hard uncertain regions [34] along the boundary. Previous study [33] also revealed that mining hard uncertain negative samples were beneficial to learn discriminative features by contrastive learning. Therefore, we design an uncertain region-aware negative sampling strategy for labeled and unlabeled data.

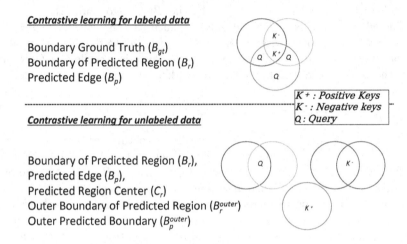

Fig. 2. Details of the proposed contrastive sampler on labeled and unlabeled data.

Given Eq. 3 for labeled data, k^+ is sampled from the predicted true-positive set $B_{gt} \cap B_p \cap B_r$; k^- is sampled from the false positive set $B_p \cap B_r \cap B_{gt}^c$; q is sampled from the false negative set $B_{gt} \cap (B_p \cap B_r)^c$, where B_{gt} is the ground truth of the boundary mask, and the superscript c denotes complement of the set. This brings the predicted false-negative (hard negative) features closer to the positive features and further away from the negative ones. For unlabeled data, we sample k^+ from $B_p \cap B_r$ of the unlabeled input. Then, we derive the outer predicted boundary mask B_p^{outer} and the region's outer boundary mask B_r^{outer} with Eq. 2, such as $B_p^{outer} = F_{sg}(-B_p)$ and $B_r^{outer} = F_{sg}(-B_r)$. Specifically, k^- are sampled from $B_p^{outer} \cap B_r^{outer}$, which are the pseudo-labeled background pixels nearest to the predicted foreground. Furthermore, inspired by [9], we construct the query q as the region center of the predicted foreground. The region center is defined as the average features of all pixels inside the region mask, excluding the boundary part. Formally, given the feature map $M \in \mathbb{R}^{C \times H \times W}$, the center region C_r of M is defined as:

$$C_r = \frac{\sum_{h,w} M(h,w) 1[P(h,w) = 1]}{\sum_{h,w} 1[P(h,w) = 1]}, \tag{4}$$

where $1[\cdot]$ is the binary indicator; $P(h,w)$ is the predicted region label at location (h,w), where $h \in [0, H]$ and $w \in [0, W]$. In the end, Eq. 3 is applied to the sampled q, k^+, and k^- for both labeled and unlabeled data, which gives L_{contr} and L_{contr}^u respectively to ensure a well-represented dedicated feature space in the contrastive learning head in Fig. 1.

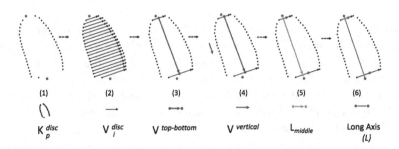

Fig. 3. The schematic diagram for long-axis (L) estimation. K_p^{disc} is the regressed vertices (disc endpoints) along the boundary. v_i^{disc} is one of the k disc vectors. $v^{top-bottom}$ is the vector from the midpoint of the disc at the top to the midpoint of the disc at the bottom. $v^{vertical}$ is the averaged unit vertical vectors of every one of k disc. L_{middle} is the length of the projection of $v^{top-bottom}$ along the $v^{vertical}$ direction. L is the length of the long-axis vector.

2.3   Differentiable Ejection Fraction Estimation of Weak Supervision

The ejection fraction is the ratio of left ventricular end-systolic volume (ESV) and end-diastolic volume (EDV), such as $LV_{EF} = (EDV - ESV)/EDV$. Area-Length [4] or Simpson methods [5] is used by previous methods [8,12] to off-line post-approximate the LV volume. This work found no significant difference in estimating LV_{EF}. And we adopted the Area-Length method [4] via the endocardial area (A) and the length of the long axis (L) in 2D echocardiography to differentiable estimate LV cavity volume, such as:

$$V_{endocardial} \approx \frac{8}{3\pi} \cdot \frac{A^2}{L}, \tag{5}$$

where A can be approximated by the number of pixels of R_r or R_p in our model without significant difference. Intuitively, the long axis length L can be directly calculated by locating the two endpoints/vertices of the long axis. However, the bottom endpoint lacks semantic and spatial information, making direct regression challenging and inaccurate [8]. To solve this, we propose to estimate the long axis via the regressed vertices (disc endpoints) along the boundary end-to-end (shown in Fig. 3). Given regressed disc endpoints K_p^{disc} in Fig. 3(1), the disc direction vector v_i^{disc} of every disc i is shown in Fig. 3(2). The long axis (L) is divided into $k + 1$ segments by each v_i^{disc}. According to the *EchoNet-Dynamic* [30] ground truth of the long axis (L) and the disc endpoints K_p^{disc}, The terminal segments protruding discs are defined as $\frac{L}{2k}$ long. And the rest of each middle segment between the top and bottom discs is defined as $\frac{L}{k}$ long. In this way, the overall length of the middle segments L_{middle} is equal to $\frac{k-1}{k}L$, such as:

$$L_{middle} = \frac{k-1}{k} \cdot L \tag{6}$$

On the other hand, L_{middle} is equal to the distance between the top and bottom disc. In order to estimate L_{middle}, we first calculate $v^{top-bottom}$ via K_p^{disc}. Specifically, $v^{top-bottom}$ is the vector from the midpoint of the disc at the top to the midpoint of the disc at the bottom, which is shown in Fig. 3(3). After that, we find the vector $v^{vertical}$ that is perpendicular to all the discs, which is shown in Fig. 3(4). In detail, $v^{vertical}$ can be calculated by averaging unit vertical vectors $v_i^{disc^\mathsf{T}}$ of every one of k disc, such as:

$$v^{vertical} = \frac{\sum_i \frac{v_i^{disc^\mathsf{T}}}{|v_i^{disc^\mathsf{T}}|}}{k}. \tag{7}$$

Thus, given $v^{top-bottom}$ and $v^{vertical}$, we can calculate L_{middle} by projecting $v^{top-bottom}$ along the $v^{vertical}$ direction, such as:

$$L_{middle} = \frac{v^{top-bottom} \cdot v^{vertical}}{|v^{vertical}|}, \tag{8}$$

and this is shown in Fig. 3(5). Finally, Given Eq. 6, and 8, we can define the long axis length L based on discs points solely, such as:

$$L = \frac{k}{k-1} \cdot \frac{v^{top-bottom} \cdot v^{vertical}}{|v^{vertical}|}, \tag{9}$$

and this is shown in Fig. 3(6). Therefore, We propose a differentiable strategy to calculate the long axis length by endpoints on discs. Mean square error (MSE) loss L_{EF} is employed to LV_{EF} prediction for supervision. Note that

3   Datasets and Implementation Details

EchoNet-Dynamic [30]: a large public dataset with 2D apical four-chamber echocardiography video sequences. Each video has about 200 frames with a resolution of 112 × 112 pixels, where experienced cardiologists manually labeled a pair of two frames (end-systole and end-diastole) in each sequence. In detail, they provided the vertex locations (disc endpoints in Sect. 2.3). We generated the region and boundary ground truth via Eq. 1 and 2. Among the 9,768 pairs of images, 8,791 image pairs were used for three-fold training and cross-validation and 977 image pairs for external testing.

Implementation: The stochastic gradient descent with a momentum of 0.9 was used to optimise the overall parameters. We trained the model around 300 epochs for all the experiments, with a learning rate of 0.06 and a step decay rate

| Input | GT | RMCNet | UAMT | UDCNet | DTC | Ours |

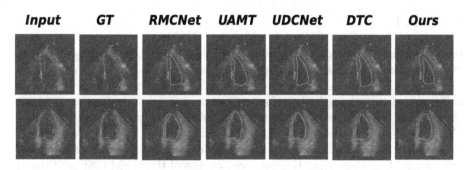

Fig. 4. Qualitative results of LV_{ED} & LV_{ES} region segmentation. We compared with *UAMT* [35], *UDCNet* [13], *DTCNet* [14] and *RMCNet* [20].

of 0.999 every 100 iterations. The batch size was set to 26. *Res2Net* [6] was used as the backbone for ours and all the compared methods. Note that we tuned all of the hyperparameters via three-fold cross-validation. All the training and testing processes were performed on a server with four *GEFORCE RTX 3090 24 GiB GPUs*. The implementation code is publicly available at https://github.com/smallmax00/Echo_lv_ef.

Loss Function: Given the supervised, unsupervised and weakly-supervised loss functions discussed in the manuscript, we defined the overall loss function as a combination of fully-supervised, semi-supervised and weakly-supervised losses. Specifically for the full-supervised loss (L_{fully}), we applied Dice loss for L_B and L_R, and mean square error loss for L_K. And L_{fully} is defined as: $L_{fully} = L_R + 0.1 * (L_B + L_K)$. Then for the unsupervised loss (L_{unsup}), we empirically added both contrastive loss terms (L_{contr} and L_{contr}^u) into it, such as: $L_{unsup} = 0.05*(L_B^u + L_R^u) + 0.01*L_{contr} + 0.005*L_{contr}^u$. In the end, we defined the weakly-supervised loss function (L_{weakly}) with ejection fraction regression loss, such as $L_{weakly} = 0.01 * L_{EF}$. Overall, our model's final loss function is defined as:

$$L_{final} = L_{fully} + \lambda * L_{unsup} + L_{weakly} \tag{10}$$

λ is adopted from [10] as the time-dependent Gaussian ramp-up weighting coefficient to account for the trade-off between the supervised, unsupervised, and weakly-supervised losses. This prevents the network from getting stuck in a degenerate solution during the initial training period. Because no meaningful predictions of the unlabeled input are obtained.

A potential concern may be the various loss functions used in this work. While our work belongs to multi-task-based semi-supervised learning, including contrastive representation learning and weakly supervised learning as well. Previous similar multi-task-based methods [19,26] also need multiple loss function terms. Thus, such a framework requires multiple supervisions from different granularity to gain sufficient regularisation and exploitation from the limited number of labeled data. Our ablation study in Table 1 in the manuscript proves that every loss terms contribute to the final model performance.

4 Results

In this section, we show qualitative (Fig. 4) and quantitative results (Table 1) of LV segmentation and LV_{EF} estimation. More qualitative and quantitative results, such as ablation study on vertices regression performance, comparison with state-of-the-art methods, cohesion and consistency visualization *etc.*, are shown in the Supplementary. Dice similarity score (*Dice*) and Hausdorff Distance (*HD*) were used as the segmentation accuracy metrics; Pearson's correlation coefficients [28] (*Corr*), average deviation *(Bias)* and standard deviation *(Std)* are used as the LV_{EF} estimation metrics. Results of the model that are trained by using labels of 10 % of the training dataset are presented.

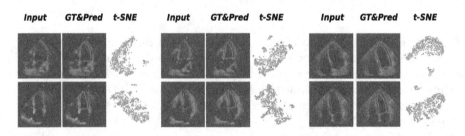

Fig. 5. *T-SNE* map of the proposed model's feature maps on region center (C_r) in green colour, boundary (B_p) in blue colour and background in red colour. The *t-SNE* results clearly show that the proposed model can learn a discriminative feature space between the background and the B_p. The C_r and B_p are expectedly near each other with a clear margin. The ground truth (*GT*) and the predictions (*Pred*) are overlaid on the input images for better visualization.

Table 1. Quantitative segmentation results of LV_{ES} and LV_{ED}, and statistical comparison of LV_{EF} estimation. The performance is reported as *Dice* (%), *HD* (mm), *Corr* and *Bias* ± *Std* (%).

Methods	LV_{ED}		LV_{ES}		LV_{EF}	
	Dice (%)	HD (mm)	Dice (%)	HD (mm)	Corr	Bias ± Std (%)
UAMT [35]	91.10 ± 0.18	3.98 ± 0.20	88.21 ± 0.18	3.99 ± 0.17	0.602	8.23 ± 3.7
URPC [16]	91.32 ± 0.15	3.66 ± 0.16	88.59 ± 0.16	3.78 ± 0.15	0.630	7.80 ± 3.9
DTCNet [14]	91.42 ± 0.13	3.77 ± 0.18	88.31 ± 0.16	3.76 ± 0.15	0.677	7.43 ± 3.8
UDCNet [13]	91.33 ± 0.15	3.87 ± 0.18	88.57 ± 0.15	3.88 ± 0.18	0.634	7.76 ± 4.4
RMCNet [20]	91.50 ± 0.17	3.89 ± 0.19	88.67 ± 0.15	3.89 ± 0.18	0.606	7.98 ± 4.2
Ours (Semi)	**92.11 ± 0.11**	**3.24 ± 0.10**	**89.25 ± 0.11**	**3.25 ± 0.13**	**0.727**	**6.86 ± 2.9**

Left Ventricular Segmentation. In Table 1, *Ours (Semi)* obtained an average 92.11 % and 89.25 % *Dice* on LV_{ED} and LV_{ES} segmentation, respectively, outperformed *UAMT* [35] by 1.1 %, outperformed *URPC* [16], *UDCNet* [13] and *RMCNet* [20] by 0.8 %, 2.5 % and 0.7 %. The *t-SNE* in Fig. 5 show that the

proposed model can learn a discriminative feature space between the background and the B_p. The C_r and B_p are expectedly near each other with a clear margin.
Clinical Evaluation. In addition to the segmentation metrics, we evaluated our method's performance via LV_{EF}. A post-process method [36] is used to generate LV_{EF} with the predicted masks for those compared methods that do not regress boundary vertices. Table 1 shows that *Ours (semi)* achieves a higher *Corr* and lower *Bias* than all the other methods. Specifically, *Ours (semi)* achieved the best performance of 0.727 *Corr*, which outperformed *URPC* [16], *UDCNet* [13] and *RMCNet* [20] by 15.4 %, 14.7 % and 20.0 %.

5   Conclusion

We propose a novel weakly/semi-supervised framework with uncertainty-aware contrastive learning. The multi-level intrinsic geometric association among vertex, boundary and region predictions are exploited in the proposed contrastive learning and weakly/semi-supervised learning pipeline. Our experiments have demonstrated that the proposed model can effectively leverage semantic region features and spatial vertex, boundary features for left ventricle segmentation and end-to-end ejection fraction estimation in 2D echocardiography.

References

1. PNPOLY- point inclusion in polygon test. https://wrf.ecse.rpi.edu/Research/Short_Notes/pnpoly.html. Accessed 01 Jun 2022
2. Cheng, B., Girshick, R., Dollár, P., Berg, A.C., Kirillov, A.: Boundary IoU: improving object-centric image segmentation evaluation. In: Proceedings of the IEEE Conference on Computer Vision and Pattern Recognition (2021)
3. Deng, K., et al.: TransBridge: a lightweight transformer for left ventricle segmentation in echocardiography. In: Noble, J.A., Aylward, S., Grimwood, A., Min, Z., Lee, S.-L., Hu, Y. (eds.) ASMUS 2021. LNCS, vol. 12967, pp. 63–72. Springer, Cham (2021). https://doi.org/10.1007/978-3-030-87583-1_7
4. Dodge, H.T., Sandler, H., Ballew, D.W., Lord, J.D., Jr.: The use of biplane angiocardiography for the measurement of left ventricular volume in man. Am. Heart J. **60**(5), 762–776 (1960)
5. Folland, E., Parisi, A., Moynihan, P., Jones, D.R., Feldman, C.L., Tow, D.: Assessment of left ventricular ejection fraction and volumes by real-time, two- dimensional echocardiography. A comparison of cineangiographic and radionuclide techniques. Circulation **60**(4), 760–766 (1979)
6. Gao, S., Cheng, M.M., Zhao, K., Zhang, X.Y., Yang, M.H., Torr, P.H.: Res2Net: a new multi-scale backbone architecture. IEEE Trans. Pattern Anal. Mach. Intell. **42**, 652–662 (2019)
7. Gu, R., et al.: Contrastive semi-supervised learning for domain adaptive segmentation across similar anatomical structures. IEEE Trans. Med. Imaging **42**(1), 245–256 (2022)
8. Guo, L., et al.: Dual attention enhancement feature fusion network for segmentation and quantitative analysis of paediatric echocardiography. Med. Image Anal. **71**, 102042 (2021)

9. Hu, H., Cui, J., Wang, L.: Region-aware contrastive learning for semantic segmentation. In: Proceedings of the IEEE/CVF International Conference on Computer Vision, pp. 16291–16301 (2021)

10. Laine, S., Aila, T.: Temporal ensembling for semi-supervised learning. In: International Conference on Learning Representations (ICLR) (2017)

11. Lazarow, J., Xu, W., Tu, Z.: Instance segmentation with mask-supervised polygonal boundary transformers. In: Proceedings of the IEEE/CVF Conference on Computer Vision and Pattern Recognition, pp. 4382–4391 (2022)

12. Li, H., Wang, Y., Qu, M., Cao, P., Feng, C., Yang, J.: EchoEFNet: multi-task deep learning network for automatic calculation of left ventricular ejection fraction in 2D echocardiography. Comput. Biol. Med. **156**, 106705 (2023)

13. Li, Y., Luo, L., Lin, H., Chen, H., Heng, P.-A.: Dual-consistency semi-supervised learning with uncertainty quantification for COVID-19 lesion segmentation from CT images. In: de Bruijne, M., et al. (eds.) MICCAI 2021. LNCS, vol. 12902, pp. 199–209. Springer, Cham (2021). https://doi.org/10.1007/978-3-030-87196-3_19

14. Luo, X., Chen, J., Song, T., Wang, G.: Semi-supervised medical image segmentation through dual-task consistency. In: Proceedings of the AAAI Conference on Artificial Intelligence, vol. 35, pp. 8801–8809 (2021)

15. Luo, X., et al.: Scribble-supervised medical image segmentation via dual-branch network and dynamically mixed pseudo labels supervision. In: Wang, L., Dou, Q., Fletcher, P.T., Speidel, S., Li, S. (eds.) Proceedings of the 25th International Conference on Medical Image Computing and Computer Assisted Intervention, MICCAI 2022, Part I, Singapore, 18–22 September 2022, pp. 528–538. Springer, Cham (2022). https://doi.org/10.1007/978-3-031-16431-6_50

16. Luo, X., et al.: Semi-supervised medical image segmentation via uncertainty rectified pyramid consistency. Med. Image Anal. **80**, 102517 (2022)

17. Meng, Y., et al.: Diagnosis of diabetic neuropathy by artificial intelligence using corneal confocal microscopy. Eur. J. Ophthalmol. **32**, 11–12 (2022)

18. Meng, Y., et al.: Bilateral adaptive graph convolutional network on CT based COVID-19 diagnosis with uncertainty-aware consensus-assisted multiple instance learning. Med. Image Anal. **84**, 102722 (2023)

19. Meng, Y., et al.: Transportation object counting with graph-based adaptive auxiliary learning. IEEE Trans. Intell. Transp. Syst. **24**(3), 3422–3437 (2022)

20. Meng, Y., et al.: Shape-aware weakly/semi-supervised optic disc and cup segmentation with regional/marginal consistency. In: Proceedings of the 25th International Conference on Medical Image Computing and Computer Assisted Intervention, MICCAI 2022, Part IV, Singapore, 18–22 September 2022. pp. 524–534. Springer, Cham (2022). https://doi.org/10.1007/978-3-031-16440-8_50

21. Meng, Y., et al.: Regression of instance boundary by aggregated CNN and GCN. In: Vedaldi, A., Bischof, H., Brox, T., Frahm, J.-M. (eds.) ECCV 2020. LNCS, vol. 12353, pp. 190–207. Springer, Cham (2020). https://doi.org/10.1007/978-3-030-58598-3_12

22. Meng, Y., et al.: Artificial intelligence based analysis of corneal confocal microscopy images for diagnosing peripheral neuropathy: a binary classification model. J. Clin. Med. **12**(4), 1284 (2023)

23. Meng, Y., et al.: CNN-GCN aggregation enabled boundary regression for biomedical image segmentation. In: Martel, A.L., et al. (eds.) MICCAI 2020. LNCS, vol. 12264, pp. 352–362. Springer, Cham (2020). https://doi.org/10.1007/978-3-030-59719-1_35

Weakly/Semi-supervised Left Ventricular Segmentation   109

24. Meng, Y., et al.: Bi-GCN: boundary-aware input-dependent graph convolution network for biomedical image segmentation. In: 32nd British Machine Vision Conference, BMVC 2021. British Machine Vision Association (2021)

25. Meng, Y., et al.: Dual consistency enabled weakly and semi-supervised optic disc and cup segmentation with dual adaptive graph convolutional networks. IEEE Trans. Med. Imaging **42**, 416–429 (2022)

26. Meng, Y., et al.: Spatial uncertainty-aware semi-supervised crowd counting. In: Proceedings of the IEEE/CVF International Conference on Computer Vision, pp. 15549–15559 (2021)

27. Meng, Y., et al.: Graph-based region and boundary aggregation for biomedical image segmentation. IEEE Trans. Med. Imaging **41**, 690–701 (2021)

28. Mukaka, M.M.: A guide to appropriate use of correlation coefficient in medical research. Malawi Med. J. **24**(3), 69–71 (2012)

29. van den Oord, A., Li, Y., Vinyals, O.: Representation learning with contrastive predictive coding. arXiv preprint arXiv:1807.03748 (2018)

30. Ouyang, D., et al.: Video-based AI for beat-to-beat assessment of cardiac function. Nature **580**(7802), 252–256 (2020)

31. Patefield, A., et al.: Deep learning using preoperative AS-OCT predicts graft detachment in DMEK. Trans. Vis. Sci. Technol. **12**(5), 14 (2023)

32. Preston, F.G., et al.: Artificial intelligence utilising corneal confocal microscopy for the diagnosis of peripheral neuropathy in diabetes mellitus and prediabetes. Diabetologia **65**, 457–466 (2022). https://doi.org/10.1007/s00125-021-05617-x

33. Robinson, J., Chuang, C.Y., Sra, S., Jegelka, S.: Contrastive learning with hard negative samples. arXiv preprint arXiv:2010.04592 (2020)

34. Shi, Y., et al.: Inconsistency-aware uncertainty estimation for semi-supervised medical image segmentation. IEEE Trans. Med. Imaging **41**(3), 608–620 (2021)

35. Yu, L., Wang, S., Li, X., Fu, C.-W., Heng, P.-A.: Uncertainty-aware self-ensembling model for semi-supervised 3D left atrium segmentation. In: Shen, D., et al. (eds.) MICCAI 2019. LNCS, vol. 11765, pp. 605–613. Springer, Cham (2019). https://doi.org/10.1007/978-3-030-32245-8_67

36. Zeng, Y., et al.: MAEF-Net: multi-attention efficient feature fusion network for left ventricular segmentation and quantitative analysis in two-dimensional echocardiography. Ultrasonics **127**, 106855 (2023)

37. Zhang, Y., Meng, Y., Zheng, Y.: Automatically segment the left atrium and scars from LGE-MRIs using a boundary-focused nnU-Net. In: Zhuang, X., Li, L., Wang, S., Wu, F. (eds.) Left Atrial and Scar Quantification and Segmentation, LAScarQS 2022. LNCS, vol. 13586, pp. 49–59. Springer, Cham (2023). https://doi.org/10.1007/978-3-031-31778-1_5

Spatial-Temporal Graph Convolutional Network for Insomnia Classification via Brain Functional Connectivity Imaging of rs-fMRI

Wenjun Zhou[1] , Weicheng Luo[1], Liang Gong[2(✉)] , Jing Ou[1] ,
and Bo Peng[1]

[1] School of Computer Science, Southwest Petroleum University, Chengdu, China
[2] Department of Neurology, Chengdu Second People's Hospital, Chengdu, China
seugongliang@hotmail.com

Abstract. Chronic Insomnia Disorder (CID) is a prevalent sleep disorder characterized by persistent difficulties in initiating or maintaining sleep, leading to significant impairment in daily functioning and quality of life. Accurate classification of CID patients is crucial for effective treatment and personalized care. However, existing approaches face challenges in capturing the complex spatio-temporal patterns inherent in rs-fMRI data, limiting their classification performance. In this study, we propose a novel approach utilizing the Spatial-Temporal Graph Convolutional Network (ST-GCN) for classification of CID patients. Our method aims to address the limitations of existing approaches by leveraging the graph convolutional framework to model the spatio-temporal dynamics in rs-fMRI data. Specifically, this method first pre-processes the raw rs-fMRI images and divides the brain into several regions of interest using a brain template. Next, it utilizes the ST-GCN network to integrate spatio-temporal features. Finally, the extracted features are utilized into a fully connected network for classification. Comparative experiment results show that the Accuracy and Specificity of the proposed method reach 98.90%, 99.08% respectively, which are better than the state-of-the-art methods.

Keywords: resting-state functional MRI · graph convolutional neural networks · functional connectivity

1 Introduction

Chronic Insomnia Disorder (CID) is a prevalent mental disorder categorized as insomnia. Patients with CID may exhibit symptoms of difficulty falling asleep and maintaining sleep [10]. Resting-state functional magnetic resonance imaging (rs-fMRI) is a non-invasive and high spatial-temporal resolution neuroimaging technique that is capable of recording the activity of various regions of the brain

© The Author(s), under exclusive license to Springer Nature Singapore Pte Ltd. 2024
Q. Liu et al. (Eds.): PRCV 2023, LNCS 14437, pp. 110–121, 2024.
https://doi.org/10.1007/978-981-99-8558-6_10

over a period of time based on the dependence of blood oxygen saturation levels [2,7]. This technique has been widely used in the diagnosis and study of Chronic Insomnia Disorder (CID) to investigate the functional connectivity patterns and aberrant brain activity associated with sleep disturbances [17]. However, the analysis of fMRI images is now often performed manually, which requires specific knowledge and is time-consuming [4]. Nowadays, deep learning, due to its superior performance, has been widely used in various fields, including applications to assist in CID diagnosis [5]. However, the application of fMRI images to deep learning for the classification of CID requires consideration of an issue: the original rs-fMRI images contain both temporal and spatial information, and existing deep learning methods used for rs-fMRI data either eliminate the information of the temporal dynamics of brain activity or overlook the functional dependency between different brain regions in a network [1].

In this paper, we propose a novel method to classify CID and Normal using the pre-processed rs-MRI data collected from the Chengdu Second People's Hospital, China. We first utilized the DPABI software [22] to preprocess the raw rs-fMRI images. Next, we employed a brain template, such as the Automated Anatomical Labeling (AAL) atlas [18], to divide the preprocessed rs-fMRI data into several brain regions. Each brain region was treated as a node in a graph, and the functional connectivity between nodes was quantified using measures such as Pearson correlation. Then, spatio-temporal graph convolutional network (ST-GCN) and fully connected network (FCN) are adopted for feature extraction and CID classification, respectively. In summary, our contributions are as follows:

1. We first apply ST-GCN to CID classification diagnosis and propose a spatial-temporal graph convolutional network for insomnia classification via brain functional connectivity imaging of rs-fMRI.
2. We introduce a new data construction method based on rs-fMRI data, which combines the feature matrix with the brain adjacency matrix based on sliding windows, forming a novel input feature.
3. The effectiveness of the method is verified by ablation experiments and comparative experiments.

The rest of paper is organized as follows. Section 2 reviews related work. In Sect. 3, we detail our method. Experimental results are presented in Sect. 4, and conclusions follow in Sect. 5.

2   Related Work

Graph convolutional networks (GCNs) have been utilized in multiple classification, prediction, segmentation and reconstruction tasks with non-structural (e.g. fMRI, EEG, iEEG) and structural data (e.g. MRI, CT). This section mainly covers application of graph learning representation on functional brain connectivity.

Attention Deficit Hperactivity Disorder: Some studies have shown that fMRI-based analysis is effective in helping understand the pathology of brain diseases such as attention-deficit hyperactivity disorder (ADHD). ADHD is a neurodevelopmental disorder characterized by symptoms of inattention, hyperactivity, and impulsivity. The model proposed by Rakhimberdina and Murata [15] was employed to differentiate adults with ADHD from healthy controls using a population graph. The model achieved an accuracy of 74.35% by constructing a graph based on gender, handedness, and acquisition site features. In a similar vein, Yao et al. [24] utilized a multi-scale tripled GCN to identify ADHD using the ADHD-200 dataset [3]. They generated functional connectivity networks at different spatial scales and ROI definitions by applying multi-scale templates to each subject. In these networks, each region of interest (ROI) was represented as a node, and the connections between ROIs were defined by the Pearson correlation of their mean time series.

Major Depressive Disorder: Major depressive disorder (MDD) is a psychiatric condition characterized by persistent feelings of sadness, reduced interest or pleasure in activities, and cognitive impairments. Yao et al. [25] employed a temporal adaptive graph convolutional network (GCN) on rs-fMRI data to leverage the temporal dynamics and capture the fluctuating patterns of brain activity, enabling effective MDD detection. The model learns a data-based graph topology and captures dynamic variations of the brain fMRI data, and outperforms traditional GCNs [12] and GATs [20] models.

Sex Classification of Brain Connectivity: Sex classification of brain connectivity refers to the process of predicting the biological sex of an individual based on patterns of brain connectivity derived from neuroimaging data. It has been observed that there are inherent differences in brain connectivity between males and females, which can be captured using advanced computational techniques such as deep learning. Deep learning models have been applied to analyze neuroimaging data, specifically functional magnetic resonance imaging (fMRI), and extract discriminative features for distinguishing between male and female brains. By leveraging large datasets, these models can learn complex patterns and relationships in brain connectivity associated with sex. Gadgil et al. [9] proposed a spatio-temporal graph convolutional network (GCN) that captures the non-stationary characteristics of functional connectivity by utilizing blood-oxygen-level-dependent (BOLD) time series data. The model achieved an accuracy of 83.7% in predicting the age and gender of healthy individuals, outperforming conventional recurrent neural network (RNN)-based methods when tested on the Human Connectome Project (HCP) dataset. Similarly, Filip et al. [6] utilized the graph attention network (GAT) architecture and employed an inductive learning strategy, incorporating the concept of a master node, to develop a graph classification framework for predicting gender on the HCP dataset [19]. To visualize the crucial brain regions associated with specific phenotypic differences, Kim et al. [11] employed a graph isomorphism network [21], which is a generalized convolutional neural network (CNN) designed to operate within the graph domain.

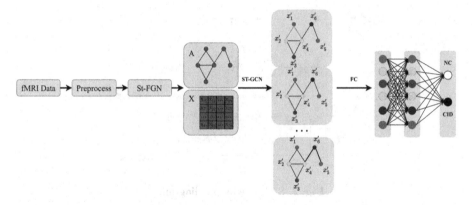

Fig. 1. Overall frame diagram, where St-FGN: Spatio-temporal functional connectivity networks; A represents the adjacency matrix of St-FGN; X represents the node feature matrix of St-FGN; x_i represents the feature vector of the ith node; x_i' represents the feature vector of the ith new node obtained after the Sptio-temporal graph convolution network

3   Methodology

Figure 1 shows the whole experimental framework. First, we performed data preprocessing, followed by the extraction of temporal and spatial features combined with the Automated Anatomical Labeling (AAL) brain template using DPABI software. The data segmentation was conducted using a sliding window. Next, connections were established between the functional networks and spatiotemporal graphs, and a spatiotemporal graph convolution (ST-GC) was defined on these graphs. Then, we presented the architecture of our spatiotemporal graph convolutional network (ST-GCN), which is specifically designed for classification based on rs-fMRI data. Finally, we outlined the importance of learning graph edges.

3.1   Data Preprocessing

All the rs-fMRI data were preprocessed using the Data Processing Assistant for Resting-State Function (DPARSF 7.0 Advanced Edition) toolkit. For each subject, the first 10 frames were discarded for magnetic saturation. The following steps were performed:

a. slice timing correction.
b. motion correction.
c. regression of nuisance signals (white matter, cerebrospinal fluid signals, and global signal), and the head-motion parameters.
d. spatial normalization to Montreal Neurological Institute space.
e. spatial smoothing using a 6 mm full-width at half-maximum Gaussian kernel.
f. linear detrend.
g. filtering the high-frequency noise.

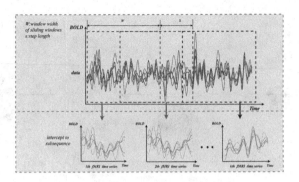

Fig. 2. Sliding window diagram

3.2   Data Augmentation

To solve the small sample size of rs-fMRI data, we use the sliding window method to expand the data, as shown in Fig. 2, where the abscissa indicates the acquisition time of the fMRI time series, and the ordinate indicates the blood oxygen signal in the brain region. For each subject, the mean rs-fMRI time series of all voxels in the kth brain ROI is defined as follows:

$$x_k = (x_{k1}, x_{k2}, ..., x_{kN}), \quad k = 1, 2, ..., M, \tag{1}$$

where M is the total number of ROI and N indicates the timestamp of the scan. The whole rs-fMRI time series is divided into K overlapping sub-segments. Each sub-rs-fMRI time series can build a spatio-temporal Graph. The value of K is calculated according to the following:

$$k = \frac{T - W}{s} + 1, \tag{2}$$

where T indicates the length of the whole rs-fMRI time series, W represents the size of sliding window, and s is the step length of each slide of the sliding window. By employing a smaller stride in the sliding window, the model gains access to a greater wealth of local information, thus enabling it to effectively capture the subtle and rapid changes within the dataset. Furthermore, the model demonstrates an enhanced capability in capturing the temporal correlations and patterns inherent in sequences. Consistent with the study [8], the sliding time window W was set to $30TR$ to obtain relatively stable functional connectivity data. According to the experimental results, in order to achieve the best results, the sliding window step size of this experiment is set to $5TR$.

3.3   Construction of Spatio-Temporal Graph

Following preprocessing, we constructed spatio-temporal graphs based on the processed rs-fMRI sequences, as shown in Fig. 3. Specifically, the rs-fMRI sequences were initially divided into N regions of interest (ROI) using the AAL brain template. Subsequently, an undirected spatiotemporal graph $\mathcal{G} = (\mathcal{V}, \mathcal{E})$ was constructed based on the rs-fMRI sequences comprising N ROIs and T frames, capturing both ROI-to-ROI connectivity and inter-frame connectivity. In this graph, $\mathcal{V} = \{v_{t,i} | t = 1, ..., T; i = 1, ..., N\}$ represents the set of nodes defined by N ROIs and T time points, while $\mathcal{E}$ comprises edges representing temporal and spatial connections. Each edge in the temporal graph connects an ROI to its corresponding ROI in the subsequent time point. In the spatial graph (associated with a time point), each edge represents the functional connectivity between two ROIs, which is determined by the brain's functional organization. The weight of the edge is defined by the functional connectivity between the respective nodes. To define this weight, we assume that the spontaneous activation of the N ROIs can be characterized by the average value of the BOLD signals within each ROI. Initially, we compute the average value for each ROI in the rs-fMRI sequence. Subsequently, the connectivity between two ROIs can be calculated using the Pearson correlation coefficient formula. Notably, this correlation is impartial to the time point.

3.4   Spatio-Temporal Graph Convolution (ST-GC)

To define an ST-GC on such graphs, consistent with [9], we denote $f_{in}(v_{ti})$ as the input feature at node v_{ti}, the spatio-temporal neighborhood $B(v_{ti})$ is defined as,

$$B(v_{ti}) = \{v_{qj} | d(v_{tj}, v_{ti}) \leq K, |q - t| \leq \lfloor \Gamma/2 \rfloor\}, \tag{3}$$

where the parameter Γ is used to adjust the range of the neighborhood graph in the temporal dimension (i.e., temporal kernel size) and K is used to determine the size of the spatial neighborhood(i.e., spatial kernel size). An ST-GC operation on node v_{ti} with respect to a convolutional kernel $w(\cdot)$ and a normalization factor Z_{ti} can then be defined as [23]

$$f_{out}(v_{ti}) := \frac{1}{Z_{ti}} \sum_{v_{qj} \in B(v_{ti})} f_{in}(v_{qj}) \cdot W(v_{qj}). \tag{4}$$

Following a similar implementation method outlined in reference [23], we utilize an approximation technique to represent the spatio-temporal convolutional kernel $W(\cdot)$ by decomposing the kernel into a spatial graph convolutional kernel $W_{SG} \in \mathbb{R}^{C \times M}$ represented in the Fourier domain and a temporal convolutional kernel $W_{TG} \in \mathbb{R}^{M \times \Gamma}$. Denote $f_t \in \mathbb{R}^{N \times C}$ as the characteristics of C-channel for N ROI at the tth frame, $f'_t \in \mathbb{R}^{N \times M}$ as the M-channel output features. Additionally, the matrix $\mathbf{A}$ is employed to capture the similarities between ROI. We first perform spatial graph convolution at each time point by [12].

$$f'_t = \Lambda^{-\frac{1}{2}} (A + I) \Lambda^{-\frac{1}{2}} f_t W_{SG}, \tag{5}$$

where Λ is a diagonal matrix of $\Lambda^{ii} = \sum_j A^{ij} + 1$. Then a temporal convolution is performed on the resulting features. Since the temporal graph has regular grid spacing, we perform a standard 1D convolution on the time series of features $f_i \in \mathbb{R}^{M \times T}$ for each node v_i by $f'_i = f_i \circledast W_{TG}$, where $\circledast$ denotes convolution in the time domain, and $f'_i \in \mathbb{R}^{M \times T}$ is the final output of ST-GC for v_i.

3.5   ST-GCN Building

In the following discussion, we present the structural details of the ST-GCN model employed for the classification of rs-fMRI time series. The model architecture consists of a singular layer of spatio-temporal graph convolutional operators. Each ST-GC layer generates 64 output channels, employing a temporal kernel size of 11. To mitigate the risks of overfitting, a feature dropout technique is employed, wherein features are randomly discarded with a probability of 50% following each ST-GCN unit. Then, global average pooling is applied after the convolutional layers to yield a 64-dimensional feature vector. Subsequently, the vector is sent to the fully connected layer for processing. Finally, a sigmoid activation function is employed to derive class probabilities. The model is trained using the Adam optimizer with a learning rate set to 0.001.

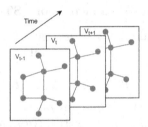

Fig. 3. Spatio-temporal map of brain regions. Each v_t indicates a frame of the connection between the brain regions at time step t.

3.6   Edge Importance Learning

To enhance the contribution of important edge nodes in the spatial graph and suppress non-important edges, we introduce a learnable weight matrix $\mathbf{M} \in \mathbb{R}^{N \times N}$. This matrix is shared across each layer of the ST-GC and is learned by replacing $(\mathbf{A} + \mathbf{I})$ in Eq. 5 with $(\mathbf{A} + \mathbf{I}) * \mathbf{M}$ (element-wise multiplication). The learnable edge importance weights allow the model to focus attention on the most relevant edges or connections in the spatiotemporal graph. This selective attention mechanism improves the model's ability to extract meaningful information and discard noise or irrelevant connections. The original proposal by Yan et al. [23] was to add a learnable mask at each layer of the spatio-temporal graph convolution. Their strategy often resulted in negative and asymmetric masks that were difficult to interpret and varied across layers, while we enforce $\mathbf{M}$ to be both positive and symmetric, making it easier to interpret.

4 Experiments

4.1 Dataset

The collection of MRI image data in this study was conducted at the Chengdu Second People's Hospital. A total of 163 participants were included in the study, with 106 individuals diagnosed with chronic insomnia forming the Chronic Insomnia Disorder (CID) group. These participants met the criteria for chronic insomnia as defined by the International Classification of Sleep Disorders (ICSD). To compare the findings with a control group, 57 healthy volunteers were recruited during the same period. These participants formed the Normal Control (NC) group and did not have a history of sleep disorders or any significant sleep complaints. As part of the comprehensive sleep assessment, the Pittsburgh Sleep Quality Index (PSQI) was utilized to evaluate sleep quality and disturbances experienced by the participants. In addition to the PSQI, the study also employed the Self-rating Anxiety Scale (SAS) and the Sleep Disturbance Scale (SDS). The SAS is a self-report scale used to measure the severity of anxiety symptoms experienced by the participants [26]. The SDS is a questionnaire specifically designed to measure the severity of sleep disturbances, including difficulties falling asleep, maintaining sleep, and early morning awakening [27]. Table 1 presents the calculated means and standard deviations of the variables under investigation.

Table 1. Demographic and scale data of all study participants. Unless otherwise noted, data are presented as mean ± standard deviation. PSQI Pittsburgh sleep quality index, SAS self-rating anxiety scale, SDS self-rating depression scale.

	CID group	HC group
Number of subjects	106	61
Gender (M/FM)	32/74	23/38
Age (year)	34.77 ± 10.52	35.15 ± 8.61
Education	15.50 ± 2.29	14.88 ± 3.55
PSQI	12.84 ± 3.16	3.55 ± 1.57
SDS	52.04 ± 10.96	38.29 ± 10.42
SAS	49.64 ± 11.61	33.72 ± 7.31

4.2 Evaluation Metrics

In order to evaluate the performance of the model, we utilize four common analysis measurements:Accuracy, Specificity, Sensitivity and F1 Score. These metrics are widely used to evaluate the quality of binary classifiers [14,16].

$$Accuracy = \frac{(TP + TN)}{(TP + TN + FP + FN)}, \tag{6}$$

$$Specificity = \frac{(TN)}{(TN + FP)}, \tag{7}$$

$$Sensitivity = \frac{(TP)}{(TP + FN)}, \tag{8}$$

$$F1 \quad Score = \frac{2 \times Precision \times Recall}{Precision + Recall}, \tag{9}$$

where TP, FP, FN and TN indicate the number of true positives, false positives, false negatives, true negatives, respectively. Precision is the proportion of true positive predictions among all positive predictions, while Recall is the proportion of true positive predictions among all actual positive instances.

4.3   Analysis of Different Sliding Window Step Size

The information derived from features within sliding windows varies at different time points, with the extent of overlap between adjacent windows determined by the chosen step size. The degree of window overlap has an impact on the smoothness of the data. Smaller overlaps in windows allow for a greater amount of localized information, enabling the model to capture subtle and rapid changes in the data more effectively. This is particularly crucial when capturing short-term variations in time series or finer details within the dataset. Conversely, larger window overlaps tend to smooth the data, making the model less sensitive to rapid changes. In the conducted ablation experiment, the only variable manipulated was the step size of the sliding windows, while all other conditions remained consistent. The selection of step size lengths, namely 5, 10, 15, and 20, was based on the extent of overlap between windows. The detailed information can be found in Table 2.

Table 2. Experimental results of different Sliding Window Step Size.

Step size	Average accuracy	SD	Average sensitivity	SD	Average specificity	SD	F1 Score
5	**98.90%**	**0.0198**	99.08%	**0.0187**	98.63%	**0.0330**	**99.08%**
10	94.13%	0.0433	96.52%	0.0329	91.49%	0.0959	90.72%
15	90.15%	0.0561	89.54%	0.0736	90.47%	0.0743	91.76%
20	87.57%	0.0601	88.88%	0.0861	87.56%	0.0880	90.25%

From the experimental results, it can be found that the performance of the model will be affected by the step size of the sliding window. Overall, when the step size is 15 or 20, the model cannot give full play to its performance, because the selected slices cannot contain enough continuous time information; as the step size decreases, the performance of the model gradually improves. When the stride is 5, our model performs best overall, ranking first in the accuracy, sensitivity, specificity, and F1 Score metrics with 98.90%, 99.08%, 98.63%, and 99.08%, respectively. Furthermore, the standard deviation of all results is less than 0.1. These results show that the dispersion of all results in cross-validation is small and the experimental results are relatively reliable.

4.4   Comparison with Other Methods

Table 3. Results for different methods.

Method	Step size	Average accuracy	Average sensitivity	Average specificity	F1 Score
LSTM [13]	20	86.77%	87.87%	84.97%	88.93%
St-LSTM [9]	20	71.21%	86.53%	46.63%	78.23%
Ours	20	87.57%	88.88%	87.56%	90.25%
	15	90.15%	89.54%	90.47%	91.76%
	10	93.58%	94.16%	96.33%	90.72%
	5	**98.90%**	**99.08%**	**98.63%**	**99.08%**

To further validate the feasibility of our proposed method, we compared our method with other deep learning methods [9,13]. We conducted experiments using a self-collected dataset. In terms of evaluation, we employed four commonly used metrics in the medical field: accuracy, sensitivity, specificity, and F1 Score. The experimental results are shown in Table 3. In order to adapt our dataset to the aforementioned network, we divided the data in the comparative experiment into segments using a sliding window with a stride of 20 TR.

Table 3 gives the results of different models for CID and Normal classification. It could be observed from Table 3 that our proposed method achieved the highest scores in accuracy, sensitivity, and specificity, with scores of 98.90%, 99.08%, and 98.63%, respectively. Among the compared models, the LSTM and St-LSTM had the score of 86.77% and 71.21% for Accuracy, which were 12.13% and 27.69% lower than our proposed method, respectively.

5   Conclusion

This article introduced an Intelligent diagnostic model for CID consisting of two main components: data processing and model construction. In the data processing phase, our approach begins by pre-processing the raw rs-fMRI data to eliminate irrelevant components and focus specifically on CID diagnosis. Subsequently, the data is segmented using sliding windows to ensure the preservation of interrelatedness and continuity between adjacent data points. Moving on to the model construction phase, we employ ST-GCN. Finally, a fully connected network is utilized for CID classification.

We conducted the ablation experiments and comparative experiments to validate the effectiveness of our proposed method. The results substantiate the efficacy of our approach in CID diagnosis. In the future, we would like to explore more sophisticated feature construction techniques, such as cortical segmentation, to further enhance the performance of our model. Additionally, we also plan to optimize our model by leveraging alternative disease datasets.

Acknowledgment. This work was supported by the National Natural Science Foundation of China under Grant 82001803.

References

1. Abrol, A., Hassanzadeh, R., Plis, S., Calhoun, V.: Deep learning in resting-state fMRI. In: 2021 43rd Annual International Conference of the IEEE Engineering in Medicine & Biology Society (EMBC), pp. 3965–3969. IEEE (2021)
2. Biswal, B.B., Kylen, J.V., Hyde, J.S.: Simultaneous assessment of flow and bold signals in resting-state functional connectivity maps. NMR Biomed. **10**(4–5), 165–170 (1997)
3. Bullmore, E., Sporns, O.: The economy of brain network organization. Nat. Rev. Neurosci. **13**(5), 336–349 (2012)
4. Carp, J.: The secret lives of experiments: methods reporting in the fMRI literature. Neuroimage **63**(1), 289–300 (2012)
5. Ching, T., et al.: Opportunities and obstacles for deep learning in biology and medicine. J. R. Soc. Interface **15**(141), 20170387 (2018)
6. Filip, A.C., Azevedo, T., Passamonti, L., Toschi, N., Lio, P.: A novel graph attention network architecture for modeling multimodal brain connectivity. In: 2020 42nd Annual International Conference of the IEEE Engineering in Medicine & Biology Society (EMBC), pp. 1071–1074. IEEE (2020)
7. Fox, M.D., Snyder, A.Z., Vincent, J.L., Corbetta, M., Van Essen, D.C., Raichle, M.E.: The human brain is intrinsically organized into dynamic, anticorrelated functional networks. Proc. Nat. Acad. Sci. **102**(27), 9673–9678 (2005)
8. Fu, Z., Du, Y., Calhoun, V.D.: The dynamic functional network connectivity analysis framework. Engineering (Beijing, China) **5**(2), 190 (2019)
9. Gadgil, S., Zhao, Q., Pfefferbaum, A., Sullivan, E.V., Adeli, E., Pohl, K.M.: Spatiotemporal graph convolution for resting-state fMRI analysis. In: Martel, A.L., et al. (eds.) MICCAI 2020, Part VII 23. LNCS, vol. 12267, pp. 528–538. Springer, Cham (2020). https://doi.org/10.1007/978-3-030-59728-3_52
10. Jaussent, I., Morin, C., Dauvilliers, Y.: Definitions and epidemiology of insomnia. Rev. Prat. **67**(8), 847–851 (2017)
11. Kim, B.H., Ye, J.C.: Understanding graph isomorphism network for rs-fMRI functional connectivity analysis. Frontiers Neuroscience **14**, 630 (2020)
12. Kipf, T.N., Welling, M.: Semi-supervised classification with graph convolutional networks. arXiv preprint arXiv:1609.02907 (2016)
13. Li, H., Fan, Y.: Brain decoding from functional MRI using long short-term memory recurrent neural networks. In: Frangi, A.F., Schnabel, J.A., Davatzikos, C., Alberola-López, C., Fichtinger, G. (eds.) MICCAI 2018, Part III 11. LNCS, vol. 11072, pp. 320–328. Springer, Cham (2018). https://doi.org/10.1007/978-3-030-00931-1_37

14. Powers, D.M.: Evaluation: from precision, recall and F-measure to ROC, informedness, markedness and correlation. arXiv preprint arXiv:2010.16061 (2020)
15. Rakhimberdina, Z., Murata, T.: Linear graph convolutional model for diagnosing brain disorders. In: Cherifi, H., Gaito, S., Mendes, J.F., Moro, E., Rocha, L.M. (eds.) COMPLEX NETWORKS 2019. SCI, vol. 882, pp. 815–826. Springer, Cham (2020). https://doi.org/10.1007/978-3-030-36683-4_65
16. Saito, T., Rehmsmeier, M.: The precision-recall plot is more informative than the ROC plot when evaluating binary classifiers on imbalanced datasets. PLoS ONE 10(3), e0118432 (2015)
17. Spiegelhalder, K., et al.: Increased EEG sigma and beta power during NREM sleep in primary insomnia. Biol. Psychol. 91(3), 329–333 (2012)
18. Tzourio-Mazoyer, N., et al.: Automated anatomical labeling of activations in SPM using a macroscopic anatomical parcellation of the MNI MRI single-subject brain. Neuroimage 15(1), 273–289 (2002)
19. Van Essen, D.C., et al.: The WU-Minn human connectome project: an overview. Neuroimage 80, 62–79 (2013)
20. Veličković, P., Cucurull, G., Casanova, A., Romero, A., Lio, P., Bengio, Y.: Graph attention networks. arXiv preprint arXiv:1710.10903 (2017)
21. Xu, K., Hu, W., Leskovec, J., Jegelka, S.: How powerful are graph neural networks? arXiv preprint arXiv:1810.00826 (2018)
22. Yan, C.G., Wang, X.D., Zuo, X.N., Zang, Y.F.: DPABI: data processing & analysis for (resting-state) brain imaging. Neuroinformatics 14, 339–351 (2016)
23. Yan, S., Xiong, Y., Lin, D.: Spatial temporal graph convolutional networks for skeleton-based action recognition. In: Proceedings of the AAAI Conference on Artificial Intelligence, vol. 32 (2018)
24. Yao, D., et al.: Triplet graph convolutional network for multi-scale analysis of functional connectivity using functional MRI. In: Zhang, D., Zhou, L., Jie, B., Liu, M. (eds.) GLMI 2019. LNCS, vol. 11849, pp. 70–78. Springer, Cham (2019). https://doi.org/10.1007/978-3-030-35817-4_9
25. Yao, D., Sui, J., Yang, E., Yap, P.-T., Shen, D., Liu, M.: Temporal-adaptive graph convolutional network for automated identification of major depressive disorder using resting-state fMRI. In: Liu, M., Yan, P., Lian, C., Cao, X. (eds.) MLMI 2020. LNCS, vol. 12436, pp. 1–10. Springer, Cham (2020). https://doi.org/10.1007/978-3-030-59861-7_1
26. Zung, W.W.: A rating instrument for anxiety disorders. Psychosom. J. Consultation Liaison Psychiatry 12, 371–379 (1971)
27. Zung, W.W.: A self-rating depression scale. Arch. Gen. Psychiatry 12(1), 63–70 (1965)

Probability-Based Nuclei Detection and Critical-Region Guided Instance Segmentation

Yunpeng Zhong, Xiangru Li[✉], Huanyu Mei, and Shengchun Xiong

School of Computer Science, South China Normal University, No. 55 West of Yat-sen Avenue, Guangzhou 510631, China
xiangru.li@qq.com

Abstract. Nuclear instance segmentation in histopathological images is a key procedure in pathological diagnosis. In this regard, a typical class of solutions are deep learning-like nuclear instance segmentation methods, especially those based on the detection of nuclear critical regions. The existing instance segmentation methods based on nuclear critical regions are still insufficient in detecting and segmenting adhesion nuclei. In this study, we proposed a Critical-Region Guided Instance Segmentation (CGIS) method to precisely segment adhesion nuclei. Specifically, CGIS embed the critical region of an instance into the original image to guide the model to segment only the target instance, and provide an accurate segmentation mask for each nucleus. To improve the accuracy of critical region detection in CGIS, we proposed a boundary-insensitive feature, Central Probability Field (CPF). The CPF is calculated based on global morphological characteristics of nuclei, and can reduce the negative effects from the unreliable nuclear boundary details on the nuclear critical regions detection. We evaluated the effectiveness of the proposed CGIS and CPF feature on the Lizard and PanNuke datasets. It is shown that the proposed CGIS method can accurately segment adhesion nuclei, and the CPF feature can efficiently detect the critical regions of nuclei.

Keywords: Nuclear instance segmentation · Computational pathology · Critical region

1 Introduction

Analyzing tissue slides is a powerful tool for cancer research and diagnosis [1]. The digitalization of tissue slices into Whole slide images (WSIs) using whole slide scanners enables permanent storage and sharing of visual data, and allows for the analysis of tissue images using some computer programs. The computer calculation and analysis process is referred to as computational pathology (CPath). In CPath, accurate nucleus identification and segmentation are critical steps in tissue pathology analysis [2]. Nuclear segmentation can be divided into semantic segmentation and instance segmentation. Semantic segmentation

© The Author(s), under exclusive license to Springer Nature Singapore Pte Ltd. 2024
Q. Liu et al. (Eds.): PRCV 2023, LNCS 14437, pp. 122–135, 2024.
https://doi.org/10.1007/978-981-99-8558-6_11

aims to predict the type label of the corresponding nucleus for each pixel, while instance segmentation not only gives the type label of the corresponding nucleus for each pixel, but also distinguishes the pixels from various nuclei. Compared with nuclear semantic segmentation, nuclear instance segmentation can provide additional information such as the average size, density, and morphology for the nuclei in a WSI. These additional information can facilitate cancer staging and prognostic assessment. However, this task is remarkably challenging. The challenging factors include the enormous quantities of nuclei, small nucleus areas, dense clustering of nuclei, overlap and adhesion between nuclei, uneven staining, blurred nuclei boundaries, and limited amount of annotated data [2,3].

Recently, Deep Learning (DL) techniques have been increasingly applied to pathology image analysis tasks, and have achieved remarkable successes. One important advancement is the development of deep learning models for nuclear instance segmentation. In instance segmentation, there are two key steps: the nucleus detection and the nucleus pixel identification. Based on the principle of the detection procedure, we categorize existing approaches into three groups: bounding box-based methods, boundary-based methods, and critical region-based methods.

The bounding box-based methods [4–10] first predict a set of rectangles, in which most of the pixels come from a single nucleus. This kind of rectangle is referred to as bounding box. Then, they process each pixel in those bounding boxes to determine whether it comes from the foreground (i.e. pixels belonging to the nuclei), to which nucleus it belongs, and type of its father nucleus, etc. However, bounding box detectors demonstrate limited performance on densely crowded and small instances such as nuclei. Furthermore, bounding box-based nuclear instance segmentation methods have significant deficiencies in localization and segmentation accuracy.

To prevent the connectivity between segmentation masks corresponding to different nuclei, the boundary-based methods [11–15] add the class of nuclear boundary to semantic segmentation task. Based on the judgment of the boundary pixels, this kind methods obtain the nucleus instance identification by connectivity analysis, and produce the instance segmentation results. These methods heavily rely on accurate boundary prediction. However, boundary prediction may fail in the cases with blurred nucleus boundaries. Moreover, boundary prediction compromises the segmentation masks of adhesion nuclei.

Critical region-based methods [3,16–18] typically generate a series of smaller representative regions (called critical regions) by thresholding some features (such as distance) to represent the detected nuclei. These critical regions are non-contiguous and are assigned instance labels through the connected component labeling algorithm. Furthermore, these critical regions are used as seeds for watershed algorithms to produce segmentation masks of nuclei. Such methods show excellent detection performance on small objects such as nuclei, and can effectively distinguish adhesion nuclei. Therefore, we chose to construct the nucleus instance segmentation algorithm based on the critical regions.

Moreover, the instance masks derived by the watershed algorithm in the critical region-based methods are not reliable and inherit some errors from semantic segmentation results (e.g. the background pixels between close nuclei are misclassified as foreground). Therefore, we proposed a Critical-Region Guided Instance Segmentation (CGIS) method to address limitations faced by the available critical region-based approaches. Specifically, we constructed a light segmentation network guided by critical region masks instead of watershed algorithm to generate instance masks. Consequently, CGIS can yield more reliable instance segmentation outcomes, particularly for adhesion nuclei. Moreover, we proposed a boundary-insensitive Central Probability Field (CPF) for nuclear critical region detection in CGIS. The CPF is calculated based on global morphology characteristics of nuclei, and can reduce the negative effects on the nuclear critical region detection from the unreliable nuclear boundary details. Key contributions of our work are:

- Proposed a Critical-Region Guided Instance Segmentation (CGIS) method.
- Proposed a boundary-insensitive Center Probability Field (CPF) for critical region detection in CGIS.
- We benchmarked our CGIS approach against other models on two public datasets, accomplishing state-of-the-art performance.

2   Related Works on Nucleus Instance Segmentation

Nuclei instance segmentation should determine which different nuclei exist in the image (called instances), judge whether each pixel belongs to a nucleus, estimate the type label of its parent nucleus for each nucleus pixel, and distinguish the pixels from different nuclei. In instance segmentation, there are two key procedures: the nucleus instance detection and the nucleus pixel identification. Based on the principle of the detection procedure, we categorize existing approaches into three groups: bounding box-based methods, boundary-based methods, and critical region-based methods. The following subsections will review and analyze the studies on nucleus instance segmentation from these three aspects.

2.1   Bounding Box-Based Methods

There are two key procedures in the bounding box-based methods [4–10]: nucleus location prediction and nucleus pixel identification. The positions of the nuclei are represented by rectangles, and these rectangles are referred to as bounding boxes. The second procedure should process each pixel in these bounding boxes to determine whether it comes from the foreground (that is, the pixel belongs to a nucleus), which nucleus (parent nucleus) it belongs to, the type of its parent nucleus, etc. Therefore, both the bounding box detection quality and pixel identification quality have important effects on the performance of such methods.

Johnson et al. [6] and Fujita et al. [10] applied the Mask R-CNN [19] to the nucleus/cell instance segmentation task, and made a series of adjustments and improvements to adapt it to such task. Yi et al. [4,5] introduced a U-net structure in the segmentation process of Mask R-CNN to improve segmentation performance. They extract ROI features from feature maps with various scales and fuse these features into the decoding path by skip connections. Cheng et al. [8] combined the SSD (Single Shot MultiBox Detector) detector with the U-Net model, achieved fast and accurate nuclear instance segmentation. Liu et al. [7,9] incorporated an additional global semantic segmentation branch based on Mask R-CNN to improve the segmentation quality.

2.2   Boundary-Based Methods

Boundary-based methods [11–15] add the class of nuclear boundary into the semantic segmentation of foreground and background. The boundary segmentation results can prevent the connectivity between foreground regions corresponding to different nuclei (especially adhesion nuclei). Based on such 3-class semantic segmentation results, this kind methods obtain nuclear instance recognition judgments by performing connectivity analysis on the foreground mask, and then obtain instance segmentation results. Therefore, the instance detection performance of this kind methods severely depend on the accuracy of nuclear boundary segmentation. The performance dependency results from the situation that missing even a few pixels on the boundary of adhesion nuclei may lead to gaps in the boundary, and the gap hinder the distinction of nuclei on either side.

Ronneberger et al. [11] and Raza et al. [13] treated boundaries as a separate category in semantic segmentation. They improve the model's sensitivity to instance boundaries using a weighted loss. Chen et al. [12] and Chen et al. [14] treated boundary prediction as a separate semantic segmentation task. In the post-processing procedure, the foreground pixels belonging to boundaries are set as background to distinguish the adhesion nuclei. He et al. [15] transformed the boundary detection into an estimation problem of the direction from each pixel to the center of its corresponding nucleus. They derived instance boundaries by calculating the cosine similarity of directions between neighboring pixels.

2.3   Critical Region-Based Methods

Critical region-based methods [3,16–18] first predict per-pixel differential features such as distance and probability through semantic segmentation models, compute a series of smaller representative regions or points to represent the nuclear instances present in the image by a threshold scheme or searching for local maximal (minimum) values. Then, the watershed algorithm are applied on the predicted feature map to complete instance segmentation.In this procedure, the critical regions are used as seeds, and the foreground mask in semantic segmentation result act as a mask. Therefore, the instance detection performance of such methods depends on the features used.

Peter et al. [20] proposed using the distance from a pixel to the nearest background pixel to represent nuclear distribution, detect the nuclear centers by locating the local maximal values, and predict the foreground mask using manually set thresholds. The method of Mahbod et al. [17] is similar to that of Peter et al., however, their foreground mask are predicted by semantic segmentation. HoVer-Net [3] predicted the normalized distances maps in the horizontal and vertical directions to the center of the nuclei for each pixel, and compute the critical regions of nuclei by thresholding the gradient of the distance map. Rashid et al. [18] proposed a scheme to detect the critical regions using a centroid map (distance from nuclear pixels to nearest nuclear boundaries) introduced an attention mechanism into the model to improve the segmentation performance.

3   CGIS Method and CPF Feature

We will introduce the proposed CGIS method in Subsect. 3.1, and then explain the proposed CPF for nuclear critical region detection in Subsect. 3.2. Finally, we elaborate on the process of nuclear classification in CGIS, including training and post-processing procedures, in Subsect. 3.3.

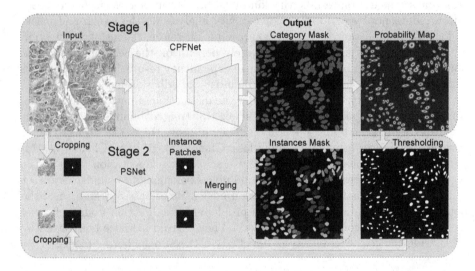

Fig. 1. CGIS processing procedure. The proposed CGIS consists of two stages, i.e., nuclear critical region detection (stage 1) and nuclear mask completion (stage 2). In the first stage, the CPF probability map and nuclear category mask are generated by CPFNet. In the second stage, some patches are cropped according to the locations of the critical regions generated by thresholding and connected component labeling and input into the PSNet to predict complete instance masks. The final output of the CGIS includes the nuclear category mask and the instance mask.

3.1   Critical-Region Guided Instance Segmentation

The process of our Critical-Region Guided Instance Segmentation is shown in Fig. 1. CGIS decomposes the nuclear instance segmentation task into two subtasks: nuclear critical region detection and nuclear mask completion. In the detection stage, we first generate the CPF feature (more about the CPF are discussed in Sect. 3.2) using a U-net style network called CPFNet. Based on the CPF features, we produce the binary critical region mask by thresholding and give a unique nucleus instance label for each critical region based on connectivity analysis. In the segmentation stage, the CGIS crops a series of image patches from the original image and the binary mask based on the location of the nuclear critical regions, and input these patches into the PSNet (Patch Segmentation Net, a light U-net style network) to predict the complete instance mask. Finally, these instance masks are recombined into one mask according to the positions where the corresponding patches are cropped.

In the cropping process, all patches are cropped using the same size (the instances at the edge of the image are handled by padding with extra pixels). There are two reasons to support us in doing so. Firstly, it is challenging to accurately determine the size of an instance only based on its critical region. Secondly, it is inefficient to use some patches with various scales in case of GPU acceleration. However, due to the translational invariance characteristics of CNNs, the PSNet cannot distinguish the target instance from the non-target ones, even if the target instance is at the center of the patch. To deal with this problem, the CGIS embeds the critical region mask of the target nucleus into the image to indicate the instance which needs to be segmented. The above-mentioned scheme is inspired by the works Hu et al. [21] and Tian et al. [22]. Specifically, we crop a patch from the binary mask that only contains the critical region of the target nucleus in the same way as we crop the original image. Then, the critical region patch is concatenated with the original image patch in the channel dimension. The critical region mask can provide a distinctive identifier for the target nucleus, help the PSNet distinguish it from non-target nuclei.

Another issue to be addressed is the pixel conflicts, i.e., two or more predicted instances cover a common pixel. The conflicts can be attributed to two factors: duplicated instance detection and inaccurate segmentation. Therefore, in the process of instance mask merging, we detect the intersection between the instance to be merged and the existing foreground mask to tackle such conflicts, including removing duplicated instances and deleting the overlapping pixels.

3.2   Central Probability Field

For critical region detection in CGIS, we proposed a boundary-insensitive Center Probability Field (CPF) to represent the probability that a pixel belongs to a nucleus. Our motivation is that the critical region detection of CGIS does not use any boundary information, so the boundary information contained in the detection feature may distract the model's attention. In addition, annotation inaccuracies congregate primarily at the nucleus boundaries, further harming the detection performance of the model (Fig. 2).

We intuitively employ the 2D Gaussian distribution to model the CPF, as this distribution only relies on a limited number of key parameters. Furthermore, the 2D Gaussian distribution can be viewed as a distance normalized in various

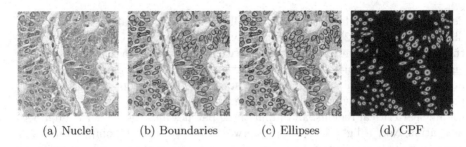

    (a) Nuclei   (b) Boundaries   (c) Ellipses   (d) CPF

Fig. 2. The generation process of the central probability field of nuclei. The nuclear boundaries are extracted by performing morphological erosion operation on the annotation masks. The elliptic parameters are estimated using the coordinates of the nuclear boundary. The CPF is calculated based on Eq. (1) and Eq. (3) with the elliptic parameters.

directions. It therefore has the potential to describe the morphological characteristics of those elongated nuclei. We approximately describe the nuclear contours using ellipses and reformulate the parameter estimation of the 2D Gaussian distribution as an ellipse-fitting problem. We employ the least squares ellipse fitting to obtain the ellipse that most closely approximates the contour of a nucleus, and take it as the contour of the 2D normal distribution at a given probability ϵ, where ϵ is a value close to 0 (such as 0.01). Given the non-uniqueness of the general equation representing an elliptic curve, the following probability function is employed:

$$\varphi_i(x, y) = \exp(\lambda^i[A^i(x - x_c^i)^2 + 2B^i(x - x_c^i)(y - y_c^i) + C^i(y - y_c^i)^2]), \quad (1)$$

where, λ^i is a coefficient to be determined, $A^i, B^i, C^i, x_c^i, y_c^i$ are the coefficients and center coordinates output by ellipse fitting. The λ^i can be solved by the following equation:

$$\varphi_i(x_c^i, y_c^i + \sqrt{\frac{F - A^i(x_c^i)^2 - B^i x_c^i y_c^i - C^i(y_c^i)^2}{C^i}}) = \epsilon. \quad (2)$$

Next, we merge the probability distributions of all nuclei in the following way:

$$P_c = \max_{i=1}^{n} \varphi_i(x, y), \quad (3)$$

where P_c is the CPF of the entire image. Merging in this way prevents the creation of peaks due to superposition on the boundary of the adhesion instance.

3.3 Nuclear Classification

Due to image disparities arising from the factors such as organ type and staining protocol, the classification process should depend on more image information rather than the instance itself. Therefore, we derive the classification results from the semantic segmentation outcomes. In this semantic segmentation task, accurate category prediction is more important than nuclear shape prediction. That is, the classification results on instance boundaries are not important. Therefore, we utilize the CPF as the weight in the loss function to drive the model's attention focus on the category estimation for the pixels in the critical regions.

$$L_{cls} = -\frac{1}{H \times W} \sum_{p} [\varphi(p) + \eta] \log \hat{P}_p^{y_p}, \tag{4}$$

where y_p denotes the category label of pixel p, $\hat{P}_p^{y_p}$ is the category probability corresponding to y_p predicted by the model, η is a smoothing constant.

During the post-processing procedure, we determine the category of an instance based on the voting results of all pixels belonging to that instance. The predicted CPF is also utilized as the weights for the pixel voting process.

$$C_i = \underset{c \in \{1, \cdots, C\}}{\mathrm{argmax}} \sum_{(i,j) \in S_i} \hat{\varphi}(i,j) \hat{p}_{ij}^c, \tag{5}$$

where, S_i is the set of pixels of the i-th instance, $\hat{\varphi}(i,j)$ is the predicted CPF, and $\hat{p}_{ij}^c$ is the predicted probability that the pixel belongs to the c-th category.

4 Experimental Verification and Analysis

4.1 Datasets and Evaluation Metrics

Datasets. We conducted experiments on two datasets, *Lizard* [23] and *PanNuke* [24]. *Lizard* is a large, high-quality H&E dataset for nuclear instance segmentation. This dataset consists of 291 well-annotated colon images with various sizes. These images are divided into three subsets for model training, validation and test. Adhering to the initial partitioning scheme, we successively designated each of three subsets as training set, validation set and test set to conduct iterative training and assessment. Therefore, this work established 6 models on *Lizard*. *PanNuke* [24] consists of 7901 H&E images with the size of 256×256. All images are divided into three subsets, and we also followed the original division for training, validation, and test.

Evaluation Metrics. We adopted the same evaluation metrics as possible in *Lizard* [23] and *PanNuke* [24]. On *Lizard*, we employed the DICE2 [3] score, Panoptic Quality (PQ) [25], multi-class PQ (mPQ), and additional F_1 score. On *PanNuke*, we employed the precision, recall, F_1, PQ and mPQ.

4.2 Parameters and Implementation Details

Generating CPF. In our experiment, we utilized the EllipseModel API from the scikit-image [26] library to conduct least-squares ellipse regression. In the cases of the ellipse regression failure, such as no solution for the equation or abnormal parameters, we replace the CPF with the mask or morphological erosion mask of that instance. On *Lizard*, the nuclei are small, therefore we used the outer boundaries of the nuclei to calculate the CPF.

Network Structure. CPFNet and PSNet are constructed upon the pre-activated Bottleneck. In CPFNet, there are two decoding paths, which are respectively responsible for predicting CPF and nuclear categories. In CPFNet, there are four downsampling operations, while PSNet has only two.

Training Configurations. For CPF regression, we use a L1 Smooth Loss, and multiply the predicted value and target value by 10 before calculation to amplify the weight of the loss. For classification, the weighted loss function in Eq. (4) is used, the η is set to 0.5. And the weights of various categories are adjusted according to the dataset to mitigate the negative effects from class imbalance. For training PSNet, we randomly select a threshold from the [0.3, 0.7] interval to generate critical region masks for training. Therefore, CPFNet and PSNet are trained independently. The size of patches are set to 48×48 (for Lizard) or 96×96 (for PanNuke).

Post-processing. During post-processing, we sequentially integrate the instance masks into a single mask. If more than half of the pixels in the instance to be merged overlap the existing foreground, the instance will be discarded; otherwise, the overlapping pixels are removed. The final category predictions of nuclei are determined by Eq. (5).

4.3   Comparisons with Other Methods

We compared our results with those of some other methods, including the baseline methods in benchmark datasets and the methods with reported performance on the same dataset literature. Quantitative comparison results are given by tables. Moreover, we visualized some of the results to facilitate intuitive analysis.

Experimental Results on *Lizard*. Our results on *Lizard* are presented in Table 1. The performance of U-net [11], Micro-Net (Microscopy Net) [13] and HoVer-Net (Horizontal and Vertical distances Net) [3] were reported in Lizard [23]. The performance test results of NC-Net (Nuclei Probability and Centroid Map Net) come from the original paper [18]. However, the division on training set and test set in [18] is different from that in this work. The PQ and mPQ scores achieved by the CGIS+CPF are 66.55% and 42.55% respectively, outperforming those of all competing methods. CGIS+CPF achieved a significant improvement in both segmentation and detection performance with a 6.1% increase in DICE2 score over Hover-Net and a 3.76% higher F_1 score than NC-Net. And our method shows excellent robustness (by comparing the standard deviation).

Table 1. Comparison Results on *Lizard*. The experimental results of U-net, Micro-Net and HoVer-Net is reported by the Lizard. The experimental results of NC-Net comes from the original paper. However, the configuration of the training data for NC-Net is different from this work and there isn't variance metric results reported for NC-Net. CGIS+CD means training with normalized center distance.

Method	F1 (%)	Dice2 (%)	PQ (%)	mPQ (%)
U-net	-	73.50 ± 2.80	51.50 ± 3.30	26.50 ± 1.30
Micro-Net	-	78.60 ± 0.40	52.20 ± 1.50	26.40 ± 1.20
HoVer-Net	-	82.80 ± 0.80	62.40 ± 1.30	39.60 ± 0.90
NC-Net	77.90	82.20	61.70	-
CGIS + CD(Ours)	78.73 ± 1.04	85.64 ± 0.64	59.59 ± 1.51	37.34 ± 1.46
CGIS + CPF(Ours)	$\mathbf{81.66 \pm 0.53}$	$\mathbf{88.90 \pm 0.19}$	$\mathbf{66.00 \pm 0.61}$	$\mathbf{42.11 \pm 1.13}$

Table 2. Comparison results on *PanNuke*. The experimental results of Dist, Mask R-CNN, Micro-Net and HoVer-Net was reported by PanNuke officials. The experimental results of NC-Net come from its original paper, where the configuration of training set and test set are different from this work. CGIS+CD means training with normalized center distance.

Method	Precision (%)	Recall (%)	F_1 (%)	PQ (%)	mPQ (%)
Dist	74.00	71.00	73.00	53.46	34.06
Mask R-CNN	76.00	68.00	72.00	55.28	36.88
Micro-Net	78.00	**82.00**	80.00	60.53	40.59
HoVer-Net	82.00	79.00	80.00	65.96	46.29
NC-Net	-	-	79.20	65.40	-
CGIS+CD (Ours)	81.24	77.07	78.22	64.48	44.36
CGIS+CPF (Ours)	**83.34**	79.92	**80.81**	**66.99**	**46.39**

To provide an intuitive demonstration of how the CGIS approach segments clustered nuclei, we visualized the CGIS segmentation outcomes. The visualizations are illustrated in Fig. 3. Here, the prediction results of HoVer-Net were yielded by our trained model. Our model can identify most of the nuclei and distinguish adhesion nuclei. For adhesion nuclear instances, CGIS can generate more reasonable segmentation boundaries.

Experimental Results on *PanNuke* Table 2 shows the comparison results on *PanNuke*. The experimental results of Dist [27], Mask R-CNN [19], Micro-Net [13] and HoVer-Net [3] were reported by PanNuke [24]. The experimental results of NC-Net come from [18], where the configuration of training set is different from ours. Except for Recall, our method outperforms other methods on other metrics and achieved 1% improvement over HoVer-Net on PQ. Owing to the small field of view of the images in *PanNuke*, the nuclei within the images appear comparatively large. Therefore, the difficulty of nuclear instance

Ground Truth   CGIS(Ours)   HoVer-Net

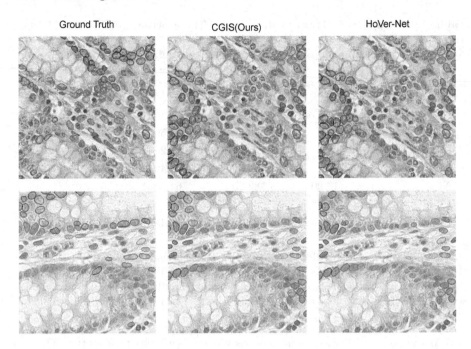

Fig. 3. Visualization of the experimental results on _Lizard_. The marked curves are the nuclear boundaries, and various colors indicate different instances.

detection is relatively low, and the improvement of CPF on detection performance is not great. Similarly, since the impact of pixel errors on larger instances is smaller, the improvement of CPF in segmentation performance is not obvious.

4.4 Ablation Study

To validate whether the proposed CPF enhances model's detection performance, we substituted CPF by the normalized center distance (CD) to train CPFNet and PSNet. On the Lizard dataset, the F_1 score of CGIS+CPF is 1% higher than that of the model trained with center distance (last two rows in Table 1). While on the PanNuke dataset, the F_1 score increased 2.59% by the CGIS+CPF (last two rows in Table 2). In addition, CGIS+CPF is more stable than CGIS+CD. Therefore, the experimental results indicate that CPF can better guide the detection for nuclear critical region.

To investigate the influences of the CPF segmentation threshold on the detection performance, we counted the matching situations between the detected critical regions and instances (i.e., whether the detected critical regions are within the instances) at various thresholds. In this experiment, the purpose of using critical regions instead of the final predicted instances is to eliminate the effects of the segmentation process. We use precision, recall, and F_1 score to describe the above-mentioned matching results. As shown in Table 3, the dependencies

Table 3. The dependences of the matching between critical regions and nuclear instances on the threshold.

Threshold	Lizard			PanNuke		
	Precision (%)	Recall (%)	F_1 (%)	Precision (%)	Recall (%)	F_1 (%)
0.30	88.89	83.21	85.93	87.95	83.05	85.42
0.35	89.19	83.98	86.48	88.24	82.76	85.41
0.40	89.28	83.99	86.53	88.52	82.40	85.34
0.45	89.47	83.74	86.49	88.77	81.96	85.22
0.50	89.58	83.31	86.31	89.10	81.45	85.10

of the various metrics on the threshold are not very significant. The optimal threshold is 0.4 on the Lizard dataset, while that on the PanNuke dataset is 0.3.

5   Conclusion

This paper proposed a Critical-Region Guided Instance Segmentation method and a boundary-insensitive Central Probability Field, and applied them respectively in nuclear instance segmentation and detection of nuclear critical regions. CGIS embeds nuclear critical regions into the image patches to guide the model generating independent segmentation masks for accurately segmenting adhesion nuclei. CPF employs the 2D Gaussian distribution to model the probability that a pixel belongs to a nucleus, indicates the nuclear distribution. The experimental results show that the proposed CGIS method outperforms other models in instance segmentation performance, and the CPF feature can effectively improve the model's detection performance for small objects.

Acknowledgment. This work is supported by the National Natural Science Foundation of China (No. 11973022, 12373108) and the Natural Science Foundation of Guangdong Province (2020A1515010710).

References

1. Kumar, N., Verma, R., Anand, D., et al.: A multi-organ nucleus segmentation challenge. IEEE Trans. Med. Imaging **39**(5), 1380–1391 (2020)
2. Nasir, E.S., Parvaiz, A., Fraz, M.M.: Nuclei and glands instance segmentation in histology images: a narrative review. Artif. Intell. Rev. **56**, 7909–7964 (2022)
3. Graham, S., Vu, Q.D., Raza, S.E.A., et al.: Hover-Net: simultaneous segmentation and classification of nuclei in multi-tissue histology images. Med. Image Anal. **58**, 101563 (2019)
4. Yi, J., Wu, P., Jiang, M., et al.: Attentive neural cell instance segmentation. Med. Image Anal. **55**, 228–240 (2019)
5. Yi, J., Wu, P., Huang, Q., et al. Context-refined neural cell instance segmentation. In: ISBI. IEEE, April 2019

6. Johnson, J.W.: Automatic nucleus segmentation with mask-RCNN. In: Arai, K., Kapoor, S. (eds.) CVC 2019. AISC, vol. 944, pp. 399–407. Springer, Cham (2020). https://doi.org/10.1007/978-3-030-17798-0_32

7. Liu, D., Zhang, D., Song, Y., et al. Nuclei segmentation via a deep panoptic model with semantic feature fusion. In: International Joint Conferences on Artificial Intelligence Organization (IJCAI), August 2019

8. Cheng, Z., Qu, A.: A fast and accurate algorithm for nuclei instance segmentation in microscopy images. IEEE Access **8**, 158679–158689 (2020)

9. Liu, D., Zhang, D., Song, Y., et al.: Panoptic feature fusion net: a novel instance segmentation paradigm for biomedical and biological images. IEEE Trans. Image Process. **30**, 2045–2059 (2021)

10. Fujita, S., Han, X.-H.: Cell detection and segmentation in microscopy images with improved mask R-CNN. In: Sato, I., Han, B. (eds.) ACCV 2020. LNCS, vol. 12628, pp. 58–70. Springer, Cham (2021). https://doi.org/10.1007/978-3-030-69756-3_5

11. Ronneberger, O., Fischer, P., Brox, T.: U-Net: convolutional networks for biomedical image segmentation. In: Navab, N., Hornegger, J., Wells, W.M., Frangi, A.F. (eds.) MICCAI 2015. LNCS, vol. 9351, pp. 234–241. Springer, Cham (2015). https://doi.org/10.1007/978-3-319-24574-4_28

12. Chen, H., Qi, X., Yu, L., et al.: DCAN: deep contour-aware networks for object instance segmentation from histology images. Med. Image Anal. **36**, 135–146 (2017)

13. Raza, S.E.A., Cheung, L., Shaban, M., et al.: Micro-Net: a unified model for segmentation of various objects in microscopy images. Med. Image Anal. **52**, 160–173 (2019)

14. Chen, S., Ding, C., Tao, D.: Boundary-assisted region proposal networks for nucleus segmentation. In: Martel, A.L., et al. (eds.) MICCAI 2020. LNCS, vol. 12265, pp. 279–288. Springer, Cham (2020). https://doi.org/10.1007/978-3-030-59722-1_27

15. He, H., Huang, Z., Ding, Y., et al.: CDNet: centripetal direction network for nuclear instance segmentation. In: ICCV. IEEE, October 2021

16. Scherr, T., Löffler, K., Böhland, M., Mikut, R.: Cell segmentation and tracking using CNN-based distance predictions and a graph-based matching strategy. PLoS ONE **15**(12), e0243219 (2020)

17. Mahbod, A., Schaefer, G., Ellinger, I., Ecker, R., Smedby, Ö., Wang, C.: A two-stage U-Net algorithm for segmentation of nuclei in H&E-stained tissues. In: Reyes-Aldasoro, C.C., Janowczyk, A., Veta, M., Bankhead, P., Sirinukunwattana, K. (eds.) ECDP 2019. LNCS, vol. 11435, pp. 75–82. Springer, Cham (2019). https://doi.org/10.1007/978-3-030-23937-4_9

18. Rashid, S.N., Fraz, M.M.: Nuclei probability and centroid map network for nuclei instance segmentation in histology images. Neural Comput. Appl. **35**, 15447–15460 (2023). https://doi.org/10.1007/s00521-023-08503-2

19. He, K., Gkioxari, G., Dollar, P., Girshick, R. Mask r-CNN. In: ICCV. IEEE, October 2017

20. Naylor, P., Laé, M., Reyal, F., Walter, T.: Segmentation of nuclei in histopathology images by deep regression of the distance map. IEEE Trans. Med. Imaging **38**, 448–459 (2018)

21. Hu, X., Tang, C., Chen, H., et al.: Improving image segmentation with boundary patch refinement. Int. J. Comput. Vis. **130**(11), 2571–2589 (2022)

22. Tian, Z., Zhang, B., Chen, H., Shen, C.: Instance and panoptic segmentation using conditional convolutions. IEEE Trans. Pattern Anal. Mach. Intell. **45**(1), 669–680 (2023)

23. Graham, S., Jahanifar, M., Azam, A., et al.: Lizard: a large-scale dataset for colonic nuclear instance segmentation and classification. In: ICCV. IEEE, October 2021

24. Gamper, J., Koohbanani, N.A., Benes, K., et al.: PanNuke dataset extension, insights and baselines. arXiv preprint arXiv:2003.10778 (2020)
25. Kirillov, A., He, K., Girshick, R., et al.: Panoptic segmentation. In: CVPR. IEEE, June 2019
26. van der Walt, S., Schönberger, J.L., Nunez-Iglesias, J., et al.: scikit-image: image processing in Python. PeerJ **2**, e453 (2014)
27. Schmidt, U., Weigert, M., Broaddus, C., Myers, G.: Cell detection with star-convex polygons. In: Frangi, A.F., Schnabel, J.A., Davatzikos, C., Alberola-López, C., Fichtinger, G. (eds.) MICCAI 2018. LNCS, vol. 11071, pp. 265–273. Springer, Cham (2018). https://doi.org/10.1007/978-3-030-00934-2_30

FlashViT: A Flash Vision Transformer with Large-Scale Token Merging for Congenital Heart Disease Detection

Lei Jiang[1], Junlong Cheng[1], Jilong Chen[1], Mingyang Gu[1], Min Zhu[1(✉)], Peilun Han[2], Kang Li[2], and Zhigang Yang[2]

[1] College of Computer Science, Sichuan University, Chengdu, China
zhumin@scu.edu.cn
[2] Department of Radiology and West China Biomedical Big Data Center, West China Hospital, Sichuan University, Chengdu, China

Abstract. Congenital heart disease (CHD) is the most common congenital malformation and imaging examination is an important means to diagnose it. Currently, deep learning-based methods have achieved remarkable results in various types of imaging examinations. However, the issues of large parameter size and low throughput limit their clinical applications. In this paper, we design an efficient, light-weight hybrid model named FlashViT, to assist cardiovascular radiologists in early screening and diagnosis of CHD. Specifically, we propose the Large-scale Token Merging Module (LTM) for more aggressive similar token merging without sacrificing accuracy, which alleviate the problem of high computational complexity and resource consumption of self-attention mechanism. In addition, we propose an unsupervised homogenous pre-training strategy to tackle the issue of insufficient medical image data and poor generalization ability. Compared with conventional pre-training strategy that use ImageNet1K, our strategy only utilizes less than 1% of the class-agnostic medical images from ImageNet1K, resulting in faster convergence speed and advanced performance of the model. We conduct extensive validation on the collected CHD dataset and the results indicate that our proposed FlashViT-S achieves accuracy of 92.2% and throughput of 3753 fps with about 3.8 million parameters. We hope that this work can provide some assistance in designing laboratory models for future application in clinical practice.

Keywords: Congenital Heart Disease Detection · Large-scale Token Merging Module · Homologous Pre-training Strategy

1 Introduction

Congenital heart disease (CHD) is the most common type of birth defect [15], which can endanger the health and life of affected children. Imaging examination is an important means to diagnose it. Recently, research on using deep learning

Q. Liu et al. (Eds.): PRCV 2023, LNCS 14437, pp. 136–148, 2024.
https://doi.org/10.1007/978-981-99-8558-6_12

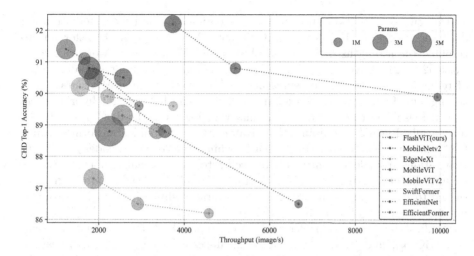

Fig. 1. Comparison of our proposed FlashViT models with light-weight CNNs and hybrid ViTs. The x-axis shows the throughput on GPU device and y-axis displays the CHD Top-1 classification accuracy. The number of parameters are mentioned for each corresponding point in the graph.

methods to assist radiologists in diagnosing CHD has become more widespread. Some works [1] use CNN and U-net to classify and segment the lesion area in ultrasound images of unborn patients with CHD, while others [27] employ Seg-CNN to classify and segment CT images of CHD patients. These methods focus on the analysis and research of CHD pathology. It is worth noting that chest X-ray plays a critical role in the early screening and diagnosis of CHD, particularly in resource-limited areas [22]. However, subtle changes in the chest X-ray of CHD patients may lead to misdiagnosis by radiologists. Therefore, combining deep learning methods with chest X-ray to assist radiologists in early screening and diagnosis of CHD is a promising direction.

Deep learning methods based on CNN have achieved surprising results in various imaging examinations, such as classification of pneumonia using ResNet [10] in [7], and the comparison of various CNN models in the classification of chest X-ray of COVID-19 patients in [19]. However, CNN model lack long-range dependencies due to inductive bias. And the issues of large parameter size and low throughput limit their clinical applications. In recent years, Transformers with global self-attention mechanism have been introduced into medical tasks [4,11,20], and have achieved competitive performance. Nevertheless, due to the quadratic computations and memory costs associated with token sequence lengths, Transformers are also difficult to deploy on edge devices with limited hospital resources.

To address this issue, numerous research endeavors have focused on reducing the length of the token sequence. Some works attempt to achieve this by replacing the original self-attention mechanism with more efficient variants, such as window self-attention [16], grid self-attention [25] and axial self-attention [8]. Other works [17,18] incorporate efficient CNN into Vision Transformers to combine the strengths of both convolution and Transformer. Moreover, some works [2,14,21] strive to improve model throughput by pruning a small number of redundant tokens, which represents a strategy that balances accuracy and throughput. In this work, we design an efficient and light-weight hybrid model named FlashViT, to assist cardiovascular radiologists in the diagnosis of CHD. Specifically, FlashViT combines the advantages of CNN and Transformer, which CNN extracts local features from images, while Transformer encodes global representations from local features using self-attention mechanism. Furthermore, we introduce the Large-scale Token Merging Module (LTM) for more aggressive similar token merging without sacrificing accuracy, which shows higher throughput versus accuracy trade-off (see Fig. 1).

Medical images have fewer numbers compared to natural images (ImageNet1K [6]), and training models on small-scale datasets can easily lead to overfitting and poor generalization capability. Transfer learning is mainly applicable to tasks where there are not enough training samples to train models from scratch, such as medical image classification and segmentation for rare or newly developed diseases [5,12,19]. Many studies have shown that pre-training on large-scale natural images and fine-tuning on small-scale target task images can result in good model performance [12,19]. However, due to the significant domain gap between medical and natural images, domain generalization of transfer learning models is often limited. In this paper, we propose an unsupervised homogenous pre-training strategy that allows the model to undergo unsupervised training on a large-scale medical image dataset, followed by supervised fine-tuning on data from the target disease. Experimental results demonstrate that our homogenous pre-training strategy, which only uses less than 1% of class-agnostic medical images from ImageNet1K, leads to faster convergence speed and superior performance compared to the pre-training strategy that typically uses ImageNet1K.

Furthermore, there are few open-source datasets related to congenital heart disease (CHD), and the data size is small. A small-scale dataset containing 110 CT images for CHD classification was released in [27], but no chest X-ray dataset related to CHD diagnosis has been found. In this study, we retrospectively collect 3466 chest X-ray images between 2018–2021, including 1358 CHD images and 2108 non-CHD images. It should be noted that this dataset is only used for research purposes in this article and will not be publicly available.

The main contributions of this paper are summarized as follows:

A. We design an efficient and light-weight hybrid model named FlashViT which combines the advantages of CNN and Transformer. Furthermore, we introduce the Large-scale Token Merging Module (LTM) for more aggressive similar token merging without sacrificing accuracy, which shows higher throughput versus accuracy trade-off.

B. We propose an unsupervised homogenous pre-training strategy to tackle the issue of insufficient medical image data and poor generalization ability. Our strategy only utilizes less than 1% of the class-agnostic medical images from

ImageNet1K, resulting in faster convergence speed and advanced performance of the model.

C. We conduct extensive validation on the collected CHD dataset and the results indicate that our proposed FlashViT-S achieves accuracy of 92.2% and throughput of 3753 fps with about 3.8 million parameters, which is 1.7% and 0.8% more accurate than MobileNetv2 (CNN-based) and MobileViT (ViT-based), 2.0× and 3.1× the throughput of above both models.

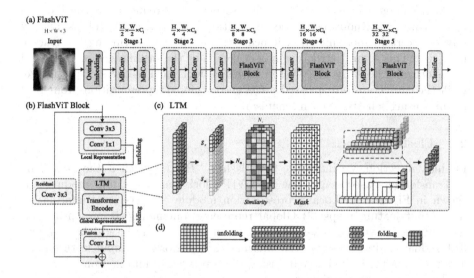

Fig. 2. Overview of FlashViT. (a) Architecture of FlashViT; (b) FlashViT block; (c) Large-scale Token Merging Module; (d) Unfolding operation and folding operation.

2   Method

In this section, we introduce the architecture of FlashViT in detail, with a focus on the LTM. The introduction of LTM allows the model to merge a large number of redundant tokens, which improves model throughput without sacrificing accuracy. Finally, we design several variants of the FlashViT models according to practical requirements to adapt the scenario of limited medical resources.

2.1   Overview

The architecture of FlashViT is illustrated in Fig. 2a. For an input image with size H × W × 3, a overlap patch embedding layer (Conv3×3, stride 2) is used to obtain a collection of initial tokens ($\frac{H}{2} \times \frac{W}{2}$) and project the channel dimension to C_1. Then, these initial tokens will pass through five stages, which include two CNN-based stages and three Transformer-based stages. For the ith stage, the dimension of the input feature map is $\frac{H}{2^i} \times \frac{W}{2^i} \times C_i$. Specifically, the CNN-based stage contains two MBConv modules (efficient convolutional module proposed in MobileNetv2 [23]), and only the first MBConv module will downsample the input feature map and change the channel dimension, while the second MBConv module further enriches the extracted feature information. The Transformer-based stage contains one MBConv module and one FlashViT Block, where the MBConv module expands the channel dimension without changing the feature map size, and the FlashViT Block utilizes a Large-scale Token Merging Module (LTM) to merge more aggressive similar tokens. LTM reduces the length of the token sequence without losing the information contained in the tokens, and by setting a certain merging ratio, it achieves the effect of shrinking the feature map size. This hybrid structure model combines the advantages of CNN and Transformer. The early CNN-based stages extract local features of the image, while the later Transformer-based stages encode global representations from local features using self-attention mechanism.

2.2   FlashViT Block

In this subsection, we introduce the FlashViT Block, which is the main part of Transformer-based stage. The FlashViT Block consists of four modules (as shown in Fig. 2b): 1) Local representation module extracts local information from input feature maps; 2) Global representation module is responsible for more aggressive similar token merging and modeling them in a global context; 3) Residual module preserves the information of the original input feature map; 4) Fusion module is in charge of fusing the two types of information.

The local representation module is composed of two convolutions. For the $F_{input} \in \mathbb{R}^{C \times H \times W}$, it will pass through a convolution with a kernel size of 3 and a stride of 1, followed by a normalization layer and SiLU activation function. This process does not change the dimensions of the feature map. Subsequently, a point-wise convolution is used to project the channel dimension to C_1, resulting in $F_{mid} \in \mathbb{R}^{C_1 \times H \times W}$. In order to adapt to the subsequent global representation module and utilize efficient self-attention, F_{mid} is converted into a token sequence $T_{grid} \in \mathbb{R}^{P \times \frac{HW}{P} \times C_1}$ using the unfolding operation (as shown in Fig. 2d), where $P = 4$ indicates the size of grid.

The global representation module comprises a LTM and a standard Transformer Encoder. For the token sequence T_{grid}, LTM merges r most similar tokens using nearest neighbor similarity algorithm to obtain a collection of merged token sequence $T_{merged} \in \mathbb{R}^{P \times \frac{HW}{4P} \times C_1}$, $(r = \frac{3HW}{4P})$. Then, the standard Transformer Encoder is used to globally model the merged token sequence for global interaction, resulting in $T_{encoded} \in \mathbb{R}^{P \times \frac{HW}{4P} \times C_1}$. Finally, the folding operation is used to restore the encoded token sequence back into a feature map $F_{encoded} \in \mathbb{R}^{C_1 \times \frac{H}{2} \times \frac{W}{2}}$.

To be concise and not introduce excessive parameters, the residual module consists of a convolution with a kernel size of 3 and a stride of 2, a normalization layer and SiLU, which downsamples the original input feature map to obtain $F_{residual} \in \mathbb{R}^{2C \times \frac{H}{2} \times \frac{W}{2}}$. The fusion module projects the channel dimension of

$F_{encoded}$ to 2C using point-wise convolution, then fuses it with $F_{residual}$, resulting in the final output feature map $F_{output} \in \mathbb{R}^{2C \times \frac{H}{2} \times \frac{W}{2}}$.

2.3 Large-Scale Token Merging Module

In this section, we provide a detailed introduction to the LTM, and its general process is shown in Fig. 2c. LTM merges a large-scale similar tokens which are redundant for model. In order to reduce the computational cost of self-attention, we place LTM before the Transformer Encoder.

Firstly, the input token sequence T_{grid} is divided into two subsets S^r and S^m with a 1:3 ratio. Subset S^r represents the set of reserved tokens while subset S^m represents the set of merged tokens. Given the two subsets, I^r and I^m are the corresponding token indices of S^r and S^m, and N^r and N^m are the number of both above, separately.

Secondly, we compute the similarity between tokens in subsets S^r and S^m, and obtain a similarity matrix $c_{i,j}$. To simplify the calculation, we measure the similarity using cosine similarity. The specific formula is as follows:

$$c_{i,j} = \frac{x_i^T x_j}{\|x_i\| \|x_j\|}, i \in I^m, j \in I^r \tag{1}$$

It should be noted that in the similarity matrix $c_{i,j}$, multiple tokens in subset S^m may be most similar to the same token in subset S^r. We then record the results in a mask matrix $M \in \mathbb{R}^{N^m \times N^r}$, which is defined as follows:

$$m_{i,j} = \begin{cases} 1, & \textit{the similarity between } x_i \textit{ and } x_j \textit{ is highest} \\ 0, & \textit{otherwise} \end{cases} \tag{2}$$

Finally, based on the mask martix M, we can easily merge tokens in subset S^m into subset S^r using efficient matrix operations. For simplicity, we use the average operation to process these tokens and obtain the final token sequence $T_{merged} \in \mathbb{R}^{P \times \frac{HW}{4P} \times C_1}$.

2.4 Architecture Variants

In this subsection, we design several variants of the FlashViT models according to practical requirements to adapt the scenario of limited medical resources. The architecture details of our model family are presented in Table 1.

Table 1. Detail settings of FlashViT variants. D_i, r_i refer to the dimension of heads and the number of merged tokens in the i-th stage.

Model	$\{D_1, D_2, D_3, D_4, D_5\}$	$\{r_3, r_4, r_5\}$	Params (M)	FLOPs (G)
FlashViT-XXS	$\{16, 24, 48, 64, 80\}$	$\{192, 48, 12\}$	0.8	0.12
FlashViT-XS	$\{32, 48, 64, 80, 96\}$	$\{192, 48, 12\}$	1.5	0.31
FlashViT-S	$\{32, 64, 96, 128, 160\}$	$\{192, 48, 12\}$	3.8	0.61

3   Experiments

To demonstrate the efficacy of our proposed approach, we conduct sufficient experiments and compare a range of highly efficient neural networks, including CNNs and ViTs.

3.1   CHD Dataset

One of the curial challenges faced by medical researchers is the limited availability of datasets. To the best of our knowledge, only a few medical image datasets related to CHD are publicly available due to privacy concerns and information restrictions. In this study, we train our model using chest X-ray images of both normal and CHD cases. We retrospectively obtain 3466 frontal chest X-ray images between 2018 and 2021, including 2108 normal images and 1358 CHD images.

1) **Data Preprocessing Stage:** All chest X-ray images are in JPEG format, and are decoded and resized to $256 \times 256 \times 3$ pixels to fit the model. The JPEG images are then organized based on their respective image classes. For the experiments, the data is divided into 80% for training, 10% for validation, and 10% for testing with an equal number of image ratios across all classes.

2) **Data Augmentation Stage:** In our experiments, we first crop all images to remove clutter from the image boundaries, allowing the model to focus more on the chest area. For the binary classification experiment (with or without CHD), to allow the model to obtain more information from the chest area, we apply center crop technique on the training, validation and test sets, and additional data augmentation technique, random horizontal flip, is used only on the training set.

3.2   Evaluations on CHD Dataset

Experiment Details. We train various models of FlashViT from scratch on the CHD dataset. We use the Pytorch framework to build the models, and train the models for 60 epochs using the cross-entropy loss function and the Adam [13] optimizer. The batch size is set to 32 and the learning rate is set to 0.0001 based on experience. We use Kaiming Initialization [9] to initialize the weights of the models.

Comparison with Light-Weight Models. As shown in Table 2, FlashViT outperforms light-weight CNNs and hybrid ViTs across different network sizes. For example, FlashViT-S achieves accuracy of 92.2% and throughput of 3753 with about 3.8 million parameters, which is 1.7% and 0.8% more accurate than MobileNetv2 (CNN-based) and MobileViT (ViT-based), 2.0× and 3.1× the throughput of above both models. In addition, we evaluate the performance of our proposed models using several commonly used indicators for classification performance. As shown in Table 3, FlashViT-S outperforms light-weight CNNs and hybrid ViTs in terms of recall, F1-score and AUC value. Our model is also comparable to the SOTA results in terms of precision and specificity. We use the

ROC curve (as shown in Fig. 3a) to demonstrate the overall performance comparison between models. According to the AUC value, except for the EdgeNeXt-S model, all models have similar performance, with FlashViT-S having a slightly higher AUC than other models.

Table 2. Performance comparison with light-weight models on CHD dataset by the way of training from scratch. Throughput is measured on a single V100 GPU with batch size 128.

Model	Type	Image size	Params (M)	FLOPs (G)	Throughput (fps)	Accuracy (%)
MobileNetv2-050	CNN	256	0.8	0.13	6685	86.5
EdgeNeXt-XXS	Hybrid	256	1.2	0.26	4573	86.2
MobileViT-XXS	Hybrid	256	1.0	0.35	2936	89.6
MobileViTv2-050	Hybrid	256	1.1	0.46	3738	89.6
FlashViT-XXS (Ours)	Hybrid	256	**0.8**	**0.12**	**9933**	**89.9**
EfficientNet-b0	CNN	256	4.0	0.54	2562	90.5
MobileNetv2-100	CNN	256	2.2	0.39	3541	88.8
SwiftFormer-XS	Hybrid	224	3.0	0.61	3356	88.8
MobileViTv2-075	Hybrid	256	2.5	1.03	2202	89.9
FlashViT-XS (Ours)	Hybrid	256	**1.5**	**0.31**	**5200**	**90.8**
EfficientNet-b1	CNN	256	6.5	0.80	1768	90.8
MobileNetv2-120d	CNN	256	4.5	0.87	1870	90.5
EdgeNeXt-S	Hybrid	256	5.3	1.25	1876	87.3
EfficientFormer-L1	Hybrid	224	11.4	1.31	2246	88.8
MobileViT-S	Hybrid	256	4.9	1.89	1219	91.4
FlashViT-S (Ours)	Hybrid	256	**3.8**	**0.61**	**3753**	**92.2**

Comparison with Pruning Models. In Table 4, we also compare FlashViT-S with some pruning models. Our model achieves higher accuracy with fewer parameters, smaller FLOPs operations, and equivalent throughput.

Table 3. Common evaluation metrics for binary classification models on the CHD test set.

Model	Precision	Recall	Specificity	F1-score	AUC
EfficientNet-b1	0.9143	0.8067	**0.9605**	0.8571	0.944
MobileNetv2-120d	0.8583	0.8655	0.9254	0.8619	0.942
EdgeNeXt-S	0.8788	0.7311	0.9474	0.7982	0.862
MobileViT-S	**0.9159**	0.8235	0.9605	0.8673	0.930
MobileViTv2-100	0.8829	0.8235	0.9430	0.8522	0.925
FlashViT-S (Ours)	0.9035	**0.8655**	0.9518	**0.8841**	**0.950**

Table 4. Performance comparison with token pruning models on CHD dataset by the way of training from scratch.

Model	Type	Image size	Params (M)	FLOPs (G)	Throughput (fps)	Accuracy (%)
EViT-DeiT-tiny	Token Pruning	224	5.5	1.08	2870	84.7
Evo-ViT	Token Pruning	224	5.5	0.73	**3993**	85.3
DynamicViT br = 0.8	Token Pruning	224	5.7	0.83	3704	84.7
Deit-T(ToMe) r = 8	Token Merging	224	5.5	0.82	3402	87.9
FlashViT-S (Ours)	Token Merging	256	**3.8**	**0.61**	3753	**92.2**

3.3   Homogenous Pre-training Strategy

Experiment Details. In this subsection, to address the issue of domain gap, we adopt an alternative unsupervised homogenous pre-training strategy using contrastive learning method SimCLR [3]. Specifically, the model is pre-trained on the NIH Chest X-ray medical dataset [26] consisting of 112,120 chest X-ray images (approximately 0.8% of the ImageNet1K dataset [6]) with a batch size of 512 and a learning rate of 0.001 over 200 epochs on a single A100 GPU. It is then fine-tuned on the supervised training set of CHD dataset.

Results. In Table 5, FlashViT uses the homogenous pre-training strategy on the NIH Chest X-ray medical dataset, while other models use the pre-training strategy on the ImageNet1K dataset. From the experimental results, our homogenous pre-training strategy enables the model to achieve greater performance improvement on relatively small medical datasets, alleviating the problem of model domain generalization.

Table 5. Performance comparison of different pre-training strategies. ImageNet indicates that the model is pre-trained on ImageNet1K. Contrastive Learning indicates that the model pre-trained on the NIH Chest X-ray dataset in a contrastive learning manner.

Model	Pretrain Strategy	Params(M)	Accuracy(%)
MobileNetv2-050	ImageNet	0.8	94.8
MobileViT-XXS	ImageNet	1.0	95.4
FlashViT-XXS (Ours)	Contrastive Learning	**0.8**	**96.8**
MobileNetv2-100	ImageNet	2.2	95.1
MobileViT-XS	ImageNet	1.9	95.7
FlashViT-XS (Ours)	Contrastive Learning	**1.5**	**97.1**
MobileNetv2-120d	ImageNet	4.5	95.7
MobileViT-S	ImageNet	4.9	96.0
FlashViT-S (Ours)	Contrastive Learning	**3.8**	**97.8**

Visualization. As shown in Fig. 3b, we extract the convolutional layer gradients of the last stage in FlashViT under two training strategies (no pre-training and homogenous pre-training), and use the Grad-CAM [24] technique to visualize the diagnosed region of CHD. By comparing the heatmaps of the second and third rows of chest X-rays, we find that the model without pre-training has biases in the observed regions and does not fully focus on the heart projection area, while the homogenous pre-training strategy makes the model more focused on the diagnosed region. The homogenous pre-training strategy provides the model with sufficient domain knowledge and alleviates the problem of domain generalization.

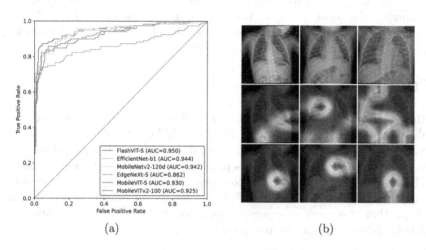

(a)   (b)

Fig. 3. (a) ROC curve of different models on the CHD test set. (b) Grad-CAM visualizations for FlashViT-S in three chest X-ray images from the test set. Vertical sets give the Original images (top row), Grad-CAM visualizations for the model trained from scratch (2nd row), and Grad-CAM visualizations for the model trained with homogenous pre-training strategy (3rd row).

3.4   Ablation Study

Table 6. Analysis for different self-attention mechanisms.

Method	Params	FLOPs	Throughput	Accuracy
Window SA	3.8	0.61	3774	91.4
Grid SA	3.8	0.61	3753	**92.2**

Table 7. Analysis for the stacking depth of Transformer Encoder.

Stacking depth	Params	FLOPs	Throughput	Accuracy
[1, 1, 1]	**3.8**	**0.61**	**3753**	**92.2**
[2, 2, 2]	4.7	0.68	3374	91.1
[3, 3, 3]	5.6	0.74	3059	91.1
[2, 4, 3]	5.8	0.72	3185	91.6

The Effectiveness of Sparse Self-attention Mechanism. We test the impact of the efficient sparse self-attention mechanisms Window SA [16] and

Grid SA [25] on the performance of the FlashViT model. The results are shown in Table 6. Due to the global token interaction, Grid SA outperforms Window SA in improving model performance. In addition, there is little difference between the two self-attention mechanisms in terms of parameters, FLOPs and throughput.

The Analysis for the Stacking Depth of Transformer Encoder. As shown in Table 7, we compare the impact of different stacking depth of Transformer Encoder on the model. Following MobileViT [18], we set [1, 1, 1], [2, 2, 2], [3, 3, 3] and [2, 4, 3] for the stacking depth in stage 3, 4 and 5, respectively. From the experimental results, despite the increase in parameters and FLOPs with increasing depth, there is no improvement in model performance. The stacking depth of [1, 1, 1] is simpler and more efficient.

4   Conclusion

In this paper, we design an efficient and light-weight hybrid model called FlashViT with a Large-scale Token Merging Module (LTM), to assist cardiovascular radiologists in early screening and diagnosis of CHD. Additionally, we propose an unsupervised homogenous pre-training strategy to tackle the issue of insufficient medical image data and poor generalization ability. Extensive validation on the collected CHD dataset demonstrate the superiority of our method. We hope that this work can provide some assistance in designing laboratory models for future application in clinical practice.

References

1. Arnaout, R., Curran, L., Zhao, Y., Levine, J.C., Chinn, E., Moon-Grady, A.J.: Expert-level prenatal detection of complex congenital heart disease from screening ultrasound using deep learning. medRxiv, pp. 2020–06 (2020)
2. Bolya, D., Fu, C.Y., Dai, X., Zhang, P., Feichtenhofer, C., Hoffman, J.: Token merging: your ViT but faster. arXiv preprint arXiv:2210.09461 (2022)
3. Chen, T., Kornblith, S., Norouzi, M., Hinton, G.: A simple framework for contrastive learning of visual representations. In: International Conference on Machine Learning, pp. 1597–1607. PMLR (2020)
4. Cheng, J., et al.: ResGANet: residual group attention network for medical image classification and segmentation. Med. Image Anal. **76**, 102313 (2022)
5. Cheng, J., Tian, S., Yu, L., Lu, H., Lv, X.: Fully convolutional attention network for biomedical image segmentation. Artif. Intell. Med. **107**, 101899 (2020)
6. Deng, J., Dong, W., Socher, R., Li, L.J., Li, K., Fei-Fei, L.: ImageNet: a large-scale hierarchical image database. In: 2009 IEEE Conference on Computer Vision and Pattern Recognition, pp. 248–255. IEEE (2009)
7. Desai, G., Elsayed, N., Elsayed, Z., Ozer, M.: A transfer learning based approach for classification of COVID-19 and pneumonia in CT scan imaging. arXiv preprint arXiv:2210.09403 (2022)
8. Dong, X., et al.: CSWin transformer: a general vision transformer backbone with cross-shaped windows. In: Proceedings of the IEEE/CVF Conference on Computer Vision and Pattern Recognition, pp. 12124–12134 (2022)

9. He, K., Zhang, X., Ren, S., Sun, J.: Delving deep into rectifiers: surpassing human-level performance on ImageNet classification. In: Proceedings of the IEEE International Conference on Computer Vision, pp. 1026–1034 (2015)

10. He, K., Zhang, X., Ren, S., Sun, J.: Deep residual learning for image recognition. In: Proceedings of the IEEE Conference on Computer Vision and Pattern Recognition, pp. 770–778 (2016)

11. Huang, X., Deng, Z., Li, D., Yuan, X.: MISSFormer: an effective medical image segmentation transformer. arXiv preprint arXiv:2109.07162 (2021)

12. Huynh, B.Q., Li, H., Giger, M.L.: Digital mammographic tumor classification using transfer learning from deep convolutional neural networks. J. Med. Imaging 3(3), 034501–034501 (2016)

13. Kingma, D.P., Ba, J.: Adam: a method for stochastic optimization. arXiv preprint arXiv:1412.6980 (2014)

14. Liang, Y., Ge, C., Tong, Z., Song, Y., Wang, J., Xie, P.: Not all patches are what you need: expediting vision transformers via token reorganizations. arXiv preprint arXiv:2202.07800 (2022)

15. Liu, Y., et al.: Global prevalence of congenital heart disease in school-age children: a meta-analysis and systematic review. BMC Cardiovasc. Disord. 20, 1–10 (2020)

16. Liu, Z., et al.: Swin transformer: hierarchical vision transformer using shifted windows. In: Proceedings of the IEEE/CVF International Conference on Computer Vision, pp. 10012–10022 (2021)

17. Maaz, M., et al.: EdgeNeXt: efficiently amalgamated CNN-transformer architecture for mobile vision applications. In: Karlinsky, L., Michaeli, T., Nishino, K. (eds.) Computer Vision, ECCV 2022 Workshops, ECCV 2022, Part VII. LNCS, vol. 13807, pp. 3–20. Springer, Cham (2023). https://doi.org/10.1007/978-3-031-25082-8_1

18. Mehta, S., Rastegari, M.: MobileViT: light-weight, general-purpose, and mobile-friendly vision transformer. arXiv preprint arXiv:2110.02178 (2021)

19. Minaee, S., Kafieh, R., Sonka, M., Yazdani, S., Soufi, G.J.: Deep-COVID: predicting COVID-19 from chest X-ray images using deep transfer learning. Med. Image Anal. 65, 101794 (2020)

20. Perera, S., Adhikari, S., Yilmaz, A.: POCFormer: a lightweight transformer architecture for detection of COVID-19 using point of care ultrasound. In: 2021 IEEE International Conference on Image Processing (ICIP), pp. 195–199. IEEE (2021)

21. Rao, Y., Zhao, W., Liu, B., Lu, J., Zhou, J., Hsieh, C.J.: DynamicViT: efficient vision transformers with dynamic token sparsification. Adv. Neural. Inf. Process. Syst. 34, 13937–13949 (2021)

22. Rashid, U., Qureshi, A.U., Hyder, S.N., Sadiq, M.: Pattern of congenital heart disease in a developing country tertiary care center: factors associated with delayed diagnosis. Ann. Pediatr. Cardiol. 9(3), 210 (2016)

23. Sandler, M., Howard, A., Zhu, M., Zhmoginov, A., Chen, L.C.: MobileNetV2: inverted residuals and linear bottlenecks. In: Proceedings of the IEEE Conference on Computer Vision and Pattern Recognition, pp. 4510–4520 (2018)

24. Selvaraju, R.R., Cogswell, M., Das, A., Vedantam, R., Parikh, D., Batra, D.: Grad-CAM: visual explanations from deep networks via gradient-based localization. In: Proceedings of the IEEE International Conference on Computer Vision, pp. 618–626 (2017)

25. Tu, Z., et al.: MaxViT: multi-axis vision transformer. In: Avidan, S., Brostow, G., Cissé, M., Farinella, G.M., Hassner, T. (eds.) Computer Vision, ECCV 2022, Part XXIV. LNCS, vol. 13684, pp. 459–479. Springer, Cham (2022). https://doi.org/10.1007/978-3-031-20053-3_27

26. Wang, X., Peng, Y., Lu, L., Lu, Z., Bagheri, M., Summers, R.M.: ChestX-ray8: hospital-scale chest X-ray database and benchmarks on weakly-supervised classification and localization of common thorax diseases. In: Proceedings of the IEEE Conference on Computer Vision and Pattern Recognition, pp. 2097–2106 (2017)
27. Xu, X., et al.: ImageCHD: a 3D computed tomography image dataset for classification of congenital heart disease. In: Martel, A.L., et al. (eds.) MICCAI 2020, Part IV 23. LNCS, vol. 12264, pp. 77–87. Springer, Cham (2020). https://doi.org/10.1007/978-3-030-59719-1_8

Semi-supervised Retinal Vessel Segmentation Through Point Consistency

Jingfei Hu[1,2,3], Linwei Qiu[1,2], Hua Wang[1,2], and Jicong Zhang[1,2,3(✉)]

[1] School of Biological Science and Medical Engineering, Beihang University, Beijing, China
`jicongzhang@buaa.edu.cn`
[2] Hefei Innovation Research Institute, Beihang University, Hefei, China
[3] Beijing Advanced Innovation Centre for Big Data-Based Precision Medicine, Beihang University, Beijing, China

Abstract. Retinal vessels usually serve as biomarkers for early diagnosis and treatment of ophthalmic and systemic diseases. However, collecting and labeling these clinical images require extensive costs and thus existing models are commonly based on extremely limited labeled data for supervised segmentation of retinal vessels, which may hinder the effectiveness of deep learning methods. In this paper, we propose a novel point consistency-based semi-supervised (PCS) framework for retinal vessel segmentation, which can be trained both on annotated and unannotated fundus images. It consists of two modules, one of which is the segmentation module predicting the pixel-wise vessel segmentation map like a common segmentation network. Otherwise, considering that retinal vessels present tubular structures and hence the point set representation enjoys its prediction flexibility and consistency, a point consistency (PC) module is introduced to learn and express vessel skeleton structure adaptively. It inputs high-level features from the segmentation module and produces the point set representation of vessels simultaneously, facilitating supervised segmentation. Meanwhile, we design a consistency regularization between point set predictions and directly predicted segmentation results to explore the inherent segmentation perturbation of the point consistency, contributing to semi-supervised learning. We validate our method on a typical public dataset DRIVE and provide a new large-scale dataset (TR160, including 160 labeled and 120 unlabeled images) for both supervised and semi-supervised learning. Extensive experiments demonstrate that our method is superior to the state-of-the-art methods.

Keywords: Retinal vessel segmentation · Point set representation · Semi-supervised learning

1 Introduction

Retinal vessels are the only complete vascular structures that can be directly observed in the human body *in vivo* in a non-invasive manner [1]. Several studies

Q. Liu et al. (Eds.): PRCV 2023, LNCS 14437, pp. 149–161, 2024.
https://doi.org/10.1007/978-981-99-8558-6_13

have shown close relationships between structural changes in retinal vessels and some systemic diseases, such as cardiovascular diseases [19], diabetic retinopathy [23], and age-related macular degeneration [1]. Retinal vessel extraction is the first step for disease diagnosis, whereas the manual annotating of retinal vessels is time-consuming and prone to human errors. With the rapid increase of requirements for medical services, automatic retinal vessel segmentation methods meeting the practical clinical needs grow essential. Recently, deep learning (DL) based methods have been applied to automatically segment retinal vessels from fundus images [2,8,22,25]. They designed and applied diversified convolutional neural networks to perform vessel segmentation. One of the most representative approaches is UNet [11], which is famous for its skip connections. Inspired by its success in medical image segmentation, some UNet variants [4,5,8] were successively proposed for retinal vessel segmentation and gained promising performance. Unfortunately, these DL-based methods are heavily thirsty for substantial images with pixel-level annotations, but are forced to be trained on the limited dataset (usually less than 20 color images) for the tremendous costs to obtain enough source images and dense labels.

Semi-supervised learning (SSL) is one of the most effective strategies to further improve the model performance by combining unlabeled data [29]. Most dominant SSL approaches are based on the generative methods [12,26], pseudo-labeling [6,9,18,20], entropy minimization [3,16], and consistency regularization [14,15,21,24]. Generative methods are not stable in the training stage and deserve a large amount of data. Both pseudo-labeling and entropy minimization methods aim to produce high-confidence results on unlabeled images, ignoring to explore the consistency lying on standalone samples. Some consistency regularization-based approaches like [10] introduce a dual task consistency by combining the level set-derived segmentation and directly semantic segmentation. However, the level set task is only harmonious for segmentation such as pancreas, and not suitable for fundus images since retinal vessels have tubular structures rather than massive textures. Furthermore, most existing supervised and semi-supervised methods usually regard the extraction of vessel structures as a regular semantic segmentation task, without the utilization of the point consistency behind their tubular structures. Different from organs or tissues in other medical image segmentation tasks, nearly all foreground areas in retinal vessels are close to boundaries. This tubular feature leads to easy segmentation mistakes of small branches but releases the flexibility of the point set representation. A type of point set structure is derived from one segmentation map while a retinal image links to a consistent topology preservation. This observation of the point consistency inspires us, which is additionally leveraged among both labeled and unlabeled data.

In this paper, we propose a novel point consistency-based semi-supervised (PCS) framework for retinal vessel segmentation. Our network mainly includes two modules, one of which is the segmentation module. It applies widely-used encoder-decoder architecture as backbone network to perform retinal vessel segmentation. Another point consistency (PC) module consisting of three heads is

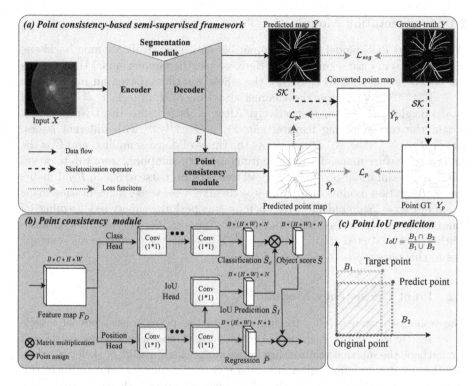

Fig. 1. Details of the proposed network. (a) is the overall structure of our point consistency-based semi-supervised framework. (b) exhibits the details of the PC module, which consists of ClassHead, PositionHead and IoUHead to obtain the classification, position and intersection over union (IoU) information for the generated points. (c) is the diagram of the point IoU predicted by IoUHead. We choose the left bottom point of a feature as the original point.

proposed, which parlays high-level features from the segmentation module to generate the point set representation of the vessel skeleton. To utilize unlabeled data, we introduce a consistency regularization between the vessel segmentation map directly generated by the segmentation module and the point set representation generated by the PC module. It can explore the inherent segmentation perturbation of the point consistency and attend to semi-supervised learning. We train and test our method on a publicly available dataset DRIVE and a large-scale private dataset TR160, which is up-to-date and the largest fundus dataset. Extensive experiments show the effectiveness of our method compared with the state-of-the-art methods.

2   Method

In this section, we introduce the details of our point consistency-based semi-supervised framework (illustrated in Fig. 1(a)) for retinal vessel segmentation.

2.1 Segmentation Module

We proposed PCS model can be compatible with almost all common backbone networks in semantic segmentation. Here, we use the famous model UNet [11] as the backbone network to elaborate the whole workflow. Given an input image $X \in \mathbb{R}^{c \times h \times w}$, where c is the channel size (for RGB image, $c = 3$), h and w are height and width, respectively. After X is fed into the UNet, we can obtain the corresponding feature map $F_D \in \mathbb{R}^{C \times H \times W}$ with different scales, where $H = \frac{h}{2^D}$, $W = \frac{w}{2^D}$. D represents the times of down-sampling, and C is the channel of feature maps. Through continuous down-sampling, from top to down, the structure information of blood vessel tree is from coarse to fine in F_D. Finally, the segmentation module output one-to-one pixel-wise vessel segmentation maps $\bar{Y}$, where $\bar{Y} = Sigmoid(F_0)$. We thereby can perform supervised learning by minimizing the supervised segmentation loss function $\mathcal{L}_{seg}$ between the ground-truth Y and predicted map $\bar{Y}$. For $\mathcal{L}_{seg}(\bar{Y}, Y)$, we use the common cross-entropy loss in this paper.

2.2 Point Consistency Module

The point consistency (PC) module extracts the point consistency of vessels by explicitly expressing each point on the vessel trees, which reconstructs and strengthens the internal relationship between the classification of skeleton points and the coordinate information. It encourages the backbone to absorb the intrinsic characteristics of retinal vessels, which further improves the vessel segmentation performance.

For a backbone, each feature vector $v_i, i \in [1, H \times W]$ in F_D^i represents the high-level connections of a region $RG^i \in \mathbb{R}^{c \times 2^D \times 2^D}$ mapped in the input space. Therefore, comprehensive information of $N = 2^D \times 2^D$ prediction points should be generated for each vector. We design three heads in the PC module to obtain the explicit adaptive structural information of the skeleton. As shown in Fig. 1(b), ClassHead is used to get the classification scores of each prediction point, and PositionHead is adopted to obtain the position information. Different from the previous work [17], the PC module not only focuses on the classification and location information of points but also focuses on the interaction information between them (depicted in Fig. 1(c)). It is important to interact with classification and locations of points. For instance, a point may have a high classification score but an extreme deviation in the location and vice versa. IoUHead is therefore added to obtain the correlation scores between classification scores and position accuracy, which can couple and coordinate classification and location dependencies.

Concretely, the feature map F_2 (we set $D = 2$ in this work) is fed into ClassHead to get the classification score $\bar{S}_c = ClassHead(F_2)$. Moreover, we input the F_2 into PositionHead to obtain the relative position information $\bar{P} = PositionHead(F_2)$. Meanwhile, IoUHead incorporates the latent features of PositionHead to get the point IoU $\bar{S}_I = IoUHead(F_2)$. $\bar{S}_I$ thus shares the same latent representations with $\bar{P}$ and facilitates classification interactions with position information. Note that we apply Sigmoid operation after ClassHead and IoUHead to normalize the results of classification to $\{0,1\}$. The final classification score $\bar{S}$ is obtained by $\bar{S} = \bar{S}_c * \bar{S}_I$. To obtain the real coordinate of the points, we simply apply the relative positions regressed by PositionHead to the center point (x_c^i, y_c^i) of a RG^i, then the coordinate of a predicted point is $\bar{P} = \{\bar{P}_{m,n}^i | \bar{P}_{m,n}^i = (x_m^i + x_c^i, y_n^i + y_c^i), m \in [1, 2^D], n \in [1, 2^D], i \in [1, H \times W]\}$. We adopt the greedy bipartite matching algorithm [17] to assign these N prediction points to the corresponding point in every RG^i of Y_p. A skeletonization operation $\mathcal{SK}$ is used to transform the pixel-wise segmentation map Y into skeletons of vessels for point set representations Y_p

$$Y_p = \mathcal{SK}(Y) = \bigcup_{t=0}^{T} ((Y \ominus tB) - (Y \ominus tB) \circ B), \tag{1}$$

where $\ominus$ and $\circ$ represent erosion operation and open operation, respectively. B indicates predefined structural elements, and $T = \max\{t | (Y \ominus tB) \neq \emptyset\}$. Then the target score $S = \{S_j^i | S_j^i \in \{0,1\}, j \in [1, N], i \in [1, H \times W]\}$ is represented as a matrix with binary values from Y_p, and we also convert it to the target position of point sets $P = \{(a, b) | Y_p(a, b) = 1, a \in [1, h], b \in [1, w]\}$.

To train this module, we use the Focal Loss [7] to deal with the unbalanced distribution for the classification score loss. Besides, the $L1$ loss is used for the relative position regression loss. The point loss $\mathcal{L}_p$ as in Eq. 2:

$$\mathcal{L}_p(\bar{Y}_p, Y_p) = -(1 - \bar{S})^\gamma Log(\bar{S}) + \alpha * MAE(P - \bar{P}), \tag{2}$$

where we set $\gamma = 2$ and $\alpha = 1$ in this work. However, Eq. 2 is only useful for the situation when the label Y is given. We propose a regularization term to leverage the point consistency from unlabeled images. In Eq. 1, it obviously projects the prediction space of segmentation to be the same as that of point set representation, naturally introducing a point-level prediction perturbation since the segmentation module focuses on pixel classifications while the PC module is devoted to tubular structures. This segmentation perturbation behind the point consistency inspires our semi-supervised learning method. For any input X with the output predicted segmentation map $\bar{Y}$ and point prediction $\bar{Y}_p$ from our two modules, we define the point regularization $\mathcal{L}_{pc}$ to enforce point consistency between these two outputs:

$$\mathcal{L}_{pc}(\bar{Y}_p, \bar{Y}) \triangleq \mathcal{L}_p(\bar{Y}_p, \hat{Y}_p), \hat{Y}_p = \mathcal{SK}(\bar{Y}) \tag{3}$$

2.3   Semi-supervised Training Through Point Consistency

Given a semi-supervised dataset $\mathcal{D}$, consisting of labeled data $(X_l, Y_l) \in \mathcal{D}_l$ and unlabeled data $X_u \in \mathcal{D}_u$, the core of our semi-supervised learning is to make full use of the point consistency between segmentation predictions and the corresponding point set representation without help of the ground truth. As shown in Fig. 1, this is achieved by applying the point regularization $\mathcal{L}_{pc}$ for unlabeled data $X_u \in \mathcal{D}_u$, i.e.

$$\mathcal{L}_{unsup}(X_u, \bar{Y}_u, \bar{Y}_{up}) = \sum_{X_u \in \mathcal{D}_u} \mathcal{L}_{pc}(\bar{Y}_{up}, \bar{Y}_u), \tag{4}$$

where $\bar{Y}_u$ and $\bar{Y}_{up}$ are outputs of our model for any input X_u. For labeled input $X_l \in \mathcal{D}_l$, we define the supervised loss $\mathcal{L}_{sup}$ by combining the segmentation loss, point loss, and point regularization

$$\mathcal{L}_{sup}(X_l, Y_l, \bar{Y}_l, \bar{Y}_{lp}) = \sum_{X_l, Y_l \in \mathcal{D}_l} \mathcal{L}_{seg}(\bar{Y}_l, Y_l) + \mathcal{L}_p(\bar{Y}_{lp}, \mathcal{SK}(Y_l)) + \mathcal{L}_{pc}(\bar{Y}_{lp}, \bar{Y}_l), \tag{5}$$

where $\bar{Y}_l$ and $\bar{Y}_{lp}$ are predicted maps corresponding to the input X_l, Y_l is label. The total semi-supervised loss is defined as

$$\mathcal{L}_{semi} = \mathcal{L}_{sup} + \lambda(s) * \mathcal{L}_{unsup}, \lambda(s) = 0.1 * e^{-5(1-\frac{s}{s_m})^2}, \tag{6}$$

where $\lambda(s)$ is the warming up function [14] to balance the influence of unlabeled data with respect to the current training step s and the maximum training step s_m.

3   Experiments

3.1   Datasets

The DRIVE [13] dataset contains 40 color fundus images with a resolution of 584×565. The set is divided into a training and a test set, both containing 20 images. The TR160 dataset is collected by us, consisting of 280 images with a resolution of 2464×2248 pixels, of which 160 images are labeled and the rest 120 images are unlabeled. In particular, the predicted area of retinal vessel was defined as the region from 0 to 1.5 optic disc diameters away from the optic disc margin for the TR160. Therefore, we cropped this area from the original image and trained graders labeled retinal vessels using computer-assisted software (ITK-SNAP). For supervised learning, 120 images with labeled for training, and the rest 40 images for testing. For semi-supervised learning, we effectively used 120 unlabeled data for training.

3.2   Implementation Details

We implemented our model using the PyTorch 1.10.0. We used the Adam optimizer to update the network parameters. And we adopted poly strategy to dynamically adjust the learning rate of each iteration with factor of $lr \times (1 - \frac{cur\_it}{max\_it})^{0.9}$, where lr, cur_it, max_it represent the base learning rate, current number of iterations, and the maximum number of iterations respectively. We trained the whole model for 2000 epochs on a single NVIDIA Tesla V100 GPU. For supervised learning, the batch size was set as 2. For semi-supervised learning, the batch size was 4, consisting of 2 labeled images and 2 unlabeled images.

Table 1. Comparison with other semi-supervised methods on the TR160 dataset. The best results are highlighted in bold.

Methods	Acc (%)	Sen (%)	Sp (%)	AUC (%)	Dice (%)
UNet	97.10	81.65	98.61	99.05	83.25
PL [6]	96.73	76.55	**98.71**	98.83	80.61
MT [14]	97.34	84.76	98.57	99.17	85.05
UAMT [24]	97.29	82.74	98.71	99.15	84.50
EM [16]	97.28	83.73	98.61	99.14	84.54
IC [15]	97.00	83.19	98.36	98.96	83.21
DAN [26]	97.12	80.86	98.71	99.03	83.25
RD [21]	97.17	83.19	98.54	99.04	83.97
PCS(Ours)	**97.55**	**89.90**	98.31	**99.22**	**86.83**

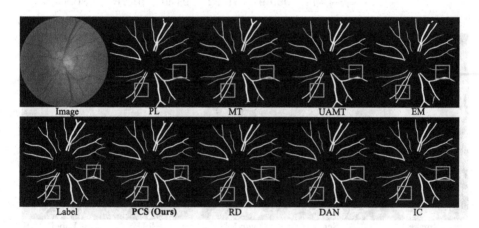

Fig. 2. Visualization of different semi-supervised segmentation methods on the TR160 dataset.

Table 2. The performance of the number of unlabeled data on TR160 dataset with the UNet backbone. The best results are highlighted in bold.

Labeled	Unlabeled	Acc (%)	Sen (%)	Sp (%)	AUC (%)	Dice (%)
120	0	97.32	86.12	**98.42**	99.18	85.15
120	30	97.43	89.85	98.18	99.18	86.25
120	60	97.50	89.75	98.26	99.20	86.56
120	90	97.51	89.33	98.31	**99.23**	86.52
120	120	**97.55**	**89.90**	98.31	99.22	**86.83**

Table 3. The effectiveness of PC module (+PC) on two datasets and three backbones for supervised learning. The best results are highlighted in bold.

Dataset	Backbone	Methods	Acc (%)	Sen (%)	Sp (%)	AUC (%)	Dice (%)
DRIVE	UNet	Seg	95.65	80.46	**97.90**	98.00	82.39
		Seg+PC	**95.69**	**82.20**	97.69	**98.20**	**82.83**
	UNet++	Seg	95.70	81.75	97.77	98.17	82.80
		Seg+PC	**95.75**	**81.98**	**97.79**	**98.25**	**83.01**
	ResUNet	Seg	**95.70**	81.86	**97.76**	98.14	82.82
		Seg+PC	**95.70**	**82.64**	97.64	**98.26**	**82.96**
TR160	UNet	Seg	97.10	81.65	**98.61**	99.05	83.25
		Seg+PC	**97.32**	**86.12**	98.42	**99.18**	**85.15**
	UNet++	Seg	97.08	79.88	**98.75**	99.02	82.83
		Seg+PC	**97.33**	**88.53**	98.20	**99.11**	**85.57**
	ResUNet	Seg	96.34	72.53	**98.66**	98.70	77.60
		Seg+PC	**96.69**	**79.22**	98.40	**98.76**	**80.87**

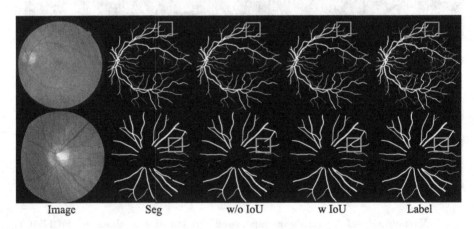

Image   Seg   w/o IoU   w IoU   Label

Fig. 3. Visualization results of UNet backbone. The first row is DRIVE and the second row is TR160 dataset.

3.3   Experimental Results

Comparison with Other Semi-supervised Methods. We quantitatively evaluate our method via five metrics, including Accuracy (Acc), Sensitivity (Sen), Specificity (Sp), Area Under Curve (AUC), and Dice. We validate our semi-supervised method on the TR160 dataset with a large amount of data, the comparison results with other semi-supervised methods are shown in Table 1. In Table 1, the first row presents the segmentation performance of UNet trained with only the labeled data, and the rest rows show the existing semi-supervised methods and our semi-supervised method (PCS). Compared with other

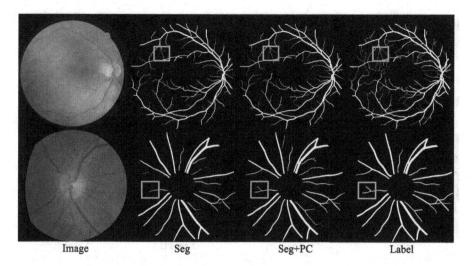

Image   Seg   Seg+PC   Label

Fig. 4. Visualization results of UNet++ backbone. The first row is DRIVE and the second row is TR160 dataset.

Table 4. Ablation experiment results of IoUHead with UNet as the backbone. The best results are highlighted in bold.

Dataset	IoUHead	Acc (%)	Sen (%)	Sp (%)	AUC (%)	Dice (%)
DRIVE	w/o	95.68	81.41	**97.80**	98.04	82.68
	w	**95.69**	**82.20**	97.69	**98.20**	**82.83**
TR160	w/o	97.16	83.57	**98.51**	99.09	83.85
	w	**97.32**	**86.12**	98.42	**99.18**	**85.15**

semi-supervised methods, our PCS achieves state-of-the-art performance with the Acc of 97.55%, Sen of 89.90%, AUC of 99.22% and Dice of 86.83%. By utilizing the unlabeled data, our PCS framework improves the vessel segmentation performance by 8.25% Sen and 3.58% Dice. Moreover, we found that the performance of PL is even lower than that of supervised method, which may be due to its failure to consider the reliability of pseudo labels. Figure 2 displays the visualization results of different semi-supervised methods, which reveals that our method has better ability to segment small blood vessels and learn the details of retinal vessels.

In addition, by observing Table 2, we can see that the semi-supervised approach consistently behaves better than the supervised learning. Therefore, the point consistency of unlabeled data is explored in our method, which gains better performance. And the more unlabeled images are employed, the better results we can get.

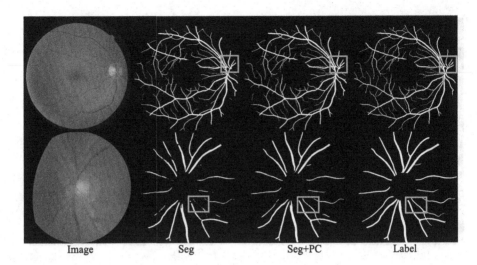

Fig. 5. Visualization results of ResUNet backbone. The first row is DRIVE and the second row is TR160 dataset.

Table 5. Results of the different rates of down-sampling (D). The best results are highlighted in bold.

Dataset	D	Acc (%)	Sen (%)	Sp (%)	AUC (%)	Dice (%)
DRIVE	1	95.49	80.01	**97.79**	97.94	81.78
	2	**95.69**	82.20	97.69	**98.20**	**82.83**
	3	95.64	**82.21**	97.64	98.16	82.68
	4	95.60	81.71	97.66	98.12	82.45
TR160	1	96.84	80.65	98.43	98.90	82.03
	2	**97.32**	86.12	98.42	**99.18**	**85.15**
	3	97.08	81.94	**98.56**	99.01	83.33
	4	96.81	**88.81**	97.60	98.85	83.29

The Effectiveness of PC for Supervised Learning. Table 3 shows the effectiveness of our PC module on two datasets and three commonly used backbone networks (UNet [11], UNet++ [28], ResUNet [27]) for supervised learning. From Table 3, we can see that our method achieves superior performance compared with backbone segmentation networks by introducing the PC module. The visualization result of vessel segmentation is shown in Fig. 3, which shows that the 'w IoU' *i.e.* our proposed supervised method produces more fine vessel segmentation results. Figure 4 and Fig. 5 show the visualization results of UNet++ and ResUNet, thanks to our point consistency module ('+PC'), our model can distinguish small vessels and branches as depicted in red boxes.

The Ablation Study of IoUHead. Table 4 shows the ablation experiment results of IoUHead with UNet as the backbone, we can see that the introduction of IoUHead improves the retinal vessel segmentation performance on two datasets with different resolutions. The experimental results demonstrate that adding IoUHead is more effective than using only two heads in the PC module.

The Effect of Different Down-Sampling Rates. We compare the vessel segmentation performance influenced by different rates of down-sampling (D) using the UNet backbone. The experimental results are shown in Table 5. By observing Table 5, we discover that $D = 2$ yields better performance in most situations, hence we set D to 2 in this study.

4 Conclusion

In this paper, we propose a novel point consistency-based semi-supervised framework for retinal vessel segmentation. Especially, the PC module can capture the point consistency of vessels by coordinating the interactive information of pixel classification score and position coordinates. In addition, our method effectively fosters the consistency regularization of unlabeled data to further improve the performance of the vessel segmentation. The experimental results qualitatively and quantitatively demonstrate the superiority of our model. Our large dataset can also pursue more new chances in retinal image analysis. In the future, we will further explore the feasibility of point consistency in other related tasks and datasets.

Acknowledgement. This work was supported by the Beijing Natural Science Foundation under Grant Z200024, in part by Hefei Innovation Research Institute, Beihang University, and in part by the University Synergy Innovation Program of Anhui Province under Grant GXXT-2019-044.

References

1. Fraz, M.M., et al.: Blood vessel segmentation methodologies in retinal images - a survey. Comput. Meth. Programs Biomed. **108**(1), 407–433 (2012)
2. Fu, H., Xu, Y., Lin, S., Kee Wong, D.W., Liu, J.: DeepVessel: retinal vessel segmentation via deep learning and conditional random field. In: Ourselin, S., Joskowicz, L., Sabuncu, M.R., Unal, G., Wells, W. (eds.) MICCAI 2016. LNCS, vol. 9901, pp. 132–139. Springer, Cham (2016). https://doi.org/10.1007/978-3-319-46723-8_16
3. Grandvalet, Y., Bengio, Y.: Semi-supervised learning by entropy minimization. In: Advances in Neural Information Processing Systems, vol. 17 (2004)
4. Guo, C., Szemenyei, M., Hu, Y., Wang, W., Zhou, W., Yi, Y.: Channel attention residual U-Net for retinal vessel segmentation. In: ICASSP, pp. 1185–1189. IEEE (2021)
5. Hu, J., Wang, H., Wang, J., Wang, Y., He, F., Zhang, J.: SA-Net: a scale-attention network for medical image segmentation. PLoS ONE **16**(4), e0247388 (2021)

6. Lee, D.H., et al.: Pseudo-label: the simple and efficient semi-supervised learning method for deep neural networks. In: Workshop on Challenges in Representation Learning, ICML, vol. 3, p. 896 (2013)
7. Lin, T.Y., Goyal, P., Girshick, R., He, K., Dollár, P.: Focal loss for dense object detection. In: Proceedings of the IEEE International Conference on Computer Vision, pp. 2980–2988 (2017)
8. Liu, Y., Shen, J., Yang, L., Yu, H., Bian, G.: Wave-Net: a lightweight deep network for retinal vessel segmentation from fundus images. Comput. Biol. Med. **152**, 106341 (2023)
9. Lokhande, V.S., Tasneeyapant, S., Venkatesh, A., Ravi, S.N., Singh, V.: Generating accurate pseudo-labels in semi-supervised learning and avoiding overconfident predictions via Hermite polynomial activations. In: Proceedings of the IEEE/CVF Conference on Computer Vision and Pattern Recognition, pp. 11435–11443 (2020)
10. Luo, X., Chen, J., Song, T., Wang, G.: Semi-supervised medical image segmentation through dual-task consistency. In: Proceedings of the AAAI Conference on Artificial Intelligence, vol. 35, pp. 8801–8809 (2021)
11. Ronneberger, O., Fischer, P., Brox, T.: U-Net: convolutional networks for biomedical image segmentation. In: Navab, N., Hornegger, J., Wells, W.M., Frangi, A.F. (eds.) MICCAI 2015. LNCS, vol. 9351, pp. 234–241. Springer, Cham (2015). https://doi.org/10.1007/978-3-319-24574-4_28
12. Souly, N., Spampinato, C., Shah, M.: Semi supervised semantic segmentation using generative adversarial network. In: Proceedings of the IEEE International Conference on Computer Vision, pp. 5688–5696 (2017)
13. Staal, J., Abràmoff, M.D., Niemeijer, M., Viergever, M.A., Van Ginneken, B.: Ridge-based vessel segmentation in color images of the retina. IEEE Trans. Med. Imaging **23**(4), 501–509 (2004)
14. Tarvainen, A., Valpola, H.: Mean teachers are better role models: weight-averaged consistency targets improve semi-supervised deep learning results. In: Advances in Neural Information Processing Systems, vol. 30 (2017)
15. Verma, V., et al.: Interpolation consistency training for semi-supervised learning. Neural Netw. **145**, 90–106 (2022)
16. Vu, T.H., Jain, H., Bucher, M., Cord, M., Pérez, P.: ADVENT: adversarial entropy minimization for domain adaptation in semantic segmentation. In: Proceedings of the IEEE/CVF Conference on Computer Vision and Pattern Recognition, pp. 2517–2526 (2019)
17. Wang, D., Zhang, Z., Zhao, Z., Liu, Y., Chen, Y., Wang, L.: PointScatter: point set representation for tubular structure extraction. In: Avidan, S., Brostow, G., Cissé, M., Farinella, G.M., Hassner, T. (eds.) Computer Vision, ECCV 2022, Part XXI. LNCS, vol. 13681, pp. 366–383. Springer, Cham (2022). https://doi.org/10.1007/978-3-031-19803-8_22
18. Wang, Y., et al.: Semi-supervised semantic segmentation using unreliable pseudo-labels. In: Proceedings of the IEEE/CVF Conference on Computer Vision and Pattern Recognition, pp. 4248–4257 (2022)
19. Wong, T.Y., Klein, R., Klein, B.E., Tielsch, J.M., Hubbard, L., Nieto, F.J.: Retinal microvascular abnormalities and their relationship with hypertension, cardiovascular disease, and mortality. Surv. Ophthalmol. **46**(1), 59–80 (2001)
20. Wu, H., Prasad, S.: Semi-supervised deep learning using pseudo labels for hyperspectral image classification. IEEE Trans. Image Process. **27**(3), 1259–1270 (2017)
21. Wu, L., et al.: R-DROP: regularized dropout for neural networks. Adv. Neural. Inf. Process. Syst. **34**, 10890–10905 (2021)

22. Wu, Y., Xia, Y., Song, Y., Zhang, Y., Cai, W.: Multiscale network followed network model for retinal vessel segmentation. In: Frangi, A.F., Schnabel, J.A., Davatzikos, C., Alberola-López, C., Fichtinger, G. (eds.) MICCAI 2018. LNCS, vol. 11071, pp. 119–126. Springer, Cham (2018). https://doi.org/10.1007/978-3-030-00934-2_14

23. Yau, J.W., et al.: Global prevalence and major risk factors of diabetic retinopathy. Diab. Care **35**(3), 556–564 (2012)

24. Yu, L., Wang, S., Li, X., Fu, C.-W., Heng, P.-A.: Uncertainty-aware self-ensembling model for semi-supervised 3D left atrium segmentation. In: Shen, D., et al. (eds.) MICCAI 2019. LNCS, vol. 11765, pp. 605–613. Springer, Cham (2019). https://doi.org/10.1007/978-3-030-32245-8_67

25. Zhang, Y., Chung, A.C.S.: Deep supervision with additional labels for retinal vessel segmentation task. In: Frangi, A.F., Schnabel, J.A., Davatzikos, C., Alberola-López, C., Fichtinger, G. (eds.) MICCAI 2018, Part II 11. LNCS, vol. 11071, pp. 83–91. Springer, Cham (2018). https://doi.org/10.1007/978-3-030-00934-2_10

26. Zhang, Y., Yang, L., Chen, J., Fredericksen, M., Hughes, D.P., Chen, D.Z.: Deep adversarial networks for biomedical image segmentation utilizing unannotated images. In: Descoteaux, M., Maier-Hein, L., Franz, A., Jannin, P., Collins, D.L., Duchesne, S. (eds.) MICCAI 2017. LNCS, vol. 10435, pp. 408–416. Springer, Cham (2017). https://doi.org/10.1007/978-3-319-66179-7_47

27. Zhang, Z., Liu, Q., Wang, Y.: Road extraction by deep residual U-Net. IEEE Geosci. Remote Sens. Lett. **15**(5), 749–753 (2018)

28. Zhou, Z., Rahman Siddiquee, M.M., Tajbakhsh, N., Liang, J.: UNet++: a nested U-Net architecture for medical image segmentation. In: Stoyanov, D., et al. (eds.) DLMIA/ML-CDS - 2018. LNCS, vol. 11045, pp. 3–11. Springer, Cham (2018). https://doi.org/10.1007/978-3-030-00889-5_1

29. Zhu, X., Goldberg, A.B.: Introduction to Semi-supervised Learning. Synthesis Lectures on Artificial Intelligence and Machine Learning, vol. 3, no. 1, pp. 1–130 (2009)

Knowledge Distillation of Attention and Residual U-Net: Transfer from Deep to Shallow Models for Medical Image Classification

Zhifang Liao, Quanxing Dong, Yifan Ge, Wenlong Liu, Huaiyi Chen, and Yucheng Song(✉) ⓘ

School of Computer Science and Engineering,
Central South University, Hunan 410083, China
214711072@csu.edu.cn

Abstract. With the widespread application of deep learning in medical image analysis, its capacity to handle high-dimensional and complex medical images has been widely recognized. However, high-accuracy deep learning models typically demand considerable computational resources and time, while shallow models are generally unable to compete with complex model. To overcome these challenges, this paper introduces a Knowledge Distillation methodology that merges features and soft labels, transfering knowledge encapsulated in the intermediate features and predictions of the teacher model to the student model. The student model imitates the teacher model's behavior, thereby improving its prediction accuracy. Based on this, we propose the Res-Transformer teacher model bases on the U-Net architecture and the ResU-Net student model incorporates Residuals. The Res-Transformer model employs dual Attention to acquire deep feature maps of the image, and subsequently employs Hierarchical Upsampling to restore the details in these feature maps. The ResU-Net model enhances stability via Residuals and recovers the loss of image information in convolution operations through optimized skip-connection. Finally, we evaluate on multiple disease datasets. The results show that the Res-Transformer achieves accuracy up to 94.3%. By applying knowledge distillation, abundant knowledge from the Res-Transformer is transferring knowledge of the Res-Transformer to the ResU-Net model, improving its accuracy up to 7.1%.

Keywords: Knowledge distillation · Medical image · U-Net · Residual · Attention

1 Introduction

In recent years, medical imaging has become an essential tool in healthcare, providing clinicians with a non-invasive means to diagnose diseases. Among various imaging methods, radiological images and histopathological images have been

Q. Liu et al. (Eds.): PRCV 2023, LNCS 14437, pp. 162–173, 2024.
https://doi.org/10.1007/978-981-99-8558-6_14

extensively used to detect and classify diseases, such as lung diseases and breast cancer [1]. However, the interpretation of these images requires expert knowledge and is time-consuming. To improve diagnostic efficiency, applying deep learning technologies to medical image analysis has attracted people's attention.Prior to deep learning, rule-based methods and knowledge-based methods were the primary approaches to medical image analysis [2]. Rule-based methods fundamentally rely on designing a fixed set of rules to parse imaging data, requiring manual design and feature selection. However, these methods are time-consuming, error-prone, and demonstrate weak adaptability to data variation [3]. Knowledge-based methods incorporate expert medical knowledge to enhance the precision and reliability of medical image analysis. However, these methods often rely on comprehensive and precise expert knowledge. Acquiring such knowledge in practice can be challenging, and encoding this knowledge into an algorithm poses a substantial task [4].

A common limitation shared by the aforementioned methods is their inability to handle the high-dimensional and highly variable nature of medical images [5], an issue effectively addressed by the advent of deep learning. In recent years, numerous studies have successfully integrated deep learning technologies with medical image analysis. For instance, Rajpurkar et al. [6] utilized a network architecture named CheXNet, a variation of the 121-layer convolutional neural network structure. Thry conducted experiments for detecting lung diseases on the ChestX-ray14 dataset and achieved results comparable to those of radiologists. Wang et al. [7] employed a patch-based classification stage of deep learning for classifying and locating tumors in breast cancer tissue slices. Combining the predictive results of the deep learning system with the diagnosis of pathologists, they increased the AUC to 0.995. Pathak et al. [8] employed a top-2 smooth loss function with cost-sensitive properties to handle noisy and imbalanced COVID-19 dataset. Despite the significant advancements made in the application of deep learning to medical image analysis, these complex models often require extensive computational resources and have a long inference time, meaning that they are unable to meet the timeliness required in clinical diagnosis.

To overcome the challenges, this paper proposes a method of knowledge distillation that combines features and soft labels, transferring the knowledge contained in the intermediate layers and prediction results of the teacher model to the student model. Two models are designed: the teacher model, a complex Res-Transformer model based on the U-Net structure, and the student model, a shallow ResU-Net model based on residual modules. The Res-Transformer teacher model uses a dual Attention to obtain deep feature maps of the images, and employs Hierarchical Upsampling and skip-connection to recover crucial details in the feature maps, aiming to achieve high-accuracy results. However, due to its complex structure, it entails an extensive inference time, making it unsuitable for using in clinical diagnosis. The ResU-Net student model, on the other hand, lowers the coupling between layers to reduce model complexity by optimizing skip-connections and ensures model stability via Residuals. With its simplified structure and reduced computational load, it is more suitable for use in clinical diagnosis. The overall architecture of the method is shown in Fig. 1.

In conclusion, the main contributions of this paper are as follows:

1. We propose a knowledge distillation method integrating features and soft labels, thus improving the performance of student model by transferring abstract feature knowledge and specific class knowledge from the teacher model.
2. We propose two U-Net-based models, the Res-Transformer teacher model, and the ResU-Net student model. The former extracts feature information from medical images using a dual attention mechanism and restores feature map details via Hierarchical Upsampling, while the latter optimizes skip-connections and introduces Residual for stability.
3. Demonstrated improvement in prediction accuracy of the student model after knowledge distillation, with promising results achieved on multiple datasets.

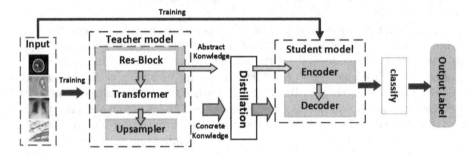

Fig. 1. The overall architecture of the proposed knowledge distillation method and the two models. Medical imaging data is first input into the teacher model for training. The knowledge in the intermediate layers and prediction results of the teacher model are then transferred to the student model through distillation.

The rest of the paper is organized as follows: Sect. 2 is the introduction to Teacher models,Student model and distillation method; Sect. 3 gives experiments and results;Sect. 4 provides a detailed discussion of the system.

2 Methods

In our proposed method of knowledge distillation, the student model is trained to mimic the intermediate layer outputs and prediction results of the teacher, learning the abstract knowledge and specific knowledge contained within, thereby improving the performance of the student model. To facilitate the learning of the knowledge encapsulated within the intermediate layers, both models are designed with a U-Net-based encoder-decoder structure. Through the encoder, the models extract image features, while the decoder restores image details, eventually generating image class information.The teacher model, with its relatively complex structure, is capable of capturing complex patterns and features within images. In contrast, the student model, although simpler in structure than the teacher model, can significantly enhance prediction accuracy via knowledge distillation.

2.1   Res-Transformer Teacher Model Based on U-Net Structure

The teacher model employes Attention-based Residual Blocks to extract critical features in encoder. After feature extraction, we tokenize the feature into a series of patches. These patches are subsequently subjected to the Transformer's encoders, enabling the weighted encoding of global context information. In decoder stage, we use a resolution-optimized approach, gradually increasing the resolution of the feature map through Hierarchical Upsampling and skip-connections with Residual Blocks. This approach ensures the restoration of detailed information in the images.

Res-Transformer Encoder Utilizes Dual Attention. In addressing the gradient vanishing problem associated with complex models [9], we introduce Residual Blocks as feature extractors in the teacher model, with three stages, each consisting of numerous Residual Blocks. Within these, the outputs of the layers are added to their original inputs, forming Residual connections. Images was encoded via this sequence of Residual blocks, generating a feature map.To enhance the attention towards global information during feature extraction, Attention are introduced within the Residual Block. This Attention computes the global significance of each feature [10], enabling the model to pay greater attention to significant portions of the image. After producing feature map, it is critical to capture the global information among different feature map sections. We tokenize the feature map x into a series of sequence of flattened 2D patches $\left\{ x_p^i \in R^{p^2 C} \mid i = 1, \ldots, N \right\}$, where each patches is of size of $P \times P$. To preserve positional information during the process, position embeddings is added to each patch as follow:

$$z_0 = [x_p^1; x_p^2; \ldots x_p^N] + E_{position} \tag{1}$$

where $E_{position}$ is the patch embedding projection. Then we input the vectors into the Transformer encoder, consists of L layers of Multihead Self-Attention (MSA) and Multi-layer Perceptron (MLP), the output of each layer are as follows:

$$z_l' = MSA(LN(z_{l-1})) + z_{l-1}, \tag{2}$$

$$z_l = MLP(LN(z_l')) + z_l' \tag{3}$$

where LN denotes layer normalization, $Z_l$is the output of the l-th Transformer encoder. This feature vector contains the global information of the image.

Hierarchical Upsampling and Skip-Connection. During the process of encoding images, continuous sampling and compression can lead to the loss of some details within the images. In medical imaging, such details often have the potential to directly contribute to global predictions. Therefore, we designed a decoder that incorporates a Hierarchical Upsampling structure. This decoder consists of multiple stages, which consists of a upsampling, a convolution block and a ReLU layer. These stages gradually elevate the output from the encoder from a lower resolution back to the full resolution of the original image. In this process, we implemented skip-connections with the intermediate feature maps of the Residual Blocks which assists in the recovery of a greater amount of detail [5]. The overall architecture of Res-Transformer is shown in Fig. 2.

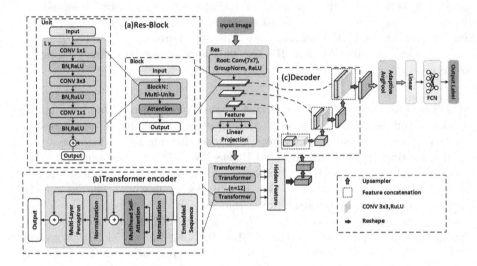

Fig. 2. The overall architecture of teacher model Res-Transformer. (a) Right part is overall architecture of three stages of Residual Blocks,which consists of multi Residual Units. The three stages respectively use 3, 4, and 9 Residual Units. (a) Left part is a single Residual Unit. (b) Architecture of a single transformer encoder. Res-Transformer contains 12 transformer encoders. (c) Architecture of hierarchical upsampling decoder.

2.2   ResU-Net Student Model Incorporates Residual

The student model is a lightweight model structured on the U-Net architecture. The model introduces Residual to design its encoder, with the aim of enhancing network stability. We recover spatial information lost in convolution operations via optimizing the way of skip-connections, which aims to improve learning efficiency and reduce the computational load of the model.

Optimized Skip-Connection to Enhance Spatial Information Recovery. The skip-connections in U-Net allow neural networks to restore spatial information lost during pooling operations, but unable to recover spatial information lost during convolution operations [11]. Moreover, skip-connections in U-Net are directly linked to convolution layers, leading to a high level of coupling within skip-connections. During the backpropagation, both skip-connection and downsampling can affect the optimization of parameters, leading to a decrease in the model's learning speed and efficiency [12]. We optimize the skip-connections between the encoder and decoder in the student model. The intermediate layer output of the encoder is directly connected to the output layer of the model, recovering the spatial information lost during convolution operations and reducing the coupling between the convolution layer and skip-connections, as shown in the Fig. 3.

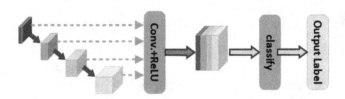

Fig. 3. The optimized skip-connection method. The outputs from each layer of the encoder are concatenated into output layer.

Residual Connection Method to Enhance Stability. The depth of a neural network is critically important for deep learning. However, due to the gradient vanishing problem, as the depth increases, there comes a point where classification begins to decline [13]. To address this issue, we employ Residual in the encoder to process the input data. The definition of the Residual is as follows:

$$x_{l+1} = F(y_l) \tag{4}$$

$$y_l = h(x_l) + F(x_l, w_l) \tag{5}$$

where x_1 and w_l represents the input and the network parameters of the l-th layer, F represents the mapping relationship, y_l is the output of the l-th layer, $h(x_l)$ and $F(y_l)$ represent the identity mapping and the ReLU function.

After extracting key features through encoder, we use multiple layers of upsampling to recover the lost details in the feature map. The feature map from the final layer of the network is then subjected to global average pooling, transforming it into a vector. The overall architecture of the ResU-Net is shown in Fig. 4.

2.3   Knowledge Distillation

Teacher model and student models mentioned employ an encoder-decoder structure. Student model learns feature knowledge via mimic output from the corresponding layers of teacher model, and learns specific knowledge from variations in the probability of the teacher model's prediction. Consequently, we introduce a method incorporating both features and soft labels for knowledge distillation.

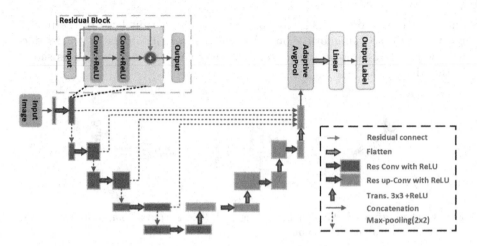

Fig. 4. The overall architecture the ResU-Net model based on Residual. The Residual Blocks shown in gray are employed in the model. Intermediate feature maps produced during the encoding are appended to the decoder's output through skip-connections. (Color figure online)

Joint Distillation of Features and Soft Labels. The distillation strategy we proposed includes two parts: abstract knowledge feature distillation and concrete knowledge soft label distillation. Feature distillation allows the student model to mimic the intermediate feature representations of the teacher model. In our method, both student and teacher models consist of three layers of encoders and decoders. The feature outputs of the intermediate layers of the encoders often harbor critical information, important to the prediction. The output from the l-th layer of the encoders in the teacher and student models is T^l and S^l, feature loss is as follows:

$$L_{feature} = \frac{1}{L} \sum_{l=1}^{L} \frac{1}{H \times W \times C} (\sum_{i=1}^{H} \sum_{j=1}^{W} \sum_{k=1}^{C} (T_{ijk}^l - S_{ijk}^l)^2) \qquad (6)$$

soft label distillation allows the student model to learn from variations in the probability of the teacher's prediction. This approach enables the student model to learn the decision boundaries of the teacher model, thereby acquiring specific class knowledge [14]. We use the distillation temperature T to soften the output $z = [z_1, z_2, ..., z_n]$ of the teacher model, using the distillation temperature T, transforming it into a smooth probability distribution $t = [t_1, t_2, ..., t_n]$,

$$t_i = \frac{e^{\frac{z_i}{T}}}{\sum_j e^{\frac{z_i}{T}}} \qquad (7)$$

the soft label loss during learning is calculated by Cross Entropy of t and output of student $s = [s_1, s, ..., s_n]$, soft label loss is as follows:

$$L_{label} = - \sum_i t_i \log(s_i) \qquad (8)$$

we use a weight coefficient α to combine these loss function, distillation loss function $L_{distillation}$ as follow:

$$L_{distillation} = (1 - \alpha)\, L_{label} + \alpha L_{feature} \tag{9}$$

Distillation Process. To enable the student model to learn from the teacher model and fit the original data, we combine the distillation loss with the original data loss. We use cross-entropy to calculate the difference between the predicted results $y = [y_1, y_2, ..., y_n]$ of the student model and the actual results $z = [z_1, z_2, ..., z_n]$ combining with the distillation loss mentioned, and we use weight coefficient β to represent the contribution of distillation loss and origin data loss.

$$Loss = \beta\, L_{distillation} + (1 - \beta)(-\sum_{i=1}^{n} Z_i \log(y_i)) \tag{10}$$

The overall structure is shown in Fig. 5.

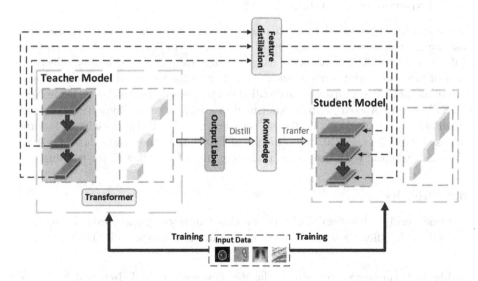

Fig. 5. Overall structure of the knowledge distillation. Teacher transfers knowledge from each layer to corresponding layer of the student, and distills the class information output into the student model.

3   Data and Experiments

3.1   Datasets

We conducted experiments on three publicly medical imaging datasets and a private dataset provided by our collaborating hospital.

COVID-19 Radiography Dataset: This dataset contains 3,616 COVID-19 positive cases, 10,192 normal cases, 6,012 cases of lung opacity, and 1,345 cases of viral pneumonia [15].

BreaKHis Dataset: This database contains 9,109 microscopic images of breast tumor tissues collected from 82 patients using different magnification factors. It includes 2,480 benign samples and 5,429 malignant samples [16].

IDC Dataset: This is a histopathological dataset originating from the scanned images of 162 full-slide Breast Cancer. It contains 277,524 image patches were extracted, with 198,738 IDC-negative and 78,786 IDC-positive patches [17].

Brain Stroke Dataset: This dataset, provided by our collaborating hospital, includes images from 28 cerebral hemorrhage patients, 47 cerebral infarction patients, and 35 healthy individuals.

3.2   Experimental Settings

We use accuracy and AUC (area under curve) to evaluate each method, and assess the resource usage of the proposed method via Parameter count, Average Inference Time (AIT), and GPU usage during training. We validated effectiveness of the proposed teacher, student models, and the proposed knowledge distillation method. We compared the distilled student method with U-Net, DenseNet, UnSTNet, and AlexNet. After experimental analysis, we set hyperparameter of distillation as following, distillation temperature T=20, the weight of feature and soft label loss $\alpha = 0.7$, the weight of distillation and origin data $\beta = 0.4$. We used Adagrad optimizer with a learning rate of $l_r = 0.0005$ for training.

3.3   Results

We first verified the effectiveness of the Res-Transformer and ResU-Net models as well as the effectiveness of knowledge distillation, as shown in Table 1.

Table 1. Performance obtained by the Res-Transformer, ResU-Net, and ResU-Net after knowledge distillation.

Models	COVID-19		BreaKHis		IDC		Brain Stroke	
	ACC	AUC	ACC	AUC	ACC	AUC	ACC	AUC
ResU-Net (student origin)	84.9	93.4	83.7	92.2	82.8	86.6	76.4	81.2
Res-Transformer (teacher)	94.3	98.4	93.7	97.3	93.7	93.4	82.9	77.2
ResU-Net (student-KD)	**91.2 (6.6%)**	**96.9**	**89.7 (7.1%)**	**94.3**	**88.1 (6.4%)**	**90.1**	**80.2 (4.5%)**	**84.1**

Table 1 shows that Res-Transformer presents the highest accuracy of 94.3%, with good performance in other datasets. The distilled student model learn from

Table 2. Comparsion with other state-of-the-art methods.

Models	Parameter count	COVID-19		BreaKHis		IDC		Brain Stroke	
		ACC	AUC	ACC	AUC	ACC	AUC	ACC	AUC
U-Net	9.526M	83.1	78.7	81.8	89.2	84.4	87.6	–	–
AlexNet [18]	36.871M	–	–	86.3	82.3	91.4	85.4	–	–
UnSTNet [19]	17.847M	90.3	97.2	90.6	93.8	86.6	89.0	–	–
DenseNet [20]	24.356M	88.7	94.1	87.2	85.6	87.7	87.4	–	–
ResU-Net(ours)	**8.032M**	**91.2**	**96.9**	**89.7**	**94.3**	**88.1**	**90.1**	**80.2**	**84.1**

the teacher model, improving its accuracy by up to 7.1%. We then compared the distilled Student model with stage-of-art deep learning models in medical imaging, as shown in Table 2.

Our method achieves 91.2% and 96.9% performance in terms of accuracy and area under the curve for the COVID-19 dataset, better than UnSTNet's 90.3% and Dense-Net's 88.7%. For the BreaKHis dataset, the accuracy reached 89.7%, better than AlexNet's 86.3%, close to UnSTNet's 90.6%, but with only 45% of its parameters. We compared the inference time and GPU usage of each model, as shown in Table 3.

Table 3. Average interface time(s) and GPU(GB) usage for all methods.

Models	COVID-19		BreaKHis		IDC		Brain Stroke	
	AIT	GPU	AIT	GPU	AIT	GPU	AIT	GPU
U-Net	0.19	3.0	0.14	3.4	0.17	0.17	–	–
AlexNet	–	–	0.51	11.1	0.48	5.6	–	–
UnSTNet	0.34	5.6	0.33	6.4	0.27	4.3	–	–
DenseNet	0.45	7.2	0.38	8.6	0.33	4.5	–	–
ResU-Net(ours)	**0.17**	**3.3**	**0.12**	**3.9**	**0.12**	**2.4**	**0.10**	**2.1**

Table 3 show that ResU-Net has the smallest inference time and GPU usage. With the similar performance, the inference time is reduced by about 51%. The loss function during COVID-19 Radiography Dataset and change in accuracy with the hyperparameter α are shown in the following Fig. 6.

From Fig. 6 (a), it can be seen that due to the guidance of the trained Res-Transformer, compared with other models, the initial training loss of ResU-Net-KD is the smallest, and its convergence speed is the fastest. Figure 6 (b) shows that when the distillation parameter α is set to 0.7, the best results are achieved.

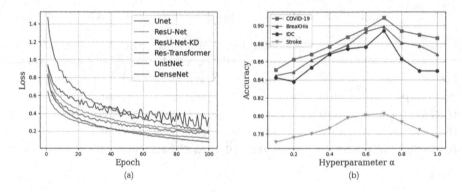

Fig. 6. The loss function during COVID-19 Radiography Dataset and change in accuracy with the hyperparameter α. (a) Losses during training on the COVID-19 Radiography Dataset for all methods used in our experiments.(b) Illustrates the relationship between the accuracy of the student model and the changes in the hyperparameter α.

4   Conclusion

In this paper, we proposed a knowledge distillation method based on joint features and soft labels, which transmits abstract and concrete knowledge from the teacher to student. Two U-Net architecture models are proposed: the Res-Transformer teacher model and the ResU-Net student model. The former is more complex and demands higher computational resources, with better performance. The latter learns from teacher model via knowledge distillation. Our experiments show that the distilled student model not only performs better, but also requires fewer computational resources and less inference time.

Acknowledgement. This work was supported by National Natural Science Foundation of China, Regional Science Fund Project, No: 72264037.

References

1. LeCun, Y., Bengio, Y., Hinton, G.: Deep learning. Nature **521**(7553), 436–444 (2015)
2. Shen, D., Wu, G., Suk, H.I.: Deep learning in medical image analysis. Ann. Rev. Biomed. Eng. **19**, 221–248 (2017)
3. Yao, L., Mao, C., Luo, Y.: Clinical text classification with rule-based features and knowledge-guided convolutional neural networks. BMC Med. Inf. Decis. Mak. **19**(3), 31–39 (2019)
4. Li, C.Y., Liang, X., Hu, Z., Xing, E.P.: Knowledge-driven encode, retrieve, paraphrase for medical image report generation. In: Proceedings of the AAAI Conference on Artificial Intelligence, vol. 33, pp. 6666–6673 (2019)
5. Tajbakhsh, N., Jeyaseelan, L., Li, Q., Chiang, J.N., Wu, Z., Ding, X.: Embracing imperfect datasets: a review of deep learning solutions for medical image segmentation. Med. Image Anal. **63**, 101693 (2020)

6. Rajpurkar, P., et al.: Chexnet: radiologist-level pneumonia detection on chest x-rays with deep learning. arXiv preprint arXiv:1711.05225 (2017)

7. Wang, D., Khosla, A., Gargeya, R., Irshad, H., Beck, A.H.: Deep learning for identifying metastatic breast cancer. arXiv preprint arXiv:1606.05718 (2016)

8. Pathak, Y., Shukla, P.K., Tiwari, A., Stalin, S., Singh, S.: Deep transfer learning based classification model for covid-19 disease. IRBM **43**(2), 87–92 (2022)

9. Rossini, N., Dassisti, M., Benyounis, K., Olabi, A.G.: Methods of measuring residual stresses in components. Mater. Des. **35**, 572–588 (2012)

10. Vaswani, A., et al.: Attention is all you need. Adv. Neural Inf. Process. Syst. **30** (2017)

11. Siddique, N., Paheding, S., Elkin, C.P., Devabhaktuni, V.: U-net and its variants for medical image segmentation: a review of theory and applications. IEEE Access **9**, 82031–82057 (2021)

12. Du, G., Cao, X., Liang, J., Chen, X., Zhan, Y.: Medical image segmentation based on u-net: a review. J. Imaging Sci. Technol. (2020)

13. He, K., Zhang, X., Ren, S., Sun, J.: Deep residual learning for image recognition. In: Proceedings of the IEEE Conference on Computer Vision and Pattern Recognition, pp. 770–778 (2016)

14. Hinton, G., Vinyals, O., Dean, J.: Distilling the knowledge in a neural network. arXiv preprint arXiv:1503.02531 (2015)

15. Rahman, T., et al.: Exploring the effect of image enhancement techniques on covid-19 detection using chest x-ray images. Comput. Biol. Med. **132**, 104319 (2021)

16. Spanhol, F.A., Oliveira, L.S., Petitjean, C., Heutte, L.: A dataset for breast cancer histopathological image classification. IEEE Trans. Biomed. Eng. **63**(7), 1455–1462 (2015)

17. Cruz-Roa, A., et al.: Automatic detection of invasive ductal carcinoma in whole slide images with convolutional neural networks. In: Medical Imaging 2014: Digital Pathology, vol. 9041, p. 904103. SPIE (2014)

18. Liu, M., et al.: A deep learning method for breast cancer classification in the pathology images. IEEE J. Biomed. Health Inf. **26**(10), 5025–5032 (2022)

19. Fan, Y., Gong, H.: An improved tensor network for image classification in histopathology. In: Pattern Recognition and Computer Vision: 5th Chinese Conference, PRCV 2022, Shenzhen, China, 4–7 November 2022, Proceedings, Part II, pp. 126–137. Springer, Heidelberg (2022). https://doi.org/10.1007/978-3-031-18910-4_11

20. Li, X., Shen, X., Zhou, Y., Wang, X., Li, T.Q.: Classification of breast cancer histopathological images using interleaved densenet with senet (idsnet). PLoS ONE **15**(5), e0232127 (2020)

Two-Stage Deep Learning Segmentation for Tiny Brain Regions

Yan Ren[1,2], Xiawu Zheng[3], Rongrong Ji[3], and Jie Chen[1,2,4(✉)]

[1] School of Electronic and Computer Engineering,
Peking University, Shenzhen, China
[2] AI for Science (AI4S)-Preferred Program,
Peking University Shenzhen Graduate School, Shenzhen, China
[3] Media Analytics and Computing Lab, Department of Artificial Intelligence,
School of Informatics, Xiamen University, Xiamen, China
[4] Peng Cheng Laboratory, Shenzhen, China
`jiechen2019@pku.edu.cn`

Abstract. Accurate segmentation of brain regions has become increasingly important in the early diagnosis of brain diseases. Widely used methods for brain region segmentation usually rely on atlases and deformations, which require manual intervention and do not focus on tiny object segmentation. To address the challenge of tiny brain regions segmentation, we propose a two-stage segmentation network based on deep learning, using both 2D and 3D convolution. We first introduce the concept of the Small Object Distribution Map (SODM), allowing the model to perform coarse-to-fine segmentation for objects of different scales. Then, a contrastive loss function is implemented to automatically mine difficult negative samples, and two attention modules are added to assist in the accurate generation of the small object distribution map. Experimental results on a dataset of 120 brain MRI demonstrate that our method outperforms existing approaches in terms of objective evaluation metrics and subjective visual effects and shows promising potential for assisting in the diagnosis of brain diseases.

Keywords: Brain Region Segmentation · Two-stage Segmentation · Small Object Distribution Map

1 Introduction

Brain diseases are often associated with damage and dysfunction in specific brain tissues or structures, posing a significant threat to people's health [1]. The structures and tissues in the brain that need to be segmented include not only large objects such as white matter and gray matter but also tiny objects such as the hippocampus and optic nerve. MRI (Magnetic Resonance Imaging) [11] is a non-invasive medical imaging technique used to generate high-resolution and high-contrast images of the internal structures of the human brain, facilitating the observation of brain regions and tissues.

Q. Liu et al. (Eds.): PRCV 2023, LNCS 14437, pp. 174–184, 2024.
https://doi.org/10.1007/978-981-99-8558-6_15

Widely used methods for brain MRI segmentation include atlas-based, intensity-based and surface-based segmentation algorithms. The atlas-based segmentation [2,3] establishes a one-to-one correspondence between the target new image and the atlas image by deformable alignment and transfers the segmentation labels from the atlas to the target image. The intensity-based methods analyze the intensity levels of the pixels/voxels to distinguish different brain structures or regions [4]. Surface-based methods include active contours [5,6] and level-set [7].

Recently, deep neural networks have demonstrated remarkable capabilities in analyzing complex patterns and extracting meaningful features from medical images. De Brebisson et al. [14] used a deep neural network approach to segment brain regions on the 2012 MICCAI Multi-Atlas Labeling Challenge brain image dataset and achieved a Dice similarity coefficient (DSC) of 72.5. DeepBrainSeg [8] automatically identify and segment brain regions by incorporating features at three different levels. Meng et al. [13] use a two-stage segmentation network to first segment liver then tumor. However, these methods do not specifically address the problem of segmenting tiny brain regions.

This paper proposes a two-stage automatic segmentation method for brain tissue in MR images. The method combines the advantages of 2D and 3D convolutions in segmentation tasks, designs a two-stage segmentation network structure, proposes the small objects distribution map, and ensures the segmentation speed of large objects while guaranteeing the fast and accurate segmentation of tiny ones . At the same time, the method uses a contrastive loss function to automatically mine difficult negative samples and adds two attention modules to effectively help generate the map. This paper collects MRI brain scan data from 120 patients, and the experiments on this dataset demonstrate the effectiveness of the proposed method.

In summary, this paper makes the following main contributions:

1. We use both 2D and 3D convolutions and propose a two-stage segmentation network structure that segment both large and tiny brain regions in MRI images.
2. We propose the **S**mall **O**bject **D**istribution **M**ap (SODM) that accurately identifies the regions of tiny objects, enabling the possibility of further fine segmentation.
3. The experimental results demonstrate that our method outperforms the other approaches on our MRI brain data set.

2 Method

2.1 Overall Workflow

Figure 1 illustrates the overall workflow of the proposed two-stage segmentation method. The first phase of the workflow is data preprocessing. The second phase is the two-stage segmentation. The preprocessed brain MRI are sliced and fed into a 2D large object segmentation network for coarse predictions. Based on the predictions, an SODM is generated, which will be explained in detail in Sect. 2.2. Using the regions of small objects on the distribution map, corresponding patches are extracted from the MRI. These patches represent the locations of small regions and are fed into a 3D small object segmentation network to obtain refined segmentation. Finally, the predictions from both networks are merged to obtain the multi-class segmentation results of the brain.

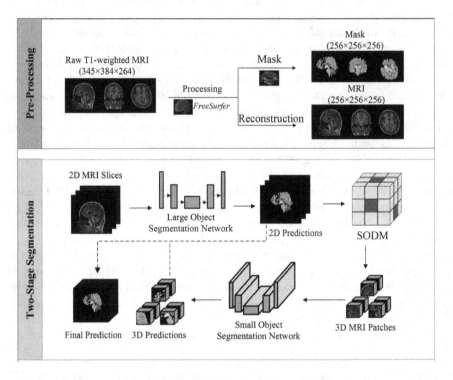

Fig. 1. Overall workflow of brain region segmentation in MRI.

2.2 Two-Stage Segmentation Network

The two-stage tiny object segmentation approach involves segmenting objects in two stages: coarse segmentation followed by fine segmentation, as shown in the Fig. 2. In the first stage, 2D convolutions operate on the image plane, which have lower computational requirements, higher efficiency, and reduced model complexity compared to 3D convolutions. This makes them suitable for fast segmentation of large objects. In the second stage, 3D convolutions are employed, taking into account the relationships between voxels in the entire volume data, for a comprehensive consideration of the relationships between adjacent voxels. 3D convolutions are better suited for precise segmentation of tiny objects. Both stages of the network are based on the nnU-Net [10] architecture, with adaptive generation of U-Net [9] network parameters depending on the input data.

In the first stage, apart from generating initial predictions for brain tissue structure segmentation, an SODM is also generated. The map finds the regions in the predictions that contain tiny objects. This map is calculated by determining the ratio of the number of voxels predicted as a specific class to the total voxels, as shown in Eq. 1.

$$\beta_i = \frac{N_i}{N_{total}}. \tag{1}$$

where N_i represents the number of voxels predicted for the ith category and N_{total} represents the overall number of non-background voxels for that image.

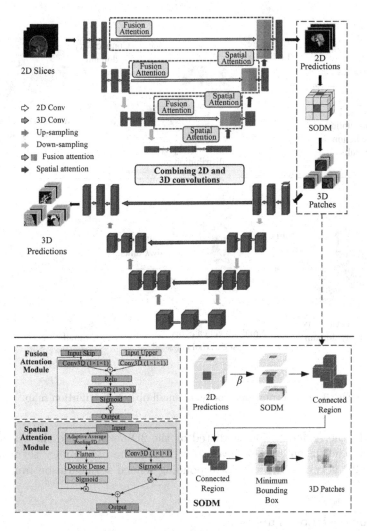

Fig. 2. Two-stage segmentation network.

As shown in Fig. 3, the first cube represents the prediction of the first-stage. In this image, each small cube represents a voxel, and each voxel has a predicted class. Next, the sum of all voxels for each class is calculated and compared to the total number of all class voxels. If the ratio is smaller than the threshold β, it is considered as a small object. Then, the minimum bounding cube enclosing all connected components of small objects is formulated, and expanded outward by k units. The corresponding region in the brain MRI is then extracted as the input patches for the second stage.

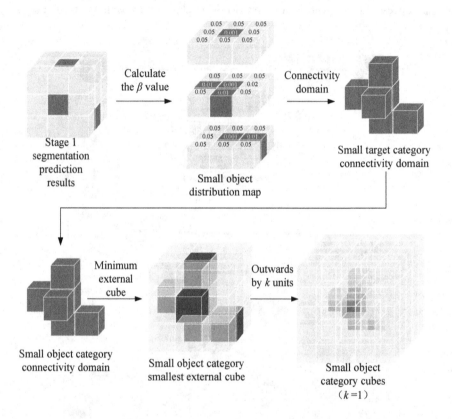

Fig. 3. The formation process of small object distribution map.

Finally, the segmentation predicted by the 2D network and the 3D network are fused. The fusion process involves replacing the preliminary segmentation from the 2D network with the corresponding refined segmentation from the 3D network.

2.3   Contrast Loss Function

We used the contrast loss function [15] in both the first stage 2-D convolutional neural network and the second stage 3-D convolutional neural network, because there are some small objects in the brain organization structure in this task, and these small objects are difficult samples in the training process, and the contrast loss function is a loss function with the self-discovery property of difficult negative samples.

Current algorithms for semantic segmentation of medical images focus on local contextual semantic information and ignore global contextual information from different samples in the training set. We use a pixel-by-pixel comparison learning loss function for structural brain organization segmentation in a fully supervised setting. The core idea is that embeddings of pixels belonging to the same semantic category are more similar than embeddings of pixels from different semantic categories.

We use a pixel-wise contrast loss function, and traditional medical image segmentation uses a pixel-wise cross-entropy loss function. The current approach is shown in Eqs. 2, 3 and 4 Starting from both regularizing the embedding space and exploring the global structure of the training data, pixel i in each image I is classified into a category $c \in C$. i^+ is a positive sample that also belongs to category c, and i^- is a negative sample that also does not belong to category c, where P_i and N_i are the pixel codes of positive and negative samples, respectively.

$$z^+ = exp\left(i \cdot \frac{i^+}{\tau}\right), \tag{2}$$

$$z^- = exp\left(i \cdot \frac{i^-}{\tau}\right), \tag{3}$$

$$L_i^{NCE} = \frac{1}{|P_i|} \sum_{i^+ \in P_i} -\log \frac{z^+}{z^+ + \sum_{i^- \in N_i} z^-}. \tag{4}$$

The advantage of using the contrast loss function is that it allows the model to have strong discriminative power and to have the ability to solve the classification of individual pixels, because it is able to disperse the distances of the embedding of different classes, making the distances of the embedding of the same classes to be clustered.

2.4 Attention Modules

To generate the SODM accurately, this study improves the first-stage network by incorporating two attention modules. The first attention module, named as spatial attention, is placed between the skip-connection layer and the decoder output layer. The second attention module is a fusion attention module added after the decoder layer, which operates in the channel-space domain to enhance feature fusion.

The structure of the spatial attention module is illustrated in the right of the Fig. 4. In the decoder layer of the first-stage network, a lightweight 3D channel-space attention module is applied at the front end of the upsampling operation. For the channel attention on the left side, assuming the size of the input is $B \times C \times W \times H \times K$, it goes through an adaptive average pooling layer, which transforms the feature map into a size of $B \times C \times 1 \times 1 \times 1$. Then, a flatten layer is applied to convert the feature map into a vector representation $B \times C$. After that, a two-layer fully connected network is used, with a Leaky ReLU activation function in the middle layer. Finally, a Sigmoid activation function is applied to obtain the weights for channel attention. These weights are multiplied with each channel of the input features, providing a weighting factor between 0 and 1 for each channel. On the right side, the spatial attention is achieved through a 3D convolution layer with a kernel size of $1 \times 1 \times 1$. The output of this convolution layer is passed through a Sigmoid activation function to obtain the weights for spatial attention. These weights are applied to each point within the feature cube of each channel, multiplying them by a value between 0 and 1. The results of both channel attention and spatial attention are then added together.

The structure of the fusion attention module is illustrated in the left of the Fig. 4. In the original U-Net [9] model, there are skip connections from the encoder to the decoder through a 3D convolutional layer. In the architecture proposed in this paper, an additional attention fusion module is introduced for the fusion of the skip connections and the upsampling output.

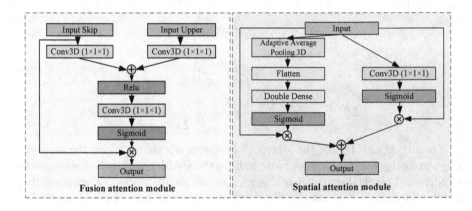

Fig. 4. Two attention modules.

By incorporating these two attention modules, it becomes effective in enhancing the segmentation of brain regions and generating the small object distribution map

3   Experiments

3.1   Dataset and Metrics

The data set for this paper contained MR head scan images from 120 patients. The patient's height ranges from 1.14 to 1.85 m, and their weight ranges from 17 to 115 kg. MRI imaging was performed using a three-dimensional t1-weighted,

10° flip angle magnetization-prepared fast gradient echo (MP-RAGE) sequence. The matrix dimension of the original t1-weighted MR images was $264 \times 384 \times 345$ (four cases are exception with dimension of $160 \times 512 \times 460$), and the voxel size was $0.67 \times 0.67 \times 0.67$ mm^3 (four cases are exception with size of $0.5 \times 0.5 \times 1$ mm^3).

Raw MRI was reconstructed as $256 \times 256 \times 256$ matrix dimensions with a voxel size of $1 \times 1 \times 1$ mm^3 and was transferred to the FreeSurfer software in a standard space to produce 44 categories of ground truth masks based on the atlas. There are 45 types of labels, and each label represents one region of the brain.

The evaluation metrics used in this paper include Dice Similarity Coefficient (DSC), Jaccard Index (JAC) [12], recall, and precision.

3.2   Comparisons Experiments

We compared the 3D U-Net [16] model without any preprocessing, the 2D nnU-Net [10] model with preprocessing and our two-stage segmentation method, and the results are shown in the Table 1 and Fig. 5.

Table 1. Comparisons of the DSC, JAC, precision and recall values on different segmentation models.

Models	Dice	Jaccard	Precision	Recall
3D U-Net [16]	11.8	7.66	18.24	11.99
2D nnU-Net [10]	80.89	72.69	80.93	81.24
Two-stage	**84.78**	**76.35**	**84.94**	**85.33**

Table 2. Performance of the two-stage segmentation network at different β values.

Method	Dice	Jaccard	Precision	Recall
2D nnU-Net [10]	80.89	72.69	80.93	81.24
$\beta = 0.02$	84.67	76.22	84.62	**85.49**
$\beta = 0.05$	84.68	76.23	84.7	85.36
$\beta = 0.1$	84.73	76.28	84.75	85.36
$\beta = 0.15$	**84.78**	**76.35**	**84.94**	85.34

It shows that our two-stage segmentation method achieves the best performance, surpassing other methods in all four evaluation metrics. The 2D nnU-Net [10] model with preprocessing techniques ranks second, showing comparable results to our method, which further demonstrates the excellent performance of the nnUN-et framework in medical image segmentation.

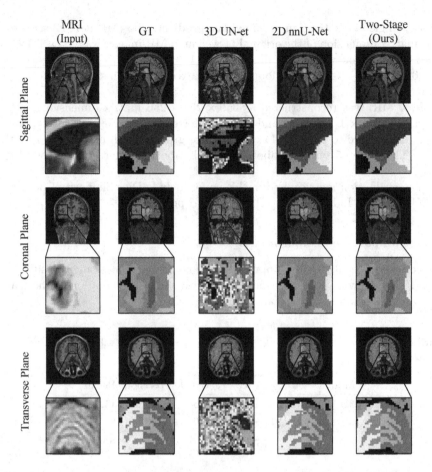

Fig. 5. Comparison of brain region segmentation results. Each figure is divided into two sections, with the top portion displaying the full brain region and the bottom part displaying a zoomed-in region, and all results are shown in the sagittal, coronal, and transverse plane perspectives..

The Fig. 5 show the comparative results of the 3D UN-et model without any preprocessing, the 2D nnU-Net model with preprocessing means and the predictions of our two-stage segmentation method in different views. Each figure is divided into two parts, showing the complete brain region at the top and the magnified region at the bottom, with all results shown in sagittal, coronal and cross-sectional fluoroscopic views. mri and labeling are located in the first two columns, respectively. First, the results of the 3D UN-et model without any preprocessing only show rough contours, while the 2D nnU-Net model with preprocessing means and our two-stage segmentation method have more accurate shapes.

We also conducted experiments on the SODM with different values of β. β represents the ratio between the number of predicted voxels for a specific class and the total number of voxels. From the comparison Table 2, it is observed that the best performance is achieved when β is set to 0.15.

4   Conclusion

To address the segmentation of tiny objects in brain regions, this paper proposes a two-stage object segmentation method based on the small object distribution map. Firstly, this method combines the advantages of 2D and 3D convolutions and designs a two-stage segmentation network structure. It introduces the concept of a small object distribution map, which ensures fast and accurate segmentation of small objects while maintaining the segmentation speed for large objects. Secondly, the method utilizes a contrastive loss function to automatically mine hard negative samples and adds two attention modules to assist in the accurate generation of the small object distribution map. This approach alleviates the scale mismatch problem between large and small objects in medical image segmentation and reduces the loss of spatial information. Finally, the effectiveness of the two-stage small object segmentation method in brain tissue segmentation is validated by comparing it with commonly used medical image segmentation methods.

Acknowledgements. This work was supported in part by the National Key R&D Program of China (No. 2022ZD0118201), Natural Science Foundation of China (No. 61972217, 32071459, 62176249, 62006133, 62271465).

References

1. Naz, F.: Human brain disorders: a review. Open Biol. J. **8**, 6–21 (2020)
2. Kuklisova-Murgasova, M., Aljabar, P., Srinivasan, L.: A dynamic 4D probabilistic atlas of the developing brain. Neuroimage **54**(4), 2750–2763 (2011)
3. Wachinger, C., Golland, P.: Atlas-based under-segmentation. In: Golland, P., Hata, N., Barillot, C., Hornegger, J., Howe, R. (eds.) MICCAI 2014. LNCS, vol. 8673, pp. 315–322. Springer, Cham (2014). https://doi.org/10.1007/978-3-319-10404-1_40
4. del Fresno, M., Vénere, M., Clausse, A.: A combined region growing and deformable model method for extraction of closed surfaces in 3D CT and MRI scans. Comput. Med. Imaging Graph. **33**(5), 69–376 (2009)
5. Kass, M., Witkin, A., Terzopoulos, D.: Snakes: active contour models. Int. J. Comput. Vision **1**(4), 321–331 (1988)
6. Moreno, J.C., Prasath, V.S., Proenca, H., Palaniappan, K.: Fast and globally convex multiphase active contours for brain MRI segmentation. Comput. Vis. Image Underst. **125**, 237–250 (2014)
7. Rivest-Hénault, D., Cheriet, M.: Unsupervised MRI segmentation of brain tissues using a local linear model and level set. Magn. Reson. Imaging **29**(2), 243–259 (2011)

8. Tan, C., Guan, Y., Feng, Z., et al.: DeepBrainSeg: automated brain region segmentation for micro-optical images with a convolutional neural network. Front. Neurosci. **14**(279), 1–13 (2020)

9. Ronneberger, O., Fischer, P., Brox, T.: U-net: convolutional networks for biomedical image segmentation. Lect. Notes Comput. Sci. **9351**, 234–241 (2015)

10. Isensee, F., Jaeger, P.F., Kohl, S.A.: nnU-Net: a self-configuring method for deep learning-based biomedical image segmentation. Nat. Methods **18**(2), 203–211 (2021)

11. Van, B.: Functional imaging: CT and MRI. Clin. Chest Med. **29**(1), 195–216 (2008)

12. Jaccard, P.: The distribution of the flora in the alpine zone. New Phytol. **11**(2), 37–50 (1912)

13. Meng, L., Zhang, Q., Bu, S.: Two-Stage liver and tumor segmentation algorithm based on convolutional neural network. Diagnostics **11**(10), 1806 (2021)

14. De Brebisson, A., Montana, G.: Deep neural networks for anatomical brain segmentation. In: Proceedings of the IEEE Conference on Computer Vision and Pattern Recognition Workshops, pp. 20–28 (2015)

15. Wang, F., Liu, H.: Understanding the behaviour of contrastive loss. In: Proceedings of the IEEE Computer Society Conference on Computer Vision and Pattern Recognition, pp. 2495–2504 (2021)

16. Çiçek, Ö., Abdulkadir, A., Lienkamp, S.S., Brox, T., Ronneberger, O.: 3D U-net: learning dense volumetric segmentation from sparse annotation. In: Ourselin, S., Joskowicz, L., Sabuncu, M.R., Unal, G., Wells, W. (eds.) MICCAI 2016. LNCS, vol. 9901, pp. 424–432. Springer, Cham (2016). https://doi.org/10.1007/978-3-319-46723-8_49

Encoder Activation Diffusion and Decoder Transformer Fusion Network for Medical Image Segmentation

Xueru Li[1], Guoxia Xu[2], Meng Zhao[1(✉)], Fan Shi[1], and Hao Wang[2]

[1] Key Laboratory of Computer Vision and System of Ministry of Education, School of Computer Science and Engineering, Tianjin University of Technology, Tianjin 300384, China
zh_m@tju.edu.cn
[2] School of Cyber Engineering of Xidian University, Xi'an 710126, China

Abstract. Over the years, medical image segmentation has played a vital role in assisting healthcare professionals in disease treatment. Convolutional neural networks have demonstrated remarkable success in this domain. Among these networks, the encoder-decoder architecture stands out as a classic and effective model for medical image segmentation. However, several challenges remain to be addressed, including segmentation issues arising from indistinct boundaries, difficulties in segmenting images with irregular shapes, and accurate segmentation of lesions with small targets. To address these limitations, we propose Encoder Activation Diffusion and Decoder Transformer Fusion Network (ADTF). Specifically, we propose a novel Lightweight Convolution Modulation (LCM) formed by a gated attention mechanism, using convolution to encode spatial features. LCM replaces the convolutional layer in the encoder-decoder network. Additionally, to enhance the integration of spatial information and dynamically extract more valuable high-order semantic information, we introduce Activation Diffusion Blocks after the encoder (EAD), so that the network can segment a complete medical segmentation image. Furthermore, we utilize a Transformer-based multi-scale feature fusion module on the decoder (MDFT) to achieve global interaction of multi-scale features. To validate our approach, we conduct experiments on multiple medical image segmentation datasets. Experimental results demonstrate that our model outperforms other state-of-the-art (SOTA) methods on commonly used evaluation metrics.

Keywords: Medical Image Segmentation · Lightweight Convolution Modulation · Encoder Activation Diffusion · Multi-scale Decoding Fusion With Transformer

---

This work was supported by the National Natural Science Foundation of China (NSFC) under grant numbers 62272342, 62020106004, 92048301.

Q. Liu et al. (Eds.): PRCV 2023, LNCS 14437, pp. 185–197, 2024.
https://doi.org/10.1007/978-981-99-8558-6_16

1   Introduction

With the development of the times, hospitals have more accurate and reliable requirements for medical image segmentation. Convolutional neural networks have gradually taken advantage of this aspect. U-Net [19] composed of encoder-decoder architecture is widely used in medical image segmentation. On this basis, researchers have proposed some feature-enhancement methods [10,15,24] to improve the segmentation performance.

Medical images often exhibit a wide range of scales for the target objects, leading to potential segmentation errors. To address this challenge, Rahman et al. [13] introduced ResPath with a residual structure in MultiResUNet to improve the connectivity problem between the encoder and decoder. Furthermore, Wang et al. [20] pioneered the use of Transformers for the connectivity between encoder and decoder layers. As the demand for better medical image segmentation increases, model architectures have become larger and more complex. To address this, Chollet et al. [6] proposed the Xception model, which introduced the concept of depth separable convolutions, significantly reducing the number of parameters and computational complexity. Similarly, in Ghost-Net [11], a linear operation is applied to one feature map to generate more similar feature maps, effectively reducing parameters.

Excessive convolution operations may result in the loss of spatial information. Azad et al. [5] proposed the CE-Net, which incorporated a digital-to-analog conversion module and a minimum mean square error module into the encoder structure to mitigate this issue. GU et al. [10] introduced the CA-Net, which utilized spatial position, channel numbers, and scale of feature maps, along with a comprehensive attention module. Additionally, Dai et al. [7] proposed MsR-EFM and MsR-DFM in MsRED to perform multi-scale feature fusion, enabling the network to adaptively learn contextual information. In addition to these advancements, researchers have proposed various improved algorithms for medical image segmentation. While these approaches [17,22] have achieved certain performance improvements, they still have certain limitations. For instance, most encoding and decoding convolution blocks suffer from a high number of parameters and computational complexity. The information extracted by the encoder layer may be insufficient, making it challenging to segment images with blurred boundaries. When restoring image space details and semantic information in the decoder layer, capturing changes in multi-scale features can be difficult. Moreover, traditional fusion methods, such as feature map concatenation [7] may suffer from redundant or missing information.

To address the aforementioned issues, we propose Encoder Activation Diffusion and Decoder Transformer Fusion Network, referred to as ADTF. We propose an LCM that seamlessly replaces the convolution layer in the encoder-decoder. Compared to the traditional convolution layer, our LCM significantly reduces the number of parameters while slightly improving segmentation performance. Additionally, the EAD retains more meaningful information, thereby improving the segmentation of blurred areas. Finally, we introduce MDFT can use the self-attention mechanism to globally interact with features of different scales. This enables better learning of correlations between different channel features, facilitating the capture of long-distance dependencies and global contextual information in complex medical images. Overall, the key contributions of this paper can be summarized as follows:

1. To enhance the feature learning of the model and reduce the amount of model parameters while training the network, a Lightweight Convolution Modulation is proposed to replace all convolutional blocks in the encoder and decoder layers.
2. To enable the encoder to extract more useful feature information to solve the activation problem of the image blur area, it is proposed to introduce the function of graph diffusion in the encoder.
3. To more effectively fuse channel features of different scales, richer global context information and long-distance dependencies can be obtained. We propose a better way to connect different decoders, using Transformer for multi-scale decoding channel information fusion.

2   Methodology

The overall frame diagram of the model is shown in Fig. 1. In order to solve the problems that still exist in medical image segmentation, we propose three solutions. Firstly, replace the encoding and decoding convolutional layers with our Lightweight Convolutional Modulation, in the MsRED-Net [7] model, we replaced the M^2F^2 in the encoder-decoder convolutional layer to generate richer feature representations. Secondly, we introduce Activation Diffusion blocks at the end of the encoding layer, which mainly solves the partial activation problem in medical image segmentation by exploiting spatial correlation. In practice,

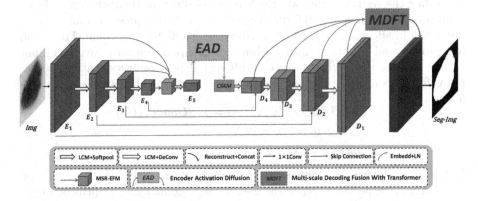

Fig. 1. The overall framework of the proposed model, the key parts of the novel model design, including 1) Lightweight Convolution Modulation(LCM), which covers many encoder and decoder layers. 2) Encoder Activation Diffusion(EAD), it is in the middle of E_5 and CSAM. 3) Multi-scale Decoding Fusion With Transformer(MDFT), which is used to connect decoder features of different scales.

we introduce learnable parameters to adjust the diffusion range and dynamically filter the noise to achieve flexible control and better adaptability, it can diffuse the semantic awareness area to the boundary area to a certain extent, he can make the generated graph can obtain clearer boundaries, which can effectively solve the problems of segmentation target is small and blurred boundary features. Finally, we use Transformer for multi-scale decoding channel information fusion, the self-attention mechanism is used to globally interact with features, and the features of different scales are better fused by learning the correlation between features, which solves the problem that ordinary fusion cannot obtain long-distance dependencies and global context information.

Furthermore, for the CSAM [18] among them, we utilize it to connect the decoder and encoder, effectively using channel and spatial attention modules for feature representation.

2.1 Lightweight Convolution Modulation

In the convolution modulation layer we designed, we try to learn from and simplify the attention mechanism [12, 25] so that it can easily replace other ordinary convolutions. Therefore, we try to use the depthwise gating mechanism to optimize the model. Specifically, three different convolution operations are performed on the input feature X. In order to optimize the conventional convolution, first use Depthwise+Sigmoid and Depthwise+Gelu to fully extract the feature information of a single channel in X and perform the dot product of the two, and then additionally use the convolution kernel size 1×1 linear to extract different channel features in X, and finally, we fuse and add the two results. The combination of Depthwise convolution and 1×1 convolution can achieve multi-scale feature fusion. Depthwise convolution can capture the spatial information of input features, and 1×1 convolution can perform information fusion between channels. By adding the results of the two, the features of different channels can be fused while preserving the spatial information, and the feature expression ability can be improved. In addition, we used GroupNorm, which divides the channel direction into groups, and then normalizes within each group. Specifically, given an input feature $X \in \mathbb{R}^{(H \times W \times C)}$, we use two different depthwise-based gating mechanisms for convolution, and then use Hadamard prod to calculate the output, as follows:

$$
\begin{aligned}
S &= DSConv_{k \times k}(X), \quad G = DGConv_{k \times k}(X) \\
SG &= S \odot G, \quad L = WX \\
OutPut &= SG + L
\end{aligned}
\tag{1}
$$

In the above formula, $\odot$ represents the Hadamard product, $DSConv_{k \times k}$ and $DGConv_{k \times k}$ respectively represent Depthwise+Sigmoid and Depthwise+Gelu with kernel size $k \times k$, W is the weight matrix of the linear layer and its convolution kernel size is 1×1. The convolutional modulation operation described above enables each spatial location (h, w) to be associated with multiple pixels within a $k \times k$ square region centered at (h, w) and the information exchange between channels can be realized through the linear layer.

2.2   Encoder Activation Diffusion

To make the encoder obtain useful information richer and dynamically expand the high-order semantic information so that it can obtain more complete segmentation regions and sharper boundaries. As shown in Fig. 2A, inspired by the graph diffusion algorithm [2,14], we propose activation diffusion for the encoder to dynamically diffuse semantically-aware regions to boundary regions.

Semantic Similarity Estimation. For the output features of the encoder, we need to construct a pair of patches to describe its semantic and spatial relationships. To achieve this, we first express the spatial information as vectors in a high-dimensional space, and each vector has potential semantic information. For this we define that v^l in the i node $\in \mathbb{R}^Q$, then we infer the semantic similarity $ESM_{i,j}^l$, where $ESM_{i,j}^l$ is defined as the cosine similarity between v_i and v_j:

$$ESM_{i,j}^l = \frac{v_i^l (v_j^l)^T}{||v_i^l|| ||v_j^l||} \tag{2}$$

where $ESM_{i,j}^l$ measures the similarity between vectors v_i and v_j. A higher value indicates greater similarity.

Dynamic Activation Diffusion. Inspired by the graph neural network, the inverse of the Laplacian matrix can be used to enhance global diffusion [3,16]. In order to represent the spatial relationship, we use v_i and v_j to construct the Laplacian matrix, the inverse of the Laplace matrix is used to describe the equilibrium state between v_i and v_j, where $A_{i,j}^l \in \mathbb{R}^{N \times N}$, which indicates whether v_i and v_j are connected. $D_{i,i}^l$ is used to represent the sum of all degrees related to v_i and $D_{j,j}^l$ is used to represent the sum of all degrees related to v_j and $D^l \in \mathbb{R}^{N \times N}$. Then we construct the Platas matrix is: $L^l = D^l - A^l$. In order to enhance the diffusion of semantic information [9], we combine $ESM_{i,j}^l$ and L^l with contextual information. We introduce a learnable parameter to dynamically adjust semantic correlation and spatial context to make the activation diffuse is more flexible for efficient information dissemination. Therefore, we define our Laplacian matrix as:

$$L^l = (D^l - A^l) \odot (\lambda ESM^l - 1) \tag{3}$$

Diffuse Matrix Approximation. It is impossible for us to directly solve the inverse of the Laplacian matrix, because we cannot guarantee that the matrix must have a positive definite shape, so we need to use Newton-Schulz Iteration to approximate the inverse transformation of the Laplacian matrix,

$$Inverse_i = \partial(L^l)^T, \quad Inverse_{i+1} = Inverse_i(2I - L^l Inverse_i) \tag{4}$$

Here, the $Inverse$ variable is obtained by multiplying a constant value ∂ by L^l, where the subscript i represents the number of iterations, the more iterations, the more accurate the obtained value, and I is the identity matrix.

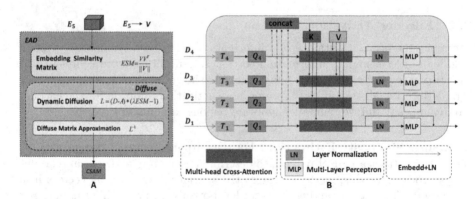

Fig. 2. This frame diagram consists of two parts. A is the structure of the Encoder Activation Diffusion (EAD). It consists of Embedding Similarity Matrix and Dynamic Diffuse Blocks. B is the Multi-scale Decoding Fusion With Transformer (MDFT).

2.3 Multi-scale Decoding Fusion with Transformer

In order to solve the semantic gap between different decoders, we propose MDFT (see Fig. 2B) to make full use of the multi-scale features of the decoder, and use Transformer for fusion. The self-attention mechanism can be used to globally interact with the feature map, and the features of different scales can be better fused by learning the correlation between features. Our model is based on Transformer [20], specific operations are as follows. First, for the four decoder outputs, D_1, D_2, D_3 and D_4 respectively, since the features of different levels have different sizes and channels, we denote it as $D_i \in \mathbb{R}^{(iC \times \frac{hw}{i})}, (i = 1, 2, 3, 4)$, map the four D_i inputs into 2D sequences through convolutional encoding (channel embeddings), still maintain the original channel size, specifically expressed as $T_i \in \mathbb{R}^{(iC \times \frac{hw}{i})}, (i = 1, 2, 3, 4)$, the specific operation is as follows:

$$
\begin{aligned}
F &= concat(T_1, T_2, T_3, T_4) \\
K &= linear(F), V = linear(F) \\
Q_i &= linear(T_i), (i = 1, 2, 3, 4)
\end{aligned}
\tag{5}
$$

Each linear here is calculated with different weights. They are based on the Query, key, and value calculated with T_i and F, where $Q_i \in \mathbb{R}^{(iC \times d)}, (i = 1, 2, 3, 4)$, $F \in \mathbb{R}^{(C_\Sigma \times d)}$, $K \in \mathbb{R}^{(C_\Sigma \times d)}$, $V \in \mathbb{R}^{(C_\Sigma \times d)}$. Specifically, the calculation method of the Muti-head Cross-Attention module and the final are:

$$
\begin{aligned}
CA_i &= M_i V^T = \sigma \left[\varphi(\frac{Q_i^T K}{\sqrt{C_\Sigma}}) \right] V^T, C_i = Drop(linear(CA_i)) \\
O_i &= T_i + MLP(LN(T_i + C_i)), (i = 1, 2, 3, 4)
\end{aligned}
\tag{6}
$$

Here, $\varphi(\cdot)$ and $\sigma(\cdot)$ represent the normalization and softmax functions respectively. The formula omits some activation functions and regularization that are essential in operations. We use more effective feature fusion and multi-scale channel cross-attention to fuse low-level features and high-level features. Semantic and resolution gaps between features to capture more complex channel dependencies make image segmentation more accurate.

3 Experiments

3.1 Datasets

We will conduct experiments on three datasets, they are PH^2, ISIC2017, and ISIC2018 datasets. PH^2 is a small dataset with a total of 200 images, and uses 80 images for training, 100 images for testing, and 20 images for validation. ISIC2017 and ISIC2018 datasets have 2150 and 2594 images and their corresponding labels respectively. Among them, ISIC2018 has become the main benchmark for algorithm evaluation in medical image segmentation, ISIC2017 and ISIC2018 datasets are randomly divided into training (70%), verification sets (10%), and test sets (20%). All datasets are re-cropped to 224×320 pixels.

3.2 Implementation Details

All our experiments were run on a single GeForce RTX3090 GPU and implemented with pytorch. We use Adam as the optimizer with an initial learning rate of 0.001 and weight decay of 0.00005, using a CosineAnnealing-WarmRestarts learning rate strategy runs 250 epochs for all networks. Because the PH^2 dataset is too small, it is easy to cause overfitting in model training, so we augment the data by using methods including image horizontal flip, cropping and rotation. In order to ensure the accuracy and reliability of the experimental data, we use five times of cross-validation to evaluate the model feasibility. Medical Segmentation Evaluation Criteria use Jaccard Index (JI), Accurary (ACC), Precision, Recall, and Dice to evaluate the model, which can be used to accurately measure the performance of the model in medical image segmentation.

3.3 Evaluation Results

In order to verify the effectiveness of our proposed model, we compare our model with state-of-the-art methods, we will make a fair comparison on the ISIC2017, ISIC2018 and PH^2 datasets in turn. As ISIC2018 has become the main benchmark for algorithm evaluation in medical image segmentation, so we will also use it to show a visual comparison with other models.

Comparsions with the State-of-arts on ISIC2017 and PH^2. The results of the two experiments are summarized in Table 1. There are more complex lesions in the ISIC2017 dataset. Obviously, the ADTF performs better than other models on complex segmentation dataset and small dataset. This is mainly due to the dynamic activation of the ISIC2017 large fuzzy lesion area by the EAD module. MDFT can improve the perception ability of the model, making it more accurate when dealing with multi-scale tasks.

Table 1. Segmentation performance of different networks on ISIC2017 and PH^2, the best results are shown in bold.

Methods	ISIC2017			PH^2			Para(M)
	Dice	ACC	JI	Dice	ACC	JI	
DeepLabv3+ [4]	87.93	96.38	80.68	93.18	96.07	87.92	37.9
DenseASPP [23]	87.88	96.05	80.73	94.13	96.61	89.38	33.7
BCDU-Net [1]	83.92	94.96	74.50	93.06	95.61	87.41	28.8
CE-Net [5]	85.61	93.51	77.54	94.36	96.68	89.62	29.0
CPF-Net [8]	88.45	95.68	81.39	94.52	96.72	89.91	43.3
CA-Net [10]	85.00	95.51	76.98	92.42	95.46	86.44	**2.7**
MsRED [7]	86.48	94.10	78.55	94.65	96.80	90.14	3.8
ADTF(ours)	**88.64**	**96.60**	**81.41**	**95.67**	**97.75**	**91.78**	8.6

Comparsions with the State-of-arts on ISIC2018. Table 2 shows the comparison between ADTF and other 11 advanced models in the ISIC2018 dataset. Among them, Dice and JI are indicators for evaluating the acquaintance of the segmentation results, indicating the degree of overlap between the predicted and real segmentation results, and Recall is the degree of under-segmentation. The indicators JI, ACC, Recall, and Dice of our ADTF have increased by 0.63%, 0.32%, 0.63%, and 0.55% respectively compared with MsRED, and achieves the best performance among the 12 networks.

Figure 3 visually compares the segmentation results of these models. These samples contain various challenges including ambiguous boundary features, small lesion objects, and irregular shapes. Most networks are always over-segmented (red area) for small objects and under-segmented (blue area) for areas with

Table 2. Segmentation performance of different networks on ISIC2018 (mean ± standard deviation), the best results are shown in bold.

Methods	Dice	JI	ACC	Precision	Recall	Para (M)
U-Net [19]	88.81 ± 0.40	81.69 ± 0.50	95.68 ± 0.29	91.31 ± 0.25	88.58 ± 0.99	32.9
U-Net++ [26]	88.93 ± 0.38	81.87 ± 0.47	95.68 ± 0.33	90.98 ± 0.53	89.10 ± 1.33	34.9
AttU-Net [21]	89.03 ± 0.42	81.99 ± 0.59	95.77 ± 0.26	91.26 ± 0.32	88.98 ± 0.67	33.3
DeepLabv3+ [4]	89.26 ± 0.23	82.32 ± 0.35	95.87 ± 0.23	90.87 ± 0.42	89.74 ± 0.72	37.9
DenseASPP [23]	89.35 ± 0.37	82.53 ± 0.55	95.89 ± 0.28	91.38 ± 0.37	89.50 ± 0.98	33.7
BCDU-Net [1]	88.33 ± 0.48	80.84 ± 0.57	95.48 ± 0.40	89.68 ± 0.59	89.13 ± 0.39	28.8
DO-Net [22]	89.48 ± 0.37	82.61 ± 0.51	95.78 ± 0.36	90.59 ± 0.71	90.36 ± 0.47	24.7
CE-Net [5]	89.59 ± 0.35	82.82 ± 0.45	95.97 ± 0.30	90.67 ± 0.52	90.54 ± 0.48	29.0
CPF-Net [8]	89.63 ± 0.42	82.92 ± 0.52	96.02 ± 0.34	90.71 ± 0.61	90.62 ± 0.48	43.3
CA-Net [10]	87.82 ± 0.63	80.41 ± 0.81	95.25 ± 0.32	90.72 ± 0.48	87.62 ± 0.61	**2.7**
MsRED [7]	89.99 ± 0.34	83.45 ± 0.41	96.19 ± 0.26	91.47 ± 0.22	90.49 ± 0.61	3.8
ADTF(ours)	**90.42 ± 0.39**	**83.60 ± 0.51**	**96.49 ± 0.19**	**92.52 ± 0.75**	**91.06 ± 0.38**	8.6

ambiguous lesions, while our ADTF shows good segmentation ability. This is mainly because EAD retains more useful information, MDFT uses the self-attention mechanism to globally interact with features and learn their correlation to better fuse features of different scales.

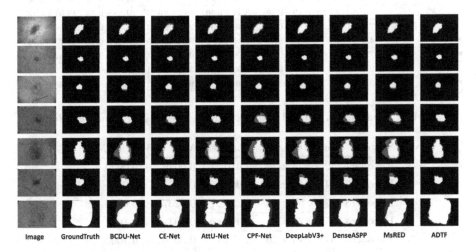

Image GroundTruth BCDU-Net CE-Net AttU-Net CPF-Net DeepLabV3+ DenseASPP MsRED ADTF

Fig. 3. Visual comparison with the state-of-the-arts on ISIC 2018. The colors white, red, and blue represent the correct segmentation, the over-segmentation, and the under-segmentation, respectively. (Color figure online)

3.4 Ablation Study

Perform Ablation Experiments on Each Module: In order to verify the effectiveness of each of our proposed modules, we step-by-step ablation experiments for the proposed three modules (see Table 3). For this purpose, we choose the ISIC2018 dataset for comparison. Compared with the Baseline, the evaluation indicators of our different models have been improved, among which, the model with only LCM has the least number of Paras (M) and Floats (G) Recall is the degree of under-segmentation, the larger the value, the more complete the segmentation. The model combined with the EAD module has a higher Recall, which is related to its dynamic activation diffusion in the blurred area of the lesion, it can obtain the more complete segmented image.

Table 3. Evaluate the segmentation performance of adding different modules. The Baseline is MsRED, the last line is ADTF. The best results are shown in bold.

LCM	EAD	MDFT	Paras	Floats	Dice	JI	ACC	Precision	Recall
			3.8	9.8	89.99 ± 0.34	83.45 ± 0.41	96.19 ± 0.26	91.47 ± 0.22	90.49 ± 0.61
✔			**3.2**	**8.0**	90.03 ± 0.29	83.29 ± 0.33	96.22 ± 0.13	91.64 ± 0.61	90.50 ± 0.80
	✔		5.1	10.1	90.19 ± 0.38	83.44 ± 0.40	96.29 ± 0.27	91.86 ± 0.27	91.01 ± 0.13
		✔	7.9	11.1	89.96 ± 0.29	83.40 ± 0.35	96.24 ± 0.23	91.85 ± 0.16	90.54 ± 0.43
✔	✔		4.5	8.4	89.91 ± 0.56	83.31 ± 0.67	96.20 ± 0.29	92.20 ± 0.47	90.73 ± 0.79
✔		✔	7.3	9.4	89.85 ± 0.39	83.29 ± 0.43	96.23 ± 0.31	91.81 ± 0.58	90.61 ± 0.76
	✔	✔	9.2	11.5	90.39 ± 0.36	83.50 ± 0.46	96.25 ± 0.31	92.51 ± 0.65	90.66 ± 1.05
✔	✔	✔	8.6	9.7	$\mathbf{90.42 \pm 0.39}$	$\mathbf{83.60 \pm 0.51}$	$\mathbf{96.49 \pm 0.19}$	$\mathbf{92.52 \pm 0.75}$	$\mathbf{91.06 \pm 0.38}$

Table 4. Comparison of LCM and convolution layer in M^2F^2 and CS^2-Net.

Stage	Feature Size	Conv(M)	M^2F^2(M)	LCM(M)
Encoder1	$32 \times 224 \times 320$	0.0104	0.0049	0.0050
Encoder2	$64 \times 112 \times 160$	0.0578	0.0061	0.0071
Encoder3	$128 \times 56 \times 80$	0.2302	0.0237	0.0265
Encoder4	$256 \times 28 \times 40$	0.9193	0.0931	0.1021
Encoder5	$512 \times 14 \times 20$	3.6736	0.3910	0.4009
CSAM	$512 \times 14 \times 20$	0.4600	0.4600	0.4600
Decoder4	$256 \times 28 \times 40$	1.9023	0.8651	0.4063
Decoder3	$128 \times 56 \times 80$	0.4760	0.1956	0.1048
Decoder2	$64 \times 112 \times 160$	0.1192	0.0495	0.0278
Decoder1	$32 \times 224 \times 320$	0.0299	0.0126	0.0078
Total		7.8787	2.1015	**1.5483**

Comparison of LCM and Other Convolutional Layer Parameters: Table 4 qualitatively shows the comparison of parameters between LCM and M^2F^2 and CS^2-Net convolutional layers. The feature sizes used are the same. Obviously, the parameters of CS^2-Net convolutional layers are much larger than M^2F^2 and LCM. LCM compared to M^2F^2 and CS^2-Net convolutional layers are reduced by 80% and 26% in all parameter comparisons, respectively.

Visual Analysis of All Models Including EAD Modules: In order to verify the effectiveness of our EAD module, we performed a visual analysis of all related models, as shown in Fig. 4, the blue area represents under-segmentation, it can be clearly seen from the figure that MsRED is difficult to completely segment the lesion area. However, the model combined with EAD can dynamically activate and diffuse the encoding area. More useful information can be extracted. It can be seen that ADTF has the least blue area and the most accurate and complete segmentation.

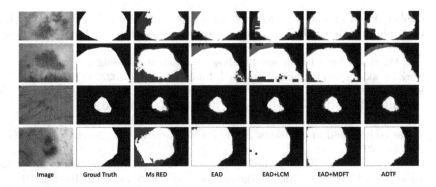

| Image | Groud Truth | Ms RED | EAD | EAD+LCM | EAD+MDFT | ADTF |

Fig. 4. All models containing EAD are compared visually with Baseline (MsRED) on ISIC2018. The colors white, blue, and red represent the correct segmentation, the under-segmentation, and the over-segmentation, respectively. (Color figure online)

4 Conclusion

In this paper, a novel model named ADTF is proposed to address the existing challenges in medical image segmentation. Specifically, the EAD dynamically activates high-order semantic information and pays more attention to image-blurred regions. The Transformer architecture is introduced to perform multi-scale channel fusion on the decoder to improve the perception ability of the model and better obtain the correlation between features. In addition, the proposed LCM adopts fewer parameters, which could be utilized to replace other convolutional layers and improve the model training ability. Furthermore, we validate the effectiveness of our model on multiple datasets and experimental results demonstrate the priority compared with other SOTA methods.

References

1. Azad, R., Asadi-Aghbolaghi, M., Fathy, M., Escalera, S.: Bi-directional convlstm u-net with densley connected convolutions. In: Proceedings of the IEEE/CVF International Conference on Computer Vision Workshops (2019)
2. Bai, H., Zhang, R., Wang, J., Wan, X.: Weakly supervised object localization via transformer with implicit spatial calibration. In: Computer Vision-ECCV 2022: 17th European Conference, Tel Aviv, Israel, 23–27 October 2022, Proceedings, Part IX, pp. 612–628. Springer, Heidelberg (2022). https://doi.org/10.1007/978-3-031-20077-9_36
3. Bruna, J., Zaremba, W., Szlam, A., LeCun, Y.: Spectral networks and locally connected networks on graphs. arXiv preprint arXiv:1312.6203 (2013)
4. Chen, L.C., Zhu, Y., Papandreou, G., Schroff, F., Adam, H.: Encoder-decoder with atrous separable convolution for semantic image segmentation. In: Proceedings of the European Conference on Computer Vision (ECCV), pp. 801–818 (2018)
5. Chen, S., Niu, J., Deng, C., Zhang, Y., Chen, F., Xu, F.: Ce-net: a coordinate embedding network for mismatching removal. IEEE Access **9**, 147634–147648 (2021)

6. Chollet, F.: Xception: deep learning with depthwise separable convolutions. In: Proceedings of the IEEE Conference on Computer Vision and Pattern Recognition, pp. 1251–1258 (2017)
7. Dai, D., et al.: MS RED: a novel multi-scale residual encoding and decoding network for skin lesion segmentation. Med. Image Anal. **75**, 102293 (2022)
8. Feng, S., et al.: CPFNET: context pyramid fusion network for medical image segmentation. IEEE Trans. Med. Imaging **39**(10), 3008–3018 (2020)
9. Gao, S., Tsang, I.W.H., Chia, L.T.: Laplacian sparse coding, hypergraph laplacian sparse coding, and applications. IEEE Trans. Pattern Anal. Mach. Intell. **35**(1), 92–104 (2012)
10. Gu, R., et al.: CA-NET: comprehensive attention convolutional neural networks for explainable medical image segmentation. IEEE Trans. Med. Imaging **40**(2), 699–711 (2020)
11. Han, K., Wang, Y., Tian, Q., Guo, J., Xu, C., Xu, C.: Ghostnet: more features from cheap operations. In: Proceedings of the IEEE/CVF Conference on Computer Vision and Pattern Recognition, pp. 1580–1589 (2020)
12. Hou, Q., Lu, C.Z., Cheng, M.M., Feng, J.: Conv2former: a simple transformer-style convnet for visual recognition. arXiv preprint arXiv:2211.11943 (2022)
13. Ibtehaz, N., Rahman, M.S.: Multiresunet: rethinking the u-net architecture for multimodal biomedical image segmentation. Neural Netw. **121**, 74–87 (2020)
14. Kondor, R.I., Lafferty, J.: Diffusion kernels on graphs and other discrete structures. In: Proceedings of the 19th International Conference on Machine Learning, vol. 2002, pp. 315–322 (2002)
15. Liu, Z., Li, X., Luo, P., Loy, C.C., Tang, X.: Semantic image segmentation via deep parsing network. In: Proceedings of the IEEE International Conference on Computer Vision, pp. 1377–1385 (2015)
16. Liu, Z., Li, X., Luo, P., Loy, C.C., Tang, X.: Deep learning markov random field for semantic segmentation. IEEE Trans. Pattern Anal. Mach. Intell. **40**(8), 1814–1828 (2017)
17. Messaoudi, H., Belaid, A., Salem, D.B.: Cross-dimensional transfer learning in medical image segmentation with deep learning. Med. Image Anal. (2023)
18. Mou, L., et al.: Cs2-net: deep learning segmentation of curvilinear structures in medical imaging. Med. Image Anal. **67**, 101874 (2021)
19. Ronneberger, O., Fischer, P., Brox, T.: U-Net: convolutional networks for biomedical image segmentation. In: Navab, N., Hornegger, J., Wells, W.M., Frangi, A.F. (eds.) MICCAI 2015. LNCS, vol. 9351, pp. 234–241. Springer, Cham (2015). https://doi.org/10.1007/978-3-319-24574-4_28
20. Wang, H., Cao, P., Wang, J., Zaiane, O.R.: Uctransnet: rethinking the skip connections in u-net from a channel-wise perspective with transformer. In: Proceedings of the AAAI Conference on Artificial Intelligence, vol. 36, pp. 2441–2449 (2022)
21. Wang, S., Li, L.: Attu-net: attention u-net for brain tumor segmentation. In: International MICCAI Brainlesion Workshop, pp. 302–311. Springer, Heidelberg (2022). https://doi.org/10.1007/978-3-031-09002-8_27
22. Wang, Y., Wei, Y., Qian, X., Zhu, L., Yang, Y.: Donet: dual objective networks for skin lesion segmentation. arXiv preprint arXiv:2008.08278 (2020)
23. Yang, M., Yu, K., Zhang, C., Li, Z., Yang, K.: Denseaspp for semantic segmentation in street scenes. In: Proceedings of the IEEE Conference on Computer Vision and Pattern Recognition, pp. 3684–3692 (2018)
24. Yuan, F., Zhang, Z., Fang, Z.: An effective CNN and transformer complementary network for medical image segmentation. Pattern Recogn. **136**, 109228 (2023)

25. Zhou, B., Wang, S., Xiao, S.: Double recursive sparse self-attention based crowd counting in the cluttered background. In: Pattern Recognition and Computer Vision: 5th Chinese Conference, PRCV 2022, Shenzhen, China, 4–7 November 2022, Proceedings, Part I, pp. 722–734. Springer, Heidelberg (2022). https://doi.org/10.1007/978-3-031-18907-4_56

26. Zhou, Z., Siddiquee, M.M.R., Tajbakhsh, N., Liang, J.: Unet++: redesigning skip connections to exploit multiscale features in image segmentation. IEEE Trans. Med. Imaging **39**(6), 1856–1867 (2019)

Liver Segmentation via Learning Cross-Modality Content-Aware Representation

Xingxiao Lin and Zexuan Ji[✉]

School of Computer Science and Engineering, Nanjing University of Science
and Technology, Nanjing 210094, China
jizexuan@njust.edu.cn

Abstract. Liver segmentation has important clinical implications using
computed tomography (CT) and magnetic resonance imaging (MRI).
MRI has complementary characteristics to improve the accuracy of medical analysis tasks. Compared with MRI, CT images of the liver are
more abundant and readily available. Ideally, it is promising to transfer
learned knowledge from the CT images with labels to the target domain
MR images by unsupervised domain adaptation. In this paper, we propose a novel framework, i.e. cross-modality content-aware representation
(CMCAR), to alleviate domain shifts for cross-modality semantic segmentation. The proposed framework mainly consists of two modules in
an end-to-end manner. One module is an image-to-image translation
network based on the generative adversarial method and representation
disentanglement. The other module is a mutual learning model to reduce
further the semantic gap between synthesis images and real images.
Our model is validated on two cross-modality semantic segmentation
datasets. Experimental results demonstrate that the proposed model outperforms state-of-the-art methods.

Keywords: Segmentation · Cross-modality · Domain adaptation ·
Liver

1 Introduce

Liver segmentation is the process of extracting the liver region from medical
images to facilitate quantitative analysis and treatment planning. The liver,
being one of the largest organs in the body, holds valuable diagnostic and thera-
peutic information. Accurate segmentation enables the detection of diseases and
assessment of treatment outcomes. Liver segmentation finds extensive applications in medical imaging, especially in the diagnosis and treatment of liver diseases.

Precise liver segmentation is essential for clinical applications that involve
measuring the volume and shape of the liver, as well as monitoring, planning,

Q. Liu et al. (Eds.): PRCV 2023, LNCS 14437, pp. 198–208, 2024.
https://doi.org/10.1007/978-981-99-8558-6_17

and performing virtual or augmented liver surgeries. This requires accurate segmentation of medical images obtained through computed tomography (CT) or magnetic resonance imaging (MRI). CT and MRI provide complementary diagnostic information on the healthy tissue and the affected area. Therefore, providing liver segmentation on both modalities is highly valuable for disease analysis and treatment [2]. Computed tomography (CT)-based radiotherapy is currently used in radiotherapy planning and is reasonably effective and magnetic resonance (MR) imaging delivers superior contrast of soft tissue compared with the CT scans [11]. CT images are more abundant clinically than MR images, but the latter contains richer quantitative diagnostic information [2].

Therefore, it is desirable to achieve unsupervised domain adaptation for transferring from a well-trained model using the source domain containing labeled CT images to the target domain containing unlabeled MR images [5]. However, due to domain shift, the unsatisfactory performance of a trained model evaluated on samples from heterogeneous domains is expressed, even if these images on sampled from heterogeneous domains [5].

There are there categories of existing adaptation methods for tackling domain shift: pixel-level appearance alignment, high-level semantics alignment, and segmentation-oriented cross-modality image translation [9]. Adaptive methods based on image-to-image translation obtain synthetic images across two modalities and then utilize abundant annotations from the source domain to assist the target domain's segmentation task. CycleGAN [12] is widely used to focus on alleviating domain shift of segmentation on the multi-modal unpaired images, which learns a reversible mapping between unpaired images based on a cycle consistency loss. CycleGAN focuses on whether the appearance of the composite image is real and ignores the correspondence of anatomical structures, which leads to mismatches between synthetic data and labels at the pixel level [9]. Therefore, it is not suitable for scenes requiring multi-modality data with high image resolution and low image noise [11].

Inspired by [11], cross-modality translation for the medical images by disentangling the representation into content and style has a very good effect. However, the effectiveness of Generative Adversarial Network is limited by the discriminator, and then the gap still remains between the synthesis target-like images and real target images [2].

In our paper, we improves the mutual-learning structure and introduces target images for training within the mutual-learning module. Moreover, we incorporate the mutual learning using pseudo-labels from images with the same content but different modalities. Comparing with prior works, the major differences of the proposed model are: (1) Introducing contrastive learning during disentangled representation to reduce the coupling between style and content, thereby enhancing the quality of target-like images. (2) Directly utilizing target images and further incorporating t2s domain images in the subsequent mutual learning stage. By using images with the same content but different modalities, the paper aims to alleviate the gap between target and target-like images and improve the overall performance. (3) Simultaneously using real labels and pseudo-labels to mitigate the issues caused by pseudo-label errors.

2   Methodology

2.1   Overview

We define image x_s from the image set of source domain X_S, $x_s \in X_S$, and image x_t from target domain X_T, $x_t \in X_T$, both domains contain the anatomical structures of the same liver. Each image in the source domain has segmentation mask y_s, $y_s \in Y_S$, while the target domain has no available masks. The proposed framework consists of a cross-modality image translation network and a peer-to-peer learning procedure utilizing self-learning technology. The translation network contains two modality-specific encoders for representation disentanglement, two content encoders with shared parameters for reconstruction, and two generators with discriminators for translating cross-modality images. The overview of the proposed framework is illustrated in Fig. 1.

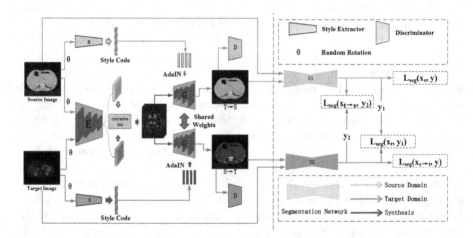

Fig. 1. Overview of the proposed framework (CMCAR). The CMCAR consists of two key components: image-to-image translation framework and peer-to-peer network. (a) On the left side of the dashed line is image-to-image translation framework is composed of a common encoder E, two generators G with shared weights, two style extractors S, and two discriminators. (b) On the left side of the dashed line is the peer-to-peer consists of two semantic segmentation utilizing self-learning methods.

2.2   Image-to-Image Network

The translation sub-network consists of encoders E_S, E_T, and E_C, generators G_S and G_T, as well as D_S and D_T for adversarial training. It should be noted that

G_S and G_T are weight-sharing. First, in the representations, disentanglement stage, each image from two domains after rotation of random angle $(\theta(x)$, noted as θ) is fed into the content encoder, and style extractor $\{s_s, s_t\}$, where (s_s, c_s) $= E_s(\theta) = (E_S^S(\theta_s^1), (E_S^C(\theta_s^2))$ and $(s_t, c_t) = E_t(\theta) = (E_T^S(\theta_t^1), (E_T^C(\theta_t^2))$.

Reconstruction Loss. The deep information would be missing in the feature disentangle process, the reconstruction loss L_{rec} is utilized to recover the original feature θ_s and θ_t with the disentangled domain-specific feature and category-specific feature, which is defined as

$$L_{rec} = ||\theta^2 - G(E^S(\theta^1), E^C(\theta^2))||_2 \qquad (1)$$

where θ^1 and θ^2 denote the original image with different rotations.

Adversarial Loss. Then we perform the translation by content code combing with style code, i.e. G_S and G_T, to produce the cross-modality translated images $g_s = G_S(s_s, c_t)$ and $g_t = G_T(s_t, c_t)$. The generators and discriminator train alternately via adversarial learning:

$$L_{adv}^T(G_T, D_T) = E_{x_t \sim X_T}[(D_T(G_T(s_t, c_s) - 1))^2] \qquad (2)$$

$$L_{dis}^T(G_T, D_T) = E_{x_t \sim X_T}[(D_T(\theta_t) - 1)^2 + (D_T(G_T(s_t, c_s))^2] \qquad (3)$$

The generator G_T generates realistic target-style images with the aim of fooling the D_T, by minimizing the loss L_{adv}^T, while D_T tries to distinguish between translated images and real images by minimizing the loss L_{dis}^T. The discriminator D_T is a binary classification network. As for D_S and G_S, L_{adv}^S and L_{dis}^T is similarly defined.

Cycle-Consistency Loss. To preserve the original contents in the transformed images, a reverse generator is usually used to impose the cycle consistency [1]. The pixel-wise cycle-consistency loss L_{cyc} is used to encourage $G_S(C(\theta_{s \rightarrow t}),$ $S_S(\theta_s)) = G_S(G_S(C(\theta_s), S_T(\theta_t)), S_S(\theta_s)) = \theta_s, G_T(C(\theta_{t \rightarrow s}), S_S(\theta_s)) = G_T(G_T$ $(C(\theta_t), S_S(\theta_s)), S_T(\theta_t)) = \theta_t$. With the adversarial loss and cycle-consistency loss, the image adaptation transforms the source images into target-like images.

Correlation Coefficient Metric. To enhance the anatomical structure similarity between realistic images and synthesized counterparts, we utilize cross-modality image synthesis to guarantee the anatomical structure consistency loss. Ideally, we assume that the method is able to be effective to enhance the structural constraint in cross-modality image synthesis [4]. Because the values of the original correlation coefficient metric range from -1 to 1, we use the square of the correlation coefficient function to constrain output values between 0 and 1 for constructing loss functions. The anatomical structure consistency loss L_{cc} is defined as:

$$L_{cc} = L_{cc}^{S \to T} + L_{cc}^{T \to S}$$

$$= \mathbb{E}_{x_s \sim X_S} \left[1 - \left(\frac{\mathrm{Cov}\,(x_{s \to t}, x_s)}{\sigma\,(x_{s \to t})\,\sigma\,(x_s)} \right)^2 \right] + \mathbb{E}_{x_t \sim X_T} \left[1 - \left(\frac{\mathrm{Cov}\,(x_{t \to s}, x_t)}{\sigma\,(x_{t \to s})\,\sigma\,(x_t)} \right)^2 \right]$$

(4)

where $\mathrm{Cov}(A', A)$ is the covariance operator between the synthesized images A' and original image A, and σ is the standard deviation.

Contrastive Loss. Apart from the correlation coefficient metric, the mutual information metric is also effective to guarantee the consistency of anatomical structures in cross-modality image synthesis [7]. To obtain high-quality and structure-consistent synthesized images, the synthesized image patches should associate with the corresponding input patches while disassociating them from any other patches. Therefore, the patch-based Noise Contrastive Estimation (NCE) operator [3] is applied to the mutual information estimation of encoded features in the share-weighted encoders.

To map the source image to the target counterpart, the shared encoder Ec extracts the last layer(l-th layer) feature map from the original images x_s and the generated images $x_{s \to t}$. The original image features and the synthesized image features are projected into the feature tensor f_s, $f_{s \to t} \in R^{S_l \times k}$ by MLP_S and MLP_T, respectively. S_l is the number of vectors and the number of spatial locations in the l-th layer. For contrastive loss of each layer, we sample one positive vector $v^{s+} \in R^K$ and one query vector $v^{sq} \in R^K$ from f_s and $f_{s \to t} \in R^{S_l \times k}$ respectively, randomly sample $S_l - 1$ negative vector $v^{s-} \in R^{(S_l-1) \times K}$. For a single location s, the contrastive loss ℓ can be expressed as the normalized temperature-scaled cross-entropy loss [3] by computing the feature similarity, which is defined as:

$$\ell\left(v^{sq}, v^{s+}, v^{s-}\right) = -\log\left(\frac{\exp\left(\mathrm{sim}\left(v^{sq}, v^{s+}\right)/\tau\right)}{\exp\left(\mathrm{sim}\left(v^{sq}, v^{s+}\right)/\tau\right) + \sum_{n=1}^{S_l-1} \exp\left(\mathrm{sim}\left(v^{sq}, v_n^{s-}\right)/\tau\right)} \right)$$

(5)

where $\mathrm{sim}(u, v) = u^T v\,/\|u\|\,\|v\|$ denotes the cosine similarity between u and v, and τ denotes a temperature parameter for scaling the feature vector distance, which is set as 0.07 [13] as the default value. In order to guarantee anatomical consistency, the aggregated contrastive loss values of all spatial locations can be defined as:

$$l_{nce} = \mathbb{E}_{x_s^S \sim X_s} \sum_{s=1}^{S_l} \ell\left((f_{s \to t})_l^{sq}, (f_s)_l^{s+}, (f_s)_l^{s-}\right)$$

(6)

where $(f)_l^*$ denotes a vector of feature tensors, and l denotes l-th layer feature of the encoder. Target images are trained with a similar method.

2.3 Peer-to-Peer Network

To address the remaining gap between synthesized target-like images and real target images, we propose a framework called peer-to-peer network. This framework builds upon the benefits of traditional knowledge distillation and deep mutual learning [5] to enable unsupervised domain adaptation. The ultimate goal is to alleviate the domain shift between target-like images and target images, as illustrated in Fig. 2. Both two peer networks U1 and U2 use the same network architecture(unshared weights), but process different imaging modalities. The network U_1 learns from the input image x_s and the network U_2 learns from the input image $x_{s \to t}$ with supervised learning method. Then the network U_1 receives the pseudo label y_2 from U_2 network as the label of $x_{t \to s}$. Similarly, the network U_2 receives the pseudo label y_1 from U_1 network as the label of x_t. The overall loss L_{p2p} of peer-to-peer learning is given as:

$$\mathcal{L}_{\text{p2p}} = \begin{cases} \mathcal{L}_{\text{seg}}^1 \left(x_s, y \right) + \mathcal{L}_{\text{seg}} \left(x_{t \to s}, y_2 \right) \\ \mathcal{L}_{\text{seg}}^2 \left(x_{s \to t}, y \right) + \mathcal{L}_{\text{seg}} \left(x_t, y_1 \right) \end{cases} \tag{7}$$

We utilize peer-to-peer learning by imitating self-learning technology to improve the performance of the desired model U_2.

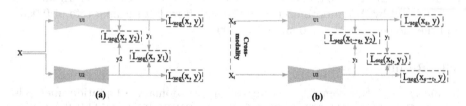

Fig. 2. The illustration of deep mutual learning and peer-to-peer learning. (a) denotes the deep mutual learning with input from the same image, and (b) denotes the proposed peer-to-peer learning with input from different modality images.

In summary, the final objective function for our framework is:

$$\begin{aligned} L_{total} =& \lambda_1 L_{rec} + \lambda_2 (L_{adv} + L_{div}) + \\ & \lambda_3 L_{cyc} + \lambda_4 L_{cc} + \lambda_5 L_{ce} + L_{p2p} \end{aligned} \tag{8}$$

3 Experiments

3.1 Dataset

In this study, we performed experiments on two publicly available challenge datasets: the LiTS-Liver Tumor Segmentation Challenge released with ISBI 2017 and the CHAOS-Combined (CT-MR) Healthy Abdominal Organ Segmentation released with ISBI 2019. The LiTs dataset contains CT images from 131 subjects, while the CHAOS challenge dataset consists of 20 subjects with CT images and 20 subjects with MR images (including both T1 and T2 sequences of MR). For this experiment, we used 20 unlabelled T2-SPIR in phase MRI images. All CT images were resized to the same size of MR images 256 × 256. The pixel value of CT images was limited within the range of -1000 to 400 (with the overflow value taken as the boundary value). Both CT and MR images were normalized to values between 0 and 1. In total, there were 151 subjects with CT images and 20 subjects with MR images (T2-SPIR), which were set as the source domain and target domain, respectively. After slicing and discarding background slices from each subject and retaining only the liver-contained images for this experiment, there were a total of 21,225 slices in the CT domain and 408 slices in the MR domain. For both datasets, we randomly selected 80% of the images as training data, while the remaining 20% of the images were used as testing data.

3.2 Setting

We implemented our network in Pytorch on an NVIDIA Tesla T4. We used SGD optimizer for the optimization of two segmentation networks with a learning rate of 0.0001 and momentum of 0.9. The encoders and generators network was trained by Adam optimizer of optimization with a learning rate of 0.0001 and the momentum $\beta1 = 0.5$ and $\beta2 = 0.999$. The style extractors and discriminators were trained by Adam optimizer of optimization with a learning rate of 0.0001 and the momentum $\beta1 = 0.0$ and $\beta2 = 0.99$.

3.3 Result

In our paper, the comparison state-of-the-art domain adaptation methods include CycleGAN [13], DTS [10], DCDA [8] DRCIT [6] and JALS [5]. DTST utilizes contrastive learning for training, but it is constrained by limited memory capacity and the number of annotated images. CycleGAN, DCDA, and DRCIT employ GAN structures, which are currently popular, to transform source domain images into target domain images, and then train using source domain labels. DCDA and DRCIT introduce disentangled representation into GAN, decoupling style and content to enhance s2t expression. However, these approaches overlook the gap between the target and target-like domains. JALS utilizes multiple networks and pseudo-labels for learning and training, but the use of pseudo-labels in multiple networks may amplify the impact of cumulative errors. In our paper, we address the aforementioned limitations by training with images of the same content but different modalities, aiming to alleviate the gap between the target domain and the target-like domains. Moreover, the proposed model can mitigate the issue of pseudo-label errors by simultaneously utilizing both pseudo-labels and real labels. By referring to the quantitative results in Table 1 and the qualitative results in Fig. 3, we observe a significant improvement in accuracy with our proposed method compared to other approaches.

Additionally, we also compared our proposed framework with the method without domain adaptation, denoted as "W/o adaptation", in which the model

Table 1. Comparison of segmentation results of several methods on CT to MRI adaptation experiment.

Methods	DSC (%)	ASSD (mm)
W/o adaptation	38.7 ± 2.3	7.52 ± 0.12
CycleGAN(2018)	57.4 ± 2.4	3.13 ± 1.48
DTST(2020)	64.8 ± 4.6	1.19 ± 0.37
DCDA(2022)	74.7 ± 4.3	1.25 ± 0.19
DRCIT(2022)	80.0 ± 2.0	0.51 ± 0.08
JALS(2022)	88.4 ± 3.6	0.32 ± 0.12
Ours	89.2 ± 2.7	0.30 ± 0.07

was trained on only source domain images and directly segmented the target domain images. The compared results of different methods are shown in Table 1. In our segmentation experiments, we employed the Dice similarity coefficient (DSC) and average symmetric surface distance (ASSD) to evaluate the agreement between the ground truth and the obtained results. A superior method is anticipated to attain a high DSC value while concurrently maintaining a low ASSD value.

Table 2. Results of ablation studies on key components of CT to MRI adaptation experiment (mean ± standard deviation).

Methods	DSC (%)	ASSD (mm)
Baseline	38.7 ± 2.3	7.52 ± 0.12
Baseline+I2I(w/o)	82.5 ± 1.6	0.80 ± 0.05
Baseline+I2I(contrastive)	84.9 ± 0.5	0.71 ± 0.03
Baseline+I2I(rotation+contrastive)	85.4 ± 0.6	0.35 ± 0.06
Proposed methods	89.2 ± 2.7	0.30 ± 0.07

Our framework consists of two main components, an image-to-image (I2I) translation network and a peer-to-peer (P2P) network, aiming to learn cross-modality content-aware representation for assisting cross-modality segmentation. To assess their effectiveness, we conducted incremental evaluations through several ablation studies, as summarized in Table 2. Initially, the baseline network trained solely on source images showed the worst performance as expected. However, significant performance improvement was observed with the introduction of the I2I network. The incorporation of image rotation and contrastive loss further enhanced the robustness and stability of the I2I translation, as illustrated in the figure above. Despite these improvements, there was still a noticeable gap between the target-like image and its real counterpart. To bridge this gap, we introduced a self-learning-based P2P network, which ultimately resulted in the best performance achieved by CMCAR.

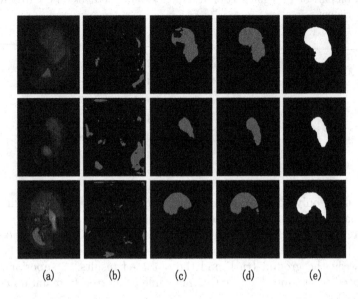

Fig. 3. Segmentation results on MRI image. From left to right, (a) Target image, (b) W/o adaptation, (c) JALS(sota), (d) CMCAR, (e) Ground Truth.

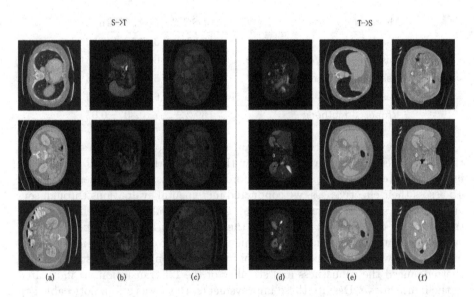

Fig. 4. Sample results of translated images. S−>T: content image (a) combined style image (b) to output translated images (c). T−>S: content image (d) combined style image (e) to output translated images (f).

The compared results of different methods are presented in Fig. 3. The figure showcases the qualitative outcomes of image-to-image translation using CMCAR in two settings: S−>T and T−>S. In the S−>T setting, one modality image is utilized as the content, while the other modality image serves as the style to generate the translated image. The roles are reversed in the T−>S setting.

In Fig. 4, we present results for no adaptation (w/o adaptation), adaptation at the output space together with cross-modality content-aware representation (CMCAR).

4   Conclusion

In this paper, we propose an unsupervised domain adaptation method that aims to minimize the negative impact of domain shift on segmentation performance. To achieve cross-modality image-to-image translation, we leverage disentangled representation and recombination of style and content codes. As future work, we introduce self-learning technology to further reduce the gap between target-like images and real target images. Our experimental results on liver MRI and CT images demonstrate that the proposed method effectively improves segmentation performance when compared to other adaptation methods.

Acknowledgements. This work was supported by National Science Foundation of China under Grants No. 62072241.

References

1. Chen, C., Dou, Q., Chen, H., Qin, J., Heng, P.A.: Synergistic image and feature adaptation: towards cross-modality domain adaptation for medical image segmentation. In: Proceedings of the AAAI Conference on Artificial Intelligence, vol. 33, pp. 865–872 (2019)
2. Chen, C., Dou, Q., Chen, H., Qin, J., Heng, P.A.: Unsupervised bidirectional cross-modality adaptation via deeply synergistic image and feature alignment for medical image segmentation. IEEE Trans. Med. Imaging **39**(7), 2494–2505 (2020)
3. Chen, T., Kornblith, S., Norouzi, M., Hinton, G.: A simple framework for contrastive learning of visual representations. In: International Conference on Machine Learning, pp. 1597–1607. PMLR (2020)
4. Ge, Y., Wei, D., Xue, Z., Wang, Q., Liao, S.: Unpaired MR to CT synthesis with explicit structural constrained adversarial learning. In: 2019 IEEE 16th International Symposium on Biomedical Imaging (ISBI) (2019)
5. Hong, J., Yu, C.H., Chen, W.: Unsupervised domain adaptation for cross-modality liver segmentation via joint adversarial learning and self-learning (2021)
6. Jiang, K., Quan, L., Gong, T.: Disentangled representation and cross-modality image translation based unsupervised domain adaptation method for abdominal organ segmentation. Int. J. Comput. Assist. Radiol. Surg. **17**(6), 1101–1113 (2022)
7. Park, T., Efros, A.A., Zhang, R., Zhu, J.-Y.: Contrastive learning for unpaired image-to-image translation. In: Vedaldi, A., Bischof, H., Brox, T., Frahm, J.-M. (eds.) ECCV 2020. LNCS, vol. 12354, pp. 319–345. Springer, Cham (2020). https://doi.org/10.1007/978-3-030-58545-7_19

8. Peng, L., Lin, L., Cheng, P., Huang, Z., Tang, X.: Unsupervised domain adaptation for cross-modality retinal vessel segmentation via disentangling representation style transfer and collaborative consistency learning. In: 2022 IEEE 19th International Symposium on Biomedical Imaging (ISBI), pp. 1–5. IEEE (2022)

9. Wang, S., Rui, L.: SGDR: semantic-guided disentangled representation for unsupervised cross-modality medical image segmentation (2022)

10. Wang, Z., et al.: Differential treatment for stuff and things: a simple unsupervised domain adaptation method for semantic segmentation. In: Proceedings of the IEEE/CVF Conference on Computer Vision and Pattern Recognition, pp. 12635–12644 (2020)

11. Wolterink, J.M., Dinkla, A.M., Savenije, M., Seevinck, P.R., Berg, C., Isgum, I.: Deep mr to ct synthesis using unpaired data. arXiv e-prints (2017)

12. Zhu, J.Y., Park, T., Isola, P., Efros, A.A.: Unpaired image-to-image translation using cycle-consistent adversarial networks. IEEE (2017)

13. Zhu, J.Y., Park, T., Isola, P., Efros, A.A.: Unpaired image-to-image translation using cycle-consistent adversarial networks. In: Proceedings of the IEEE International Conference on Computer Vision, pp. 2223–2232 (2017)

Semi-supervised Medical Image Segmentation Based on Multi-scale Knowledge Discovery and Multi-task Ensemble

Yudie Tu, Xiangru Li$^{(\boxtimes)}$, Yunpeng Zhong, and Huanyu Mei

School of Computer Science, South China Normal University,
Guangzhou 510631, China
xiangru.li@qq.com

Abstract. The high cost of manual annotation for medical images leads to an extreme lack of annotated samples for image segmentation. Moreover, the scales of target regions in medical images are diverse, and the local features like texture and contour of some images (such as skin lesions and polyps) are often poorly distinguished. To solve the above problems, this paper proposes a novel semi-supervised medical image segmentation method based on multi-scale knowledge discovery and multi-task ensemble, incorporating two key improvements. Firstly, to detect targets with various scales and focus on local information, a multi-scale knowledge discovery framework (MSKD) is introduced and discovers multi-scale semantic features and dense spatial detail features from cross-level (image and patches) inputs. Secondly, by integrating the ideas of multi-task learning and ensemble learning, this paper leverages the knowledge discovered by MSKD to perform three subtasks: semantic constraints for the target regions, reliability learning for unsupervised data, and representation learning for local features. Each subtask is treated as a weak learner focusing on learning unique features for a specific task in the ensemble learning. Through three-task ensemble, the model achieves multi-task feature sharing. Finally, comparative experiments are conducted on datasets for skin lesions, polyps, and multi-object cell nucleus segmentation, indicate the superior segmentation accuracy and robustness of the proposed method.

Keywords: Medical image segmentation · Semi-supervised learning · Multi-scale knowledge discovery · Multi-task ensemble strategy

1 Introduction

Medical image segmentation plays a crucial role in assisting diagnosis and improving diagnostic efficiency. In recent years, with the rise of deep learning and convolutional neural networks, many researchers investigated the medical image segmentation problem based on this kind technologies [1,2]. However, due

© The Author(s), under exclusive license to Springer Nature Singapore Pte Ltd. 2024
Q. Liu et al. (Eds.): PRCV 2023, LNCS 14437, pp. 209–222, 2024.
https://doi.org/10.1007/978-981-99-8558-6_18

to high dependence of deep learning on vast data volume and the strong difficulty of acquiring authoritative medical images with pixel-by-pixel labels, medical image segmentation methods based on semi-supervised learning [3] attract much attention. These methods aim to extract more knowledge from unlabeled samples to improve segmentation accuracy.

The available semi-supervised medical image segmentation (SSMIS) algorithms based on deep learning can be divided into four main categories, including pseudo label-based methods [4], self-ensembling-based methods [5], entropy minimization-based methods [6], and consistency regularization-based methods [7]. These methods exploit unlabeled data by constructing pseudo-labels for network training or by implementing consistency learning [8–10] on the predictions of the network. Among them, consistency learning-based methods discover the intrinsic semantic information from unlabeled data by a semantic consistency constraint in case of unsupervised perturbations (SCCUP). The SCCUP constraint means that a label-free image and its perturbed sample generated by adding some perturbations should share a common semantic segmentation result. Since it is difficult to determine the optimal perturbation for various tasks, the SCCUP constraint may lead to some suboptimal results in applications. Therefore, Verma et al. [11] abandoned the adversarial perturbations to the input data and proposed interpolation consistency, which guides the training direction of the network by keeping the semantic consistency among the interpolated data.

However, these approaches lack the measure of the disparity between different pixels (features) and also ignore the inter-pixel connection on features. Therefore, to strengthen the mining and utilization of the inter-pixel connection information on features and improve the representational ability of the network, Zhao et al. proposed CLCC (Contrastive Learning and Consistency Constraint) [12] based on the cross-level contrastive learning [13,14]. Nevertheless, there are some still unresolved issues, namely: 1) available methods only capture the information at a single layer but neglect the extraction of representative information in the hierarchical deep network architecture. As the network deepens, local features such as image textures and contours are diluted or even lost. This kind deficiencies are passed layer by layer and accumulated to the output, ultimately damage the segmentation prediction; 2) these methods struggle to extract more effective knowledge for samples with varying target scales; 3) the weights of sub-loss term in existing models' training are usually set as constant, such fixed and empirically selected weights constrain meaningful learning of the model, lead to poor robustness and interpretability.

To deal with the above-mentioned issues, this paper proposes a semi-supervised medical image segmentation method based on multi-scale knowledge discovery and multi-task ensemble. Firstly, we construct a Multi-Scale Knowledge Discovery (MSKD) framework that can adapt to the targets with various scales, capture local information, and exploit rich cross-level multi-scale semantic features and dense spatial detail features. Secondly, we propose a parameter hard-sharing-based multi-task ensemble strategy (MTES) based on the ideas of multi-task learning and ensemble learning. Specifically, based on the knowledge extracted by MSKD, MTES establishes three subtasks: semantic constraint on target regions, unsupervised data reliability learning, and representation learning for local features. Each subtask is treated as a weak learner that focuses on learning a distinctive set of features. By three-task ensemble, the model can share features between multiple tasks, enhances its capability to transfer from representation to category, to eliminate the uncertainty in predicting unlabeled data, and to perceive texture, contours, and other local detail features. Finally, comparative experiments are conducted between our method and mainstream semi-supervised medical image segmentation algorithms on datasets related to skin lesions, polyps, and multi-target cell nucleus segmentation. The experimental results demonstrate the superior segmentation accuracy and robustness of our method. Ablation studies further validate the effectiveness of our proposed method.

2   Related Works on SSMIS

The general idea of available semi-supervised medical image segmentation algorithms based on deep learning is to fully exploit the information from unlabeled data. A prevalent approach is to impose semantic consistency constraints between the prediction results from each sample and its perturbed sample. Perturbations can be achieved by data augmentation operations such as rotation, grayscale transformation, or the addition of Gaussian noise. For instance, Perone et al. [15] applied the idea of Mean-Teacher (MT) [9] to improve bone marrow gray matter segmentation accuracy. Some other scholars used rotational invariance [16] or elastic invariance [10] to construct perturbed samples, improved the segmentation performance by enforcing consistency between each original sample and the corresponding perturbed sample. In addition to adding perturbations to the input, some different consistency learning methods have been proposed based on this idea.

For example, Luo et al. [17] introduced the Uncertainty Rectified Pyramid Consistency (URPC) model, which encourages the consistency between the prediction results from different output layers of the same network. Xu et al. [18] developed a dual uncertainty-guided mixing consistency network to enhance the accuracy of 3D image segmentation. Zhang et al. proposed DTML [19] to incorporate geometric shape constraints by enforcing cross-task consistency constraints among networks with shared parameters. Zhang et al. [20] achieved significant performance improvement by incorporating the boundary information of target regions from unlabeled data through task-level consistency learning. Despite the promising results achieved by these methods, the usual consistency learning-based semi-supervised segmentation algorithms often rely on the assumption that all samples are from similar distributions. However, in reality, medical sample sets are highly complex, and the model that lack of representation capability may lead to uncontrolled predictions. Contrastive learning has shown significant advantages in learning features from unlabeled images and improving the model's representation capability. Therefore, this scheme has been investigated by scholars for semi-supervised medical image segmentation tasks. For example, Wang et al. [21] incorporated image-level contrast learning into the MT model, improved the segmentation accuracy of the heart and the prostate. To address the problem of ambiguous characteristics and limited semantic information in image blocks, Shi et al. [22] introduced a cross-image pixel contrast learning scheme into the model to learn intra-block features. More recently, CLCC [12] employed contrast learning on both whole-image features and patch features to extract effective information embedded in unlabeled data.

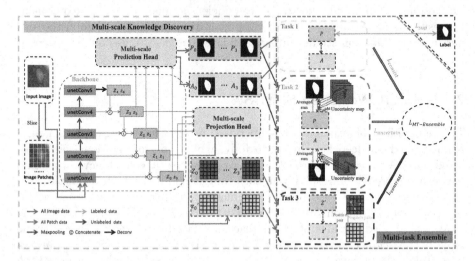

Fig. 1. Overview of the proposed method, which consists of two modules: Multi-scale Knowledge Discovery framework and Multi-task Ensemble strategy. The arrows with different colors represent different data processing flows.

3 Proposed Method

In this section, we detailedly introduce the proposed method (Fig. 1). It consists of two main modules: the Multi-scale Knowledge Discovery framework (MSKD) and the Multi-task Ensemble strategy (MTES). MSKD is composed of an extended backbone network, a multi-scale prediction head, and a multi-scale projection head. It is designed to convert input images into two types of knowledge: semantic segmentation probability maps and dense spatial detail features. These two types of knowledge are further input into the MTES. MTES performs three parallel tasks: semantic constraints on target regions, unsupervised data reliability learning, and representation learning of local features. These three tasks aim to enhance the network's ability to map representations to categories, eliminate uncertainty in predictions from unlabeled data, and improve discriminative power across different representations. By generating different loss feedback and utilizing the backpropagation mechanism, these tasks jointly train parameters of the model to share features. Moreover, MTES guides the MSKD to learn more meaningful and reliable knowledge, to promote the learning for different types of features and improve the segmentation performance of the proposed method. The implementation details of the two modules will be described respectively in Sect. 3.1 and Sect. 3.2. To describe the work in this paper more clearly, we first define some of the mathematical notations used later (listed in Table 1).

Table 1. Some of the used variables and their definitions.

Variables	Definitions
X_i	The i-th image of the input
Y_i	Pixel-by-pixel label of X_i (only for labeled data)
x_j	The j-th patch of X_i, containing n^2 patches in total
s	Representing the scale, which ranges from 0 to 3 in this paper
P_s	Segmentation probability map for X_i at scale s
A_s	Segmentation probability map obtained by concatenating the segmentation probability maps of all patches x_j at scale s
Z'_s	Spatial dense feature maps for X_i at scale s
z'_s	Spatial dense feature map obtained by concatenating the spatial dense feature maps of all patches x_j at scale s

3.1 Multi-scale Knowledge Discovery

The key to semantic segmentation lies in extracting both semantic features and spatial detail features. Semantic features reflect the information of the mapping from representation to pixel categories, while spatial detail features contribute to refining object boundaries. To achieve this, we propose a framework for discovering these two types of knowledge. Firstly, to fully exploit the knowledge at multiple scales, we extend the decoder side of the backbone network. The input data X_i passes through the extended backbone network to generate multi-scale feature maps Z_s. To confine the receptive field within a limited space and enhance the network's sensitivity to small target regions, we incorporate patches x_j of the input image as inputs to the backbone for generating multi-scale feature maps z_s. The multi-scale feature maps Z_s and z_s are then fed into the multi-scale prediction head (PreH) and multi-scale projection head (ProH), which produce multi-scale semantic features (P_s and A_s) and dense spatial detail features (Z'_s and z'_s).

Specifically, the PreH consists of a 1×1 convolutional layer and a sigmoid function, is to generate segmentation probability maps at multiple scales. Furthermore, we upsampled the low-resolution probability maps to the original image resolution, and represent the result as P_s and A_s. By performing interpolation upsampling, the details such as contours and textures can be better captured and represented, thereby improving the segmentation accuracy. At the same time, resampling the segmentation probability maps to the same resolution also facilitates the implementation of subsequent MTES. The ProH, as an improvement based on the projection head, enables different-scale feature maps to be mapped to a new representation space of the same dimension, which exhibits better discriminability for learning local feature representations. The projection head used in this paper consists of three convolutional layers, each followed by a BN layer and a ReLU layer, enhancing the model's robustness and non-linear expressive power. The dense spatial detail features generated by the ProH from the original feature maps Z_s and z_s can be represented as Z_s' and z_s', respectively. The aforementioned process is formally described as follows:

$$P_s = I_s(Pre_s(f_s(X_i))), \tag{1}$$

$$A_s = I_s(Pre_s(\bigcup_{j=1}^{n^2} f_s(x_j))), \tag{2}$$

$$Z_s' = Pro_s(f_s(X_i)), \tag{3}$$

$$z_s' = Pro_s(\bigcup_{j=1}^{n^2} f_s(x_j)), \tag{4}$$

where $f(\cdot)$ denotes the extended backbone network, $I(\cdot)$ denotes the bilinear interpolation operation, $Pre(\cdot)$ denotes the PreH, $Pro(\cdot)$ denotes the ProH, and s represents the scale, which ranges from 0 to 3. The symbol $\bigcup$ denotes the concatenation operation.

3.2 Multi-task Ensemble Strategy

In this paper, we focus on enhancing the mapping capability from features to categories, eliminating the negative effects from the noises generated in predicting the unlabeled data, and improving the discriminative power of different representations. To achieve this, we parallelly perform three tasks: (1) semantic constraints on target regions, (2) reliable learning from unlabeled data, and (3) representation learning of local features. These tasks are further integrated to train a common model, enabling feature sharing between the tasks and achieving more accurate segmentation.

Specifically, **Task 1** focuses on enforcing semantic consistency between segmentation probability maps P_s and A_s generated at the same scale from cross-level inputs. Then such semantic consistency is achieved by $L_{consist}$ and propagated hierachically across multiple scales. For labeled data, the segmentation probability map P_s is directly supervised by the ground truth (GT) (L_{sup}), ensuring a strong mapping from features to categories. So A_s can benefit such mapping from $L_{consist}$. By implementing Task 1, the network can improve the mapping capability from features to categories, while also enhancing its prediction's spatial invariance.

Task 2 addresses the issues of unreliable predictions or noise in the segmentation probablity maps of unlabeled data in Task 1. Directly enforcing $L_{consist}$ on unlabeled data may lead to model collapse or loss of details in Task 1. To alleviate this problem, Task 2 employs the KL divergence to quantify the dissimilarity between feature maps (P_s, A_s) and their benchmarks P_{avg} and A_{avg}, which are computed by averaging the feature maps across scales. This divergence metric then serves as a proxy for uncertainty estimation. Based on the uncertainty estimation maps K_{P_s} and K_{A_s}, we design a loss $L_{uncertain}$ that emphasizes high-confidence pixel predictions and ignores low-confidence ones.

Task 3 introduces contrastive learning $L_{contrast}$ between cross-level dense spatial detail features Z'_s and z'_s. By comparing the similarity of features with various scales at the same pixel position k and their differences with features at other pixel positions j, the network becomes more attentive to and emphasizes local features such as textures and contours. This enhances the perception of local features. Moreover, as Task 3 focuses on local feature, it may inadvertently overlook the global background, which can be beneficial for segmentation tasks with complex backgrounds.

Furthermore, to facilitate collaboration among the three tasks, the idea of ensemble learning (EL) is introduced. Each subtask is treated as a weak learner within EL and their weak feedbacks are ensembled by learnable weight coefficients α, β, γ. This integration fusion results in a comprehensive and meaningful strong feedback, referred to as the multi-task ensemble entropy loss. The formula for this loss is represented as Eq. (5):

$$L_{MT-Ensemble} = \alpha \cdot \underbrace{\frac{1}{S}\sum_{s=0}^{S-1}\left[\frac{1}{V}\sum_{v}^{V}(A_s^v - P_s^v)^2\right]}_{L_{consist}} + \beta \cdot \frac{1}{S}\sum_{s=0}^{S-1}\left[\frac{\sum_{v=1}^{V}\left(K_{P_s}^v\right)^2 + \sum_{v=1}^{V}\left(K_{A_s}^v\right)^2}{V}\right.$$

$$\left. + \frac{\sum_{v=1}^{V}\left(P_s^v - P_{avg}^v\right)^2 \cdot \omega_{P_s}^v}{\sum_{v=1}^{V}\omega_{P_s}^v} + \frac{\sum_{v=1}^{V}\left(A_s^v - A_{avg}^v\right)^2 \cdot \omega_{A_s}^v}{\sum_{v=1}^{V}\omega_{A_s}^v}\right]$$

$$+ \gamma \cdot \underbrace{\frac{1}{S \cdot n^2}\sum_{s=0}^{S-1}\left[\sum_{j=1}^{n^2}\log\left(1 + \frac{\sum_{k \neq j}exp\left((Z_s)'_j \cdot (Z_s)'_k/\tau\right)}{exp\left((Z_s)'_j \cdot (z_j)'_s/\tau\right)}\right)\right]}_{L_{contrast}},$$

(5)

where the unspecified term refers to $L_{uncertain}$, it's only used for unlabeled data. V represents the total number of pixels, $w_{P_s}^v = e^{-K_{P_s}^v}$, $w_{A_s}^v = e^{-K_{A_s}^v}$.

$$L_{total} = \underbrace{\frac{1}{2}\sum_{s=0}^{S-1}[Dice(P_s, Y_i) + CrossEntropy(P_s, Y_i)]}_{L_{sup}} + L_{MT-Ensemble},$$

(6)

where Y_i denotes the label of X_i.

Based on the aforementioned multi-task integrated entropy loss, our model is capable of simultaneously considering the mapping ability from representations to categories, the capability to eliminate the noises generated from predictions of unlabeled data, and the discriminative power among different representations during the learning process. Consequently, it obtains a result that is optimal on these three aspects, learns a network model with high segmentation accuracy and robustness.

4   Experiments and Analysis

4.1   Datasets and Implementation Details

We evaluate our method on three publicly available datasets. (1) **ISIC2018** [23] consists of 2594 skin lesion images with pixel-level labels. We randomly select 60% (1556 images) for training, and the remaining 40% is divided equally for validation and testing. For the training set, we use 10% (156 images) as labeled data and the rest as unlabeled data [12]. (2) **Kvasir-SEG** [24] has 1000 colon polyp images. We randomly select 600 images for training, 200 images for validation, and the rest for testing. Similar to [12], we use 20% of the training samples as labeled data. (3) **MoNuSeg2018** [25] contains 50 cell nucleus images with a resolution of 1000×1000. We randomly select 60% for training, 20% for validation, and the remaining 20% for testing. From the training set, we use 20% (6 images) as labeled data.

The experiments are conducted using Python and Pytorch framework on a Windows 10 operating system, with an NVIDIA RTX 3060Ti GPU and 16 GB RAM. To ensure fairness, all methods involved in the experiments are based on the U-Net network. Data augmentation operations, such as random cropping, horizontal and vertical flipping, and random rotation, are applied to the training samples of the network. The images are randomly cropped to a resolution of 320×320, 320×320, and 512×512, corresponding to ISIC2018, Kvasir-SEG, and MoNuSeg2018, respectively. The initial learning rate of the proposed method is set to 1e-3, with a weight decay of 1e-2. The learning rate is adjusted using the cosine decay strategy, and the training process is set to 200 epochs. Due to hardware limitations, the mini-batch size for each training iteration is set to 4. Each batch consists of 2 labeled images and 2 unlabeled images. The network parameters are updated iteratively using the Adamw optimizer.

To quantitatively evaluate the segmentation performance and the effectiveness of network training, we utilize the Dice coefficient (Dice), mean intersection over union (mIoU), and mean absolute error (MAE), following the evaluation metrics used in [12].

4.2   Comparisons with State-of-the-Art Methods

To validate the effectiveness and robustness of the proposed segmentation method, we perform a series of comparison experiments and visualize the segmentation results on three datasets. The compared methods include U-Net [1], URPC [17], CLCC [12] and DTML [19].

Experiments on ISIC2018. We first evaluate the performance of our proposed method on ISIC2018 dataset, and present the quantitative and qualitative results in Table 2 and Fig. 2. The following observations are made: (1) Our proposed method outperforms all the other SOTA methods in terms of both evaluation metrics, Dice and mIoU. Although the MAE value is slightly higher than that of CLCC, it may attribute to the similarity between some skin lesion images and the skin background. This similarity could introduce some deviations in the binarization calculation, and furthessr affects the MAE-based evaluation results. (2) Our method significantly improves the segmentation performance of skin lesion boundary, especially in the cases with high similarities between the skin lesion boundary and the background. Compared to other methods, our method exhibits evident superiorities on capturing local information such as texture and contours.

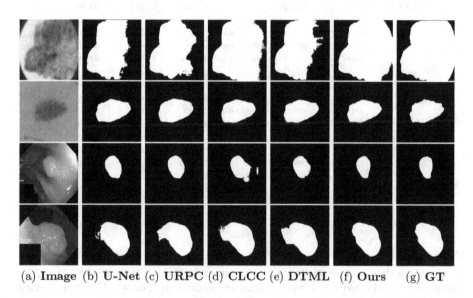

(a) **Image** (b) **U-Net** (c) **URPC** (d) **CLCC** (e) **DTML** (f) **Ours** (g) **GT**

Fig. 2. Qualitative comparisons on ISIC2018 (the first two rows) and Kvasir-SEG (the last two rows), with 10% and 20% labeled data for model training, respectively.

Experiments on Kvasir-SEG. Similarly, we conduct a series of comparative experiments on the polyp segmentation dataset Kvasir-SEG (Table 2, Fig. 2). It is shown that our method outperforms other SOTA methods with an increase of over 2.5% on the Dice metric and over 2.8% improvement on the mIoU evaluation metric compared to CLCC. Moreover, the mean absolute error significantly decreases. All three evaluation metrics consistently demonstrate the effectiveness of the method proposed in this paper. The visualization comparison chart (Fig. 2) can more intuitively show the superiority of this method, especially in the case in which the target area is not clearly distinguished from the background area. Our proposed method can handle it well and segment the shape of polyps more finely and accurately, while the other methods are relatively prone to the disturbances from be disturbed by the background and noise, which result in error segmentation.

Table 2. Quantitative comparison between our method and other semi-supervised segmentation methods on three public datasets. To ensure fairness, all methods use the same U-Net backbone. For each dataset, the first row present the results of a fully supervised baseline method. The bold items indicate the best performance on this dataset. ↑ denotes the higher the better and ↓ denotes the lower the better.

Method		Data used		Metrics		
		Labeled	Unlabeled	Dice (%) ↑	mIoU (%) ↑	MAE(%) ↓
ISIC2018	U-Net [1]	10%(156)	0	80.761	72.141	6.634
	URPC [17]	10%(156)	90%(1400)	77.340	68.069	7.243
	CLCC [12]	10%(156)	90%(1400)	81.847	73.410	**5.893**
	DTML [19]	10%(156)	90%(1400)	81.580	72.853	6.479
	Ours	10%(156)	90%(1400)	**83.196**	**74.518**	6.029
Kvasir-SEG	U-Net [1]	20%(120)	0	77.921	68.489	6.094
	URPC [17]	20%(120)	80%(480)	65.690	55.436	8.146
	CLCC [12]	20%(120)	80%(480)	79.087	69.630	6.010
	DTML [19]	20%(120)	80%(480)	79.243	69.492	6.200
	Ours	20%(120)	80%(480)	**81.595**	**72.431**	**5.570**
MoNuSeg2018	U-Net [1]	20%(6)	0	67.595	52.846	18.901
	URPC [17]	20%(6)	80%(24)	62.243	46.321	20.191
	CLCC [12]	20%(6)	80%(24)	65.335	49.975	17.902
	DTML [19]	20%(6)	80%(24)	69.439	54.974	16.210
	Ours	20%(6)	80%(24)	**69.860**	**55.481**	**15.755**

Experiments on MoNuSeg2018. To demonstrate the generalizability of the proposed method on the multi-target region segmentation task, a series of experiments are conducted on a cell nucleus segmentation dataset. Compared with the previous two datasets, this dataset has a smaller number of images, which increases the risk of overfitting during the network training. The experiments on this dataset help to verify the robustness of our method. We conduct a quantitative comparison (Table 2) and a qualitative segmentation comparison (Fig. 3). As shown in Table 2, our proposed method outperforms the suboptimal method DTML on the cell nucleus segmentation dataset with more complex segmentation tasks. This paper achieves a performance gain of 0.79% on the Dice evaluation metric and 0.947% on the mIoU evaluation metric compared to DTML. Additionally, there is a decrease of 0.874% on the MAE metric.

4.3   Ablation Studies

To further verify the significance of each task in this method, ablation studies are conducted on all three datasets, using the same dataset settings as the previous

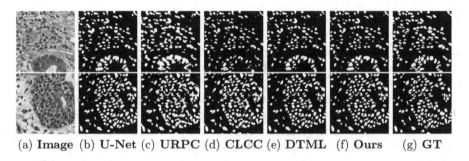

(a) **Image** (b) **U-Net** (c) **URPC** (d) **CLCC** (e) **DTML** (f) **Ours** (g) **GT**

Fig. 3. Qualitative comparisons on MoNuSeg2018, with 20% labeled data involved in model training.

comparison experiments. Specifically, based on the U-Net backbone and multi-scale knowledge discovery framework (MSKD), we make sequential additions to Task1, Task2 and Task3 to verify the effectiveness of each task step by step. When performing single or double tasks, the training losses are weighted with fixed weights, following the settings described in previous work [12]. Moreover, as MSKD is designed to serve MTES, conducting ablation studies on MSKD alone is not feasible. The results of the ablation studies on three datasets are summarized in Table 3.

Table 3. Ablation study of multi-scale knowledge discovery framework (MSKD) and three tasks (T1, T2 and T3) on three datasets. The last row corresponds to the proposed method in this paper.

MSKD	T1	T2	T3	ISIC2018 Metrics(%)			Kvasir-SEG Metrics(%)			MoNuSeg2018 Metrics(%)		
				Dice↑	mIoU↑	MAE↓	Dice↑	mIoU↑	MAE↓	Dice↑	mIoU↑	MAE↓
✓	✓			82.116	73.726	6.140	80.895	71.938	5.820	67.648	52.937	18.708
✓	✓	✓		80.781	72.140	6.820	77.825	68.144	6.057	67.068	52.086	18.552
✓	✓		✓	82.261	73.899	6.232	81.297	**72.563**	5.627	68.733	54.269	18.647
✓	✓	✓	✓	**83.196**	**74.518**	6.029	81.595	72.431	**5.570**	**69.860**	**55.481**	**15.755**

The experimental results show that after adopting MSKD framework and performing Task1, significant performance gains over the backbone networks are achieved on all three datasets, demonstrating the effectiveness and necessity of MSKD and Task1. It can be found that adding Task2 alone on the basis of MSKD+Task1 is ineffective and even reduces the performance of the method. We speculate that this is because the fixed weight loss calculation rule weakens Task2's ability to correct high uncertain samples, making this task unable to play a role, and may even affect the segmentation results due to the noise of the uncertainty estimation itself. Moreover, as we envisioned, after adding Task3 on the basis of MSKD+Task1, the segmentation performance on the three datasets is improved to some extent due to the model's better ability to perceive local feature details. At the same time, from Table 3, it can be seen that based on the MSKD framework, when executing the three tasks in parallel, the Dice and MAE values obtained on the three datasets are superior to other schemes. For the mIoU value, the proposed method in this paper achieves the second best result on the Kvasir-SEG dataset, while the best results are achieved on the other two datasets. Overall, through ablation experiments of each task on the three datasets in this section, it further explains the necessity and effectiveness of each task. Ultimately, by integrating the three tasks to work together, the segmentation performance of the method is improved.

5   Conclusion

In this study, we propose a novel semi-supervised medical image segmentation method based on multi-scale knowledge discovery (MSKD) and multi-task ensemble strategy (MTES). Through MSKD, our method can capture rich multi-scale global semantic features and dense spatial detail features, which are then applied in MTES. The MTES with some learnable weights involves parallel execution of three tasks: semantic constraint learning on target regions, unsupervised data reliability learning, and representation learning of local features. These tasks are trained in parallel based on backpropagation and loss feedback, achieve feature sharing and guide the learning of MSKD. The synergy between MSKD and MTES enables our proposed method to perform excellently in segmenting target with diverse scales. It also enhances the perception of local features such as texture and contours, while suppressing noise, results in more accurate and fine-grained segmentation. Experimental results on three public datasets validate the robustness and accuracy of our method in semi-supervised medical image segmentation tasks, demonstrate its potentials in clinical medicine.

Acknowledgment. This work is supported by the National Natural Science Foundation of China (No. 11973022, 12373108) and the Natural Science Foundation of Guangdong Province (2020A1515010710).

References

1. Ronneberger, O., Fischer, P., Brox, T.: U-Net: convolutional networks for biomedical image segmentation. In: Navab, N., Hornegger, J., Wells, W.M., Frangi, A.F. (eds.) MICCAI 2015. LNCS, vol. 9351, pp. 234–241. Springer, Cham (2015). https://doi.org/10.1007/978-3-319-24574-4_28
2. Isensee, F., Jaeger, P.F., Kohl, S.A., et al.: nnu-net: a self-configuring method for deep learning-based biomedical image segmentation. Nat. Methods 18(2), 203–211 (2021)
3. Zhu, X., Goldberg, A.B.: Introduction to semi-supervised learning. Synth. Lect. Artif. Intell. Mach. Learn. 3(1), 1–130 (2009)
4. Chen, X., Yuan, Y., Zeng, G., et al.: Semi-supervised semantic segmentation with cross pseudo supervision. In: Proceedings of the IEEE/CVF Conference on Computer Vision and Pattern Recognition, pp. 2613–2622 (2021)

5. Li, X., Yu, L., Chen, H., et al.: Transformation-consistent self-ensembling model for semisupervised medical image segmentation. IEEE Trans. Neural Netw. Learn. Syst. **32**(2), 523–534 (2020)
6. Vu, T.H., Jain, H., Bucher, M., et al.: Advent: adversarial entropy minimization for domain adaptation in semantic segmentation. In: Proceedings of the IEEE/CVF Conference on Computer Vision and Pattern Recognition, pp. 2517–2526 (2019)
7. Fan, Y., Kukleva, A., Dai, D., et al.: Revisiting consistency regularization for semi-supervised learning. Int. J. Comput. Vision **131**(3), 626–643 (2023)
8. Ouali, Y., Hudelot, C., Tami, M. Semi-supervised semantic segmentation with cross-consistency training. In: Proceedings of the IEEE/CVF Conference on Computer Vision and Pattern Recognition, pp. 12674–12684 (2020)
9. Tarvainen, A., Valpola, H.: Mean teachers are better role models: weight-averaged consistency targets improve semi-supervised deep learning results. Adv. Neural Inf. Process. Syst. **30** (2017)
10. Bortsova, G., Dubost, F., Hogeweg, L., Katramados, I., de Bruijne, M.: Semi-supervised medical image segmentation via learning consistency under transformations. In: Shen, D., et al. (eds.) MICCAI 2019. LNCS, vol. 11769, pp. 810–818. Springer, Cham (2019). https://doi.org/10.1007/978-3-030-32226-7_90
11. Verma, V., Kawaguchi, K., Lamb, A., et al.: Interpolation consistency training for semi-supervised learning. arXiv preprint arXiv:1903.03825 (2019)
12. Zhao, X., Fang, C., Fan, D.J., et al.: Cross-level contrastive learning and consistency constraint for semi-supervised medical image segmentation. In: 2022 IEEE 19th International Symposium on Biomedical Imaging (ISBI), pp. 1–5. IEEE (2022)
13. Chen, T., Kornblith, S., Norouzi, M., et al.: A simple framework for contrastive learning of visual representations. In: International Conference on Machine Learning, pp. 1597–1607. PMLR (2020)
14. Xiao, T., Liu, S., De Mello, S., et al.: Learning contrastive representation for semantic correspondence. Int. J. Comput. Vision **130**(5), 1293–1309 (2022)
15. Perone, C.S., Cohen-Adad, J.: Deep semi-supervised segmentation with weight-averaged consistency targets. In: Stoyanov, D., et al. (eds.) DLMIA/ML-CDS - 2018. LNCS, vol. 11045, pp. 12–19. Springer, Cham (2018). https://doi.org/10.1007/978-3-030-00889-5_2
16. Li, X., Yu, L., Chen, H., et al.: Semi-supervised skin lesion segmentation via transformation consistent self-ensembling model. arXiv preprint arXiv:1808.03887 (2018)
17. Luo, X., et al.: Efficient semi-supervised gross target volume of nasopharyngeal carcinoma segmentation via uncertainty rectified pyramid consistency. In: de Bruijne, M., et al. (eds.) MICCAI 2021. LNCS, vol. 12902, pp. 318–329. Springer, Cham (2021). https://doi.org/10.1007/978-3-030-87196-3_30
18. Xu, C., Yang, Y., Xia, Z., et al.: Dual uncertainty-guided mixing consistency for semi-supervised 3d medical image segmentation. IEEE Trans. Big Data **9**, 1156–1170 (2023)
19. Zhang, Y., Zhang, J.: Dual-task mutual learning for semi-supervised medical image segmentation. In: Ma, H., et al. (eds.) PRCV 2021. LNCS, vol. 13021, pp. 548–559. Springer, Cham (2021). https://doi.org/10.1007/978-3-030-88010-1_46
20. Zhang, Y., Jiao, R., Liao, Q., et al.: Uncertainty-guided mutual consistency learning for semi-supervised medical image segmentation. Artif. Intell. Med. **138**, 102476 (2023)
21. Wang, K., Zhan, B., Zu, C., et al.: Semi-supervised medical image segmentation via a tripled-uncertainty guided mean teacher model with contrastive learning. Med. Image Anal. **79**, 102447 (2022)

22. Shi, J., Gong, T., Wang, C., Li, C.: Semi-supervised pixel contrastive learning framework for tissue segmentation. IEEE J. Biomed. Health Inf. **27**(1), 97–108 (2023)
23. Codella, N.C., Gutman, D., Celebi, M.E., et al. Skin lesion analysis toward melanoma detection: a challenge at the 2017 international symposium on biomedical imaging (ISBI), hosted by the international skin imaging collaboration (ISIC). In: 2018 IEEE 15th international symposium on biomedical imaging (ISBI 2018), pp. 168–172. IEEE (2018)
24. Jha, D., et al.: Kvasir-SEG: a segmented polyp dataset. In: Ro, Y.M., et al. (eds.) MMM 2020. LNCS, vol. 11962, pp. 451–462. Springer, Cham (2020). https://doi.org/10.1007/978-3-030-37734-2_37
25. Kumar, N., Verma, R., Anand, D., et al.: A multi-organ nucleus segmentation challenge. IEEE Trans. Med. Imaging **39**(5), 1380–1391 (2019)

LATrans-Unet: Improving CNN-Transformer with Location Adaptive for Medical Image Segmentation

Qiqin Lin[1], Junfeng Yao[1,2,3(✉)], Qingqi Hong[1,3,4(✉)], Xianpeng Cao[1],
Rongzhou Zhou[1], and Weixing Xie[1]

[1] Center for Digital Media Computing, School of Film, School of Informatics,
Xiamen University, Xiamen 361005, China
{yao0010,hongqq}@xmu.edu.cn

[2] Key Laboratory of Digital Protection and Intelligent Processing of Intangible
Cultural Heritage of Fujian and Taiwan,
Ministry of Culture and Tourism, Xiamen, China

[3] Institute of Artificial Intelligence, Xiamen University, Xiamen 361005, China

[4] Hong Kong Centre for Cerebro-Cardiovascular Health Engineering (COCHE),
Hong Kong, China

Abstract. Convolutional Neural Networks (CNNs) and Vision Transformers (ViTs) have been widely employed in medical image segmentation. While CNNs excel in local feature encoding, their ability to capture long-range dependencies is limited. In contrast, ViTs have strong global modeling capabilities. However, existing attention-based ViT models face difficulties in adaptively preserving accurate location information, rendering them unable to handle variations in important information within medical images. To inherit the merits of CNN and ViT while avoiding their respective limitations, we propose a novel framework called LATrans-Unet. By comprehensively enhancing the representation of information in both shallow and deep levels, LATrans-Unet maximizes the integration of location information and contextual details. In the shallow levels, based on a skip connection called SimAM-skip, we emphasize information boundaries and bridge the encoder-decoder semantic gap. Additionally, to capture organ shape and location variations in medical images, we propose Location-Adaptive Attention in the deep levels. It enables accurate segmentation by guiding the model to track changes globally and adaptively. Extensive experiments on multi-organ and cardiac segmentation tasks validate the superior performance of LATrans-Unet compared to previous state-of-the-art methods. The codes and trained models will be available soon.

Keywords: Medical image segmentation · Transformer · Location information · Skip connection

The paper is supported by the Natural Science Foundation of China (No. 62072388), the industry guidance project foundation of science technology bureau of Fujian province in 2020 (No. 2020H0047), the Fujian Sunshine Charity Foundation, and ITC-InnoHK.

Q. Liu et al. (Eds.): PRCV 2023, LNCS 14437, pp. 223–234, 2024.
https://doi.org/10.1007/978-981-99-8558-6_19

1   Introduction

Medical image segmentation aims to provide reliable evidence for clinical diagnosis and pathological research by identifying and highlighting anatomical or pathological changes in images, assisting doctors in making more accurate diagnoses. With the development of deep learning, convolutional neural networks (CNNs) have been widely used in various medical image segmentation tasks [1]. Since 2015, the U-Net structure has dominated this field, and the U-Net based on Vision Transformer (ViT) gained prominence in early 2021.

U-Net [2] is a CNN-based model that extracts features through consecutive downsampling in the encoder and then gradually utilizes the features output by the encoder in the decoder through skip connections. This allows the network to obtain features of different granularities for better segmentation. U-Net++ [3], ResUNet [4], DenseUNet [5], R2U-Net [6], and UNet3+ [7] have followed this approach and become representative successors. Despite the impressive achievements, the effectiveness of U-Net in capturing long-range feature dependencies is still limited due to the inherent limitations of the CNN receptive field. Moreover, this global interaction is crucial in medical image segmentation [2] as the segmentation of different regions heavily relies on contextual information. To alleviate this problem, various efforts have been made to expand the receptive field by incorporating spatial pyramid augmentation [8], favorable sampling strategies [9,10], or different attention mechanisms [11–13]. However, all these methods still face challenges in effectively modeling dynamic contextual relationships between arbitrary pixels, indicating significant room for improvement.

Inspired by the tremendous success of Transformers [14] in natural language processing, attention mechanisms have also sparked significant discussions in the computer vision (CV) community [15,16]. Vision Transformer (ViT) [15] is the first purely transformer-based image recognition model. TransUNet [17] utilizes CNNs to extract features and then feeds them into the self-attention of Transformers for modeling long-range dependencies. MT-UNet [18] based on External Attention captures potential correlations in the dataset more effectively by blending transformer modules and learning internal affinity and inter-sample relationships. Swin-Unet [19] is the first pure transformer-based U-shaped network. While these methods are promising to some extent, they still fail to produce satisfactory results in medical image segmentation tasks for the following reasons: 1) Existing medical image segmentation models fail to timely track variations in organ shape and location, thereby impeding efficient and accurate segmentation. 2) Although attention mechanisms are widely used in most models, they do not effectively preserve accurate global location information. This limitation is amplified in medical image segmentation, as precise localization and fine details are crucial for small object segmentation. 3) Although the downsampling operation in CNNs exhibits powerful local feature encoding capabilities, it also leads to the loss of detailed information. The skip connections in the U-Net architecture partially alleviate the issue by restoring the contextual information lost during the encoder's downsampling process. However, it is evident that a simple skip connection scheme is insufficient to model global multi-scale contexts. The semantic gap caused by skip connections restricts further improvements in segmentation performance.

Building upon these insights, we propose a new framework named LATrans-Unet that aims to inherit the advantages of Transformer and convolution while addressing their limitations, to achieve effective medical image segmentation. In neural networks, shallow features contain rich local details, while deep features typically carry more global information. Therefore, in LATrans-Unet, we leverage the phenomenon of spatial suppression from neural science theory [20] when processing shallow features. This phenomenon entails that neurons with the highest information content inhibit the activity of surrounding neurons. Based on this, we propose the SimAM-skip structure to recover the information lost in CNNs. SimAM-skip focus on the details of each pixel, identifying the boundary between important and minor pixels, which aligns with the essence of segmentation. We then integrate and enhance the important pixels, connecting them to the corresponding decoder parts, thereby addressing the semantic gap. When dealing with deep features of long sequences, we propose Location-Adaptive Attention (LAA) to compensate for the limitations of Transformer in capturing location information. LAA utilizes scale factors to capture the information inherent in weight adjustments during training. These weight adjustments represent the variations in organ shape and location on the image. Subsequently, through two separate pooling operations, we obtain inter-channel information of deep features while preserving their location awareness along both horizontal and vertical directions. Finally, we embed the varying weight information into the location information, adaptively guiding the model to timely perceive regional changes of each organ from a global perspective. LATrans-Unet enhances the feature representation of both shallow and deep information comprehensively, thereby improving the performance of the neural network.

In summary, the main contributions of this work are as follows: (1) We propose LATrans-Unet, which inherits the advantages of Transformer and CNNs in medical image segmentation. (2) The proposed Location-Adaptive Attention can not only capture information between channels but also adaptively capture and adjust direction perception and location-sensitive information. (3) For the first time, we integrate neural science theory into skip connections and design SimAM-skip to bridge the semantic gap between the encoder and decoder. (4) We conducted extensive experiments on multi-organ and cardiac segmentation tasks and performed ablation studies to demonstrate the effectiveness of LATrans-Unet.

2   Method

Figure 1 illustrates the overall architecture of the proposed LATrans-Unet. It inherits the advantages of Transformer and convolution while avoiding their respective limitations. In practice, given a medical image input $\mathbf{X} \in \mathbb{R}^{H \times W \times C}$, we first pass it through convolution and pooling operations for feature extraction. Simultaneously, The SimAM-skip structure is designed to fuse and enhance contextual image information from different scales during the encoding of shallow information. Resolutions of $\{\frac{1}{2}, \frac{1}{4}, \frac{1}{8}\}$ compared to the original resolution. Then, we connect the strengthened information to the corresponding part of the decoder, which helps reduce the semantic gap between the encoder and decoder. When dealing with deep information of long sequences, the Location-Adaptive Attention module (LAA) extracts variance-weighted information and location-aware information from deep-level features, and applies them complementarily to enhance the representation of the regions of interest. Following the typical design of TransUNet [17], the decoder performs cascaded upsampler on the extracted multi-level features. Lastly, our LATrans-Unet accurately predicts the pixel-level segmentation masks at the resolution of $H \times W$. Detailed explanations of each component will be provided in the following sections.

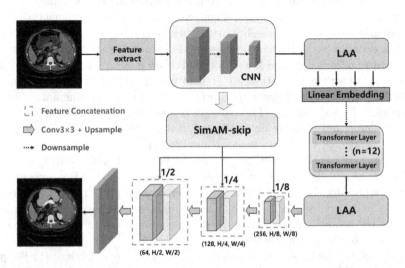

Fig. 1. An overview of the proposed LATrans-Unet framework. The Location-Adaptive Attention (LAA) is used to adaptively perceive the locations of important information within deep-level features.

2.1   Encoder-Decoder Architecture

We follow the encoder-decoder architecture of TransUNet [17]. Specifically, the feature maps obtained after consecutive downsampling by CNNs are reshaped into a sequence of flattened 2D patches $\left\{ \mathbf{x}_t^i \in \mathbb{R}^{P^2 C} \mid i = 1, \cdots, N \right\}$. Each patch has a size of $P \times P$, and the total number of patches is $N = \frac{HW}{P^2}$. The vectorized patches $\mathbf{x}_t$ are then mapped to a latent d-dimensional embedding space, resulting in the original embedding sequence z_0. The transformed patches pass through a 12-layer Transformer to extract global information and patch merging layers. The decoder consists of a cascaded upsampler to achieve full-resolution recovery of $H \times W$. Each block within the decoder includes a $2\times$ upsampling operator, a 3×3 convolutional layer, and a ReLU layer in sequential order.

2.2 Location-Adaptive Attention

The Transformer can easily capture global dependencies. However, the self-attention mechanism in the Transformer does not fully consider the importance of location information in capturing global regions of deep features and overlooks important information related to weight changes during training. This leads to suboptimal performance when dealing with highly variable medical image shapes and locations of organs. To overcome these limitations, we propose Location-Adaptive Attention (LAA) to capture long-range dependencies while preserving accurate location information. Additionally, we adaptively embed weight changes into the location information, enabling the model to quickly adapt to changes in organ shapes and locations from different angles, enhancing the representation of the objects of interest. As shown in the Fig. 2, we first encode each channel using two pooling kernels, $(1, W)$ and $(H, 1)$, along the horizontal and vertical directions, respectively. Therefore, the output at width w can be formulated as:

$$f_w(w) = \frac{1}{H} \sum_{0 \leq i < H} x(i, w) \tag{1}$$

Similarly, the output at height h can be formulated as:

$$f_h(h) = \frac{1}{W} \sum_{0 \leq j < W} x(h, j) \tag{2}$$

where f_w and f_h represent the feature map outputs of the image along the width and height directions, respectively. These two transformations enable LAA to effectively capture distant relationships in one spatial dimension while accurately preserving location information in the other spatial dimension. To fully utilize the captured location information from the aforementioned process, f_w and f_h are concatenated together to form $f \in \mathbb{R}^{C \times (H+W) \times 1}$. Then, we perform batch normalization [21] and activation operations on f. Additionally, we incorporate the scaling factor from batch normalization [21] to measure the variance of the input X and indicate their significance, as:

$$L = BN(B) = \alpha \frac{B - \mu_{\mathcal{B}}}{\sqrt{\sigma_{\mathcal{B}}^2 + \epsilon}} + \beta \tag{3}$$

where $\mu_{\mathcal{B}}$ and $\sigma_{\mathcal{B}}$ are the mean and standard deviation of the input with a batch size of B, respectively. α and β are trainable affine transformation parameters. We calculate the importance weight $W_\alpha = \alpha_i / \sum_{j=0} \alpha_j$ using the trainable α to timely capture the variations in the shape and location of organs in the image. To adapt the location perception according to the changing weight W_α, we split $\mathbf{f}$ into two tensors, $\mathbf{q}_w \in \mathbb{R}^{C \times 1 \times W}$ and $\mathbf{q}_h \in \mathbb{R}^{C \times H \times 1}$, along the spatial dimension. Then $\mathbf{q}_w$ and $\mathbf{q}_h$ are transformed into tensors with the same number of channels as W_α by using two 1×1 convolutional transformations. Finally, we embed the weight variation information into the location information to enhance representation and suppress unimportant information, generating the LAA. The output of our LAA can be formulated as:

$$\mathbf{Y} = \mathbf{X} \cdot \sigma \left[F_h \left(\mathbf{q}_h \right) \times F_w \left(\mathbf{q}_w \right) \times W_\alpha \right] \tag{4}$$

where F_w and F_h are two 1×1 convolutional transforms. σ is a sigmoid function.

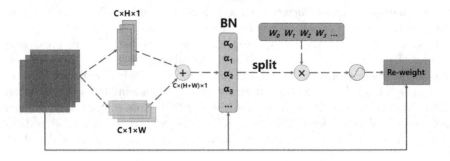

Fig. 2. Detail of the Location-Adaptive Attention.

2.3   SimAM-Skip Structure

The CNNs in the encoder serve as feature extractors, generating feature maps for the input of the Transformer. The hybrid CNN-Transformer encoder performs better than simply using a pure Transformer as the encoder [17]. However, the continuous downsampling process in the hybrid CNN-Transformer encoder leads to the loss of crucial contextual information, which can negatively impact the performance of medical image segmentation. In fact, shallow features contain rich local details. To enhance these local details, We design the SimAM-skip structure, which takes into account the interdependencies and spatial suppression relationships among pixels in medical images, highlighting important pixel information, i.e., the organ's location, and distinguishing it from less significant information, aligning with the essence of segmentation. Specifically, to identify the segmentation location, the importance of each pixel needs to be evaluated. Drawing inspiration from the neural science theory proposed by SimAM [20], we incorporate the phenomenon of spatial suppression, wherein neurons with maximum information content inhibit the activity of surrounding neurons. In other words, neurons exhibiting pronounced spatial suppression effects should be assigned higher priority in visual processing. Therefore, we obtain three different resolution feature maps $\{\mathbf{x}_k \mid k = 1, 2, 3\}$ generated during the downsampling process of CNNs, considering each pixel in the image as an independent neuron. The linear separability between a specific pixel p and other pixels is simulated through the following energy function:

$$e_k^p = \frac{4\left(\hat{\sigma_k}^2 + \lambda\right)}{(p - \hat{\mu}_k)^2 + 2\hat{\sigma_k}^2 + 2\lambda} \tag{5}$$

where $\hat{\mu}_k = \frac{1}{N}\sum_{i=1}^{N} p_k^i$ and $\hat{\sigma}_k^2 = \frac{1}{N}\sum_{i=1}^{N}\left(p_k^i - \hat{\mu}_k\right)^2$ are the mean and variance of all pixel in the channel. Equation (5) indicates that a lower energy e_k^p value represents a greater difference between pixel p and its surrounding pixels, indicating its importance in medical segmentation. To better integrate these important pixels, we utilize a scaling operator instead of addition for feature refinement, as:

$$y_k = \sigma\left(\frac{1}{E}\right) \odot x_k \tag{6}$$

where E groups all e_k^p across channel and spatial dimensions. σ is a sigmoid function. To the best of our knowledge, this work is the first attempt to incorporate the neural science theory into the skip connections of medical image segmentation, and it has achieved promising results.

3   Experiments

3.1   Dataset

Synapse Multi-organ Segmentation Dataset (Synapse): The dataset comprises a total of 30 abdominal CT scans, consisting of 3779 axial contrast-enhanced clinical images. Each CT encompasses volumes in range of 85–198 slices of 512×512 pixels, with a voxel spatial resolution of $([0.54$–$0.54] \times [0.98$–$0.98] \times [2.5$–$5.0])\,\mathrm{mm}^3$. Following [17,22] a random split of 18 cases for training and 12 cases for testing, we utilize the average Dice Similarity Coefficient (DSC) and average Hausdorff Distance (HD) as evaluation metrics to assess the performance of our method on 8 abdominal organs: aorta, gallbladder, spleen, left kidney, right kidney, liver, pancreas, and stomach.

Automated Cardiac Diagnosis Challenge (ACDC): The ACDC dataset contains 100 MRI scans with ground truths for the left ventricle (LV), right ventricle (RV) and myocardium (Myo). The dataset is divided into 70 training samples, 10 validation samples, and 20 testing samples. Similar to [17], we utilize the average DSC as the evaluation metric for our method on this dataset.

3.2 Implementation Details

For our experiments, we implement LATrans-Unet using the PyTorch framework and train it on an NVIDIA RTX 3090 GPU. To diversify the training set and ensure an unbiased training strategy, we apply simple data augmentation techniques such as random rotation and flipping. The input resolution and patch size P are set as 224×224 and 16. Furthermore, we use the pretrained weights on ImageNet [23] to initialize the parameters of the model. The models are trained using the SGD optimizer with a learning rate of 0.01, momentum of 0.9, and weight decay of 1e-4. The default batch size is set to 24.

3.3 Evaluation Results

Experiment Results on Synapse Dataset. The comparison of the proposed LATrans-Unet with previous state-of-the-art methods on the Synapse dataset is presented in Table 1. The experimental results demonstrate the effectiveness and generalizability of our approach, achieving the highest segmentation accuracy of 80.81% (DSC↑). This indicates better consistency between the segmentation results produced by our approach and the ground truth. The segmentation results of different methods on the Synapse dataset are shown in Fig. 3. Our approach accurately detects and classifies organ instances, with minor variations in the segmentation contours. In particular, medical images of organs such as pancreas, stomach, and kidney exhibit high variability in terms of location and shape, which poses challenges for segmentation. From Table 1 and Fig. 3, it can be observed that our approach significantly improves the segmentation of these organs compared to previous state-of-the-art methods. This is attributed to the ability of LATrans-Unet to fuse and enhance contextual semantic information at the shallow level and capture timely changes in organ shape and location at the deep level, effectively embedding them as location information and generating stronger feature representations.

Table 1. Segmentation accuracy of different methods on the Synapse multi-organ CT dataset. DSC of each single class is also presented. All results of our method are averaged over five runs.

Methods	DSC↑	HD↓	aorta	gallbladder	Kidney(L)	Kidney(R)	liver	pancreas	spleen	stomach
V-Net [24]	68.81	–	75.34	51.87	77.10	80.75	87.84	40.05	80.56	56.98
DARR [22]	69.77	–	74.74	53.77	72.31	73.24	94.08	54.18	89.90	45.96
R50 U-Net [17]	74.68	36.87	87.74	63.66	80.60	78.19	93.74	56.90	85.87	74.16
U-Net [2]	76.85	39.70	89.07	**69.72**	77.77	68.60	93.43	53.98	86.67	75.58
R50 Att-UNet [17]	75.57	36.97	55.92	63.91	79.20	72.71	93.56	49.37	87.19	74.95
Att-UNet [25]	77.77	36.02	**89.55**	68.88	77.98	71.11	93.57	58.04	87.30	75.75
R50 ViT [17]	71.29	32.87	73.73	55.13	75.80	72.20	91.51	45.99	81.99	73.95
MT-UNet [18]	78.59	26.59	87.92	64.99	81.47	77.29	93.06	59.46	87.75	76.81
Swin-Unet [19]	79.13	**21.55**	85.47	66.53	83.28	79.61	**94.29**	56.58	**90.66**	76.60
TransUNet [17]	77.48	31.69	87.23	63.13	81.87	77.02	94.08	55.86	85.08	75.62
LATrans-Unet (ours)	**80.81**	26.47	86.83	67.74	**84.70**	**81.16**	93.85	**64.89**	87.19	**80.09**

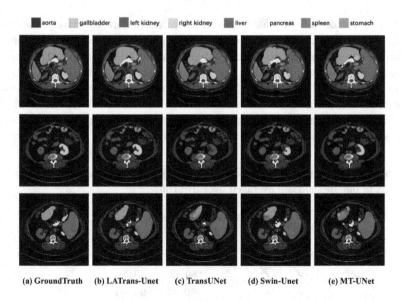

Fig. 3. Visualization results of different methods on the Synapse dataset. From left to right: (a) Ground Truth, (b) LATrans-Unet, (c) TransUNet, (d) Swin-Unet, (e) MT-UNet. Our method predicts less false positive and keep finer information.

Experiment Results on ACDC Dataset. Similar to the Synapse dataset, our proposed LATrans-Unet is trained on the ACDC dataset for medical image segmentation. The experimental results are summarized in Table 2. Despite using MR mode images as input, LATrans-Unet demonstrates outstanding performance with an accuracy of 90.40% (DSC↑), highlighting the strong generalization ability and robustness of our approach.

Table 2. Segmentation accuracy of different methods on the ACDC dataset. All results of our method are averaged over five runs.

Methods	DSC↑	RV	Myo	LV
R50 U-Net [17]	87.55	87.10	80.63	94.92
R50 Att-UNet [17]	86.75	87.58	79.20	93.47
R50 ViT [17]	87.57	86.07	81.88	94.75
Swin-Unet [19]	90.00	88.55	85.62	**95.83**
TransUNet [17]	89.71	88.86	84.53	95.73
LATrans-Unet (ours)	**90.40**	**89.15**	**86.99**	95.05

3.4 Ablation Study

We conducted further ablation studies to investigate the impact of various factors on LATrans-Unet. The results are summarized in Table 3. After incorporating the Location-Adaptive Attention (LAA), the limitations of Transformer in perceiving location information and self-adaptation were effectively addressed, leading to a significant improvement in performance. In addition, by applying the SimAM-skip structure to restore the contextual information lost during the downsampling process of CNNs, the semantic gap between the encoder and decoder is alleviated, leading to further improvement in segmentation performance. These experiments validate the effectiveness of LAA and SimAM-skip. Ablation studies demonstrate that each individual component of LATrans-Unet significantly contributes to improving its performance.

Table 3. Ablation study of two medical image segmentation tasks on DSC metrics. All results of our method are averaged over five runs.

LAA	SimAM-skip	Synapse	ACDC
✗	✗	77.48	89.50
✗	✓	78.50	89.91
✓	✗	80.02	90.09
✓	✓	**80.81**	**90.40**

3.5   Discussion

As is well-known, model pre-training has a substantial impact on the performance of Transformer-based models. In our study, we directly utilize the pretrained weights of Transformer backbone [15] and ResNet-50 [26], trained on ImageNet [23], to initialize the encoder and decoder networks. However, this approach may not be optimal. To address this limitation, we plan to compare and analyze the effects of different initialization methods on the final results, exploring more advanced strategies.

4   Conclusions

In this study, we introduce a novel architecture called LATrans-Unet for precise medical image segmentation. Our approach harnesses the capability of CNNs to capture detailed local information and the global modeling ability of Transformer. Additionally, by considering the distinct characteristics of shallow and deep-level information, we comprehensively enhance the information expression of different levels, maximizing the integration of location information and contextual details. Extensive experiments validate the advantages of LATrans-Unet, showcasing its exceptional performance and generalization capability.

References

1. Wang, R., Lei, T., Cui, R., Zhang, B., Meng, H., Nandi, A.K.: Medical image segmentation using deep learning: a survey. IET Image Proc. **16**(5), 1243–1267 (2022)

2. Ronneberger, O., Fischer, P., Brox, T.: U-Net: convolutional networks for biomedical image segmentation. In: Navab, N., Hornegger, J., Wells, W.M., Frangi, A.F. (eds.) MICCAI 2015. LNCS, vol. 9351, pp. 234–241. Springer, Cham (2015). https://doi.org/10.1007/978-3-319-24574-4_28

3. Zhou, Z., Rahman Siddiquee, M.M., Tajbakhsh, N., Liang, J.: UNet++: a nested U-net architecture for medical image segmentation. In: Stoyanov, D., et al. (eds.) DLMIA/ML-CDS -2018. LNCS, vol. 11045, pp. 3–11. Springer, Cham (2018). https://doi.org/10.1007/978-3-030-00889-5_1

4. Xiao, X., Lian, S., Luo, Z., Li, S.: Weighted res-unet for high-quality retina vessel segmentation. In: 2018 9th International Conference on Information Technology in Medicine and Education (ITME), pp. 327–331. IEEE (2018)

5. Li, X., Chen, H., Qi, X., Dou, Q., Fu, C.W., Heng, P.A.: H-DenseUNet: hybrid densely connected UNET for liver and tumor segmentation from CT volumes. IEEE Trans. Med. Imaging 37(12), 2663–2674 (2018)

6. Alom, M.Z., Hasan, M., Yakopcic, C., Taha, T.M., Asari, V.K.: Recurrent residual convolutional neural network based on u-net (r2u-net) for medical image segmentation. arXiv preprint arXiv:1802.06955 (2018)

7. Huang, H., et al.: Unet 3+: A full-scale connected UNET for medical image segmentation. In: ICASSP 2020–2020 IEEE International Conference on Acoustics, Speech and Signal Processing (ICASSP), pp. 1055–1059. IEEE (2020)

8. Zhao, H., Shi, J., Qi, X., Wang, X., Jia, J.: Pyramid scene parsing network. In: Proceedings of the IEEE Conference on Computer Vision and Pattern Recognition, pp. 2881–2890 (2017)

9. Chen, L.C., Papandreou, G., Schroff, F., Adam, H.: Rethinking atrous convolution for semantic image segmentation. arXiv preprint arXiv:1706.05587 (2017)

10. Gu, Z., et al.: CE-NET: context encoder network for 2d medical image segmentation. IEEE Trans. Med. Imaging 38(10), 2281–2292 (2019)

11. Dai, Y., Gieseke, F., Oehmcke, S., Wu, Y., Barnard, K.: Attentional feature fusion. In: Proceedings of the IEEE/CVF Winter Conference on Applications of Computer Vision, pp. 3560–3569 (2021)

12. Hou, Q., Zhou, D., Feng, J.: Coordinate attention for efficient mobile network design. In: Proceedings of the IEEE/CVF Conference on Computer Vision and Pattern Recognition, pp. 13713–13722 (2021)

13. Zhu, L., Wang, X., Ke, Z., Zhang, W., Lau, R.W.: Biformer: vision transformer with bi-level routing attention. In: Proceedings of the IEEE/CVF Conference on Computer Vision and Pattern Recognition, pp. 10323–10333 (2023)

14. Vaswani, A., et al.: Attention is all you need. Adv. Neural Inf. Process. Syst. 30 (2017)

15. Dosovitskiy, A., et al.: An image is worth 16×16 words: transformers for image recognition at scale. arXiv preprint arXiv:2010.11929 (2020)

16. Zheng, S., et al.: Rethinking semantic segmentation from a sequence-to-sequence perspective with transformers. In: Proceedings of the IEEE/CVF Conference on Computer Vision and Pattern Recognition, pp. 6881–6890 (2021)

17. Chen, J., et al.: Transunet: transformers make strong encoders for medical image segmentation. arXiv preprint arXiv:2102.04306 (2021)

18. Wang, H., et al.: Mixed transformer u-net for medical image segmentation. In: ICASSP 2022–2022 IEEE International Conference on Acoustics, Speech and Signal Processing (ICASSP), pp. 2390–2394. IEEE (2022)

19. Cao, H., et al.: Swin-unet: Unet-like pure transformer for medical image segmentation. In: European Conference on Computer Vision, pp. 205–218. Springer, Heidelberg (2022). https://doi.org/10.1007/978-3-031-25066-8_9

20. Yang, L., Zhang, R.Y., Li, L., Xie, X.: Simam: a simple, parameter-free attention module for convolutional neural networks. In: International Conference on Machine Learning, pp. 11863–11874. PMLR (2021)
21. Ioffe, S., Szegedy, C.: Batch normalization: accelerating deep network training by reducing internal covariate shift. In: International Conference on Machine Learning, pp. 448–456. PMLR (2015)
22. Fu, S., Lu, Y., Wang, Y., Zhou, Y., Shen, W., Fishman, E., Yuille, A.: Domain adaptive relational reasoning for 3D multi-organ segmentation. In: Martel, A.L., et al. (eds.) MICCAI 2020. LNCS, vol. 12261, pp. 656–666. Springer, Cham (2020). https://doi.org/10.1007/978-3-030-59710-8_64
23. Deng, J., Dong, W., Socher, R., Li, L.J., Li, K., Fei-Fei, L.: Imagenet: a large-scale hierarchical image database. In: 2009 IEEE Conference on Computer Vision and Pattern Recognition, pp. 248–255. IEEE (2009)
24. Milletari, F., Navab, N., Ahmadi, S.A.: V-net: fully convolutional neural networks for volumetric medical image segmentation. In: 2016 Fourth International Conference on 3D vision (3DV), pp. 565–571. IEEE (2016)
25. Oktay, O., et al.: Attention u-net: learning where to look for the pancreas. arXiv preprint arXiv:1804.03999 (2018)
26. He, K., Zhang, X., Ren, S., Sun, J.: Deep residual learning for image recognition. In: Proceedings of the IEEE Conference on Computer Vision and Pattern Recognition, pp. 770–778 (2016)

Adversarial Keyword Extraction and Semantic-Spatial Feature Aggregation for Clinical Report Guided Thyroid Nodule Segmentation

Yudi Zhang[1,2,3], Wenting Chen[4], Xuechen Li[1,2,3], Linlin Shen[1,2,3(✉)], Zhihui Lai[1], and Heng Kong[5]

[1] College of Computer Science and Software Engineering,
Shenzhen University, Shenzhen, China
zhangyudi2021@email.szu.edu.cn

[2] AI Research Center for Medical Image Analysis and Diagnosis,
Shenzhen University, Shenzhen, China

[3] National Engineering Laboratory for Big Data System Computing Technology,
Shenzhen University, Shenzhen, China
llshen@szu.edu.cn

[4] City University of Hong Kong, Hong Kong SAR, China

[5] The Department of Breast and Thyroid Surgery,
BaoAn Central Hospital of Shenzhen, BaoAn District, Shenzhen, Guangdong, China

Abstract. Existing thyroid nodule segmentation methods are primarily developed based on ultrasound images, which generally neglects the clinical reports that include rich semantic information for nodules. However, current text guided segmentation methods for natural images are not applicable to the image-report thyroid nodule dataset, due to the many-to-one correspondence between images and reports in current data. To this end, we propose a clinical report guided thyroid nodule segmentation framework with Adversarial Keyword Extraction (AKE) module to extract keywords from reports and Semantic-Spatial Feature Aggregation (SSFA) module to integrate reports into the segmentation model. To alleviate the many-to-one correspondence issue, we devise the AKE module to highlight the keywords about current ultrasound images from clinical reports with a keywords mask, which adopts adversarial learning to encourage the mask generator to mask out the useful descriptions to boost segmentation performance. We further propose the SSFA module to effectively and efficiently map semantic information from reports to each pixel of spatial features, so as to emphasize the target regions. Moreover, we manually collect a clinical Reports Assisted Thyroid Nodule segmentation dataset (RATN), which includes the ultrasound images, the pixel-wise nodule segmentation annotation, and the clinical reports. Extensive experiments have been conducted on the RATN dataset, and the results prove the effectiveness and computational efficiency of the proposed method over the existing methods. Code and data are available at https://github.com/cvi-szu.

© The Author(s), under exclusive license to Springer Nature Singapore Pte Ltd. 2024
Q. Liu et al. (Eds.): PRCV 2023, LNCS 14437, pp. 235–247, 2024.
https://doi.org/10.1007/978-981-99-8558-6_20

Keywords: Thyroid Nodule Segmentation · Clinical Report · Adversarial Keyword Extraction · Feature Aggregation · Ultrasound Image

1   Introduction

Thyroid cancer is the most common malignancy of the endocrine system, and its incidence has increased by 211% in the past two decades in the US [1]. The malignant nodules responsible for thyroid cancer are more likely to exhibit irregular borders when compared to benign nodules, emphasizing the importance of thyroid nodule segmentation in order to accurately observe the nodule border and further classify the nodule [12]. To assist doctors with diagnosis and further treatment, numerous automatic algorithms [6,8,9,14,15,17,18,21,25] for thyroid nodule segmentation have been widely investigated, and the deep learning (DL) based methods [2,6,15,18,21,25] achieve great performance. However, DL-based methods generally ignore the clinical reports paired with the ultrasound images, which can provide additional semantic information (e.g. the characteristics of tissues, location, size, and shape of nodules) to help diagnosis [13]. Hence, it is beneficial to leverage the reports to promote the performance of thyroid nodule segmentation.

In recent years, several text guided segmentation methods [7,11,19,24] have been proposed to utilize relevant descriptions of input images to provide semantic information for image segmentation. These methods adopt the transformer [11,24] or convolutional neural network (CNN) [7,19] as the backbone and fuse the visual features with textual features through the attention mechanism. Despite their significant performance for natural images, they still meet two main challenges for clinical report guided thyroid nodule segmentation. Firstly, existing methods are not applicable to the thyroid nodule dataset with special characteristics. Specifically, the ultrasound doctors usually write an overall clinical report to describe all the view of nodules, leading to the many-to-one correspondence between images and reports. It implies that each report includes descriptions of different ultrasound images. However, current text guided methods are mainly designed for image-report pairs with one-to-one correspondence. Thus, it is necessary to extract the corresponding description for each image from reports and leverage it to facilitate nodule segmentation.

Another challenge is that current text guided segmentation methods [11,19, 24] require a large-scale image-text dataset to train their models. Such methods typically employ an attention mechanism [22] to fuse visual and textual features at different scales, resulting in models with the large number of parameters. However, obtaining large-scale precise annotations for thyroid nodules can be time-consuming, and labor-intensive [10]. The current public dataset is thus usually small. When trained on a small-scale dataset, these methods may encounter the overfitting issue that can compromise the capability of model generalization, leading to reduced segmentation performance. Therefore, it is crucial to develop an effective text guided segmentation method with fewer parameters, which can be effectively trained using the small-scale thyroid nodule dataset.

To address the aforementioned challenges, we propose a novel clinical report guided thyroid nodule segmentation framework, which includes an **Adversarial Keyword Extraction (AKE)** module to extract useful descriptions and a **Semantic-Spatial Feature Aggregation (SSFA)** module for cross-modal feature fusion. Aiming to obtain the corresponding description for each ultrasound image, **Adversarial Keyword Extraction module (AKE)** module utilizes adversarial learning to learn a mask that highlights the relevant descriptions for each image. Concretely, it employs a mask generator to predict a keyword mask and a BERT-tiny [20] model as text encoder to extract masked report features. To mask out more related descriptions, we train the text encoder by minimizing the distance between the original and masked report features, and optimize the mask generator by using adversarial learning to maximize the distance conversely, which encourages the keyword mask to cover more crucial information about thyroid nodules. Moreover, to avoid the overfitting issue for the small-scale dataset, we introduce a **Semantic-Spatial Features Aggregation (SSFA)** to effectively and efficiently integrate reports to the segmentation model, which maps the semantic information to each pixel of spatial features to emphasize the target regions. More importantly, we manually collect a clinical Reports Assisted Thyroid Nodule segmentation dataset (RATN), which, to the best of our knowledge, is the first thyroid nodule segmentation dataset that provides the clinical reports, ultrasound images and precise annotations for each patient. This dataset can facilitate future research on clinical report guided thyroid nodule segmentation. Extensive experiments prove the effectiveness of our method and our superiority to existing methods, indicating the necessity of exploiting the clinical reports for thyroid nodule segmentation.

2   Method

As shown in Fig. 1, we propose a clinical report guided thyroid nodule segmentation framework with Adversarial Keyword Extraction (AKE) module and Semantic-Spatial Feature Aggregation (SSFA) module, which aims to extract useful semantic information to assist the nodule segmentation task. To acquire crucial semantic information, AKE adopts a mask generator to estimate a keyword mask that highlights the relevant keywords for segmentation, and utilizes the text encoder to extract reports features. To supervise the learning of keyword extraction, we employ adversarial learning to play a min-max game between the mask generator and text encoder to make the generated mask to cover more keywords. Moreover, to efficiently and effectively integrate the masked reports into the segmentation model, SSFA is proposed to map the semantic information to spatial features at the pixel level and emphasize the nodule regions.

2.1   Adversarial Keyword Extraction (AKE)

Since current text guided segmentation methods [7,11,19,24] are mainly designed for the paired image-text dataset, these methods are not applicable

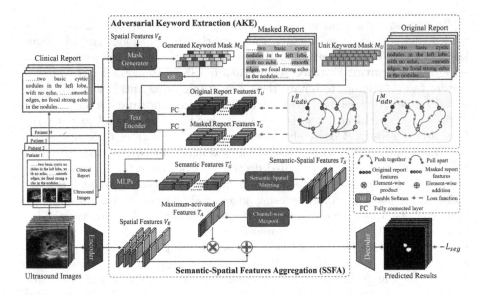

Fig. 1. The overview of clinical report guided thyroid nodule segmentation framework with Adversarial Keyword Extraction (AKE) and Semantic-Spatial Feature Aggregation (SSFA) modules.

to thyroid nodule datasets, where each report is related to several ultrasound images for each patient. Thus, to highlight the relevant descriptions for each image from one report, we propose an Adversarial Keyword Extraction (AKE) to play a min-max game between mask generator and text encoder to learn a keyword mask that covers the related texts.

Specifically, the mask generator, a two-layer transformer followed by a fully connected layer, takes a clinical report, and the spatial features for an ultrasound image encoded by the encoder as input, and estimates a keyword mask M_G. To discrete the generated keyword mask to a one-hot vector, we apply a Gumbel-Softmax [5] to M_G as a differentiable approximation. Then, we perform element-wise multiplication between M_G and the tokens for the clinical report. With the masked tokens, text encoder followed by a fully connected layer produces the corresponding masked report features T_G. Similarly, the original report features T_U are computed with the unit keyword mask M_U and the original report, where M_U is an all-ones vector.

To supervise the learning of generating masks, we employ adversarial learning to play a min-max game between the mask generator and text encoder, which encourages the generated keyword mask to cover as many keywords as possible. When optimizing text encoder, we minimize the distance between the original and masked report features by adopting contrastive representation learning. We treat the original report features and the masked ones as positive pairs and any other pairs as negative pairs. With batch samples number N, we denote the original and masked report features as $\{T_U^i\}_{i=1}^N$ and $\{T_G^i\}_{i=1}^N$, respectively. To reduce the dimension, we use a MLP layer $g(\cdot)$ to transform all the report features to the new ones $Z = \{g(T_U^1), \ldots, g(T_U^N), g(T_G^1), \ldots, g(T_G^N)\} = \{z_1, \ldots, z_{2N}\}$. The contrastive loss is defined as:

$$\ell_{i,j} = -\log \frac{\exp\left(\mathrm{sim}\left(z_i, z_j\right)/\tau\right)}{\sum_{k=1}^{2N} \mathbb{1}_{[k \neq i]} \exp\left(\mathrm{sim}\left(z_i, z_k\right)/\tau\right)}, \tag{1}$$

where sim indicates the cosine similarity. To narrow the gap between the original report features and the masked ones, we minimize the contrastive loss for all the positive pairs, which is calculated as:

$$L_{adv}^T = \sum_{i=1}^{N} \frac{1}{2N} \left(\ell_{i,i+N} + \ell_{i+N,i}\right). \tag{2}$$

With the L_{adv}^T, we can force text encoder to extract more representative features from the clinical reports. Conversely, when optimizing the mask generator, we maximize the distance between the original report features and the masked ones, in order to encourage the mask generator to mask out more keywords and construct the new masked reports different from the original ones. Thus, the adversarial loss for mask generator $L_{adv}^M = -L_{adv}^T$ is the opposite to L_{adv}^T. By adopting adversarial learning for the AKE module, we can make the text encoder capable of extracting representative features and guide mask generator to highlight the keywords related to nodules.

2.2 Semantic-Spatial Features Aggregation (SSFA)

Existing text guided segmentation methods [11,19,24] with attention mechanisms at different scales, comprising numerous parameters, require the large-scale dataset to train their models. However, thyroid nodules segmentation datasets are relatively small, which may make these methods overfitting to these datasets. Hence, to effectively and efficiently fuse the image with report features, we introduce a Semantic-Spatial Feature Aggregation (SSFA) module to map the semantic information to spatial features and further emphasize target regions.

Given the masked report features $T_G \in \mathbb{R}^{L \times E}$ extracted by AKE, an MLP block produces the condensed semantic features $T_G' \in \mathbb{R}^{L \times HW}$, where E denotes the length of word embeddings, H and W are the height and width of the spatial features V_E extracted by image encoder, and L represents the number of token. Afterward, we introduce semantic-spatial mapping to map each word embeddings to each pixel of spatial features. Specifically, the semantic-spatial mapping first adopts an MLP layer to transform the semantic features and then uses a reshape operation to obtain the semantic-spatial features $T_S \in \mathbb{R}^{L \times H \times W}$, where T_S has the same feature size as the spatial features. It maps the semantic information for each word to each pixel of features in the space.

To avoid the noise brought by unrelated descriptions from reports, the channel-wise Maxpooling is proposed to filter out the embeddings with a weak response and mainly retain those with a strong response along the channel dimension. Concretely, we channel-wise pick the maximum value for each pixel in semantic-spatial features and obtain the maximum-activated features $T_A \in \mathbb{R}^{H \times W}$, which shows the regions of interest from masked reports. Subsequently, to aggregate the spatial and semantic information, we perform an element-wise product between spatial features and maximum-activated features, and an element-wise addition to implement a residual path in order to maintain the detailed information in V_E. The final aggregated output V_F is computed as followed: $V_F = (V_E \otimes T_A) \oplus V_E$. With the SSFA module, we can efficiently and effectively integrate the mask report features to image features with very least parameters.

2.3   The Full Objective Functions

We train the proposed framework in three stages to alternatively train the segmentation model, mask generator and text encoder. In the first stage, we freeze the parameters of the mask generator and text encoder, and optimize the segmentation model with dice loss L_{Dice} and binary cross-entropy loss L_{BCE}, which are defined as follows:

$$L_{Dice} = 1 - \frac{2\sum_{i=1}^{N} p_i q_i}{\sum_{i=1}^{N} p_i^2 + \sum_{i=1}^{N} q_i^2}, \tag{3}$$

$$L_{BCE} = -\frac{1}{M} \sum_{i=1}^{M} (q_i \log p_i + (1 - q_i) \log (1 - p_i)). \tag{4}$$

where P denotes the predicted probability map, Q represents the ground truth label, and M is the total pixel number in Q. $p_i \in P$ is the predicted value of the ith pixel in P, and $q_i \in Q$ is the ground truth of the ith pixel in Q. The overall segmentation loss is $L_{seg} = \frac{1}{2}(L_{Dice} + L_{BCE})$.

During the second training stage, we fix the parameters of the segmentation model and mask generator of AKE, and update the text encoder of AKE with L_T. Besides L_{adv}^T, we adopt L_{seg} to encourage text encoder to include more representative and semantic features for segmentation, $L_T = L_{adv}^T + \lambda L_{seg}$. In the last stage, we freeze the segmentation model and text encoder, and optimize the mask generator with $L_M = L_{adv}^M + \lambda L_{seg}$, where L_{seg} forces mask generator to cover more keywords for better segmentation results. λ is a hyper-parameter.

3   Experiment

Dataset. We manually collect a clinical Reports Assisted Thyroid Nodule segmentation dataset (**RATN**), which can facilitate the research on clinical report guided nodule segmentation. This dataset includes 1,522 patient cases with 3,169 ultrasound images in total. Each case includes a clinical report and multiple ultrasound images ranging from 1 to 12, which include different views for thyroid nodules. In addition, the experienced radiologists provide pixel-level thyroid nodule segmentation labels for each ultrasound image. We split RATN dataset

Table 1. Quantitative comparison of the proposed method with existing methods.

Methods	Modalities	Dice (%)	IoU (%)	ACC (%)	FLOPs (G)	Param (M)
U-Net [16] (2015)		82.49	74.81	98.45	7.59	13.66
HRNet [23] (2020)	Image	82.77	75.21	98.61	33.45	62.80
MANet [3] (2020)		83.06	75.49	98.53	8.31	20.67
TransUnet [2] (2020)		82.18	74.19	98.43	45.95	88.91
UNeXt [21] (2022)		81.74	74.52	98.38	0.79	1.40
LAVT [24] (2022)	Image	75.61	63.57	97.64	331.40	501.04
LTS [7] (2021)	+	82.34	74.04	98.49	217.24	150.53
Ours	Report	**84.20**	**77.13**	**98.61**	**8.70**	**21.46**

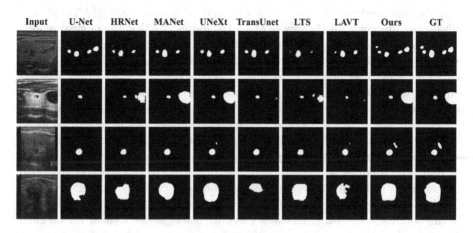

Fig. 2. Qualitative comparison of the proposed method with existing methods.

into a training set with 2,538 images and 1,215 reports from 1,215 patients, and a test set with 631 images and 307 reports from 307 patients. The resolution of ultrasound images is 224 × 224. The maximum length of reports is 256. The RATN dataset is to be released for public access.

Implementation Details. We adopt the MANet [3] backbone with a pre-trained ResNet18 [4] as encoder and a pre-trained BERT-tiny [20] model to extract report features in the AKE module due to its computational efficiency. We alternatively train the segmentation model with L_{seg}, text encoder with L_T and mask generator with L_M. Our method is implemented with the PyTorch library on NVIDIA V100. Adam optimizer is utilized to optimize the model for a maximum of 200 iterations with the learning rate of 1×10^{-3}. A warm-up strategy is used with 0.1 warm-up ratio. λ is 0.1. Batch size is 8. The source code is to be released.

3.1   Comparison with the State-of-the-Arts

Table 1 shows the segmentation performance comparison between the proposed method and related methods on the RATN dataset. Compared to medical image

Table 2. Ablation study results of the AKE module and image-report fusion modules (i.e. SSFA and PWAM).

Combination			Modalities	Dice (%)	IoU (%)	ACC (%)
PWAM	SSFA	AKE				
			Image	82.49	74.81	98.45
✓			Image+Report	78.96	71.92	98.22
	✓			82.87	75.11	98.50
	✓	✓		**83.67**	**75.69**	**98.56**

segmentation methods [2,3,16,21,23], our method surpasses the MANet [3] by a substantial margin with Dice score of 1.14%, indicating that the clinical reports can improve the thyroid nodule segmentation performance. Among text guided segmentation methods [7,24], our method achieves the best performance, exceeding LTS [7] by a significant margin with Dice score of 1.86%, which suggests our superiority to existing text guided methods. Moreover, the computational demand of our method is extremely lower than those of text guided methods, and only slightly higher than those of segmentation methods using single modality. This implies that our method is more effective and efficient than current text guided segmentation methods for the small-scale dataset.

Figure 2 visualizes segmentation results predicted by our method and other segmentation methods. Apparently, our method can precisely segment out more nodules with irregular shapes. In the first row, our method is able to find out all the nodules, while other methods can only detect some of them with regular shapes available in the training set. When learning from training data, the medical image segmentation methods may encounter inductive bias, leading to weak capability of model generalization on the nodules with special shapes. Although LAVT [24] and LTS [7] use the clinical reports as guidance, the segmentation results are similar to the previous methods due to noise brought by the irrelevant description in clinical reports. When integrating the keywords about nodules from clinical reports, our method can capture more nodules with irregular shapes, indicating that our method has a stronger capability of generalization on the nodules with special shapes.

3.2 Ablation Study

Effectiveness of SSFA and AKE. In Table 2, we conduct detailed ablation studies on the RATN dataset to prove the effectiveness of AKE module and SSFA module. We adopt the commonly used U-Net [16] as the baseline model. When integrating reports into the baseline model, the segmentation performance is improved by 0.38%, indicating the effectiveness of SSFA for semantic and spatial feature fusions. Current text guided segmentation methods [11,19,24] mainly adopt the attention mechanism for image-text feature fusion. In particular, PWAM proposed by LAVT [24] degraded the performance (Dice) by a large

Table 3. Quantitative results with different segmentation backbones.

Methods	Dice (%)	IoU (%)	ACC (%)	FLOPs (G)	Param (M)
U-Net [16] (2015)	82.49	74.81	98.45	7.59	13.66
+Ours	**83.67**	**75.69**	**98.56**	7.97	14.45
HRNet [23] (2020)	82.77	75.21	98.61	33.45	62.80
+Ours	**83.65**	**77.68**	**98.79**	33.83	63.59
MANet [3] (2020)	83.06	75.49	98.53	8.31	20.67
+Ours	**84.20**	**77.13**	**98.61**	8.70	21.46

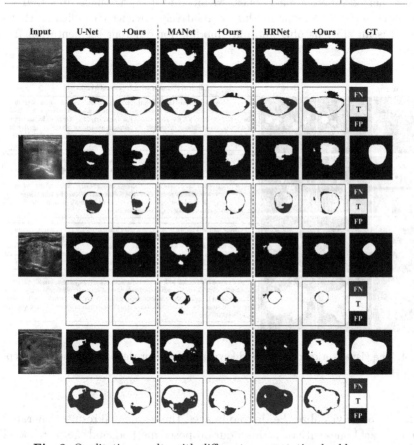

Fig. 3. Qualitative results with different segmentation backbones.

margin, i.e. 3.53%. It means that PWAM is not as effective as our SSFA module when integrating the semantic information to the network on small dataset. Finally, with both SSFA and AKE, the Dice score is substantially improved by 1.18%, to as high as 83.67%, validating the effectiveness of AKE for keyword extraction.

Effectiveness of Our Method Across Different Backbones. To further investigate the effectiveness and generalization of the proposed method across different backbones, we plug the proposed AKE and SSFA into different encoder-decoder-based segmentation backbones, such as U-Net [16], HRNet [23], and MANet [3]. Table 3 shows the segmentation performance of different segmentation backbones with or without our proposed method. Apparently, the segmentation performance of these backbones with our method is improved significantly by 1.18%, 0.88%, and 1.14% in Dice score for U-Net [16], HRNet [23] and MANet [3], respectively, implying the effectiveness and generalization of our method across different backbones. Meanwhile, the increment in FLOPs and parameters brought by our method is relatively low, which indicates that our method is an efficient plug-and-play module for different segmentation frameworks.

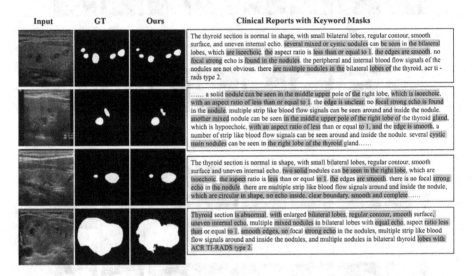

Fig. 4. The visualization of generated keyword masks for three example clinical reports and ultrasound images.

In Fig. 3, we visualize the thyroid images, the segmented results generated by different backbones with or without our proposed method, and the corresponding error maps. In each error map, the regions are filled in red, blue, and white, and these colors represent the different segmentation results as False Positive (FP), True Positive (TP), True positive, and True negative (T), respectively. When injecting the semantic information through our proposed method, the segmentation backbones can more accurately segment the nodule area with the blurry texture and make the segmented results more complete, rather than splitting the tumors into scattered areas. This implies that our method is applicable to various segmentation backbones, demonstrating its generalization capability.

3.3   Visualization of Generated Keyword Masks

To visualize the generated keyword masks of AKE, Fig. 4 displays the ultrasound images, the ground-truth nodules and generated labels, and the clinical reports highlighted by the generated keyword masks in yellow. In each test case, AKE pays attention to the descriptions of thyroid nodules, such as the type (e.g., cystic, mixed and solid), echo (e.g., isoechoic, strong), edge (e.g., smooth), which are regarded as the important visual characteristics to distinguish the nodule regions by the experienced radiologists. In addition, the generated keyword masks also highlight descriptions of the number and position of nodules. For instance, in the hard cases from 1st and 2nd rows, 'multiple nodules' and 'middle upper pole of the right lobe' are emphasized to facilitate the nodule segmentation. These illustrate that our method can extract the keywords about the nodules and effectively utilize them to boost the performance of nodule segmentation.

4   Conclusion

We introduce a clinical report guided thyroid nodule segmentation framework with an AKE module to extract keywords from reports and a SSFA module to integrate reports to image features. To highlight keywords for each image, AKE module generates a keyword mask by using adversarial learning to mask out critical information for segmentation. To integrate reports to image features, a SSFA module is advanced to map semantic information from reports to each pixel of spatial features. Moreover, we collect the RATN dataset to facilitate future research for clinical report guided nodule segmentation. Our extensive experiments on the RATN dataset show our superiority to existing methods in terms of effectiveness, generalization performance and computational efficiency.

Acknowledgement. This work was supported in part by the Natural Science Foundation of China under Grant 82261138629 and 62272319; and Guangdong Basic and Applied Basic Research Foundation under Grant 2023A1515010688, 2021A1515220072 and 2023A1515010677; and Shenzhen Municipal Science and Technology Innovation Council under Grant JCYJ20220531101412030, JCYJ20220530155811025, and JCYJ20220818095803007.

References

1. Acuña-Ruiz, A., Carrasco-López, C., Santisteban, P.: Genomic and epigenomic profile of thyroid cancer. Best Pract. Res. Clin. Endocrinol. Metab. **37**(1), 101656 (2023)
2. Chen, J., et al.: TransUNet: transformers make strong encoders for medical image segmentation. ArXiv (2021)
3. Fan, T., Wang, G., Li, Y., Wang, H.: MA-Net: a multi-scale attention network for liver and tumor segmentation. IEEE Access **8**, 179656–179665 (2020)
4. He, K., Zhang, X., Ren, S., Sun, J.: Deep residual learning for image recognition. In: CVPR, pp. 770–778 (2016)

5. Jang, E., Gu, S., Poole, B.: Categorical reparameterization with Gumbel-Softmax. ArXiv (2016)
6. Jin, Z., Li, X., Zhang, Y., Shen, L., Lai, Z., Kong, H.: Boundary regression-based reep neural network for thyroid nodule segmentation in ultrasound images. Neural. Comput. Appl. **34**, 1–10 (2022)
7. Jing, Y., Kong, T., Wang, W., Wang, L., Li, L., Tan, T.: Locate then segment: a strong pipeline for referring image segmentation. In: CVPR, pp. 9858–9867 (2021)
8. Kaur, J., Jindal, A.: Comparison of thyroid segmentation algorithms in ultrasound and scintigraphy images. Int. J. Comput. Appl. **50**(23), 24–27 (2012)
9. Kollorz, E.N., Hahn, D.A., Linke, R., Goecke, T.W., Hornegger, J., Kuwert, T.: Quantification of thyroid volume using 3-D ultrasound imaging. IEEE Trans. Med. Imaging **27**(4), 457–466 (2008)
10. Li, Z., Zhou, S., Chang, C., Wang, Y., Guo, Y.: A weakly supervised deep active contour model for nodule segmentation in thyroid ultrasound images. Pattern Recognit. Lett. **165**, 128–137 (2023)
11. Li, Z., et al.: LViT: language meets vision transformer in medical image segmentation. ArXiv (2022)
12. Ma, J., Wu, F., Jiang, T., Zhao, Q., Kong, D.: Ultrasound image-based thyroid nodule automatic segmentation using convolutional neural networks. Int. J. Comput. Assist. Radiol. Surg. **12**, 1895–1910 (2017)
13. Monajatipoor, M., Rouhsedaghat, M., Li, L.H., Jay Kuo, C.C., Chien, A., Chang, K.W.: BERTHop: an effective vision-and-language model for chest X-ray disease diagnosis. In: Wang, L., Dou, Q., Fletcher, P.T., Speidel, S., Li, S. (eds.) MICCAI 2022. LNCS, vol. 13435, pp. 725–734. Springer, Cham (2022). https://doi.org/10.1007/978-3-031-16443-9_69
14. Mylona, E.A., Savelonas, M.A., Maroulis, D.: Automated adjustment of region-based active contour parameters using local image geometry. IEEE Trans. Cybern. **44**(12), 2757–2770 (2014)
15. Pan, H., Zhou, Q., Latecki, L.J.: SGUNet: semantic guided UNet for thyroid nodule segmentation. In: ISBI, pp. 630–634 (2021)
16. Ronneberger, O., Fischer, P., Brox, T.: U-Net: convolutional networks for biomedical image segmentation. In: Navab, N., Hornegger, J., Wells, W.M., Frangi, A.F. (eds.) MICCAI 2015. LNCS, vol. 9351, pp. 234–241. Springer, Cham (2015). https://doi.org/10.1007/978-3-319-24574-4_28
17. Savelonas, M.A., Iakovidis, D.K., Legakis, I., Maroulis, D.: Active contours guided by echogenicity and texture for delineation of thyroid nodules in ultrasound images. IEEE Trans. Inf. Technol. Biomed. **13**(4), 519–527 (2008)
18. Tang, Z., Ma, J.: Coarse to fine ensemble network for thyroid nodule segmentation. In: Shusharina, N., Heinrich, M.P., Huang, R. (eds.) MICCAI 2020. LNCS, vol. 12587, pp. 122–128. Springer, Cham (2021). https://doi.org/10.1007/978-3-030-71827-5_16
19. Tomar, N.K., Jha, D., Bagci, U., Ali, S.: TGANet: text-guided attention for improved polyp segmentation. In: Wang, L., Dou, Q., Fletcher, P.T., Speidel, S., Li, S. (eds.) MICCAI 2022. LNCS, vol. 13433, pp. 151–160. Springer, Cham (2022). https://doi.org/10.1007/978-3-031-16437-8_15
20. Turc, I., Chang, M.W., Lee, K., Toutanova, K.: Well-read students learn better: the impact of student initialization on knowledge distillation. ArXiv 13 (2019)
21. Valanarasu, J.M.J., Patel, V.M.: UNeXt: MLP-based rapid medical image segmentation network. In: Wang, L., Dou, Q., Fletcher, P.T., Speidel, S., Li, S. (eds.) MICCAI 2022. LNCS, vol. 13435, pp. 23–33. Springer, Cham (2022). https://doi.org/10.1007/978-3-031-16443-9_3

22. Vaswani, A., et al.: Attention is all you need. NIPS **30**, 5998–6008 (2017)
23. Wang, J., et al.: Deep high-resolution representation learning for visual recognition. IEEE TPAMI **43**(10), 3349–3364 (2020)
24. Yang, Z., Wang, J., Tang, Y., Chen, K., Zhao, H., Torr, P.H.: LAVT: language-aware vision transformer for referring image segmentation. In: CVPR, pp. 18155–18165 (2022)
25. Zhang, Y., Lai, H., Yang, W.: Cascade UNet and CH-UNet for thyroid nodule segmentation and benign and malignant classification. In: Shusharina, N., Heinrich, M.P., Huang, R. (eds.) MICCAI 2020. LNCS, vol. 12587, pp. 129–134. Springer, Cham (2021). https://doi.org/10.1007/978-3-030-71827-5_17

A Multi-modality Driven Promptable Transformer for Automated Parapneumonic Effusion Staging

Yan Chen[1], Qing Liu[1], and Yao Xiang[2,3(✉)]

[1] School of Computer Science and Engineering, Central South University, Changsha, China
chenyan504@csu.edu.cn
[2] Ningbo No. 2 Hospital, Ningbo, China
yao.xiang.csu@gmail.com
[3] Electrical Engineering and Computer Science, Ningbo University, Ningbo, China

Abstract. Pneumonia is a prevalent disease, and some pneumonia patients develop parapneumonic effusion. Among patients with pneumonia, the mortality rate of patients with parapneumonic effusion is higher than that of patients without parapneumonic effusions. Parapneumonic effusion is classified as the uncomplicated stage or complicated stage. Patients with complicated parapneumonic effusion necessitate pleural drainage, while patients with uncomplicated parapneumonic effusion only require antibiotic treatment. Consequently, staging parapneumonic effusion plays a crucial role in reducing mortality among pneumonia patients. The previous method employs convolutional neural networks to extract features from CT slice and graph neural networks for classifying the slice-level feature sequence in parapneumonic effusion staging. However, we argue that transformers outperform graph neural networks in feature sequence classification. Thus, we integrate transformers into our method. Distinguishing between uncomplicated and complicated parapneumonic effusions based solely on CT slices is challenging due to their similar appearance. Therefore, we incorporate additional information in the form of a prompt to aid in the staging process. This allows us to propose a promptable model for parapneumonic effusion staging that incorporates multimodal information. Our method surpasses previous approaches, achieving impressive results with an F1-score of 92.72 and an AUC of 96.67.

Keywords: Deep learning · Medical image processing · Pneumonia effusion staging

1 Introduction

Parapneumonic effusion (PPE) is an accumulation of exudative pleural fluid associated with pneumonia [10], which occurs in up to 50% of all cases of community-acquired pneumonia [7]. Generally, from the view of pathophysiology, PPE progresses from the uncomplicated PPE (UPPE) stage to the complicated PPE (CPPE) stage and finally develops to empyema [6]. The treatment for PPE varies at different stages. Usually, treatment with antibiotics alone for UPPE is likely to be adequate while CPPE requires chest tube drainage [8]. Thus, the identification of the PPE stage is of great significance for doctors to select the treatment strategy.

Computed tomography (CT) has demonstrated its effectiveness in diagnosing diseases related to pleural infections, such as parapneumonic effusions [21,24,25]. CT has also been demonstrated as an effective tool for diagnosing CPPE [25]. In a study conducted by Porcel et al. [21], it is found that certain CT findings are associated with a higher probability of CPPE. These findings included increased attenuation of the extrapleural fat, the presence of a split pleura sign, pleural thickening, and the presence of microbubbles. Conversely, the absence of pleural contrast enhancement was identified as the only radiological finding that decreased the probability of CPPE. Currently, PPE staging in clinical practice mainly relies on radiologists' expertise, which is time-consuming and labor-intensive. Therefore, it is urgent to develop automated method with 3D CT images for PPE staging.

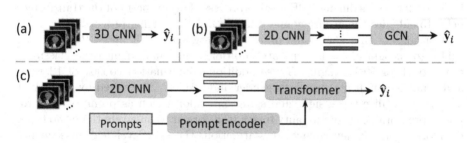

Fig. 1. (a) Works [3,34] that apply 3D CNNs directly on CT images. (b) UG-GAT [13] which utilizes a graph neural network. (c) Our method employs transformer and incorporates signs and symptoms information.

The most intuitive way for automated PPE staging with 3D CT images is to formulate it as a classification problem and directly train a 3D convolutional neural network (CNN) classifier as shown in Fig. 1(a), similar to other 3D image based classification tasks, e.g. hepatocellular carcinoma classification with MRI images [3,34]. However, 3D CNNs usually suffer from the intrinsic defect of over-fitting and has high training overhead. Instead, Hao et al. [13] proposed uncertainty-guided graph attention network for parapneumonic effusion diagnosis (UG-GAT). To effectively leverage the 3D information from CT slices while maintaining a compact model size, the UG-GAT approach adopts a two-step process. Firstly, it extracts features using a 2D CNN on slice-level. Subsequently, these slice-level features are used as graph nodes to construct a graph neural network specifically designed for PPE staging, as shown in Fig. 1 (b). In order to mitigate the noise caused by lesion-free slices, Monte Carlo Dropout (MC Dropout) is adopted to estimate uncertainty for each slice-level feature as the weight of the graph node. Despite achieving state-of-the-art performance, UG-GAT has certain limitations. One notable flaw lies in their graph convolutional network, where direct information interaction occurs solely between adjacent slices. This restricted interaction may hinder the model's ability to capture long-range dependencies. Furthermore, UG-GAT derives the weights of nodes in their graph convolutional network using MC Dropout. However, these weights inferred by MC Dropout may not accurately represent the true importance of slices.

In this paper, we propose a novel PPE staging method that utilizes a transformer for PPE staging. In our method, CT visual information and sign-symptom indicators are integrated. Particularly, we consider the CT slices of a case as a sequence, then the task of staging patients with PPE is regarded as sequence classification. Given the transformer's ability to effectively model long-range dependencies [27], it presents a superiority in tackling sequence-related problems [27]. Hence, we employ the transformer architecture for PPE staging. Transformer leverages the attention mechanism to capture the relationships between features within the sequence. In this way, the token in the sequence that contributes more to the prediction will be highlighted, which mitigate the noise introduced by lesion-free slices during PPE staging process. This property of the transformer helps to enhance the accuracy and reliability of the staging process.

Considering that the visual differences between UPPE and CPPU are too subtle to recognize, we incorporate additional information into the classification model for PPE staging. These additional information consist of 12 signs and symptoms indicators such as gender and age, etc. Motivated by Kirillov et al. [17], we utilize these signs and symptoms information as prompts and propose a promptable cross-modality fusion transformer to integrate these two types of information. By leveraging cross-attention, the slice-level feature sequence and indicator prompts interact with each other, facilitating more accurate PPE staging. This approach allows for a comprehensive analysis by combining the visual information extracted from CT slices with the relevant indicators. The transformer-based fusion model enhances the understanding and integration of these different modalities, leading to improved accuracy in PPE staging.

In summary, our contributions are listed as follows:

- We propose a multi-modality driven promptable transformer method which utilizes both visual features from CT image encoder and signs and symptoms features from a promtable encoders for automated PPE staging.
- We propose a cross-modality fusion transformer for multi-modality feature fusion in which self-attention modules are employed to mitigate the impact of noise from lesion-free slices and cross-attention modules are employed to fuse the two modality features. In this way, discriminative and powerful features are extracted for more accurate classification of pneumonia effusion.
- Our proposed method surpasses previous deep learning approaches in PPE staging, achieving an impressive F1-score of 90.72 and an AUC of 96.78.

2   Related Works

In this section, we first review previous classification methods with CT images for disease detection. Then considering that consecutive CT images of a CT volume are very similar to time sequence videos, we also give a brief literature review regarding video classification.

2.1   Disease Detection Methods with CT Images

The automated identification of PPE stages with CT images belongs to the field of 3D image analysis. Inspired by the remarkable progress in natural scene image classification achieved by 2D CNNs, an abundant of CNN-based methods have been proposed for disease detection with CT images.

Considering the computational efficiency and small-scale datasets, Roth et al. degrade the 3D medical image classification task to a 2.5D image classification [23] and propose to randomly sample three images from three orthogonal views and aggregate them to train a 2D CNN model for classification tasks, e.g., colonic polyp detection with 3D CT images. Rather than feeding the classification network with whole CT images, considering that the regions containing distinguishable information about the disease, i.e., coronavirus disease 2019 (COVID-19), only account for a small ratio, Xu et al. propose to train the disease detection model with abnormal candidate regions detected by a segmentation model [32]. Similar pipeline is also adopted in [29,33] for COVID-19 detection. To accurately segment lesion areas, a large-scale well-annotated database with CT volumes are required. However, creating such a database is high cost as the annotation work is tedious and demands expertise knowledge. To bypass this issue, an alternative way is to design a 3D CNN to classify CT in an end-to-end manner [30,34]. However, the optimization of 3D CNNs is still a challenge and prone to overfitting on small-scale dataset.

Alternatively, many works solve 3D CT classification by 2D plus 1D framework [5,11,14]. Ilse et al. [15] formulate this problem as multiple instance learning (MIL) task. They use a CNN to instance embeddings from 2D slices and aggregate them by an attention pooling. Recently, graph convolutional network (GCN) [16] is introduced for 3D medical image classification tasks. These works treat slice-level features from 2D CNN as nodes and construct a GCN for classification [4,13]. Hao et al. propose a deep learning method UG-GAT for PPE staging [13], where a ResNet is adopted for extracting features from CT slices and a graph neural network is utilized to aggregate slice-level features and make predictions on PPE staging. We argue that transformer is a better choice for aggregating slice-level features of CT.

2.2   Classification Methods with Time Sequence Videos

Early methods for video classification are based on CNNs. Carrera et al. [2] propose I3D, which expand the convolutional and pooling layers of Inception-V1 from 2D to 3D. To balance accuracy and speed, some methods factorize 3D convolutions into a combination of 2D convolutions and 1D convolutions [22,26, 31]. Xie et al. replace certain 3D convolutions in I3D with 2D convolutions [31]. They demonstrate that it is more capable and useful to model semantically-rich and high-level features deep in the network via 3D convolutions.

Since the transformer has been very successful in the field of natural language processing [27] and 2D image processing [9,18], numerous methods have introduced transformer to the field of video classification. Neimark et al. [20] proposed VTN, which employs a pure transformer architecture for video classification. Liu et al. [19] utilized the swin transformer [18], a popular architecture in the realm of 2D images, for video classification tasks [19]. Arnab et al. [1] utilize transformer to classify frame-level sequences extracted from video frames. The work of Arnab et al. demonstrates the effectiveness of the 2D+1D framework in 3D image classification tasks.

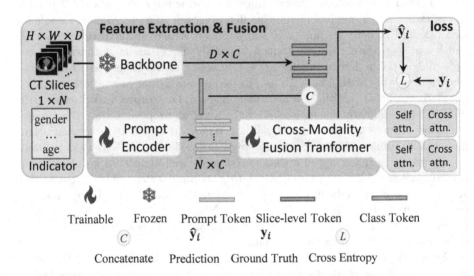

Fig. 2. Our network architecture which comprises three main components: a backbone, a prompt encoder, and a cross-modality fusion transformer. The backbone of our architecture utilizes a pretrained CNN to extract slice-level features from CT slices. The prompt encoder is designed to encode signs and symptoms indicators. The cross-modality fusion transformer module enables the fusion of features extracted from the CT slices by the backbone and the encoded indicators from the prompt encoder.

3   Method

The architecture of our method is depicted in Fig. 2. Our method consists of a CNN-based backbone which is responsible for extracting slice-level features from the CT slices, a promptable encoder which encodes the sign and symptom information, and a cross-modality fusion transformer to capture the interdependencies among these slice-level features and promptable features.

3.1 CNN-Based Slice-Level Feature Extraction

Convolutional neural networks (CNNs) have achieved remarkable successes in various computer vision tasks due to their strong generalization and inductive bias capabilities, thus we employ a CNN (such as ResNet) as a feature extractor to extract slice-level features from CT slices. The input CT volume is defined as having dimensions $H \times W \times D$, where D represents the number of slices and each slice contains $H \times W$ pixels. To process the CT slices, we feed each slice into the well pretrained CNN and extract a 1-dimensional feature vector after applying a global average pooling layer. We denote the feature vector as $\mathbf{f} \in \mathbb{R}^{1 \times C}$, where C denotes the dimensionality of slice-level feature. In our implementation, we directly extract these slice-level features with the CNN-based backbone model provided by Hao et al. [13].

3.2 Prompt Encoder

As the visual differences between UPPE and CPPE are subtle, only relying on visual information from CT images is still challenge to distinguish the PPE stages. To increase the disorientation of the features, we draw inspiration from the work of Kirillov et al. [17] and introduce additional information in the form of prompts for the PPE staging task. Our additional information consists of 12 indicators related to signs and symptoms, including gender, age, fever, cough, sputum, nausea and vomiting, chest tightness, shortness of breath, body mass index, white blood cell count, lymphocyte count, neutrophil count, and C-reactive protein. Each indicator is quantitatively represented as a numeric value, allowing for the integration of these vital clinical parameters into the PPE staging process.

To incorporate these indicators into our model, we first normalize each indicator to a range between 0 and 1. We then employ Fourier feature mapping γ, to encode the input indicators into C-dimensional embedding. Specifically, the function γ maps input indicator $\mathbf{p} \in [0,1]$ to a higher dimensional embedding $\hat{\mathbf{p}} \in \mathbb{R}^{1 \times C}$ with a set of sinusoids. We then concatenate these embedding into a sequence and pass it through a multi-layer perceptron as our prompt encoding. The output of the prompt encoder, denoted as $\mathbf{P} \in \mathbb{R}^{N \times C}$, where N is the number of indicators and C is the dimensionality of the prompt encoding.

It is worth noting that not all cases will have signs and symptoms indicators. In such instances, the prompt encoder is not applicable. In these case, the slice-level features only engage in cross-attention with themselves, allowing for information interaction within the sequence. This ensures that the model can effectively handle cases without indicators. By incorporating prompts based on signs and symptoms indicators, we enhance the model's ability to distinguish between UPPE and CPPE.

3.3 Cross-Modality Fusion Transformer

To effectively combine the slice-level features with the prompt encoding and mitigate the influence of noise from lesion-free slices, we introduce a cross-modality fusion transformer module. This module comprises two essential components: self-attention blocks and cross-attention blocks.

Self-attention is employed to enhance the representation of the slice-level feature sequence. By leveraging self-attention, the model can effectively capture the dependencies among the slices, allowing for better feature extraction and reducing the impact of noise caused by lesion-free slices. Cross-attention is utilized to fuse the slice-level feature sequence with the prompt encodings. This enables effective information interaction and integration between the visual information from CT and the supplementary information from signs and symptoms indicators. The cross-attention mechanism allows the model to attend to relevant cues from both modalities, enhancing the overall accuracy and performance of PPE staging.

Self-attention. The self-attention mechanism models long-range dependencies of sequence and calculates relationship between features in sequence. For the slice-level feature sequence from CT, the weight of the lesion-free slices that bring noise is naturally reduced. For the prompt encodings, the weight of indicators that are more valuable for predicting the stage of PPE is increased. With self-attention, the classification results will be more accurate.

In detail, we normalize the slice-level features $\mathbf{f}$ of a case and concatenate these features to obtain the feature of CT, represented as $\mathbf{S}_{in} \in \mathbb{R}^{D \times C}$, where D represents the number of slices of the patient, and C represents the number of dimensions. We treat $\mathbf{S}_{in}$ as a sequence of slice-level features and feed it into a self-attention block. We also concatenate the prompt encodings, expressed as $\mathbf{P} \in \mathbb{R}^{N \times C}$, where N represents the number of prompts. Similarly, we regard $\mathbf{P}$ as a prompt encoding sequence and input it into a self-attention block. In the self-attention block, the slice-level feature sequence is dot produced with learnable weights to obtain query, key, and value, which are represented as $\mathbf{Q}$, $\mathbf{K}$, and $\mathbf{V}$, respectively. The output $\mathbf{S}_{out}$ of the self-attention block is obtained via

$$\mathbf{S}_{out} = \text{norm}\left(\mathbf{Q} \cdot \mathbf{K}^{\top}\right) \times \mathbf{V} + \mathbf{S}_{in} . \tag{1}$$

Cross-Attention. Since self-attention only performs information interaction within the sequence, we introduce cross-attention to enable information interaction between the slice-level feature sequence and the prompt encoding sequence. Similar to self-attention, cross-attention calculates query, key, and value for slice-level feature sequence and prompt encoding sequence. We define the query, key, and value of the slice-level feature sequence as $\mathbf{Q}_s$, $\mathbf{K}_s$, and $\mathbf{V}_s$, respectively. Correspondingly, query, key, and value are defined as $\mathbf{Q}_p$, $\mathbf{K}_p$, and $\mathbf{V}_p$ for prompt encoding sequence. We make $\mathbf{Q}_s$ and $\mathbf{K}_p$ perform dot product as the attention from prompt encodings to slice-level features. We denote the input sequence as $\mathbf{S}_{in}$ output sequence as $\mathbf{S}_{out}$, then the output of the cross-attention block is calculated via

$$\mathbf{S}_{out} = \text{norm}\left(\mathbf{Q}_s \cdot \mathbf{K}_p^{\top}\right) \times \mathbf{V}_p + \mathbf{S}_{in} . \tag{2}$$

In turn prompt sequences also capture information from slice-level feature sequences via another cross-attention block.

4   Experiments

4.1   Setting and Implementation

Data Description. In this study, our CT data comes from Hao et al. [13] and our signs and symptoms indicators are from a corresponding hospital. Our data consist of 99 UPPE cases and 99 CPPE cases. We have signs and symptoms indicators for 37 UPPE cases and 41 CPPE cases, respectively. All data only include patient-level labels. For convenience, we directly use the slice-level features provided by Hao et al [13].

Experiments Setting. We employ ResNet for slice-level feature extraction. And the slice-level feature is the feature after global average pooling, represented as $\mathbf{f} \in \mathbb{R}^{1 \times 512}$. After the concatenation, the number of channels will be adjusted to 192 through the fully connected layer to match ViT-tiny. Due to the small amount of data, we set the layer of the cross-attention fusion transformer to 2. For a fair comparison with Hao et al. we set the batch size to 1. The learning rate is set to 1e−5 and adjusted by cosine annealing.

Metrics. To evaluate the proposed method, a comparison is made between our method and several approaches previously proposed for medical image classification using 3D data. Some of these methods [3,34] took 3D volumetric data as input and apply 3D CNNs with different architectures to extract features for classification. [4,12,13] extract features using CNN and then utilized a classifier to aggregate information and obtain the prediction results.

Following commonly-used metrics in the classification task, the weighted sensitivity (SEN), specificity (SPE), and accuracy (ACC) are calculated via

$$SEN = \sum_{i=1}^{N_c} \mathrm{w}_i \frac{TP_i}{TP_i + FN_i}, \tag{3}$$

$$SPE = \sum_{i=1}^{N_c} \mathrm{w}_i \frac{TN_i}{TP_i + FP_i}, \tag{4}$$

$$ACC = \sum_{i=1}^{N_c} \mathrm{w}_i \frac{TN_i + TP_i}{TN_i + FP_i + FN_i + TP_i}. \tag{5}$$

where TP_i, TN_i, FP_i, and FN_i denote the true positive, true negative, false positive, and false negative values for the i-th category, respectively, and the w_i represents the percentage of cases whose ground truth labels are i. N_c denotes the number of total classes.

To trade the sensitivity off the specificity, F1-score (F1) is also calculated.

Besides, we calculate the area under curve which is defined based on all possible pairs of SEN and 1-SPE obtained by changing the threshold performed on the classification scores.

4.2 Results

In this section, we present a comparison between our method and other 3D medical image classification methods. Table 1 displays the results of staging PPE using signs and symptoms indicators. Our method demonstrates superior performance compared to the previous state-of-the-art method by Hao et al. [13] in PPE staging. Specifically, we achieve a SEN of 93.68, SPE of 91.57, ACC of 92.63, F1-score of 92.72, and AUC of 96.67. Furthermore, our method outperforms works by Hao et al. [4,13] that employ GCN [16,28] for feature fusion. This finding highlights the superiority of using a transformer [9] to aggregate slice-level features. We attribute this improvement to the transformer's ability to model long-range dependencies.

Additionally, our method surpasses works by Zhou et al. [34] and Chang et al. [3] that directly apply 3D CNN on CT volumes. Methods utilizing a 2D+1D framework [4,12,13] outperform 3D CNN approaches [3,34], thereby confirming the effectiveness of the 2D+1D framework. To further validate our method, we

Table 1. Compared with other methods in classification of 99 UPPE, 99 CPPE.

Methods	SEN	SPE	ACC	F1-score	AUC
Chang et al. (3D ResNet) [3]	61.62	85.87	73.74	70.11	78.61
Zhou et al. (3D SE-DenseNet) [34]	62.63	79.80	71.21	68.51	75.65
Hao et al. (2D CNN+ConvLSTM) [12]	66.67	92.93	79.80	76.74	85.85
Hao et al. (Bayesian CNN+UG-GAT) [13]	80.82	93.31	86.88	86.02	90.96
Ours (CNN+Transformer)	**93.68** ± 3.94	**91.57** ± 5.36	**92.63** ± 1.97	**92.72** ± 1.81	**96.67** ± 2.09

Table 2. Compared with other methods in classification of 37 UPPE, 41 CPPE. All PPE are with signs and symptoms information included.

Methods	SEN	SPE	ACC	F1-score	AUC
Chang et al. (3D ResNet) [3]	62.66 ± 5.33	66.53 ± 3.29	62.66 ± 5.33	65.86 ± 2.90	77.85 ± 6.32
Zhou et al. (3D SE-DenseNet) [34]	66.57 ± 23.57	76.00 ± 24.80	73.33 ± 8.43	69.69 ± 14.49	87.71 ± 8.93
Chen et al. (I2GCN) [4]	**87.94** ± 6.79	79.01 ± 12.57	82.66 ± 3.26	83.88 ± 4.54	83.40 ± 4.03
Hao et al. (Bayesian CNN+UG-GAT) [13]	85.00 ± 14.57	**91.42** ± 11.42	88.00 ± 4.98	87.79 ± 6.23	93.21 ± 5.46
Ours (CNN+Transformer)	87.50 ± 11.18	**91.42** ± 11.42	**89.33** ± 5.33	**89.56** ± 5.35	**97.14** ± 3.49

Table 3. Ablation study of indicator encoder in staging of UPPE and CPPE. Data consist of 99 UPPE cases and 99 CPPE cases.

Indicators encoder	SEN	SPE	ACC	F1-score	AUC
–	88.42 ± 11.24	89.47 ± 5.76	88.94 ± 5.36	88.56 ± 6.21	92.13 ± 4.98
Fully-connected	**93.68 ± 3.94**	90.52 ± 5.15	92.10 ± 2.35	92.24 ± 2.24	96.39 ± 2.53
Prompt Encoder	**93.68 ± 3.94**	**91.57 ± 5.36**	**92.63 ± 1.97**	**92.72 ± 1.81**	**96.67 ± 2.09**

constructed a sub-dataset in which all cases contain signs and symptoms features. As indicated in Table 2, our method consistently outperforms all previous methods in this subset as well.

Table 4. Ablation study of different cross-modality fusion transformer layers.

Layers	SEN	SPE	ACC	F1-score	AUC
1	**93.68 ± 5.36**	**93.68 ± 5.36**	**93.68 ± 3.72**	**93.68 ± 3.70**	93.68 ± 2.43
2	**93.68 ± 3.94**	91.57 ± 5.36	92.63 ± 1.97	92.72 ± 1.81	**96.67 ± 2.09**
3	**93.68 ± 6.13**	90.52 ± 5.15	92.10 ± 3.33	92.18 ± 3.41	96.50 ± 2.40
4	**93.68 ± 5.15**	91.57 ± 5.36	92.63 ± 3.86	92.70 ± 3.78	96.45 ± 2.56

4.3 Ablation Study

To showcase the effectiveness of our method, we conducted a series of ablation experiments. In Table 3, we explored various approaches to encode signs and symptoms indicators. Notably, we compared the performance of the prompt encoder to that of a fully-connected layer in mapping the sign symptom indicators into a high-dimensional vector. The results clearly demonstrate that the prompt encoder outperforms the fully-connected layer in this task.

We conducted a PPE staging experiment solely using signs and symptoms indicators, which validates their effectiveness. The results show significant performance with an F1 score of 74.56 and an AUC of 79.28. These findings underscore the predictive power of signs and symptoms indicators in accurately staging PPE, emphasizing the importance of incorporating additional information beyond visual cues in the staging process.

Layers of Cross-Modality Transformer. We examined the impact of the number of layers in the cross-modality fusion transformer on the staging results. As illustrated in Table 4, we evaluated different numbers of layers and found that setting the number of layers to 2 strikes a balance between model stability and effectiveness.

5   Conclusion

In this paper, we propose a novel approach for PPE staging. Our method introduces self-attention to address the noise caused by lesion-free slices in CT. To incorporate signs and symptoms indicators into the network, we employ a prompt encoder and fuse the prompt encodings with slice-level features using cross-attention. The results demonstrate that our method achieves state-of-the-art performance on the parapneumonic effusion staging task.

Acknowledgements. This work was supported in part by the Ningbo Clinical Research Center for Medical Imaging (No. 2021L003) and Funded by the Project of NINGBO Leading Medical&Health Discipline, Project Number: No. 2022-S02.

References

1. Arnab, A., et al.: ViViT: a video vision transformer. In: ICCV, pp. 6836–6846 (2021)
2. Carreira, J., Zisserman, A.: Quo Vadis, action recognition? A new model and the kinetics dataset. In: CVPR, pp. 6299–6308 (2017)
3. Chang, K., et al.: Residual convolutional neural network for the determination of IDH status in low- and high-grade Gliomas from MR imaging. Clin. Cancer Res. **24**(5), 1073–1081 (2018)
4. Chen, Z., et al.: Instance importance-aware graph convolutional network for 3D medical diagnosis. Media **78**, 102421 (2022)
5. Chikontwe, P., et al.: Dual attention multiple instance learning with unsupervised complementary loss for COVID-19 screening. Media **72**, 102105 (2021)
6. Choudhury, S.R.: Pediatric Surgery. Springer, Cham (2018). https://doi.org/10.1007/978-981-10-6304-6
7. Corcoran, J.P., Wrightson, J.M., Belcher, E., DeCamp, M.M., Feller-Kopman, D., Rahman, N.M.: Pleural infection: past, present, and future directions. Lancet Respir. Med. **3**(7), 563–577 (2015)
8. Davies, H.E., Davies, R.J., Davies, C.W.: Management of pleural infection in adults: British thoracic society pleural disease guideline 2010. Thorax **65**(Suppl 2), ii41–ii53 (2010)
9. Dosovitskiy, A., et al.: An image is worth 16×16 words: transformers for image recognition at scale. In: ICLR (2021)
10. Hamm, H., et al.: Parapneumonic effusion and empyema. Eur. Respir. J. **10**(5), 1150–1156 (1997)
11. Han, Z., et al.: Accurate screening of COVID-19 using attention-based deep 3D multiple instance learning. TMI **39**(8), 2584–2594 (2020)
12. Hao, H., et al.: Angle-closure assessment in anterior segment OCT images via deep learning. Media **69**, 101956 (2021)
13. Hao, J., et al.: Uncertainty-guided graph attention network for parapneumonic effusion diagnosis. Media **75**, 102217 (2022)
14. He, K., et al.: Synergistic learning of lung lobe segmentation and hierarchical multi-instance classification for automated severity assessment of COVID-19 in CT images. PR **113**, 107828 (2021)
15. Ilse, M., et al.: Attention-based deep multiple instance learning. In: ICML, pp. 2127–2136 (2018)

16. Kipf, T.N., et al.: Semi-supervised classification with graph convolutional networks. In: ICLR (2017)
17. Kirillov, A., et al.: Segment anything. arXiv preprint arXiv:2304.02643 (2023)
18. Liu, Z., et al.: Swin transformer: hierarchical vision transformer using shifted windows. In: ICCV, pp. 10012–10022 (2021)
19. Liu, Z., et al.: Video swin transformer. In: CVPR, pp. 3202–3211 (2022)
20. Neimark, D., et al.: Video transformer network. In: ICCV, pp. 3163–3172 (2021)
21. Porcel, J.M., et al.: Computed tomography scoring system for discriminating between parapneumonic effusions eventually drained and those cured only with antibiotics. Respirology **22**(6), 1199–1204 (2017)
22. Qiu, Z., Yao, T., Mei, T.: Learning spatio-temporal representation with pseudo-3D residual networks. In: ICCV, pp. 5533–5541 (2017)
23. Roth, H.R., et al.: Improving computer-aided detection using convolutional neural networks and random view aggregation. TMI **35**(5), 1170–1181 (2016)
24. Stark, D., Federle, M., Goodman, P., Podrasky, A., Webb, W.: Differentiating lung abscess and empyema: radiography and computed tomography. Am. J. Roentgenol. **141**(1), 163–167 (1983)
25. Svigals, P.Z., Chopra, A., Ravenel, J.G., Nietert, P.J., Huggins, J.T.: The accuracy of pleural ultrasonography in diagnosing complicated parapneumonic pleural effusions. Thorax **72**(1), 94–95 (2017)
26. Tran, D., Wang, H., Torresani, L., Ray, J., LeCun, Y., Paluri, M.: A closer look at spatiotemporal convolutions for action recognition. In: CVPR, pp. 6450–6459 (2018)
27. Vaswani, A., et al.: Attention is all you need. In: NeurIPS, vol. 30 (2017)
28. Veličković, P., et al.: Graph attention networks. In: ICLR (2018)
29. Wang, S., et al.: A deep learning algorithm using CT images to screen for Corona virus disease (COVID-19). Eur. Radiol. **31**(8), 6096–6104 (2021)
30. Wang, X., et al.: A weakly-supervised framework for COVID-19 classification and lesion localization from chest CT. TMI **39**(8), 2615–2625 (2020)
31. Xie, S., Sun, C., Huang, J., Tu, Z., Murphy, K.: Rethinking spatiotemporal feature learning: speed-accuracy trade-offs in video classification. In: Ferrari, V., Hebert, M., Sminchisescu, C., Weiss, Y. (eds.) ECCV 2018. LNCS, vol. 11219, pp. 318–335. Springer, Cham (2018). https://doi.org/10.1007/978-3-030-01267-0_19
32. Xu, X., et al.: A deep learning system to screen novel coronavirus disease 2019 pneumonia. Engineering **6**(10), 1122–1129 (2020)
33. Zhang, K., et al.: Clinically applicable AI system for accurate diagnosis, quantitative measurements, and prognosis of COVID-19 pneumonia using computed tomography. Cell **181**(6), 1423–1433.e11 (2020)
34. Zhou, Q., et al.: Grading of hepatocellular carcinoma using 3D SE-DenseNet in dynamic enhanced MR images. Comput. Biol. Med. **107**, 47–57 (2019)

Assessing the Social Skills of Children with Autism Spectrum Disorder via Language-Image Pre-training Models

Wenxing Liu[1], Ming Cheng[1], Yueran Pan[1], Lynn Yuan[3], Suxiu Hu[3], Ming Li[1,2(✉)], and Songtian Zeng[3(✉)]

[1] School of Computer Science, Wuhan University, Wuhan 430072, China
[2] Data Science Research Center, Duke Kunshan University, Kunshan 215316, China
ming.li369@duke.edu
[3] Shenzhen Fumi Health Technology Ltd., Co., Shenzhen 518000, China
songtian.zeng@umb.edu

Abstract. Autism Spectrum Disorder (ASD) is a neurodevelopmental disorder that has gained global attention due to its prevalence. Clinical assessment measures rely heavily on manual scoring conducted by specialized physicians. However, this approach exhibits subjectivity and challenges in regions without sufficient medical resources. This study presents paradigms designed to automatically evaluate various aspects of social skills in children with ASD, utilizing our multiview and multimodal behavior system. Moreover, we propose a new pipeline to predict autism-related social skill scores using the language-image pre-training model. Our multimodal behavioral database comprises 12 subjects (511 videos with labeled social skill scores). Finally, we achieve 81.46% accuracy on the paradigm success prediction task and 69.23% accuracy on the social skill ability scoring task. The results demonstrate that the language-image pre-training model can effectively introduce domain knowledge into video assessment tasks.

Keywords: Autism Spectrum Disorder · Behavior · Language-Image Pre-training Models · Social Skills Assessment

1 Introduction

Autism Spectrum Disorder (ASD) is a prevalent neurodevelopmental disorder that commonly emerges during early childhood and significantly impacts social communication throughout an individual's lifespan [1]. The global estimate indicates a population of at least 78 million individuals affected by ASD [2].

The core characteristic of autism is a lack of social engagement and participation, along with restricted and repetitive behaviors [3]. Presently, ASD diagnosis

This research is funded in part by the Science and Technology Program of Guangzhou City (202007030011), DKU Synear and Wang-Cai Seed Grant and Shenzhen Fumi Health Technology Ltd.

primarily relies on behavior assessments, such as ADOS-G (Autism Diagnostic Observation Schedule-Generic) [4], and its revision ADOS2 [5]. Doctors typically complete a series of pre-designed social games called paradigms with children. Then, they observe the child's performance through recorded video, including social communication skills, attention span, and body movements, for 10 to 20 h per assessment case [6]. The diagnosis is based on specialized scales and heavily relies on the subjective judgment and clinical experience of the physician [7].

Recently, the rapid development of artificial intelligence [8] and sensing technologies [9] show the potential to improve ASD assessment further. There are mainly two types of AI-based automatic assessment approaches. The first is rule-based methods. This type analyzes behavioral, social reflections, head behaviors, and eye patterns in a semi-structured paradigm and manually designs a scoring rubric from the clinical point of view to assess the child's behavioral abilities [10–12]. In this way, the paradigm design and accuracy of pattern recognition modules significantly influence the assessment performance. The second is the raw-video-based method. These methods directly use deep learning to process visual features and directly predict the labels [13,14], suffering from the availability of training data and lack of interpretability.

This study introduces a novel methodology that utilizes language-image pre-training models to assess the social skills of ASD through paradigm videos. First, we adopt a dedicated data collection system, ensuring the presentation of standardized audiovisual stimuli and the recording of multimodal behavioral data [12]. Then, we design a series of social skill assessment paradigms inspired by the ADOS2 protocol and a behavior assessment coding rubric for each paradigm. Furthermore, we use text prompts, designed by autism domain knowledge, to guide a language-image model [15] to extract video features highly correlated with those prompts. Finally, machine learning algorithms are employed to predict the social skill ability scores of children with ASD. To our knowledge, this work is one of the first methods using the language-image pre-training model to assess the social skills of children with ASD. The contributions can be summarized as follows:

- To automatically assess the social skill of children with autism, we design and collect data for nine social skill assessment paradigms covering three areas: language, cognition, and attention.
- The proposed language-image method performs better in the social skill ability score prediction task than the rule-based and raw-video-based methods.

2   Related Works

2.1   Behavior Signal Processing System

We use a standardized platform that includes the stimulation, collection, analysis, modeling, and interpretation of human behavioral data for computer-aided ASD diagnosis [12]. We utilize the proposed assessment environment and audiovisual analysis algorithms for rule-based approaches. Moreover, we design nine new paradigms and the associated rubric logics targeting children's different social skills.

2.2 Language-Image Pre-training Models

In recent years, pre-trained multimodal models [16–18] have achieved impressive performance on many downstream tasks, such as Image-Text Retrieval [19], Image Captioning [20], Visual Question Answering [21], etc. The contrastive language-image per-training (CLIP) [16] uses contrast learning to build connections between images and text. The vision-and-language transformer (VILT) [17] blends visual and textual inputs with a single transformer structure. To address the misalignment between visual and textual features in the semantic space, the align before fuse (ALBEF) [18] introduces a contrastive loss to align the image and text representations before fusing. However, these aforementioned models require substantial training resources, making fine-tuning them on downstream small-scale datasets challenging. In autism, the limited availability of medical data poses difficulties in supporting fine-tuning large-scale models. Our proposed method is inspired by Bootstrapping Language-Image Pre-training (BLIP) [22] and BLIP2 [15]. We utilize the lightweight Q-former to bridge the visual and language models for retrieving text-guided image features, facilitating the use of domain knowledge.

3 Methodology

3.1 Paradigm Design

Paradigms comprise a meticulously referenced series of interactive games that reflect children's social skills in language, cognition, and attention. In this study, we design a set of nine paradigms, as shown in Fig. 1. When experiments start, participants are encouraged to engage in a warm-up phase of unstructured play, allowing the children to relax and become familiar with the therapist. To reassure parents, they are invited to observe their child's performance through videos outside the assessment room. Following the warm-up phase, the child and the therapist assume seated positions on opposite sides of the table, engaging in face-to-face interaction.

Imitative Saying. This paradigm assesses children's capacity to replicate the pronunciation of two-word objects [5]. The procedure is depicted in Fig. 1 a). (1) The therapist employs a slide controller to display images. A pre-selected picture, such as chopsticks, chili, mango, eggplant, or zebra, is projected onto the wall in front of the child. (2) The therapist articulates the name of the picture distinctly and audibly, only once. (3) If the child reproduces the picture's name within 3 s, it is considered a correct response, and the counter is added by 1. Conversely, if the child fails to respond, utters a different word, or articulates unclearly, it is deemed an incorrect behavior, and do not add to the counter

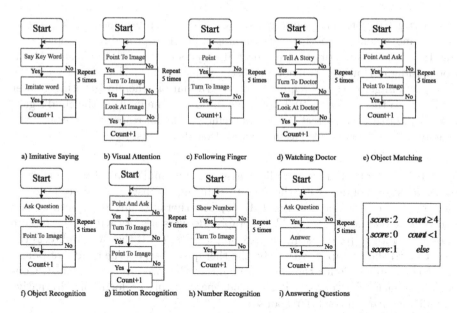

Fig. 1. The procedure of the social assessment paradigms: Imitative Saying (IS), Visual Attention (VA), Following Finger (FF), Watching Doctor (WD), Object Matching (OM), Object Recognition (OR), Emotion Recognition (ER), Number Recognition (NR), Answering Questions (AQ). The words in red denote the doctor's actions, while the words in blue indicate the child's actions. Fundamental behavioral operations encompass looking, speaking, pointing, and so forth. The social assessment computer rules to determine the social skill ability score are enclosed within the box located in the lower right corner. (Color figure online)

(4) Repeat steps (1)–(3) five times for different pictures independently, and get this paradigm score according to assessment rubric in Fig. 1. In the rule-based automatic method, If the count is greater than or equal to 4, the computer determines the social skill ability score as 2. If the count is less than 1, the score is set as 0. In other cases, the score is considered 1.

Visual Attention. This paradigm assesses the child's focus and engagement with visual stimuli, including pictures, objects, or educational material [5]. The procedure is depicted in Fig. 1 b). (1) The therapist puts the hands on the table during the initial phase, and the target image is shown on the walls. (2) The therapist raises a hand to indicate the target picture without verbal cues and then lowers the hand. (3) The therapist observes the child's reaction within 5 s. If the child shifts their attention to the picture and maintains looking for more than five seconds, the test is considered successful, and the counter is incremented by 1. On the contrary, incorrect behaviors encompass failure to redirect attention towards the picture or inability to maintain continuous observation. (4) Each of the nine paradigms is repeated 5 times.

Following Finger. This paradigm assesses the child's responsiveness in tracking the movement of another person's finger [5]. The procedure is depicted in Fig. 1 c). It is noteworthy that the overall process aligns with the Visual Attention paradigm. However, two notable distinctions exist: (1) The orientation of the therapist's finger does not require prior determination, thereby granting greater flexibility to the therapist. 2) There is no imposed time limit on the child's gaze toward the target. Incorrect behaviors include a lack of response, exceeding the time limit, or diverting attention toward other directions.

Watching Doctor. This paradigm assesses the child's capacity to maintain continuous eye contact with the therapist during an instructional scenario [5]. The procedure is depicted in Fig. 1 d). (1) The therapist unveils a picture book and puts it on the chest. (2) The therapist delivers a story within the picture book at a natural pace, lasting approximately 10 s. (3) If the child redirects the gaze toward the therapist within 3 s, the therapist observes the child's response. A correct response is for the child to maintain the gaze on the therapist. Conversely, a lack of response, short visual engagement, or focus on alternative stimuli constitute incorrect responses.

Object Matching. This paradigm is utilized to assess the child's capability for accurately matching pictures of objects [5]. The procedure is depicted in Fig. 1 e). (1) The therapist puts the hands on the table and employs a slide controller to show a picture. The target item is presented on the wall in front of the child. Two option pictures are presented on each side of the child wall. (2) The therapist raises a hand, indicates the wall containing the target, and then poses the question, "Please observe this picture and indicate which image corresponds to it." (3) If the child turns or points to a wall housing the correct object within 3 s, the behavior is considered successful, and the counter is added by 1. On the contrary, the incorrect behavior is that the child does not respond or points to the wrong picture.

Object Recognition. This paradigm assesses children's proficiency in identifying familiar objects [5]. The procedure is depicted in Fig. 1 f). This procedure aligns with Object Matching, with the distinction lying in the second step: The therapist raises a hand and indicates the wall containing the target while posing the question, "Which one corresponds to the corn (grapes, spoons, shoes, puppies)?".

Emotion Recognition. This paradigm assesses children's proficiency in recognizing the four fundamental emotions: happiness, sadness, anger, and fear. The procedure is shown in Fig. 1 g) [5]. It is evident that the overall process closely resembles object matching. The general framework, involving presenting objects, posing questions, and awaiting responses, remains consistent. However, the content of the pictures and questions differs from that of object matching.

The protocol is specifically designed to evaluate the child's emotional cognition, prompting the therapist's questions such as "Which one is happy?", "Which one is sad?", "Which one is angry?", "Which one is scared?".

Number Recognition. This paradigm assesses children's aptitude in recognizing single-digit numbers [5]. The procedure is depicted in Fig. 1 h). The overall process for the paradigm is still consistent with Emotion Recognition. The overall paradigm remains consistent with Emotion Recognition. A brief description of the procedure is as follows: (1) The therapist sequentially presents a number of pictures on the walls facing the child. (2) Without any prompting, the child's successful performance is determined if he or she correctly articulates the number within 3 s. Conversely, failure to respond or providing irrelevant answers are considered incorrect behaviors.

Answering Questions. This paradigm assesses the child's ability to provide verbal or physical responses to questions [5]. The procedure is depicted in Fig. 1 i). (1) The therapist asks the child, "Do you want to drink water?". As the rounds progress, the questions are modified accordingly (e.g., "Do you want to read a book?", "Do you want to go to the bathroom? "). (2) If the child can answer questions using verbal or body language, such as saying "Yes/No," nodding, or shaking their head, it is considered as successful. On the contrary, the incorrect reaction is responding or saying irrelevant words.

In summary, nine paradigms are designed to assess the language, cognitive, and attention abilities of children with ASD. Language paradigms include **Imitative Saying** and **Answering Questions**. Cognitive paradigms include **Object Matching**, **Object Recognition**, **Emotion Recognition**, and **Number Recognition**. Attention paradigms consist of **Visual Attention**, **Finger Tracking**, and **Teacher Observation**. Consistency has been maintained in the design of count and score ranges, with higher scores indicating stronger abilities that align more closely with typical development. Our proposed behavior signal processing system captures the ASD paradigm videos, better reflecting the social skills characteristics of children with ASD. This approach proves more relevant when extracting video features compared to using raw video data.

3.2 Language-Image Based Method

As illustrated in Fig. 2, the fusion feature extraction framework consists of three phases. (1) We first extract video frame features through the image encoder VIT-22B [23], which stands as an excellent visual model in terms of parameter count and similarity to resembles human visual perception (relying less on texture and more on the shape). (2) The Querying Transformer (Q-former) [15], an integral component in BLIP2, generates query features that interact with the text features of the prompt through the attention layer and query video features through the cross-attention layer. To summarize, Q-former allows for querying video features that exhibit semantic relevance to the text prompt, referred to

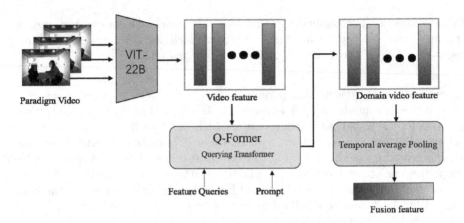

Fig. 2. Overview of language-image pre-training model base fusion feature extraction framework.

as domain video features. (3) we employ a simple temporal fusion method for domain video features. Temporal Average Pooling (TAP) averages the features across the temporal channels. Hence, we obtain fused features that potentially represent the semantics of the entire video.

In experiments, we directly employ the pre-trained VIT-22B and Q-former models for inference. Each paradigm video is encoded as a 768-dimension fused feature. Subsequently, we employ the support vector machine (SVM) with the linear kernel as a classifier to predict the social skill paradigms count and ability score in a leave-one-subject-out manner. An essential aspect of this method lies in the prompt design: The American Psychiatric Association's Diagnostic and Statistical Manual Fifth Edition (DSM-5) provides standardized criteria to help diagnose ASD [24]. We design text prompts regarding the DSM-5 to integrate domain knowledge with video features, as shown in Table 1.

4   Experimental Results

4.1   Database

We recruit 12 participants, aged between 3 and 6 years, consisting of seven individuals diagnosed with autism and five typically developing children. Informed consent was obtained from the parents of all children prior to commencing the formal assessment. We created a dataset containing 9 paradigms for 511 paradigm videos (58 for IS, 60 for VA, 36 for FF, 60 for WD, 60 for OM, 60 for OR, 60 for ER, 57 for NR, and 60 for AQ). Each paradigm video has 8 views, and our method uses only one view when extracting fusion features.

Table 1. Details of DSM-5 Prompts.

Index	Prompt Context
1	The child of Deficits in social-emotional reciprocity
2	The child of Deficits in nonverbal communicative behaviors used for social interaction
3	The child of Deficits in developing, maintaining, and understanding relationships
4	The child of Stereotyped or repetitive motor movement
5	The child of Insistence on sameness, inflexible adherence to routines, or ritualized patterns of verbal or nonverbal behavior
6	The child of Highly restricted, fixated interests that are abnormal in intensity or focus
7	The child of hyper- or hyporeactivity to sensory input or unusual interest in sensory aspects of the environment

To minimize subjectivity, we engaged three professional therapists as evaluators to score the collected video database during the manual labeling process. Each evaluator underwent a training session to familiarize themselves with the rubrics for coding paradigm performance. Additionally, they independently reviewed each video recording in the database, and the majority vote from the three evaluators was used as the ground truth for paradigm successes and ability scores. Our use of data is approved by our Institutional Review Board (IRB).

4.2 Results

We adopt two metrics to evaluate our automatic assessment methods, namely Paradigm Success Accuracy (P-Acc) and Ability Score Accuracy (A-Acc). P-Acc, representing the performance of a two-class classification task, signifies the accuracy rate of an individual single-round paradigm success assessment. A-Acc, representing the performance of a three-class classification task, indicates the accuracy rate of predicting the child's social skill ability scores.

We conduct comparative experiments on the proposed dataset with rule-based and raw video-based methods. The rule-based method uses the Multiview and Multimodal Behavior Transcription (MMBT) system to recognize fundamental human behaviors from recorded data [12]. The raw video-based method extracts features directly from raw videos, and also selects VIT-22B as the feature extractor and SVM as the classifier to ensure the fairness of the experiment.

As the number of video recordings in each paradigm is small, leave-one-subject-out cross validation [25] is adopted to obtain the accuracy estimation for two classification tasks. In our leave-one-subject-out cross-validation, we leave one child's data as the test data and other children's data as the train data for each paradigm. The data for each paradigm is independent, and the training and testing process does not use data from other paradigms.

Table 2. The performance of our proposed pre-training model based method compared to the rule-based and raw-video-based methods. P-Acc and A-Acc denote the Paradigm Success Accuracy and Ability Score Accuracy, respectively.

Paradigm	Rule-Based		Raw-Video-Based		Ours	
	P-Acc	A-Acc	P-Acc	A-Acc	P-Acc	A-Acc
IS	88.33%	75.00%	51.72%	29.31%	84.48%	65.51%
VA	85.56%	75.00%	41.66%	53.33%	70.00%	78.33%
FF	86.25%	66.67%	83.33%	22.22%	91.66%	81.67%
WD	80.56%	75.00%	72.22%	38.88%	88.88%	66.67%
OM	80.00%	33.33%	43.33%	21.67%	70.00%	48.33%
OR	81.67%	66.67%	38.33%	31.67%	71.67%	66.66%
ER	80.00%	75.00%	61.67%	30.00%	80.00%	58.33%
NR	74.03%	33.33%	82.45%	63.16%	96.49%	84.21%
AQ	83.33%	66.67%	70.00%	40.00%	80.00%	73.33%
average	82.20%	62.96%	60.52%	36.69%	81.46%	69.23%

Table 2 compares the results of our proposed method and two baselines from both P-Acc and A-Acc perspectives. From the P-Acc viewpoint, our method achieves an average accuracy of 81.46% across the 9 paradigms, which approaches the 82.20% accuracy achieved by the rule-based method. Notably, the rule-based method utilizes multiple additional high-precision sensors (RGB-D cameras) and high-accuracy behavior signal processing re-trained models [12] (human detection, gaze, head pose, hand gesture, speech recognizer, etc.) Our proposed method achieves a comparable level of accuracy utilizing the RGB video data solely, without using the paradigm design rubrics. In addition, our proposed method improves accuracy by 20.94% compared to the raw-video-based method. Regarding the A-Acc perspective, the rule-based method only attains a score of 62.96% due to the constraints imposed by manually defined ability scoring rules, despite having a higher paradigm accuracy rate. Our method achieves the highest average ability accuracy of 69.23% across the 9 paradigms, outperforming the two baselines by 6.27% and 32.54%, respectively.

The highest P-Acc for our proposed method is 96.49% on the Number Recognition (NR) paradigm, and the highest A-Acc for our proposed method is 81.67% on the Visual Attention (VA) paradigm. The reason for this result may be related to the design of the prompt, where our approach introduces attention-related prompts, e.g. (6) in Table 1. The ability to focus is a common feature of both paradigms. In the NR paradigm, children need to attend to the positions of numbers, whereas, in the VA paradigm, their attention is directed toward the positions of pictures.

4.3 Discussion

In this paper, we propose a language-image pre-training model based behavior assessment method, which incorporates domain knowledge inspired prompts

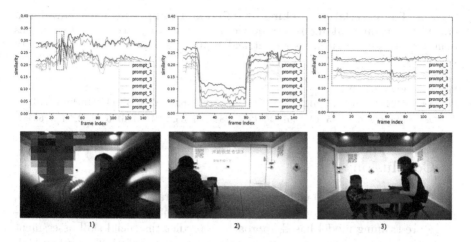

Fig. 3. Three instances of prompt similarity curves corresponding to real scenes. The Similarity meaning is the cosine similarity between image and prompt features. The images in the first row are the prompt similarity curves crossing time, and the second row is one real scene image in the red box. (Color figure online)

in the feature extraction process. The method assesses the similarity between prompt features and image features for each video frame, allowing for the retrieval of features that capture prompt semantics. Consequently, two essential questions arise: 1) Does the change in similarity correspond to events in real-life scenarios? 2) Can the content of the prompt partially explain the paradigm?

To address the first question, Fig. 3 provides three examples. In scenario 1), the similarity curve of the red box exhibits a sharp decline, indicating a realistic situation where children suddenly cover the camera with their hands. In scenario 2), the similarity curve of the red box reaches a significantly low level for a period, corresponding to instances where the child moves away from the camera scene. The smooth variation observed in scenario 3) aligns with children's sustained attention to pictures in the VA paradigm. The child's disappearance in the actual scenario leads to a decrease in correlation, and the continued focus of the child leads to a stable value in correlation. This illustrates that the similarity between prompts and pictures reflects the true variability of the scene. Furthermore, children with ASD generally exhibit higher similarity average value compared to normal children. These observations validate the effectiveness of the Language-Image method in extracting domain-specific features since our prompts are common symptoms of ASD.

To address the second question, we analyze the correlation rankings of different prompts for each paradigm, as presented in Table 3. The content of the prompt contributes to explaining the paradigm to some extent. For instance, the child's attention is a shared observation across the VA, FF, WD, and OR paradigms. Among all the prompts, prompt 6 stands out as the most representative in terms of autistic attention. Consequently, prompt 6 exhibits the highest correlation, as anticipated.

Table 3. The correlation ranking of different Prompts for each paradigm. The Correlation ranking is organized in descending order. P-order means prompt ordering in paradigm assessment, and A-order means prompt ordering in ability score assessment.

Paradigm	IS	VA	FF	WD	OM	OR	ER	NR	AQ
P-Order	5164237	2543671	6124735	1657324	3672145	6574213	5732614	2315467	4726153
A-Order	1476325	6273145	6123457	4671235	2845376	6521347	3746125	6532174	7263451

5 Conclusion

This work designs nine social skills assessment paradigms covering language, cognition, and attention domains. Additionally, we introduced a novel language-image pre-training model based approach to predict the social skill assessment labels. Our method achieves a paradigm success accuracy of 81.46% and an ability score accuracy of 69.23% on our dataset. Notably, our proposed method outperforms the rule-based and raw-video-based methods by incorporating domain knowledge into the text prompts while relying solely on the RGB video data. It has great potential to avoid manually designing the social skill paradigm assessment rubrics and complex systems built upon many behavior signal processing modules.

References

1. Lord, C., et al.: The Lancet Commission on the future of care and clinical research in autism. Lancet **399**(10321), 271–334 (2022)
2. Baio, J., et al.: Prevalence of autism spectrum disorder among children aged 8 years-autism and developmental disabilities monitoring network. MMWR Surveill. Summ. **67**(6), 1 (2018)
3. Harris, J.: Leo Kanner and autism: a 75-year perspective. Int. Rev. Psychiatry **30**(1), 3–17 (2018)
4. Lord, C., et al.: The Autism Diagnostic Observation Schedule-Generic: a standard measure of social and communication deficits associated with the spectrum of autism. J. Autism Dev. Disord. **30**, 205–223 (2000)
5. Gotham, K., Risi, S., Pickles, A., Lord, C.: The Autism Diagnostic Observation Schedule: revised algorithms for improved diagnostic validity. J. Autism Dev. Disord. **37**, 613–627 (2007)
6. Falkmer, T., Anderson, K., Falkmer, M., Horlin, C.: Diagnostic procedures in autism spectrum disorders: a systematic literature review. Eur. Child Adolesc. Psychiatry **22**, 329–340 (2013)
7. Taylor, L.J., et al.: Brief report: an exploratory study of the diagnostic reliability for autism spectrum disorder. J. Autism Dev. Disord. **47**, 1551–1558 (2017)
8. de Belen, R.A.J., Bednarz, T., Sowmya, A., Del Favero, D.: Computer vision in autism spectrum disorder research: a systematic review of published studies from 2009 to 2019. Transl. Psychiatry **10**(1), 333 (2020)
9. Winoto, P., Chen, C.G., Tang, T.Y.: The development of a Kinect-based online socio-meter for users with social and communication skill impairments: a computational sensing approach. In: Proceedings of ICKEA, pp. 139–143 (2016)

10. Bovery, M., Dawson, G., Hashemi, J., Sapiro, G.: A scalable off-the-shelf framework for measuring patterns of attention in young children and its application in autism spectrum disorder. IEEE Trans. Affect. Comput. **12**(3), 722–731 (2019)

11. Wang, Z., Liu, J., He, K., Xu, Q., Xu, X., Liu, H.: Screening early children with autism spectrum disorder via response-to-name protocol. IEEE Trans. Industr. Inf. **17**(1), 587–595 (2019)

12. Cheng, M., et al.: Computer-aided autism spectrum disorder diagnosis with behavior signal processing. IEEE Trans. Affect. Comput. **14**(4), 2982–3000 (2023)

13. Li, J., Zhong, Y., Han, J., Ouyang, G., Li, X., Liu, H.: Classifying ASD children with LSTM based on raw videos. Neurocomputing **390**, 226–238 (2020)

14. Negin, F., Ozyer, B., Agahian, S., Kacdioglu, S., Ozyer, G.T.: Vision-assisted recognition of stereotype behaviors for early diagnosis of autism spectrum disorders. Neurocomputing **446**, 145–155 (2021)

15. Li, J., Li, D., Savarese, S., et al.: BLIP-2: bootstrapping language-image pre-training with frozen image encoders and large language models. arXiv preprint arXiv:2301.12597 (2023)

16. Radford, A., Kim, J.W., Hallacy, C., et al.: Learning transferable visual models from natural language supervision. In: Proceedings of ICML, pp. 8748–8763 (2021)

17. Kim, W., Son, B., Kim, I.: ViLT: vision-and-language transformer without convolution or region supervision. In: Proceedings of ICML, pp. 5583–5594 (2021)

18. Li, J., Selvaraju, R., Gotmare, A., Joty, S., Xiong, C., Hoi, S.C.H.: Align before fuse: vision and language representation learning with momentum distillation. Adv. Neural. Inf. Process. Syst. **34**, 9694–9705 (2021)

19. Cao, M., Li, S., Li, J., et al.: Image-text retrieval: a survey on recent research and development. arXiv preprint arXiv:2203.14713 (2022)

20. Stefanini, M., Cornia, M., Baraldi, L., et al.: From show to tell: a survey on deep learning-based image captioning. IEEE Trans. Pattern Anal. Mach. Intell. **45**(1), 539–559 (2021)

21. Lin, Z., Zhang, D., Tao, Q., et al.: Medical visual question answering: a survey. Artif. Intell. Med. **143**, 102611 (2023)

22. Li, J., Li, D., Xiong, C., Hoi, S.: BLIP: bootstrapping language-image pre-training for unified vision-language understanding and generation. In: Proceedings of ICML, pp. 12888–12900 (2022)

23. Dehghani, M., Djolonga, J., Mustafa, B., et al.: Scaling vision transformers to 22 billion parameters. arXiv preprint arXiv:2302.05442 (2023)

24. American Psychiatric Association D, American Psychiatric Association: Diagnostic and statistical manual of mental disorders. American Psychiatric Association, Washington, DC (2013)

25. Wong, T.T.: Performance evaluation of classification algorithms by k-fold and leave-one-out cross validation. Pattern Recogn. **48**(9), 2839–2846 (2015)

PPS: Semi-supervised 3D Biomedical Image Segmentation via Pyramid Pseudo-Labeling Supervision

Xiaogen Zhou, Zhiqiang Li, and Tong Tong$^{(\boxtimes)}$

College of Physics and Information Engineering, Fuzhou University,
Fuzhou, People's Republic of China
ttraveltong@gmail.com

Abstract. Although deep learning models have demonstrated impressive performance in various biomedical image segmentation tasks, their effectiveness heavily relies on a large amount of annotated training data, which can be costly to acquire. Semi-supervised learning (SSL) methods have emerged as a potential solution to mitigate this challenge by leveraging the abundance of unlabeled data. In this paper, we propose a highly effective SSL method for 3D biomedical image segmentation, called Pyramid Pseudo-Labeling Supervision (PPS). The PPS comprises three segmentation networks, forming a pyramid-like network structure. To enforce consistency in the outputs of the unlabeled data, we introduce a novel rectified pyramid consistency (RPC) loss. The PPS learns from the plentiful unlabeled data by minimizing the RPC loss, which ensures consistency between the pyramid predictions and the cycled pseudo-labeling knowledge among the three segmentation networks. Additionally, weak data augmentation is applied to perturb the inputs, further enhancing the consistency of the unlabeled data outputs. Experimental results demonstrate that our method achieves state-of-the-art performance on two publicly available 3D biomedical image datasets.

Keywords: Semi-supervised biomedical image segmentation ·
Rectified pyramid consistency

1 Introduction

Brain tumors are a prevalent neurological disorder, classified into primary brain tumors, brain-derived tumors, and brain metastatic tumors. Among primary malignancies, gliomas, which exhibit varying degrees of aggressiveness, account for over 50% of all primary central nervous system tumors, making them the most common type of brain tumors. High-grade gliomas exhibit rapid growth and aggressive spread, resulting in a median survival of two years or less, even with immediate medical intervention [5]. Low-grade gliomas have an average survival of approximately seven years, but they inevitably progress to high-grade gliomas and ultimately lead to death.

Q. Liu et al. (Eds.): PRCV 2023, LNCS 14437, pp. 272–283, 2024.
https://doi.org/10.1007/978-981-99-8558-6_23

Additionally, atrial fibrillation (AF) is the most prevalent cardiac arrhythmia among elderly individuals. Late gadolinium-enhanced (LGE) magnetic resonance imaging (MRI) has gained widespread use in visualizing the extent and distribution of cardiac scars. The segmentation of the left atrium (LA) and scars from LGE MRI provides reliable information for clinical diagnosis and treatment stratification [11]. Manual labeling of the LA is time-consuming; thus, there is a high demand for automatic segmentation methods. However, the development of automated LA segmentation remains challenging due to issues such as poor image quality, varying shapes of the LA, thin boundaries, and strong noise originating from surrounding tissues.

Automatic segmentation of the brain and left atrium (LA) regions from different MRI modalities is essential for assessing treatment effectiveness before and after interventions. While deep convolutional neural networks (DCNNs) have made significant advancements in biomedical image segmentation [3,4,10,14,18, 19,21–26], their performance heavily relies on a large amount of annotated data. However, the acquisition and annotation of such extensive datasets are time-consuming and expensive. To address this issue, considerable efforts have been devoted to leveraging additional information from abundant unlabeled data to alleviate the problem.

Recently, various DCNN-based approaches have been proposed to reduce the labeling cost and achieve impressive performance in biomedical image segmentation with limited labeled data. For example, self-supervised learning methods [27] aim to train a model in a self-supervised manner using unlabeled data to learn essential information for knowledge transfer. Semi-supervised learning (SSL) methods [7,12,13] generate high-confidence segmentation predictions by training on a combination of limited labeled samples and abundant unlabeled data. Weakly supervised learning frameworks [2,16] learn from bounding boxes, scribbles, or image-level tags for image segmentation, reducing the need for pixelwise labeled data and the associated annotation burden. In this work, we adopt semi-supervised learning to achieve 3D image segmentation.

To overcome the limitations of the Teacher-Student structure-based models, we propose a novel pyramid consistency regularization framework, referred to as the Teacher-Student-Teacher network structure. Unlike most Teacher-Student models that employ an exponential moving average (EMA) strategy during training, which tightly couples the weights, our proposed method avoids the use of EMA. Consequently, the weights of the three roles in our framework are not tightly coupled, allowing each role to learn its own knowledge. Specifically, we introduce a novel rectified pyramid consistency (RPC) loss to ensure consistency among the outputs of unlabeled data. The proposed method learns from unlabeled data by minimizing the RPC loss between the pyramid pseudo labels and the cycled pseudo-labeling information among the three segmentation networks. Experimental results demonstrate that our method significantly improves the performance of SSL on two 3D biomedical image benchmarks. The main contributions of our study can be summarized as follows:

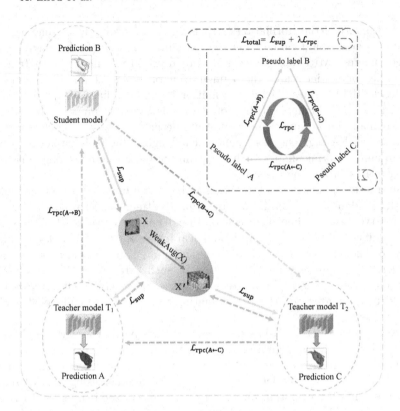

Fig. 1. Illustration of the architecture of the proposed PPS. The PPS comprises three segmentation models, forming a pyramid-shaped network structure. A novel rectified pyramid consistency (RPC) loss is introduced to enforce consistency among the outputs of unlabeled data from the three segmentation models. The three models are initialized differently, and weak data augmentation is applied to perturb their inputs.

1. We propose a novel pyramid network structure for semi-supervised 3D biomedical image segmentation. The proposed method aims to overcome the limitations of the Teacher-Student structure-based networks by incorporating an additional teacher network.
2. We introduce a novel rectified pyramid consistency (RPC), which promotes consistency among the outputs of unlabeled data and facilitates the transfer of cycled pseudo-labeling knowledge among the three segmentation networks.
3. We conduct comprehensive experiments on two publicly available 3D MRI datasets to evaluate the performance of our proposed framework. The experimental results demonstrate the feasibility and superiority of our approach.

2 Method

2.1 Overview

An overview of the proposed PPS architecture is presented in Fig. 1. Unlike most existing consistency-based semi-supervised learning (SSL) methods, our PPS simultaneously learns three segmentation models instead of using the Teacher-Student structure. The definition of our proposed PPS is as follows.

2.2 Pyramid Pseudo-Labeling Supervision

Data Definition. In our semi-supervised biomedical image segmentation task, we are provided with a labeled dataset $X^l = (x_i^l, y_i^l)_{i=1}^{N_l}$ and an unlabeled dataset $X^u = (x_j^u, y_j^u)_{j=1}^{N_u}$ for training. Here, x, y, N_l, and N_u represent an image, the corresponding label, and the number of labeled and unlabeled data, respectively.

Pyramid Network Structure. Drawing inspiration from the fact that a pyramid network structure exhibits geometric stability, we construct the PPS using one student model and two teacher models. The overall architecture resembles a pyramid shape. These models can be defined as follows:

$$T_1 = f(X; \theta_1) \tag{1}$$

$$S = f(X; \theta_2) \tag{2}$$

$$T_2 = f(X; \theta_3) \tag{3}$$

$$X_{ssl} = max([T_1, S, T_2]) \tag{4}$$

where T_1 represents the first Teacher model, S represents the Student model, and T_2 represents the second Teacher model. These models share the same network structure but are initialized with different weights, denoted as θ_1, θ_2, and θ_3. X_{ssl} represents the final semi-supervised learning (SSL) segmentation prediction of our proposed method, which is obtained by selecting the maximum output from the three segmentation models (T_1, S, and T_2). The logical illustration of our proposed method is as follows:

$$X \rightarrow WeakAug(X) \rightarrow \begin{cases} T_1 \rightarrow Y_1 \rightarrow P_1, \\ S \rightarrow Y_2 \rightarrow P_2, \\ T_2 \rightarrow Y_3 \rightarrow P_3. \end{cases} \tag{5}$$

where Y_i (for $i \in 1, 2, 3$) represents the i-th segmentation prediction. P_i (for $i \in 1, 2, 3$) is the corresponding pseudo label, which is utilized as an additional supervision signal to train the other two segmentation networks.

2.3 Loss Function

Supervised Loss: The proposed method was trained by minimizing a supervised loss $\mathcal{L}_{sup}$ among three segmentation models from limited labeled data. The supervised loss $\mathcal{L}_{sup}$ is formulated as follows:

$$\mathcal{L}_{sup} = \frac{1}{|D^L|} \sum_{X \in D^L} \frac{1}{w \times h} \sum_{i=0}^{w \times h} (\mathcal{L}_{ce}(Y_i^k, p_i^k) + \mathcal{L}_{dice}(Y_i^k, p_i^k)), \tag{6}$$

where $\mathcal{L}_{ce}$ is the cross-entropy loss function, and $\mathcal{L}_{dice}$ is the Dice coefficient loss function. The p_i^k is the ground truth and the k_{th} number of the models (i.e., $k \in [1, 2, 3]$). w and h denote the width and height of the input image.

Rectified Pyramid Consistency Loss: The rectified pyramid consistency (RPC) loss is tri-directional. Specifically, these losses are like a pyramid structure (see Fig. 1). The RPC loss $\mathcal{L}_{rpc}$ on the unlabeled data is defined as:

$$\mathcal{L}_{rpc} = \frac{1}{|D^U|} \sum_{X \in D^U} \frac{1}{w \times h} \sum_{i=0}^{w \times h} (\mathcal{L}_{ce}(Y_i^1, p_i^2) + \mathcal{L}_{ce}(Y_i^2, p_i^3) + \mathcal{L}_{ce}(Y_i^3, p_i^1)), \tag{7}$$

The total loss is defined as:

$$\mathcal{L}_{total} = \mathcal{L}_{sup} + \lambda \mathcal{L}_{rpc} \tag{8}$$

where $\mathcal{L}_{sup}$ is a hybrid loss by combining cross-entropy loss and dice loss. Following [12], we employ time-dependent Gaussian warming up function $\lambda(t) = e^{(-5(1-\frac{t}{t_{max}})^2)}$ to control the balance between the supervised loss and the unsupervised consistency loss, where t denotes the current training step and t_{max} is the maximum training step.

3 Experiments and Results

Dataset: We conducted experiments on two public biomedical image datasets to assess the performance of our proposed PPS. Specifically, we evaluated the effectiveness of the proposed PPS by comparing it with several state-of-the-art methods on the BraTS2019 and LA2018 datasets. The BraTS2019 dataset [8] used for training consists of 335 scans, each containing four modalities (FLAIR, T1, T1ce, and T2), with an isotropic resolution of $1\,mm^3$. In this study, we focused on SSL segmentation of whole brain tumors using FLAIR images. The dataset was randomly divided into 250, 25, and 60 samples for training, validation, and testing, respectively.

Additionally, we evaluated the performance of our proposed PPS on the LA2018 dataset, which is the 2018 Atrial Segmentation Challenge dataset [14]. This dataset comprises 100 gadolinium-enhanced MR imaging scans along with corresponding left atrium (LA) segmentation masks for training and validation. The scans have an isotropic resolution of $0.625 \times 0.625 \times 0.625\,mm^3$. We split the 100 scans into 80 samples for training, 10 samples for validation, and 10 samples for testing.

Table 1. Quantitative results of brain tumour semi-supervised segmentation between our proposed PPS and the other methods on the BraTS2019 dataset.

Methods	# Scan Used		Metrics			
	Labelled data	Unlabeled data	Dice ↑	JI ↑	95HD	ASD
			(%)	(%)	(voxel)	(voxel)
V-Net [9]	25(10%)	0	80.09	68.12	22.43	7.53
V-Net [9]	50(20%)	0	83.58	76.06	22.09	7.33
V-Net [9]	250(All)	0	88.51	83.82	7.52	1.81
MT [12]	25(10%)	225	81.7	75.12	13.28	3.56
EM [13]	25(10%)	225	82.35	75.75	14.7	3.68
UAMT [15]	25(10%)	225	83.93	5.43	15.81	3.27
DAN [17]	25(10%)	225	82.5	73.48	15.11	3.79
DTC [6]	25(10%)	225	85.06	76.55	14.47	3.74
CPS [1]	25(10%)	225	85.03	77.53	10.27	3.26
DTSC-Net [20]	25(10%)	225	85.12	**77.83**	10.15	3.16
PPS	25(10%)	225	**85.62**	77.53	**9.26**	**2.7**
MT [12]	50(20%)	200	85.03	78.72	11.8	1.89
EM [13]	50(20%)	200	84.82	80.21	12.37	3.21
UAMT [15]	50(20%)	200	85.05	81.14	12.31	3.03
DAN [17]	50(20%)	200	84.63	81.21	8.96	2.34
DTC [6]	50(20%)	200	86.37	81.65	8.97	2.1
CPS [1]	50(20%)	200	87.12	81.48	9.48	2.16
DTSC-Net [20]	50(20%)	200	87.22	81.58	8.41	2.13
PPS	50(20%)	200	**87.68**	**82.51**	**7.95**	**1.83**

Implementation Details and Evaluation Metrics: In this study, all approaches were implemented using PyTorch on a Ubuntu 18.04 desktop with an NVIDIA GTX 2080TI GPU. The V-Net [9] was employed as the backbone segmentation network, which was consistent across all methods for a fair comparison. A dropout rate of 0.3 was utilized. The SGD optimizer with a weight decay of $1e^{-4}$ and a momentum of 0.9 was employed, along with Eq. 8 as the final loss function. The poly learning rate scheme was adopted, where the initial learning rate l_i was multiplied by $(1 - \frac{t}{t_{max}})^{0.9})$, with l_i set to 0.1 and t_{max} set to 60000 for the BraTS2019 dataset. The batch size was set to 4 for all compared approaches, with half of the data being labeled and the other half unlabeled. Randomly cropped patches were used as input to the network, with a patch size of 96 × 96 × 96 for the BraTS2019 dataset. For the LA2018 dataset, the input data was randomly cropped to a size of 112 × 112 × 80. Random cropping, flipping, and rotation were applied to augment the training dataset and prevent overfitting. During the testing phase, four metrics were employed to quantitatively evaluate the segmentation performance: Dice coefficient (Dice),

Jaccard Index (JI), 95% Hausdorff Distance (HD95), and Average Surface Distance (ASD).

Table 2. Quantitative results of left atrium semi-supervised segmentation between our proposed PPS and the other methods on the LA2018 dataset.

Methods	# Scan Used		Metrics			
	Labelled data	Unlabeled data	Dice ↑	JI ↑	95HD	ASD
			(%)	(%)	(voxel)	(voxel)
V-Net [9]	8(10%)	0	79.99	68.12	21.11	5.48
V-Net [9]	16(20%)	0	86.03	76.06	14.26	3.51
V-Net [9]	80(All)	0	91.14	83.82	5.75	1.52
MT [12]	8(10%)	72	85.54	75.12	13.29	3.77
EM [13]	8(10%)	72	85.91	75.75	12.67	3.31
UAMT [15]	8(10%)	72	81.89	71.23	15.81	3.8
DAN [17]	8(10%)	72	84.25	73.48	13.84	3.36
DTC [6]	8(10%)	72	86.57	76.55	14.47	3.74
CPS [1]	8(10%)	72	87.32	77.23	10.70	2.86
DTSC-Net [20]	8(10%)	72	87.68	77.53	10.26	2.70
PPS	8(10%)	72	**88.28**	**78.43**	**10.10**	**2.60**
MT [12]	16(20%)	64	88.23	78.72	10.64	2.73
EM [13]	16(20%)	64	88.45	80.21	14.14	3.72
UAMT [15]	16(20%)	64	88.88	81.24	7.51	2.26
DAN [17]	16(20%)	64	87.52	81.31	9.01	2.42
DTC [6]	16(20%)	64	89.42	81.35	7.32	2.1
CPS [1]	16(20%)	64	89.61	80.98	7.61	3.26
DTSC-Net [20]	16(20%)	64	90.15	82.11	**7.23**	1.83
PPS	16(20%)	64	**91.06**	**82.51**	7.33	**1.72**

Comparison with State-of-the-Art Methods: We conducted a comparative analysis of the proposed method with other SSL methods, including MT [12], EM [13], UAMT [15], CPS [1], DAN [17], DTC [6], DTSC-Net [20], and a fully supervised learning method, V-Net [9]. The implementations of these approaches are publicly available, enabling fair comparisons on the same datasets. The results of the quantitative evaluation are presented in Table 1 and Table 2, where our proposed PPS was compared with the other models using 10% and 20% of labeled training samples on two benchmark datasets. Our method demonstrated superior performance in terms of Dice and ASD metrics, particularly with 20% labeled data. On the BraTS2019 dataset, the Dice score improvement was approximately 3.92% and 2.65% compared to MT [12] with the Student-Teacher structure, and an increase of about 0.5% and 0.46% compared to DTSC-Net [20] with the dual-teacher structure. Notably, the most significant improvement was achieved on the challenging LA2018 dataset.

Additionally, we have presented visualizations of the semi-supervised segmentation results in our experiments. Figure 2 presents the visualized left atrium segmentation results obtained by our proposed method and the other methods on the LA2018 dataset. Similarly, Fig. 3 shows the visualized brain tumor segmentation results generated by our proposed method and the other methods on the BraTS2019 dataset. It is evident that our proposed PPS achieves higher accuracy and fewer false-positive regions compared to the other comparison methods.

Table 3. Ablation studies for each component of our proposed PPS for left atrium segmentation on the LA2018 dataset.

10% Labelled Scan Used	
Methods	Dice (%)
Base	79.99
Base + Weak Data Augmentation	83.88
Base + Weak Data Augmentation + RPC loss	**88.28**
20% Labelled Scan Used	
Base	86.03
Base + Weak Data Augmentation	87.56
Base + Weak Data Augmentation + RPC loss	**91.06**

3.1 Ablation Analysis

Impact of the Different Number of Teacher Models Used in Our Pyramid Structure: To validate the effectiveness of our proposed pyramid structure, we investigated the effect of different numbers of teacher models, ranging from No-Teacher to Nine-Teacher. Table 4 presents the performance of our proposed PPS with varying numbers of teacher models used in the pyramid structure on the BraTS2019 dataset. In order to analyze the effectiveness of our proposed Teacher-Student-Teacher scheme, we incrementally added teacher models from 0 to 9. The best results were obtained when the number of teacher models was 2, corresponding to our proposed Teacher-Student-Teacher structure. The Dice scores ranged from 80.09% to 85.62% when 10% labeled data was used, as shown in Table 4. However, as the number of teacher models increased beyond 2, the Dice scores decreased, indicating the superiority of our proposed pyramid network structure. The pyramid network structure provides geometric stabilization and yields more accurate results compared to other structures.

Impact of Perturbations of Weak Data Augmentation in Our PPS: Ablation studies evaluating the impact of weak data augmentation perturbation on semi-supervised segmentation are presented in Table 3. Compared to the case without weak data augmentation, our proposed PPS achieved more accurate left atrium segmentation results than the Base model. This demonstrates that the perturbations introduced by the weak data augmentation strategy are beneficial for 3D semi-supervised medical image segmentation.

Impact of RPC Loss Strategy for Semi-supervised Learning: We investigated the impact of incorporating the proposed rectified pyramid consistency (RPC) loss on the segmentation performance. The results are presented in Table 3, revealing that our proposed method with the RPC loss achieved the highest Dice score compared to other scenarios. Specifically, when comparing our proposed PPS with the RPC loss to the case without the RPC loss ('Base+Weak

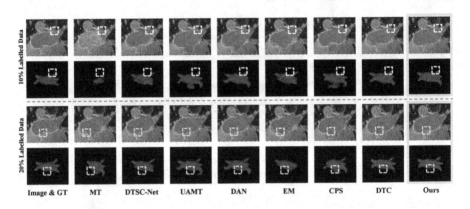

Fig. 2. Visualization of left atrium segmentation results was obtained by our proposed method and the other methods on the LA2018 dataset. We used yellow boxes to highlight our results. (Color figure online)

Table 4. Ablation studies for different the number of Teacher models used in our PPS structure on the BraTS2019 dataset.

Teacher models	10% Labelled data	20% Labelled data
	Dice (%)	Dice (%)
No-Teacher	80.09	83.58
One-Teacher	81.81	85.03
Two-Teacher (Ours)	**85.62**	**87.68**
Three-Teacher	82.53	83.82
Four-Teacher	83.19	82.78
Five-Teacher	82.54	83.67
Six-Teacher	79.87	83.14
Seven-Teacher	82.72	83.26
Eight-Teacher	80.83	82.34
Nine-Teacher	82.64	81.97

Data Augmentation'), we observed a noticeable improvement of 4.4% and 3.5% in the Dice score for the labeled ratios of 10% and 20%, respectively. This clearly demonstrates the effectiveness of our proposed RPC loss in enhancing the performance of the semi-supervised segmentation task.

Furthermore, we have visualized the ablation study of the semi-supervised segmentation results generated by each component of our proposed PPS on the BraTS2019 and LA2018 datasets. In comparison to the Base model (i.e., V-Net [9]), our proposed PPS, which incorporates weak data augmentation and the RPC loss, produces more accurate results with fewer over-/under-segmentation regions.

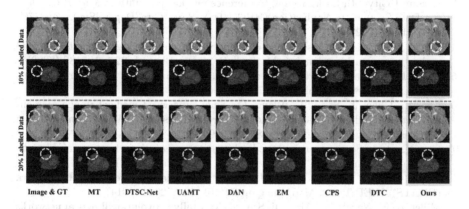

Fig. 3. Visualization of brain tumour segmentation results was generated by our proposed method and the other methods on the BraTS2019 dataset. We used yellow boxes to highlight our results. (Color figure online)

4   Conclusion

In this study, we propose a novel pyramid network structure for semi-supervised 3D biomedical image segmentation. The pyramid network structure is designed to generate pyramid pseudo labels and facilitate the transfer of cycled pseudo-labeling knowledge among three segmentation models. Additionally, a novel pyramid unsupervised consistency loss is introduced to enforce consistency in the outputs of unlabeled data among the three segmentation networks during training. The effectiveness of our approach is validated on two publicly available biomedical image datasets, and experimental results demonstrate its superiority over several state-of-the-art methods. Future work will focus on semi-supervised domain adaptation segmentation on multimodal images.

Acknowledgment. This work was supported by National Natural Science Foundation of China under Grant 62171133, in part by the Science and Technology Innovation Joint Fund Program of Fujian Province of China under Grant 2019Y9104, the Health Science and Technology Program of Fujian Province of China under Grant 2019-1-33, and the Industry-University-Research Cooperation Program of Fujian Province of China under Grant 2022H6006.

References

1. Chen, X., Yuan, Y., Zeng, G., Wang, J.: Semi-supervised semantic segmentation with cross pseudo supervision. In: Proceedings of the IEEE/CVF Conference on Computer Vision and Pattern Recognition, pp. 2613–2622 (2021)
2. Gao, F., et al.: Segmentation only uses sparse annotations: unified weakly and semi-supervised learning in medical images. Med. Image Anal. **80**, 102515 (2022)
3. Lin, X., Zhou, X., Tong, T., Nie, X., Li, Z.: SG-Net: a super-resolution guided network for improving thyroid nodule segmentation. In: 2022 IEEE 24th International Conference on High Performance Computing & Communications; 8th International Conference on Data Science & Systems; 20th International Conference on Smart City; 8th International Conference on Dependability in Sensor, Cloud & Big Data Systems & Application (HPCC/DSS/SmartCity/DependSys), pp. 1770–1775. IEEE (2022)
4. Lin, X., et al.: A super-resolution guided network for improving automated thyroid nodule segmentation. Comput. Methods Programs Biomed. **227**, 107186 (2022)
5. Louis, D.N., et al.: The 2007 who classification of tumours of the central nervous system. Acta Neuropathol. **114**, 97–109 (2007)
6. Luo, X., Chen, J., Song, T., Wang, G.: Semi-supervised medical image segmentation through dual-task consistency. In: Proceedings of the AAAI Conference on Artificial Intelligence, vol. 35, pp. 8801–8809 (2021)
7. Luo, X., et al.: Semi-supervised medical image segmentation via uncertainty rectified pyramid consistency. Med. Image Anal. **80**, 102517 (2022)
8. Menze, B.H., et al.: The multimodal brain tumor image segmentation benchmark (BRATS). IEEE Trans. Med. Imaging **34**(10), 1993–2024 (2014)
9. Milletari, F., Navab, N., Ahmadi, S.A.: V-Net: fully convolutional neural networks for volumetric medical image segmentation. In: 2016 Fourth International Conference on 3D Vision (3DV), pp. 565–571. IEEE (2016)
10. Nie, X., et al.: N-Net: a novel dense fully convolutional neural network for thyroid nodule segmentation. Front. Neurosci. **16**, 872601 (2022)
11. Njoku, A., et al.: Left atrial volume predicts atrial fibrillation recurrence after radiofrequency ablation: a meta-analysis. EP Europace **20**(1), 33–42 (2018)
12. Tarvainen, A., Valpola, H.: Mean teachers are better role models: eight-averaged consistency targets improve semi-supervised deep learning results. In: Advances in Neural Information Processing Systems, vol. 30 (2017)
13. Vu, T.H., Jain, H., Bucher, M., Cord, M., Pérez, P.: ADVENT: adversarial entropy minimization for domain adaptation in semantic segmentation. In: Proceedings of the IEEE/CVF Conference on Computer Vision and Pattern Recognition, pp. 2517–2526 (2019)
14. Xiong, Z., et al.: A global benchmark of algorithms for segmenting the left atrium from late gadolinium-enhanced cardiac magnetic resonance imaging. Med. Image Anal. **67**, 101832 (2021)
15. Yu, L., Wang, S., Li, X., Fu, C.-W., Heng, P.-A.: Uncertainty-aware self-ensembling model for semi-supervised 3D left atrium segmentation. In: Shen, D., et al. (eds.) MICCAI 2019. LNCS, vol. 11765, pp. 605–613. Springer, Cham (2019). https://doi.org/10.1007/978-3-030-32245-8_67
16. Zhang, D., Chen, B., Chong, J., Li, S.: Weakly-supervised teacher-student network for liver tumor segmentation from non-enhanced images. Med. Image Anal. **70**, 102005 (2021)

17. Zhang, Y., Yang, L., Chen, J., Fredericksen, M., Hughes, D.P., Chen, D.Z.: Deep adversarial networks for biomedical image segmentation utilizing unannotated images. In: Descoteaux, M., Maier-Hein, L., Franz, A., Jannin, P., Collins, D., Duchesne, S. (eds.) Medical Image Computing and Computer Assisted Intervention–MICCAI 2017: 20th International Conference, Quebec City, QC, Canada, 11–13 September 2017, Proceedings, Part III 20, pp. 408–416. Springer, Cham (2017). https://doi.org/10.1007/978-3-319-66179-7_47

18. Zheng, H., Zhou, X., Li, J., Gao, Q., Tong, T.: White blood cell segmentation based on visual attention mechanism and model fitting. In: 2020 International Conference on Computer Engineering and Intelligent Control (ICCEIC), pp. 47–50. IEEE (2020)

19. Zhong, Z., Wang, T., Zeng, K., Zhou, X., Li, Z.: White blood cell segmentation via sparsity and geometry constraints. IEEE Access 7, 167593–167604 (2019)

20. Zhou, X., Li, Z., Tong, T.: DTSC-Net: semi-supervised 3D biomedical image segmentation through dual-teacher simplified consistency. In: 2022 IEEE International Conference on Bioinformatics and Biomedicine (BIBM), pp. 1429–1434. IEEE (2022)

21. Zhou, X., Li, Z., Tong, T.: DM-Net: a dual-model network for automated biomedical image diagnosis. In: Tang, H. (eds.) International Conference on Research in Computational Molecular Biology, RECOMB 2023. LNCS, vol. 13976, pp. 74–84. Springer, Cham (2023). https://doi.org/10.1007/978-3-031-29119-7_5

22. Zhou, X., et al.: CUSS-Net: a cascaded unsupervised-based strategy and supervised network for biomedical image diagnosis and segmentation. IEEE J. Biomed. Health Inform. 27(5), 2444–2455 (2023)

23. Zhou, X., et al.: Leukocyte image segmentation based on adaptive histogram thresholding and contour detection. Curr. Bioinform. 15(3), 187–195 (2020)

24. Zhou, X., et al.: H-Net: a dual-decoder enhanced FCNN for automated biomedical image diagnosis. Inf. Sci. 613, 575–590 (2022)

25. Zhou, X., Tong, T., Zhong, Z., Fan, H., Li, Z.: Saliency-CCE: exploiting colour contextual extractor and saliency-based biomedical image segmentation. Comput. Biol. Med. 154, 106551 (2023)

26. Zhou, X., Wang, C., Li, Z., Zhang, F.: Adaptive histogram thresholding-based leukocyte image segmentation. In: Pan, J.S., Li, J., Tsai, P.W., Jain, L. (eds.) Advances in Intelligent Information Hiding and Multimedia Signal Processing: Proceedings of the 15th International Conference on IIH-MSP in Conjunction with the 12th International Conference on FITAT, 18–20 July 2020, Jilin, China, vol. 2, pp. 451–459. Springer, Cham (2020). https://doi.org/10.1007/978-981-13-9710-3_47

27. Zhuang, X., Li, Y., Hu, Y., Ma, K., Yang, Y., Zheng, Y.: Self-supervised feature learning for 3D medical images by playing a Rubik's cube. In: Shen, D., et al. (eds.) Medical Image Computing and Computer Assisted Intervention-MICCAI 2019: 22nd International Conference, Shenzhen, China, 13–17 October 2019, Proceedings, Part IV 22, vol. 11767, pp. 420–428. Springer, Cham (2019). https://doi.org/10.1007/978-3-030-32251-9_46

A Novel Diffusion-Model-Based OCT Image Inpainting Algorithm for Wide Saturation Artifacts

Bangning Ji[1], Gang He[1,2(✉)], Zhengguo Chen[2], and Ling Zhao[2(✉)]

[1] School of Computer Science and Technology, Southwest University of Science and Technology, Mianyang 621010, China
ganghe@swust.edu.cn
[2] NHC Key Laboratory of Nuclear Technology Medical Transformation (MIANYANG CENTRAL HOSPITAL), Mianyang, China
zhaolingssaa@163.com

Abstract. Saturation artifacts in Optical Coherence Tomography (OCT) images will affect the image quality and reduce the accuracy of clinical diagnosis. Recently, the researcher proposed various OCT image inpainting algorithms for saturation artifacts, and these algorithms were limited to .oct format files only (spectral data) or simple interpolation algorithms, which led to the failure of the best performance on wide saturation artifacts. In this paper, a novel image inpainting model based on a generative model (diffusion model) is proposed, which can recover degraded regions in OCT images. Experimental results show that the average PSNR and SSIM values outperformed existing approaches. Besides, the classification models, vision transformer (ViT), for OCT images were implemented to compare the accuracy difference before and after the proposed image inpainting algorithm. The proposed algorithm presents a promising solution for better OCT image inpainting methods.

Keywords: Optical Coherence Tomography · Image inpainting · Diffusion model

1 Introduction

Optical Coherence Tomography (OCT) is commonly used for assessing eye diseases such as macular degeneration, glaucoma, and diabetic macular edema [3] by adopting low-coherence light to capture micrometer resolution images from biological tissue. However, in some cases, OCT images may be degraded due to light signal overflow in the sensor, and this phenomenon is called saturation artifacts. Saturation artifacts in OCT images will cause semantic information loss, and the analysis accuracy for these OCT images will be dramatically decreased in many situations, including image identification or segmentation [21]. A typical optical setup of OCT is given in Fig. 1.

Different kinds of inpainting algorithms were proposed by several works to overcome the previous problems, which can be divided into three subcategories:

Q. Liu et al. (Eds.): PRCV 2023, LNCS 14437, pp. 284–295, 2024.
https://doi.org/10.1007/978-981-99-8558-6_24

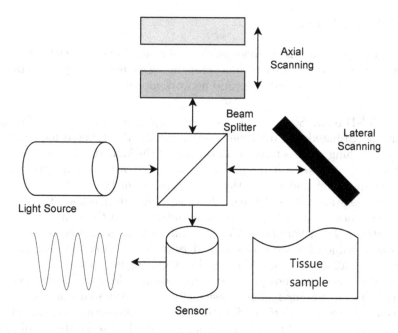

Fig. 1. Typical optical setup of OCT

sequential-based, CNN-based, and GAN-based methods [6]. Moreover, a recent generative model called the diffusion model demonstrated impressive performance lately [2].

Unlike GANs, diffusion models are generative models based on like-hood estimation, which have recently been proven to generate high-quality images [4]. More specifically, as an algorithm inspired by nonequilibrium thermodynamics, a diffusion model is a parameterized Markov chain trained using variational inference to produce samples matching the data after a finite time.

In this paper, a novel diffusion-model-based OCT image inpainting method for wide saturation artifacts is proposed. To train the proposed model, the OCT image dataset [10] with 83,484 samples was used (We train a diffusion model on the OCT dataset and recover saturation artifacts in the sampling step). The comparison between our model and recent works is implemented to validate the proposed algorithm. The Peak Signal Noise Ratio (PSNR) and Structure Similarity Index Measure (SSIM) are utilized to measure the performance of the proposed inpainting method.

As the artifacts on OCT images will decrease the accuracy of classification models, artifacts with different positions or widths may lead to varying accuracies for these models. Therefore, the influence of the artifact's parameters (the width and position) on the classification accuracy is presented in the paper.

2   Related Work

Existing image inpainting methods exploited low-level features within the input image, such as sequential-based image inpainting or missing region recovering, and most of them are based on neural network models.

Sequential-Based. Sequential-based image inpainting methods can be classified into patch-based and diffusion-based methods. Patch-based methods copy the well-matching patches from the intact part of the image to the missing region. Liu et al. [13] proposes a patch-based method to recover OCT images with the saturated spectrum and proves that the proposed method outperforms cubic spline interpolation (SI) and Euler's Elastica (EE) [16] method. Xu et al. [18] introduces a novel inpainting algorithm by investigating the sparsity of natural image patches. Total Variation (TV) [7] regularization is an effective inpainting technique capable of recovering sharp edges under some conditions. Similarly, Zhang et al. [20] applies the TV method to OCT image inpainting.

Diffusion-based methods propagate image content from the boundary(intact region) to fill the missing region. These methods perform well when the missing regions are narrow but perform dissatisfiedly when inpainting wide artifacts or texture-rich structures. Li et al. [11] proposes a method to locate the diffusion-based inpainted regions in digital images. In addition, fractional-order nonlinear diffusion is also used to image inpainting by G et al. [17].

CNN-Based. CNN-based methods, such as Shift-net [19] proposed by Yan and the method with Recurrent Feature Reasoning (RFR) module proposed by Li et al. [12], choose encoder-decoder structures as the backbone network. Despite performing better than traditional image inpainting methods, CNN-based methods need a relatively large dataset which is challenging to obtain.

GAN-Based. Generative adversarial networks (GANs) [8] are a common technique in computer vision, which trades off diversity for fidelity and produces high-quality samples. GAN-based methods are proven to be helpful for image inpainting: Chen et al. [1] proposes a GAN-based image inpainting method that contains GAN structures and a pyramid strategy. Zhao et al. [22] proposes a Sinogram Inpainting Network (SIN) based on GAN to reconstruct CT images.

Diffusion-Model-Based. As a powerful generative model, diffusion models have been proven to perform amazingly in image generation. Denoising Diffusion Probabilistic Models (DDPM) [9] is a milestone in diffusion models. After that, Nichol et al. [15] demonstrates that a specific cosine noise schedule can improve log-likelihoods and proposes Improved Denoising Diffusion Probabilistic models (IDDPM). Dhariwal et al. [4] propose an IDDPM-based model to condition reverse the noising process. Interestingly, Lugmayr et al. [14] offers a diffusion-model-based image inpainting method, which only works for natural images, inspired us to apply a diffusion model to inpaint OCT images with wide artifacts.

3   Proposed Work

Two stages inpainting method is proposed in our work: the training process and the inference process.

3.1   Training Process

An unconditional DDPM on the OCT image dataset was trained to generate the degraded part. Such as other generative models, DDPM also learns the distribution of training images and generates samples that follow the distribution learned from the training process. DDPM could be divided into two processes. The overview of DDPM is illustrated in Fig. 2.

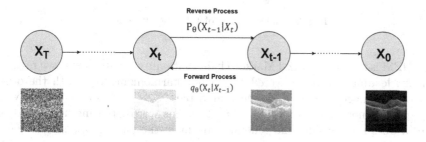

Fig. 2. The forward and reverse process in training

In the forward process, DDPM adds Gaussian noise to raw image X_0 until X_0 transforms to Gaussian noise $X_T \sim N(0,1)$, and the accumulated noise denotes as $\bar{\alpha}_t$:

$$q(X_t|X_0) = N(X_t; \sqrt{\bar{\alpha}_t}X_0, (1 - \bar{\alpha}_t)I) \tag{1}$$

The reverse process predicts the parameters μ_θ and Σ_θ with a U-Net model and sample image from those parameters:

$$p(X_t - 1|X_t) = N(X_t - 1; \mu_\theta(X_t, t), \Sigma_\theta(X_t, t)) \tag{2}$$

3.2   Inference Process

The degraded region is recovered with our method in the inference process. Raw image X combine with degraded region $X^{degraded}$ and intact region X^{intact}. The noise was added gradually to the intact region to get X_t^{intact}:

$$X_t^{intact} = N(\sqrt{\bar{\alpha}_t}X_0, (1 - \bar{\alpha}_t)I) \tag{3}$$

Obtain the $X_t^{degraded}$ with X_{t+1}, μ and Σ which predicted from neural network:

$$X_t^{degraded} = N(\mu_\theta(X_{t+1}, t + 1), \Sigma_\theta(X_{t+1}, t + 1)) \tag{4}$$

Combine the X_t^{intact} and $X_t^{degraded}$ (5):

$$X_t = X_t^{intact} + X_t^{degraded} \qquad (5)$$

The overview of our method is illustrated in Fig. 3:

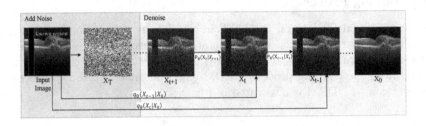

Fig. 3. The overview of Inference process

DDPM is a parameterized Markov chain that only exploits the known region's context, leading to the recovered region not harmonizing well with the other part of the image. For example, generated part $X_{t-1}^{degraded}$ is only related with X_t, and has not relation with X_{t-1}^{intact}. In other words, the method based on DDPM does not consider the generated result in the past step, and this defect will become serious with increasing steps. As reported by Lugmayr *et al.* [14], a resampling process will alleviate this problem. Lugmayr *et al.* [14] diffuse the output X_{t-1} back to X_t with formula (1) in resampling process. However, some information incorporated in the generated region $X_{t-1}^{degraded}$ is still preserved in $X_t^{degraded}$, which leads generated region being more harmonizing. The overview of this process is shown in Fig. 4

3.3 Experiments

Experimental Setup. To evaluate the performance of our image inpainting method, we use the retinal OCT dataset [10], which captures high-resolution cross-sections of the retinas of living patients. This dataset contains 84,484 OCT images divided into four categories: choroidal neovascularization (CNV), diabetic macular edema (DME), drusen, and normal (Fig. 5).

The number of images in the dataset is 83,484 images (37,205 with CNV, 11,348 with DME, 8,616 with drusen, and 26,315 normal) in the train set and 1,000 images (250 with CNV, 250 with DME, 250 with drusen, and 250 with normal) in the test set.

Our proposed method is evaluated on OCT images with synthetic degraded parts (our paper does not focus on artifact detection). The synthetic degraded parts are a series of lines with different positions and widths, shown in Fig. 6 Fig. 7.

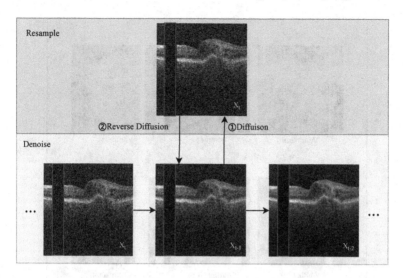

Fig. 4. The overview of resampling process

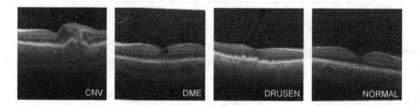

Fig. 5. Four categories OCT images

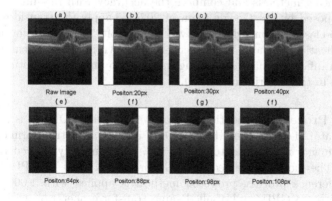

Fig. 6. Different positions artifacts

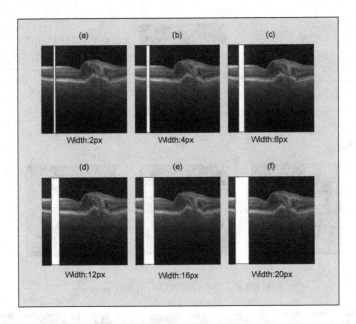

Fig. 7. Different width artifacts

3.4   Evaluation Metrics

The performance of our method is evaluated on three evaluation metrics: classification tasks, PSNR, and SSIM. Inpainted image is fed to a classification network Vision Transformer (ViT) [5] to evaluate the performance of different image inpainting methods and compare the accuracy with raw images. ViT is a model proposed by the Google team in 2020 to apply the transformer module to image classification, which evolved a milestone in the transformer-based image classification method. ViT's overview is shown in Fig. 8. Accuracy is calculated with different image inpainting methods to analyze inpainting method performance in the experiment section.

Inpainting Process. As mentioned above, a pretrained DDPM is trained to be the generator of our method. For the OCT images, some transform operation is applied: images are resized to 128 × 128, and randomly flipped to reduce overfitting. The model is trained on a GeForce RTX 2080 Ti GPU in 70,000 loops. In inference process, proposed method is applied with 1,000 diffusion steps on the same GPU, and applied same transform operation with training process.

3.5   Result

To confirm the impact of artifact with different width and different positions, oct images' classification accuracy is calculated in different condition (Fig. 9).

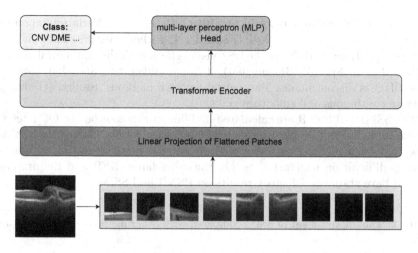

Fig. 8. Model overview of ViT

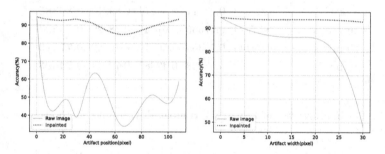

Fig. 9. The classification accuracy in different condition

As the artifact gets wider and closer to the middle of the image, the accuracy rate decreases (The center of the image contains more features and information). In the following article, our experiment is applied with the same position positioned 20 pixels from the left of the image.

As shown in Fig. 10 is a comparison between different methods for synthetic artifacts with different widths (8 pixels, 16 pixels, 20 pixels, 30 pixels), and the black masks with a red border in Fig. 10 (b-1 4) are used to emulate the saturation artifact in OCT images. Euler's Elastica inpainting method (EE) [16] only generate some blurry block in Fig. 10(c-1 4). The total Variation inpainting method (TV) creates a gray block (Fig. 2 d-1 4) without texture. The coherence Transport(CO) method shows the best visual effect with the non-neural network method, but in Fig. 10(e-3) and (e-4), it is not difficult to find that border of the artifact is discontinuous. Two neural network methods' results, (f) and (g), generate continuous and smooth regions.

The SSIM and PSNR are calculated in different methods on the OCT testing test in Table 1 and Table 2 to measure the program results. Table 1 and Table 2 show that each method performs well on narrow synthetic artifacts, but the result is difficult on wide artifacts. On the other hand, RFR and the proposed method have stably satisfactory results on PSNR and SSIM.

Table 1. The improvement of SSIM compared with other methods on synthetic artifact.

Artifact Width	TV	CO	EE	RFR	Proposed
8 pixels	0.82543	0.87411	0.84274	0.89215	**0.89389**
16 pixels	0.67974	0.77555	0.67747	0.78229	**0.79327**
20 pixels	0.64145	0.73364	0.56813	0.76442	**0.77119**
30 pixels	0.53892	0.65738	0.42683	0.72214	**0.74393**

Table 2. The improvement of PSNR(dB) compared with other methods on synthetic artifact.

Artifact Width	TV	CO	EE	RFR	Proposed
8 pixels	25.2768	31.5637	28.923	30.913	**32.7848**
16 pixels	19.999	24.5062	21.8656	26.5532	**28.7093**
20 pixels	18.6978	21.8654	19.49	25.5357	**27.5717**
30 pixels	16.8139	17.4055	17.8684	23.5257	**25.5961**

In Table 3, to quantitatively evaluate the inpainting quality, the accuracy of the classification model is compared with previous works. The methods based on neural networks increase the accuracy of classification results because they recover more semantic information.

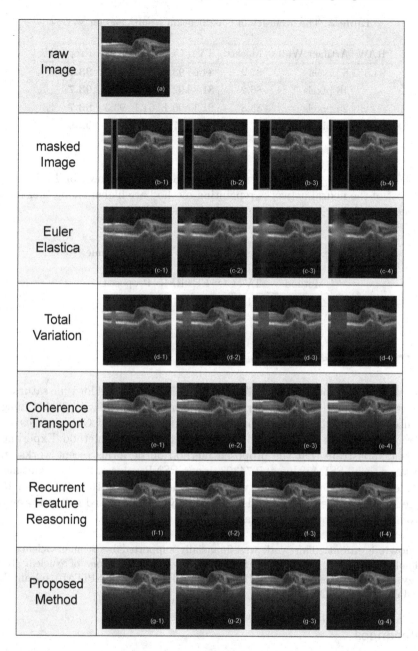

Fig. 10. The comparison of different methods (columns one to four represent: 8pixels, 16pixels, 20pixels, and 30pixels)

Table 3. The accuracy of classification results (%) on ViT.

RAW	Artifact Width	Masked	TV	CO	EE	RFR	Proposed
94.5	8 pixels	87.9	84.6	93.6	93.5	93.5	**93.7**
	16 pixels	86.2	81.8	92.6	71.4	93.1	**93.7**
	20 pixels	85.8	81.1	90.1	61.1	93.3	**93.7**
	30 pixels	47.5	66.9	81	62.3	92.4	**92.6**

The comparison results in terms of inference time is shown in Table 4. The time cost of the neural networks repainting method are higher than other methods because of their higher computational overhead.

Table 4. The comparison results in terms of inference time (second).

Method	TV	CO	EE	RFR	Proposed
Time Cost	<1	16	14	27	92

4   Conclusions

A novel diffusion-model-based OCT image inpainting method for wide saturation artifacts is proposed in this paper. The OCT image dataset with 83,484 images was used to train the diffusion model. Degraded regions of OCT images with different parameters are simulated to evaluate the proposed method. Experiment results demonstrate that the proposed method outperforms recent works, and achieves a mean SSIM value of 0.74393, mean PSNR value of 25.5961, and mean accuracy of classification models with 92.6%, increased by 10%, 28%, and 15% to the average of the four methods, respectively. The proposed method presents a promising solution for OCT image impainting.

Acknowledgement. This work was financially supported by Sichuan Science and Technology Program (NO. 2020YFS0454), NHC Key Laboratory of Nuclear Technology Medical Transformation (MIANYANG CENTRAL HOSPITAL) (Grant No. 2021HYX024, No. 2021HYX031)

References

1. Chen, Y., Hu, H.: An improved method for semantic image inpainting with GANs: progressive inpainting. Neural Process. Lett. **49**, 1355–1367 (2019)
2. Croitoru, F.A., Hondru, V., Ionescu, R.T., Shah, M.: Diffusion models in vision: a survey. IEEE Trans. Pattern Anal. Mach. Intell. **45**, 10850–10869 (2023)
3. De Carlo, T.E., Romano, A., Waheed, N.K., Duker, J.S.: A review of optical coherence tomography angiography (OCTA). Int. J. Retina Vitreous **1**, 1–15 (2015)

4. Dhariwal, P., Nichol, A.: Diffusion models beat GANs on image synthesis. Adv. Neural. Inf. Process. Syst. **34**, 8780–8794 (2021)
5. Dosovitskiy, A., et al.: An image is worth 16x16 words: transformers for image recognition at scale. arXiv preprint arXiv:2010.11929 (2020)
6. Elharrouss, O., Almaadeed, N., Al-Maadeed, S., Akbari, Y.: Image inpainting: a review. Neural Process. Lett. **51**, 2007–2028 (2020)
7. Getreuer, P.: Total variation inpainting using split Bregman. Image Process. On Line **2**, 147–157 (2012)
8. Goodfellow, I., et al.: Generative adversarial networks. Commun. ACM **63**(11), 139–144 (2020)
9. Ho, J., Jain, A., Abbeel, P.: Denoising diffusion probabilistic models. Adv. Neural. Inf. Process. Syst. **33**, 6840–6851 (2020)
10. Kermany, D.S., et al.: Identifying medical diagnoses and treatable diseases by image-based deep learning. Cell **172**(5), 1122–1131 (2018)
11. Li, H., Luo, W., Huang, J.: Localization of diffusion-based inpainting in digital images. IEEE Trans. Inf. Forensics Secur. **12**(12), 3050–3064 (2017)
12. Li, J., Wang, N., Zhang, L., Du, B., Tao, D.: Recurrent feature reasoning for image inpainting. In: Proceedings of the IEEE/CVF Conference on Computer Vision and Pattern Recognition, pp. 7760–7768 (2020)
13. Liu, H., Cao, S., Ling, Y., Gan, Y.: Inpainting for saturation artifacts in optical coherence tomography using dictionary-based sparse representation. IEEE Photon. J. **13**(2), 1–10 (2021)
14. Lugmayr, A., Danelljan, M., Romero, A., Yu, F., Timofte, R., Van Gool, L.: Repaint: inpainting using denoising diffusion probabilistic models. In: Proceedings of the IEEE/CVF Conference on Computer Vision and Pattern Recognition, pp. 11461–11471 (2022)
15. Nichol, A.Q., Dhariwal, P.: Improved denoising diffusion probabilistic models. In: International Conference on Machine Learning, pp. 8162–8171. PMLR (2021)
16. Shen, J., Kang, S.H., Chan, T.F.: Euler's elastica and curvature-based inpainting. SIAM J. Appl. Math. **63**(2), 564–592 (2003)
17. Sridevi, G., Srinivas Kumar, S.: Image inpainting based on fractional-order nonlinear diffusion for image reconstruction. Circuits Syst. Signal Process. **38**, 3802–3817 (2019)
18. Xu, Z., Sun, J.: Image inpainting by patch propagation using patch sparsity. IEEE Trans. Image Process. **19**(5), 1153–1165 (2010)
19. Yan, Z., Li, X., Li, M., Zuo, W., Shan, S.: Shift-Net: image inpainting via deep feature rearrangement. In: Ferrari, V., Hebert, M., Sminchisescu, C., Weiss, Y. (eds.) Computer Vision – ECCV 2018. LNCS, vol. 11218, pp. 3–19. Springer, Cham (2018). https://doi.org/10.1007/978-3-030-01264-9_1
20. Zhang, X., Chan, T.F.: Wavelet inpainting by nonlocal total variation. Inverse Probl. Imaging **4**(1), 191–210 (2010)
21. Zhang, Y.y., Xie, D.: Detection and segmentation of multi-class artifacts in endoscopy. J. Zhejiang Univ. Sci. B **20**(12), 1014 (2019)
22. Zhao, J., Chen, Z., Zhang, L., Jin, X.: Unsupervised learnable sinogram inpainting network (SIN) for limited angle CT reconstruction. arXiv preprint arXiv:1811.03911 (2018)

Only Classification Head Is Sufficient for Medical Image Segmentation

Hongbin Wei[1], Zhiwei Hu[2], Bo Chen[2], Zhilong Ji[2], Hongpeng Jia[1],
Lihe Zhang[1(✉)], and Huchuan Lu[1]

[1] Dalian University of Technology, Dalian, China
{weihongbin,22109042}@mail.dlut.edu.cn, {zhanglihe,lhchuan}@dlut.edu.cn
[2] Tomorrow Advancing Life, Beijing, China
{huzhiwei3,chenbo2,jizhilong}@tal.com

Abstract. Medical image segmentation is a pivotal research domain that has garnered widespread attention in contemporary medical diagnostics. In pursuit of enhancing network efficacy, researchers have taken great efforts to develop various well-designed decoders. Unfortunately, due to the limited medical training data, the issues of underfitting and overfitting frequently arise. To this end, we undertake plentiful experiments to decouple the encoder and decoder components, and obtain a critical finding that excessively complex decoders impede the encoder's potentiality of feature extraction. Inspired by some remarkable image generation work, we devise a straightforward segmentation network, which incorporates a pre-trained encoder backbone network and a pixel classification head. Our network not only ensures adequate feature decoding ability but also maximizes feature representation capability of the backbone. Experimental results on four datasets of three tasks show the outstanding performance against the state-of-the-art methods. The source code will be publicly available at https://github.com/weihongbin/CHNet

Keywords: Medical image segmentation · Encoder-decoder decoupling · Classification head

1 Introduction

Medical image segmentation aims to separate objects of interest from the surrounding environment in medical images, such as human tissues, organs, and pathological regions. It has broad applications in medical research, clinical diagnosis, pathological analysis, and assisted surgery, and is of significant importance.

Due to patient privacy concerns, acquiring medical images is difficult, and medical images need to be annotated by professional doctors, which results in expensive labelling cost. As a result, the scale of current medical image datasets is generally small. The lack of enough training data makes it difficult for neural

Q. Liu et al. (Eds.): PRCV 2023, LNCS 14437, pp. 296–308, 2024.
https://doi.org/10.1007/978-981-99-8558-6_25

networks to be sufficiently trained. Many natural image segmentation methods cannot be directly transferred to medical segmentation field, which severely limits its development.

Currently, UNet [24] and its variants [8,13,15,17–19] dominate medical segmentation field. They utilize the encoder-decoder architecture. As shown in Fig. 1 (a), the UNet adopts a symmetrical encoder-decoder structure, which extracts and reconstructs features through multiple downsampling and upsampling operations. The encoder can obtain features of different scales, which go through upsampling and skip-connection operations into the decoder. These features are fused to obtain segmentation results. Researchers believe that this direct connection method is too rough to fully tap into the potential of features at different levels. Therefore, various feature enhancement modules have been designed to strengthen feature expression ability, as shown in Fig. 1 (b). A large amount of research shows that using an encoder that has been pre-trained on a large-scale dataset such as ImageNet can significantly improve the performance of downstream tasks. Therefore, the common paradigm is to design innovative decoder based on a pre-trained encoder and then fine-tune the whole model on medical images for domain adaptation.

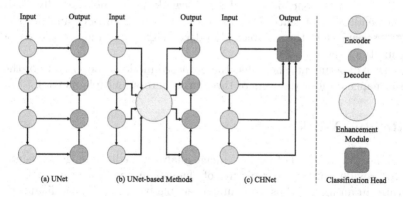

Fig. 1. Illustration of different medical image segmentation architectures

In an encoder-decoder structured neural network, the overall performance is determined by both the feature extraction ability of the encoder and the feature inferring ability of the decoder. In our study, we decouple the network to investigate the two components separately, and find that the encoders of identical structure, which are training with different decoders, exhibit different feature extraction capabilities, which can be attributed to two factors: 1) different inferring abilities of the decoders, and 2) the influence of the decoder on the feature extraction ability of the encoder during coupled training.

In image generation field, some studies [3,28,29] simultaneously generate images and their ground truths. The generation of ground truths is accomplished by using a dedicated pixel classification head. Specifically, the input of the classification head is the intermediate features of the generated image, and the output is supervised by a small number of ground truths. These approaches have achieved remarkable results, which indicates that the feature inferring ability of the pixel classification head is sufficient in scenarios where training data is limited.

Recently, the development of encoders has progressed from convolutional neural networks (CNNs) to transformers, which have increasingly powerful feature extraction capabilities. Moreover, a plethora of pre-training strategies have been introduced to learn prior knowledge so that the extracted features already contain rich semantic information. We designed a simple classification head decoder, as shown in Fig. 1 (c), which not only possesses sufficient feature inferring capabilities but also maximizes the feature extraction capabilities of the encoder. Our model has a simple and flexible structure, with few parameters and computational requirements. Most importantly, it achieves remarkably high performance. Overall, our contributions can be summarized as follows:

- We proposed a simple yet effective segmentation network (CHNet) with a classification head as the decoder, which can sufficiently learn representation capacity of the whole network on the limited training data.
- We decoupled the encoder-decoder framework, and experimentally analyzed the dependencies of feature representation between encoder and decoder. It was observed that the complex decoder design can suppress the encoding capability of the encoder.
- Extensive experiments show that the proposed model achieves state-of-the-art performance on four datasets of three tasks across multiple metrics.

2   Related Work

As an important method of medical image analysis, medical image segmentation aims to provide a more clear picture of changes in anatomical or pathological structures in an image, thus providing a reliable basis for clinical diagnosis and early diagnosis of diseases. It plays a crucial role in computer-aided diagnosis and intelligent medical treatment, greatly improving the efficiency and accuracy of diagnosis.

Feature extraction for medical images is more difficult than for normal RGB images, as the former often suffers from blur, noise and low contrast. Traditional medical image segmentation algorithms rely heavily on human prior knowledge and have insufficient generalization capabilities, making it difficult to obtain satisfactory results. With the rapid development of deep learning techniques in recent years, convolutional neural networks [20–22,30–34] have successfully achieved hierarchical feature representation of images, which has become a hot research topic in computer vision.

U-Net [24] is widely used in medical image segmentation and has become the benchmark for most medical image segmentation tasks. U-Net uses skip connections to combine the high-level semantic feature maps from the decoder

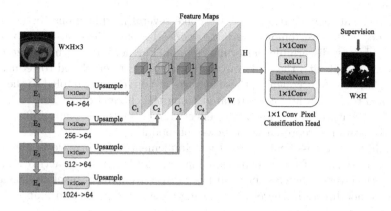

Fig. 2. The overall architecture of the proposed CHNet.

and corresponding low-level detailed feature maps from the encoder. It is widely believed that the success of U-Net depends on the U-shape structure, and many U-Net-based models have been proposed. Atten-UNet [19] embeds an attention gate in each transition layer between encoder and decoder blocks, which automatically learns to focus on target structures of different shapes and sizes. UNet++ [35] introduces nesting and dense skip connections to reduce the semantic gap between encoders and decoders. Although reasonable performance can be achieved, nested network structures are too complex to check for sufficient information at full scale. In each feature fusion, UNet3+ [13] aggregates feature maps at all scales using comprehensive skip connections to make more complete use of the full-scale feature information.

CNN can only capture local information, while transformer excels at direct global relationship modeling. Recently, the Transformer architecture has been successful in many tasks. Some works [7,12] explore its effectiveness for the medical vision tasks. NTNet [12] is a simple but powerful hybrid transformer architecture, which integrates self-attention into CNN to enhance medical image segmentation. Another representative transformer-based model is TransUNet [7], which has the advantages of both transformers and U-Net. It encodes tokenized image patches from CNN feature maps into an input sequence to extract global context, and utilizes the low-level CNN features via a u-shaped hybrid architectural design.

3 The Proposed Method

3.1 Overall Architecture

Our design principles are simplicity and effectiveness, as illustrated in Fig. 2. It is primarily composed of four encoder blocks and a single pixel classification head. Following the works [10,34], Res2Net-50 was selected as the backbone for feature extraction, yielding four layers of feature representations. Additionally, to suit medical segmentation task, we make several modifications to the backbone. Specifically, we remove the last encoder block of Res2Net-50, as it does not provide any discernible performance gain. Furthermore, to minimize computational complexity, the 1×1 convolutional layers are employed to reduce the feature channels. Subsequently, these features are directly upsampled to the original image resolution and concatenated along the channel dimension to form a 256×W×H feature map. This design effectively mitigates semantic information loss and boosts the representation power of modal.

In addition, a classification head was implemented for pixel-level classification, which consists of 1×1 convolutional layers, ReLU non-linear activation, and BatchNorm layers. The classification head directly conducts pixel classification along channel direction, thereby generating a complete prediction map.

3.2   Loss Function

The total loss function can be formulated as follows:

$$L_{total} = L_{IoU}^{w} + L_{BCE}^{w}, \tag{1}$$

where L_{IoU}^{w} and L_{BCE}^{w} represent the weighted IoU [27] loss and binary cross entropy (BCE) [27] loss, which restrict the prediction map in terms of the global structure and local details. They have been widely adopted in segmentation tasks. We use the same definitions as in [10,23] and their effectiveness has been validated in these works.

4   Encoder-Decoder Decoupling Analyses

Typically, the encoder and decoder of a model are trained jointly in an end-to-end manner, without attention to their individual performance or the coupling effect. To investigate this issue, we conduct a series of decoupling experiments using U-Net [24], Atten-UNet [19], and the proposed CHNet. To ensure fairness,

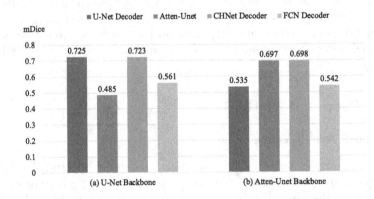

Fig. 3. Feature Inferring capability of different decoders.

the backbone architectures of the three networks are identical, and the only variation is the decoder. Firstly, the three networks are trained to reach the optimal performance, respectively. And then the models are decoupled to obtain three backbones of different parameters. After freezing each backbone, different decoders are separately connected for decoupling research. Here, all reported results are conducted on the COVID-19 Lung dataset [1,2]. They also have the same tendencies on the other datasets and tasks.

4.1 Feature Inferring Capability of Decoder

The backbones of U-Net and Atten-UNet are combined with the decoders of CHNet and FCN-32 s [16] to form new networks, respectively. After the backbones are all frozen, we retrain the decoders, which are randomly initialized as done in their original end-to-end learnt models. As shown in Fig3 (a), the CHNet decoder and U-Net decoder perform similarly based on the U-Net backbone. Meanwhile, our decoder and Atten-UNet decoder also behave the same way in Fig. 3 (b). These results indicate that the decoder of classification head has similar feature inferring capability to the complex U-Net and Atten-UNet decoders. The poor performance of FCN decoder, Atten-UNet decoder in Fig. 3 (a) and U-Net decoder in Fig. 3 (b) indicates the bad fitting ability of these decoders. In addition, when these decoders load the parameters of the optimal end-to-end models as initialization, and then are further trained based on the frozen backbone, the results are just the same.

4.2 Feature Extraction Capability of Encoder

In order to evaluate the encoder, we choose a simplest decoder FCN-32 s [16] and combine it with the backbones of CHNet, U-Net and Atten-UNet to form new networks, respectively. Similarly, all the backbones are frozen and we retrain the decoder. Because this decoder is very light, its own influence on the final performance can be ignored. Thus, the prediction results mainly reflect the abilities

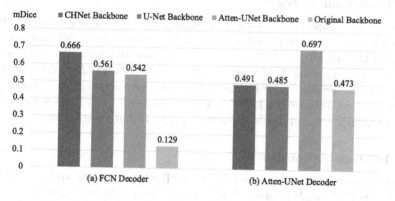

Fig. 4. Feature extraction capability of different encoders.

of different backbones. From Fig. 4 (a), we can see that the extraction ability of CHNet backbone is much better than those of U-Net and Atten-UNet, which actually reflects that the complex designs of U-Net and Atten-UNet decoders suppress their own backbones. In addition, we implement the same experiments on the Atten-UNet decoder, as shown in Fig. 4 (b). Here, the original backbone only loads ImageNet pre-trained parameters. The relatively complex decoder is able to well fit the gaps between the backbones (even though the backbone does not see medical image data) and achieves similar results except the self-coupled backbone. The results in Fig. 4 (b) also mean that a complexly structured decoder is more easy to deeply couple its own backbone.

5   Experiments

5.1   Datasets

We verify the effectiveness of the proposed framework on three medical segmentation tasks, which cover diverse data modalities, such as computed tomography (CT), ultrasound imaging, and color colonoscopy imaging.

COVID-19 Lung Infection. Few publicly available COVID-19 lung CT datasets are suitable for infection segmentation. To have relatively sufficient samples for training, we merged two datasets [1,2] to obtain 1,277 high-quality CT images. We divide them into 894 training images and 383 testing ones.

Breast Ultrasound Segmentation. The BUSI [25] is a common dataset for breast ultrasound segmentation, which consists of 780 images acquired from 600 female patients, including 133 normal cases, 437 benign tumors, and 210 malignant tumors. We compared with two well-established task-specific methods [5,6] and performed a four-fold cross-validation on this dataset.

Polyp Segmentation. Two benchmark datasets ClinicDB [4] and Kvasir [14] are used. We adopt the same training set as [10], that is, 550 samples from the ClinicDB and 900 samples from the Kvasir are used for training. And the remaining 62 images and 100 images are used for testing.

5.2   Evaluation Metrics

There are many popular metrics used in different medical segmentation branches. Following [11], five metrics are employed for quantitative evaluation, including mean Dice ($mDice$), $Precision$, $Recall$, S-measure (S_α) and mean absolute error(MAE). Following [6], $Jaccard$, $Precision$, $Recall$, and $Dice$ are more commonly used for breast tumor segmentation. For polyp segmentation, mean Dice ($mDice$), mean IoU ($mIoU$), the weighted F-measure (F_β^ω), S-measure (S_α), E-measure (E_ϕ^{max}) and mean absolute error(MAE) are widely used.

5.3   Implementation Details

Our model is implemented based on the PyTorch framework and trained on a single 3090 GPU with mini-batch size 16. We resize the inputs to 352 × 352. Random horizontally flipping and random rotate augmentation are used.

Table 1. Quantitative comparisons on the COVID-19 Lung dataset. Top 2 scores are highlighted in red and blue, respectively. "†" represents the medicine-specific method.

Methods	Backbone	$mDice \uparrow$	$Precision \uparrow$	$Recall \uparrow$	$S_\alpha \uparrow$	$MAE \downarrow$
U-Net[24]	R2-50	0.725	0.744	0.810	0.819	0.010
Atten-UNet[19]	R2-50	0.697	0.720	0.803	0.799	0.013
UTNet[12]	R-50 + ViT-B16	0.735	0.782	0.786	0.836	0.007
TransUnet[7]	R-50 + ViT-B16	0.710	0.770	0.776	0.831	0.007
Inf-Net†[11]	R2-50	0.783	0.774	0.852	0.843	0.007
BCS-Net†[9]	R2-50	0.763	0.775	0.763	0.840	0.007
Ours	R2-50	0.800	0.816	0.830	0.846	0.006

Table 2. Quantitative comparisons on the breast ultrasound dataset.

Methods	Backbone	$Jaccard \uparrow$	$Precision \uparrow$	$Recall \uparrow$	$Dice \uparrow$
U-Net[24]	R2-50	65.73±1.49	81.25±1.00	74.53±1.95	74.65±1.29
Atten-UNet[19]	R2-50	65.70±0.88	79.27±1.60	77.03±1.05	75.06±1.14
UTNet[12]	R-50 + ViT-B16	67.46±1.78	79.88±1.22	74.82±1.95	74.41±1.39
TransUnet[7]	R-50 + ViT-B16	71.47±0.98	81.66±1.52	80.78±1.63	79.00±0.79
SKU-Net†[5]	SKs	64.48±2.37	75.37±3.22	78.56±3.27	74.03±2.21
NU-Net†[6]	Deeper U-Net	68.86±1.99	78.90±2.26	82.48±2.14	77.79±1.88
Ours	R2-50	72.47±0.62	81.93±0.38	82.94±0.95	80.30±0.57

Table 3. Quantitative comparisons on the polyp segmentation dataset ClinicDB.

Methods	Backbone	$mDice \uparrow$	$mIoU \uparrow$	$F_\beta^\omega \uparrow$	$S_\alpha \uparrow$	$E_\phi^{max} \uparrow$	$MAE \downarrow$
U-Net[24]	R2-50	0.890	0.839	0.877	0.920	0.954	0.011
Atten-UNet[19]	R2-50	0.909	0.864	0.899	0.937	0.962	0.009
UTNet[12]	R-50 + ViT-B16	0.860	0.818	0.856	0.910	0.963	0.017
TransUnet[7]	R-50 + ViT-B16	0.847	0.798	0.831	0.907	0.920	0.020
PraNet†[10]	R2-50	0.899	0.849	0.896	0.936	0.963	0.009
SANet†[26]	R2-50	0.916	0.859	0.909	0.939	0.971	0.012
Ours	R2-50	0.926	0.882	0.915	0.946	0.969	0.009

Table 4. Quantitative comparisons on the polyp segmentation dataset Kvasir.

Methods	Backbone	$mDice \uparrow$	$mIoU \uparrow$	$F_\beta^\omega \uparrow$	$S_\alpha \uparrow$	$E_\phi^{max} \uparrow$	$MAE \downarrow$
U-Net[24]	R2-50	0.855	0.790	0.830	0.880	0.918	0.042
Atten-UNet[19]	R2-50	0.883	0.823	0.861	0.900	0.931	0.037
UTNet[12]	R-50 + ViT-B16	0.862	0.803	0.843	0.886	0.911	0.042
TransUnet[7]	R-50 + ViT-B16	0.869	0.816	0.847	0.899	0.920	0.040
PraNet†[10]	R2-50	0.898	0.840	0.885	0.915	0.944	0.030
SANet†[26]	R2-50	0.904	0.847	0.892	0.915	0.949	0.028
Ours	R2-50	0.911	0.864	0.900	0.924	0.950	0.026

For the optimizer, we adopt the SGD, and the momentum and weight decay are set as 0.9 and 0.0005, respectively. Warm-up and linear decay strategies are used to adjust the learning rate. For any medical image sub-tasks, the above training strategy is same. The difference among these models is only in the number of training epochs due to different convergence speeds. Specifically, the number of training epochs settings in the polyp segmentation, breast tumor segmentation and COVID-19 Lung Infection are 50, 100 and 200, respectively.

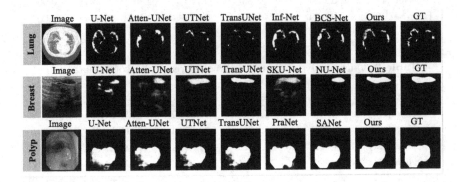

Fig. 5. Visual comparison of different medicine-general and medicine-specific methods.

5.4 Comparisons with State-of-the-Art Methods

For the purpose of a comprehensive and unbiased comparison, we not only compare our approach with medicine-specific methods, but also with medicine-general methods, such as U-Net [24], Attention U-Net [19], UTNet [12], and TransUNet [7]. To be fair, we retrain these general-purpose methods using the same training data and settings as the proposed model. For different tasks, we use their common evaluation metrics. The results are shown in Tables 1, 2, 3, and 4.

Table 1 shows performance comparison on the COVID-19 datasets. Compared to the second best method Inf-Net, our CHNet achieves an improvement of 1.7% in terms of mDice and the score has increased to 0.8. Table 2 shows that on the breast ultrasound dataset, our CHNet achieves the best preformance in terms of all four metrics. Moreover, our model is more stable than the competitors, which can be verified by the standard deviations. In Table 3 and Table 4, our model consistently achieves the best segmentation performance on the polyp segmentation dataset ClinicDB and Kvasir and especially obtains significant improvement on mIoU metric.

Figure 5 shows that the qualitative comparison with other methods. It can be seen that our results have greater advantages in terms of detection accuracy, object completeness, and contour sharpness across different image modalities.

Table 5. The FLOPs and parameters of different methods and their backbone.

Metrics	Backbone	U-Net	Atten-UNet	UTNet	TransUNet	Ours
FLOPs (GB)↓	∼ 4.8	∼ 17.5	∼ 9.7	∼ 27.1	∼ 24.7	∼ 7.1
Params (MB)↓	∼ 8.7	∼ 17.4	∼ 19.2	∼ 12.6	∼ 34.0	∼ 8.8

5.5 FLOPs and Parameters

In Table 5, we list FLOPs and parameters of different medicine-general methods and their backbone Res2Net50. It can be seen that the encoder and decoder of U-Net have used approximate amounts of parameter. The parameter number for Transformer-based TransUNet is even larger.

Compared to these U-Net-based models, our model achieves the best performance against both FLOPs and parameter amount metrics. The majority of parameters in the proposed method are derived from the backbone. The proposed model is efficient and its decoder is only about 0.15M.

5.6 Ablation Study

From Table 6, it can be seen that our classification head decoder has very strong feature expression ability. And only using 1×1 convolutional layer can achieve the best results, while using 3×3 convolutional layer performs slightly worse. The baseline adopts the FCN-32 s [16] decoder.

With similar performance, 1×1 convolutional layers can save a lot of computational resources compared to 3×3 convolutional layers. Usually, applying a convolutional kernel with larger perception will cause the compulational complexity to increase in the square mangificence.

Table 6. Ablation experiments on the COVID-19 Lung dataset.

Metrics	$mDice$	E_ϕ^{max}	F_β^ω	S_α
Baseline (FCN-32s)	0.666	0.601	0.496	0.799
CHNet (3×3 Conv)	0.796	0.761	0.653	0.843
CHNet (1×1 Conv)	0.800	0.765	0.655	0.846

6 Conclusion

In this paper, we propose a simple yet effective classification head network (CHNet) for medical image segmentation task, whose decoder is quite different from previous models. By conducting encoder-decoder decoupling experiments, we find that a decoder with complex structure will cause the feature extraction capability of the encoder to be under-utilized. Moreover, the decoupling experiments demonstrate that the feature expression capability of the classification head is similar to the well-designed decoders of other models, even though CHNet has very few parameters. More importantly, the feature extraction capability of the backbone is stronger for the end-to-end trained CHNet. Extensive experimental results on four datasets of three tasks demonstrate that the proposed model outperforms various state-of-the-art methods.

Acknowledgements. This work was supported by the National Natural Science Foundation of China # 62276046 and the Liaoning Natural Science Foundation # 2021-KF-12-10.

References

1. Covid-19 CT lung and infection segmentation dataset. https://zenodo.org/record/3757476 (2020)
2. Covid-19 CT segmentation dataset. https://medicalsegmentation.com/COVID19/ (2020)
3. Baranchuk, D., Voynov, A., Rubachev, I., Khrulkov, V., Babenko, A.: Label-efficient semantic segmentation with diffusion models. In: International Conference on Learning Representations (2022). https://openreview.net/forum?id=SlxSY2UZQT
4. Bernal, J., et al.: WM-DOVA maps for accurate polyp highlighting in colonoscopy: validation vs. saliency maps from physicians. Comp. Med. Imaging Graph. **43**, 99–111 (2015)
5. Byra, M., et al.: Breast mass segmentation in ultrasound with selective kernel U-Net convolutional neural network. Biomed. Signal Process. Control **61**, 102027 (2020)
6. Chen, G.P., Li, L., Dai, Y., Zhang, J.X.: NU-net: an unpretentious nested U-Net for breast tumor segmentation. arXiv preprint arXiv:2209.07193 (2022)
7. Chen, J., et al.: TransUNet: transformers make strong encoders for medical image segmentation. arXiv preprint arXiv:2102.04306 (2021)
8. Çiçek, Ö., Abdulkadir, A., Lienkamp, S.S., Brox, T., Ronneberger, O.: 3D U-Net: learning dense volumetric segmentation from sparse annotation. In: Ourselin, S., Joskowicz, L., Sabuncu, M.R., Unal, G., Wells, W. (eds.) MICCAI 2016. LNCS, vol. 9901, pp. 424–432. Springer, Cham (2016). https://doi.org/10.1007/978-3-319-46723-8_49
9. Cong, R., et al.: BCS-Net: boundary, context, and semantic for automatic COVID-19 lung infection segmentation from CT images. IEEE Trans. Instrum. Meas. **71**, 1–11 (2022)
10. Fan, D.-P., et al.: PraNet: parallel reverse attention network for polyp segmentation. In: Martel, A.L., et al. (eds.) MICCAI 2020. LNCS, vol. 12266, pp. 263–273. Springer, Cham (2020). https://doi.org/10.1007/978-3-030-59725-2_26
11. Fan, D.P., et al.: Inf-Net: automatic COVID-19 lung infection segmentation from CT images. IEEE Trans. Med. Imaging **39**(8), 2626–2637 (2020)
12. Gao, Y., Zhou, M., Metaxas, D.N.: UTNet: a hybrid transformer architecture for medical image segmentation. In: de Bruijne, M., Cattin, P.C., Cotin, S., Padoy, N., Speidel, S., Zheng, Y., Essert, C. (eds.) MICCAI 2021. LNCS, vol. 12903, pp. 61–71. Springer, Cham (2021). https://doi.org/10.1007/978-3-030-87199-4_6

13. Huang, H., et al.: UNet 3+: a full-scale connected UNet for medical image segmentation. In: ICASSP 2020-2020 IEEE International Conference on Acoustics, Speech and Signal Processing (ICASSP), pp. 1055–1059. IEEE (2020)
14. Jha, D., et al.: Kvasir-SEG: a segmented polyp dataset. In: Ro, Y.M., et al. (eds.) MMM 2020. LNCS, vol. 11962, pp. 451–462. Springer, Cham (2020). https://doi. org/10.1007/978-3-030-37734-2_37
15. Jha, D., et al.: ResUNet++: an advanced architecture for medical image segmentation. In: 2019 IEEE International Symposium on Multimedia (ISM), pp. 225–2255. IEEE (2019)
16. Long, J., Shelhamer, E., Darrell, T.: Fully convolutional networks for semantic segmentation. In: Proceedings of the IEEE Conference on Computer Vision and Pattern Recognition, pp. 3431–3440 (2015)
17. Mehta, S., Mercan, E., Bartlett, J., Weaver, D., Elmore, J.G., Shapiro, L.: Y-Net: joint segmentation and classification for diagnosis of breast biopsy images. In: Frangi, A.F., Schnabel, J.A., Davatzikos, C., Alberola-López, C., Fichtinger, G. (eds.) MICCAI 2018. LNCS, vol. 11071, pp. 893–901. Springer, Cham (2018). https://doi.org/10.1007/978-3-030-00934-2_99
18. Milletari, F., Navab, N., Ahmadi, S.A.: V-Net: fully convolutional neural networks for volumetric medical image segmentation. In: 2016 Fourth International Conference on 3D Vision (3DV), pp. 565–571. IEEE (2016)
19. Oktay, O., et al.: Attention U-Net: learning where to look for the pancreas. arXiv preprint arXiv:1804.03999 (2018)
20. Pang, Y., Zhao, X., Xiang, T.Z., Zhang, L., Lu, H.: Zoom in and out: a mixed-scale triplet network for camouflaged object detection. In: CVPR, pp. 2160–2170 (2022)
21. Pang, Y., Zhao, X., Zhang, L., Lu, H.: Multi-scale interactive network for salient object detection. In: CVPR, pp. 9413–9422 (2020)
22. Pang, Y., Zhao, X., Zhang, L., Lu, H.: CAVER: cross-modal view-mixed transformer for bi-modal salient object detection. IEEE TIP (2023)
23. Qin, X., Zhang, Z., Huang, C., Gao, C., Dehghan, M., Jagersand, M.: BASNet: boundary-aware salient object detection. In: Proceedings of the IEEE/CVF Conference on Computer Vision and Pattern Recognition, pp. 7479–7489 (2019)
24. Ronneberger, O., Fischer, P., Brox, T.: U-Net: convolutional networks for biomedical image segmentation. In: Navab, N., Hornegger, J., Wells, W.M., Frangi, A.F. (eds.) MICCAI 2015. LNCS, vol. 9351, pp. 234–241. Springer, Cham (2015). https://doi.org/10.1007/978-3-319-24574-4_28
25. Shin, S.Y., Lee, S., Yun, I.D., Kim, S.M., Lee, K.M.: Joint weakly and semi-supervised deep learning for localization and classification of masses in breast ultrasound images. IEEE Trans. Med. Imaging 38(3), 762–774 (2018)
26. Wei, J., Hu, Y., Zhang, R., Li, Z., Zhou, S.K., Cui, S.: Shallow attention network for polyp segmentation. In: de Bruijne, M., et al. (eds.) MICCAI 2021. LNCS, vol. 12901, pp. 699–708. Springer, Cham (2021). https://doi.org/10.1007/978-3-030-87193-2_66
27. Wei, J., Wang, S., Huang, Q.: F^3net: fusion, feedback and focus for salient object detection. In: Proceedings of the AAAI Conference on Artificial Intelligence, vol. 34, pp. 12321–12328 (2020)
28. Wu, Z., et al.: Synthetic data supervised salient object detection. In: Proceedings of the 30th ACM International Conference on Multimedia, pp. 5557–5565 (2022)
29. Zhang, Y., et al.: DatasetGAN: efficient labeled data factory with minimal human effort. In: Proceedings of the IEEE/CVF Conference on Computer Vision and Pattern Recognition, pp. 10145–10155 (2021)

30. Zhao, X., et al.: M^{2}SNet: multi-scale in multi-scale subtraction network for medical image segmentation. arXiv preprint arXiv:2303.10894 (2023)

31. Zhao, X., Pang, Y., Zhang, L., Lu, H.: Joint learning of salient object detection, depth estimation and contour extraction. IEEE TIP **31**, 7350–7362 (2022)

32. Zhao, X., Pang, Y., Zhang, L., Lu, H., Zhang, L.: Suppress and balance: a simple gated network for salient object detection. In: ECCV, pp. 35–51 (2020)

33. Zhao, X., Pang, Y., Zhang, L., Lu, H., Zhang, L.: Towards diverse binary segmentation via a simple yet general gated network. arXiv preprint arXiv:2303.10396 (2023)

34. Zhao, X., Zhang, L., Lu, H.: Automatic polyp segmentation via multi-scale subtraction network. In: de Bruijne, M., et al. (eds.) MICCAI 2021. LNCS, vol. 12901, pp. 120–130. Springer, Cham (2021). https://doi.org/10.1007/978-3-030-87193-2_12

35. Zhou, Z., Rahman Siddiquee, M.M., Tajbakhsh, N., Liang, J.: UNet++: a nested U-Net architecture for medical image segmentation. In: Stoyanov, D., et al. (eds.) DLMIA/ML-CDS -2018. LNCS, vol. 11045, pp. 3–11. Springer, Cham (2018). https://doi.org/10.1007/978-3-030-00889-5_1

Task-Incremental Medical Image Classification with Task-Specific Batch Normalization

Xuchen Xie[1,4], Junjie Xu[1,4], Ping Hu[3], Weizhuo Zhang[1,4], Yujun Huang[1,4], Weishi Zheng[1,4], and Ruixuan Wang[1,2,4(✉)]

[1] School of Computer Science and Engineering,
Sun Yat-sen Univerisity, Guangzhou, China
wangrulx5@mall.sysu.edu.cn
[2] Peng Cheng Laboratory, Shenzhen, China
[3] School of Information Science and Engineering, Xinjiang University, Urumqi, China
[4] Key Laboratory of Machine Intelligence and Advanced Computing, MOE,
Guangzhou, China

Abstract. Currently, multiple intelligent diagnosis systems are developed independently partly due to the difficulty in collecting training data for all tasks of disease diagnosis. It would be better if an intelligent diagnosis system can incrementally learn more diagnosis tasks as what general practitioners have done. Existing approaches to such a task-incremental learning problem still more or less suffer from the catastrophic forgetting issue, i.e., system performance on old tasks is often gradually decreased when the system incrementally learns new tasks. Here a simple but effective approach is proposed to solve the catastrophic forgetting of old task knowledge for task-incremental learning. Specifically, when training a convolutional neural network (CNN) classifier to incrementally learn more tasks, the kernels in the task-shared CNN feature extractor are initially learned or fine-tuned from the first task and then fixed through subsequent learning tasks, and only the task-specific batch normalization parameters in the CNN feature extractor and the new task-specific classifier head are learned and stored for each new task. Empirical evaluation on medical image datasets supports that the task-specific batch normalization provides an effective solution for an intelligent system to incrementally learn multiple tasks of disease diagnosis. The source code will be released publicly.

Keywords: Task-incremental learning · Task-specific batch normalization · Intelligent diagnosis

Supplementary Information The online version contains supplementary material available at https://doi.org/10.1007/978-981-99-8558-6_26.

1   Introduction

Deep learning techniques, particularly convolutional neural networks (CNNs), have been widely applied to the intelligent diagnosis of various diseases based on medical images [8, 26, 37]. However, current intelligent diagnosis systems can each only help diagnose a specific set of diseases due to difficulty in collecting training data for all disease diagnosis tasks at once. This results in independent systems for each disease, which causes difficulties in managing multiple systems for medical centers and requires medical staff to spend more time learning to use various systems. One solution to this issue is to develop a single intelligent system which can incrementally learn more and more tasks, with each task for the diagnosis of a specific set of diseases. To limit system size growth, the underlying classifier is often assumed to have a task-shared feature extractor but multiple task-specific classifier heads, with each classifier head for one specific task. During inference, the task identification is available and users know which classifier head should be applied for any new data.

Multiple approaches have been recently proposed to enable the classifier to have the task-incremental learning (TIL) ability. One approach is by regularization on changes in model parameters which are crucial for old knowledge when learning new tasks, such as the elastic weight consolidation (EWC) method [22], the memory aware synapses (MAS) method [3], and RWalk [6]. Another approach tries to distill the knowledge of old tasks from the old classifier to the new classifier when the (new) classifier learns a new task, such as the well-known learning without forgetting (LwF) [25]. This approach can better keep old knowledge when a small set of old samples are stored for new task learning [29, 30].

However, both regularization-based and distillation-based approaches have the difficulty in balancing between learning new knowledge and keeping old knowledge when more and more new tasks need to be incrementally learned. To make the classifier more flexibly learn new knowledge, another approach tries to add new kernels, layers, or even sub-networks when learning new tasks [17, 27]. By adopting this expansion-oriented strategy, the classifier gains the capacity to integrate newfound information, enriching its learning capabilities. However, this strategy requires adjustments to the classifier's structure, mainly in its feature extraction part. While this enhances growth, it also demands more memory for the enlarged network. Moreover, it is challenging to determine when and where to add new network components.

In this study, different from all previous approaches, a simple approach is proposed to effectively solve the catastrophic issue by making use of the batch normalization (BN) function in popular CNN models. Specifically, built on an initial feature extractor which is well-trained in advance or from the first task, the kernels in the feature extractor of the classifier are fixed through all subsequent learning tasks, and only the two parameters in BN per kernel are learned together with the task-specific classifier head when updating the classifier to learn each new task. Since the number of BN parameters is very limited compared to kernel parameters, the learned task-specific BN parameters can be directly stored together with the learned task-specific classifier head for each new task with

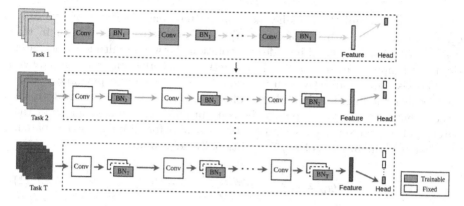

Fig. 1. Task-specific batch normalization for TIL. The kernels in the CNN feature extractor are learned or fine-tuned in the first task. For new tasks, our approach fixes kernels in the feature extractor, and learns and stores only task-specific batch normalization parameters in the feature extractor and the new task-specific classifier head.

negligible additional memory usage. In this way, the learned knowledge for each task is well kept without forgetting, and the fixed kernels plus the task-specific BN and task-specific classifier head can be applied during inference for prediction of any new task-identified image. Extensive evaluations on medical image datasets confirmed the effectiveness of the proposed approach.

2   Methodology

The problem of interest is for a classifier to continually learn more and more tasks over time. At each time, the classifier learns a new task for the classification of a set of new classes, and the classes learned at different times are often assumed to be non-overlapped, e.g., each task corresponds to the diagnosis of diseases from a specific body organ or tissue. Specifically, no data of any previously learned task is kept and only the training data of the new task are available during learning the new task. Traditionally, different tasks share the same feature extractor of the classifier but have task-specific classifier heads. During inference, for any new (test) input image, the task identification of the image is assumed to be known, and therefore the shared feature extractor and the corresponding task-specific head are combined to predict the class of the input. Different from all existing TIL strategies and inspired by recent progress in generating various styles of realistic images by tuning BN parameters with generative adversarial networks [13,14,35], a task-specific BN strategy is proposed for task-incremental learning of a CNN classifier (Fig. 1), where different tasks share the same feature extractor except for task-specific BN parameters in the feature extractor.

2.1   Preliminary: Batch Normalization (BN)

Suppose a CNN classifier consists of a feature extractor and a classifier head. In the CNN model, the feature extractor $f(\cdot)$ often consists of multiple convolutional layers and the classifier head $h(\cdot)$ often consists of one or two fully connected layers with final softmax output. Every convolutional layer in the feature extractor in general includes convolutions between multiple kernels and the layer's input feature maps, a BN function per kernel and the subsequent element-wise nonlinear activation function. Given a batch of input images during classifier training, for the j-th convolutional kernel at the l-th convolutional layer, denote by $z_{l,j}$ the output of the convolutional layer for an image in the batch, and $Z_{l,j}$ the set of convolutional layer outputs for all images in the batch, then batch normalization of $z_{l,j}$ is traditionally defined as

$$ q_{l,j}(z_{l,j}) = \gamma_{l,j} \frac{z_{l,j} - \mu(Z_{l,j})}{\sigma(Z_{l,j})} + \beta_{l,j}, \tag{1} $$

where $\mu(Z_{l,j})$ and $\sigma(Z_{l,j})$ represent respectively the mean and standard deviation of all the elements in $Z_{l,j}$, and $\gamma_{l,j}$ and $\beta_{l,j}$ are the to-be-learned parameters of the BN function. Note that BN is kernel-specific, i.e., each kernel has a unique BN with specific parameters $(\gamma_{l,j}, \beta_{l,j})$.

2.2 Task-Specific Batch Normalization for TIL

When a CNN classifier is incrementally updated to handle multiple classification tasks, most existing TIL strategies assume that all these tasks share the same feature extractor $f(\cdot)$ but have task-specific classifier head $h_t(\cdot)$ for the t-th task. That means, both kernel parameters and BN parameters of all kernels in the feature extractor are incrementally updated over tasks, and all tasks share the same set of updated kernel parameters and BN parameters. Such change in parameters of the feature extractor is the origin of catastrophic forgetting in TIL, causing downgraded performance on previously learned tasks.

To alleviate or ideally avoid the catastrophic forgetting issue, it is necessary for the feature extractor to be updated as small as possible or not updated at all. However, keeping the feature extractor unchanged would probably make the classifier not learn well for new tasks, because in this case only the new task-specific classifier head is learnable and the fixed or little-changed feature extractor may have less representation ability to extract discriminative features for new tasks. Inspired by the proposed StyleGAN model which can be trained to generate various realistic images by tuning the BN parameters in the model [13,14], we hypothesize that for an initially well-trained feature extractor, fixing kernel parameters and only tuning batch normalizing parameters may be sufficient for the feature extractor to learn to represent discriminative features for new tasks. Also, since the number of kernels (e.g., 4,800 in ResNet18 [9]) in a feature extractor is much less than the number of kernel parameters (e.g., around 11 millions in ResNet18), the learned BN parameters for each task can be directly stored together with the learned task-specific classifier head with negligible additional memory usage. In this way, the incrementally updated classifier assures avoiding the catastrophic forgetting issue, because all the new parameters (i.e., task-specific BN parameters of all fixed kernels, and the model parameters in the new task-specific classifier head) appearing in the updated classifier for the t-th task are stored and accessible even if the classifier is further updated over more new tasks at later time. That means, after the classifier is incrementally updated to learn T tasks, for any new (test) input image from the t-th task where $t \in \{1, 2, \ldots, T\}$, the classifier can be applied to predict the class of the image based on the well-trained kernels (with parameters $\boldsymbol{\theta}_1$) at the first task and task-specific BN (with parameters denoted by $\boldsymbol{\omega}_t$) for the t-th task. In contrast, existing approaches apply the final updated kernels (with parameters $\boldsymbol{\theta}_T$) and BN (with parameters $\boldsymbol{\omega}_T$) at the last (T-th) task to all T tasks during inference. In short, the difference between existing approaches and our approach in feature extractor (not including classifier head) can be summarized below,

$$\text{Previous approaches} : (\boldsymbol{\theta}_T, \boldsymbol{\omega}_T) \text{ for all the learned } T \text{tasks}; \tag{2}$$

$$\text{Our approach} : (\boldsymbol{\theta}_1, \boldsymbol{\omega}_t) \text{ for the } t\text{-th task}, \forall t \in \{1, 2, \ldots, T\}. \tag{3}$$

In our approach, since parameters of all kernels in the feature extractor are fixed across all tasks, it is crucial to have well-trained kernels at the first task. The kernels can be obtained by (1) training the feature extractor from the scratch using the training set of the first task (Strategy I), (2) using a fixed pre-trained feature extractor particularly when the pre-trained feature extractor is well trained using a large-scale public dataset (e.g., ImageNet) or other relevant dataset (Strategy II), or (3) fine-tuning a pre-trained feature extractor using the training set of the first task (Strategy III).

While our TIL strategy can prevent the updated classifier from forgetting knowledge of old tasks, one potential new issue is that the classifier might be limited in learning new knowledge from new tasks due to the relatively smaller number of learnable model parameters. However, empirical evaluation suggests that the learnable model parameters together with well-trained and fixed kernels in the feature extractor may be sufficient for the classifier to learn new knowledge for each new task. This suggests only tuning batch normalization parameters in the well-trained feature extractor may be sufficient to adapt it to the new task, enabling it to learn and extract discriminative features specific to the new task effectively.

3   Experiments

3.1   Experimental Setup

Multiple public medical image datasets were collected to form two sets of multiple tasks for comprehensive evaluation of our task-incremental learning method. The first set is single-modality (i.e., histopathology), which is a subset of multiple publicly available histopathology datasets, including seven tasks (Table 1 first 7 rows). The second set is multi-modality and contains seven tasks as well, with each task being single-modality (Table 1, last 7 rows). The second set corresponds to the original (mostly high-resolution) version of a subset of MedMNIST-v2 [36].

Table 1. Statistics of the single-modality and multi-modality sets of tasks.

Tasks	Imaging Modality	Classes	Image Size	Training	Validation	Test
Stomach (ST) [11]	Histopathology	4	512×512	630	90	180
Colon (CO) [5]	Histopathology	2	768×768	630	90	180
Lung (LU) [5]	Histopathology	2	768×768	630	90	180
Breast(BT) [12]	Histopathology	2	50×50	630	90	180
Colorectal Polyps (CP) [34]	Histopathology	2	224×224	630	90	180
Lymph Node (LN) [4,33]	Histopathology	2	96×96	630	90	180
Oral Cavity (OC) [18]	Histopathology	2	2048×1538	630	90	180
Blood (BL) [1]	Blood Cell Microscope	8	360×363	11,959	1,712	3,421
Derma (DE) [7,32]	Dermatoscope	7	600×450	7,014	1,001	2,000
Breast (BR) [2]	Breast Ultrasound	3	500×500	544	78	158
Pneumonia (PN) [19,20]	Chest X-Ray	2	[384, 2916]×[127, 2713]	4,710	522	625
Path (PA) [15,16]	Colon Pathology	9	224×224	17,997	1,994	1,431
OCT (OC) [19,20]	Retinal OCT	4	[384, 1536]×[277, 512]	19,493	2,163	196
Tissue (TI) [28]	Kidney Cortex Microscope	8	32×32	33,093	4,728	9,456

Table 2. Selection of multi-modality medical datasets. We excluded certain datasets (last five rows) and collected the corresponding original datasets from the remaining sub-datasets in MedMNIST-v2, forming multi-modality sets.

Tasks	Corresponding MNIST-Set	Original Dataset Type	Task Type	Selected	Images Evaluated
Blood (BL)	BloodMNIST	2D	Multi-Class	Yes	All
Derma (DE)	DermaMNIST				
Breast (BR)	BreastMNIST				
Pneumonia (PN)	PneumoniaMNIST				
Path (PA)	PathMNIST	2D	Multi-Class	Yes	20%
OCT (OC)	OCTMNIST				
Tissue (TI)	TissueMNIST				
–	ChestMNIST	2D	Multi-Label	No	–
–	RetinaMNIST	2D	Ordinal Regression	No	–
–	OrganAMNIST	3D	Multi-Class	No	–
	OrganBMNIST				
	OrganCMNIST				

Table 2 presents the details of the selection process for the multi-modality set. We excluded datasets that were originally in 3D format, as well as ordinal regression tasks and multi-label tasks. Additionally, for datasets with a significantly larger amount of data, we randomly selected 20% of the data for evaluation. All the images were resized to 224×224 pixels. The gray images were converted to three-channel images by duplicating the single channel to the other two.

By default, ResNet18 was used as the classifier backbone following previous studies in continual learning and the first kernel training strategy (Strategy I) was adopted to obtain the kernel parameters of the feature extractor. When incrementally learning a new task, Adam optimizer [21] (batch size 64) was adopted. The initial learning rate was 1e-3 and then divided by 10 at the 50-th and the 75-th epoch respectively. Each training lasted for 100 epochs, with training convergence consistently observed. The task-specific BN and task-specific classifier head with best classification performance on the task's validation set was selected for evaluation. After learning a new task, the mean class recall

(MCR) over all learned classes so far was calculated considering the class imbalance in the test dataset across learned tasks. Note that MCR is equivalent to class accuracy on balanced dataset which is widely adopted in continual learning studies. For each order of tasks to be incrementally learned by the classifier, the average and standard deviation of MCRs over five runs were reported.

3.2   Performance on Task-Incremental Learning

Our method was compared with several representative baselines developed for task-incremental learning, including RWalk [6], MAS [3], EWC [22], LwF [25], CF-IL [29]. We also compared our method with two more methods, iCaRL [30] and UCIR [10], which were originally developed for class-incremental learning (with a single classifier head for all learned classes), but were also used for task-incremental learning by knowing which part of the output is used during inference. As in original studies, we preserved 20 images per learned old class for iCaRL and UCIR, and generated 20 synthetic images per learned old class for CF-IL. All other baselines, as well as our method, did not retain any old-class images. Similar amount of effort was put into tuning each baseline. In addition, an upper-bound result was also reported by training a unique classifier for each task and then collecting the multiple classifiers for prediction during inference (Figs. 2 and 3, 'Joint').

Figures 2 and 3 show that our method ('Ours') outperforms all the baselines over the course of incrementally learning seven tasks respectively on the two sets, regardless of the choice of the first task and the appearing order of these tasks (Figs. 2 and 3, first two subfigures). In Fig. 2 (Left), the performance of all baselines was reduced substantially after learning the second task ('OC') probably because the change in imaging modality from the first task to the second task makes it challenging for a single classifier to work well for two imaging modalities. The third task ('BL') is relatively easier to be learned and therefore the average performance over all learned classes was boosted after incrementally learning this task. The superior performance over multi-modality TIL tasks by our method is probably from the strong learning ability of the thousands BN parameters in the feature extractor, where the task-specific BN parameters adapt the original feature extractor to the new task of another modality and the task-specific classifier head helps discriminate classes within the new task. Furthermore, Fig. 2 (Right) demonstrates classifier performance on two tasks (i.e., first one 'PA' and third one 'BL') over the course of TIL, which shows that the classifier from our method can keep classification performance for each of the tasks over TIL, while the classifiers from the baselines showed downgraded performance on old tasks. Similar results were observed on the single-modality set (Fig. 3). Furthermore, on the single-modality set, our method achieves performance that is very close to upper-bound ('Joint'). All these results directly confirm the effectiveness of our method in keeping old knowledge over TIL.

3.3   Ablation and Sensitivity Studies

An ablated study was performed by replacing the task-specific BN with task-shared BN, i.e., all incrementally learned tasks share the same BN for each kernel, and BN parameters were fixed after the first task (Fig. 4, 'Shared BN (fixed)') or updated together with a new task-specific classifier head when the classifier incrementally learns a new task (Fig. 4, 'Shared BN (updated)'). As shown in Fig. 4, the two ablated versions resulted in downgraded performance over the course of TIL, regardless of the appearing order of tasks.

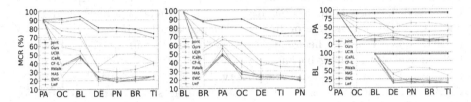

Fig. 2. Task-incremental learning performance on the multi-modality set. Left and Middle: TIL with two different task orders. Right: performance on old tasks 'PA' and 'BL' (y-axis) over the TIL course. x-axis: tasks with a specific order. The default kernel learning strategy (Strategy I) was adopted for each baseline at the first task.

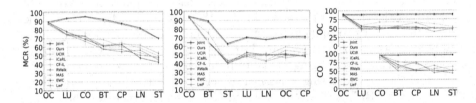

Fig. 3. Task-incremental learning performance on the single-modality set.

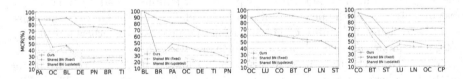

Fig. 4. Ablation study of task-specific batch normalization. Left two: TIL performance for two different orders of tasks on the multi-modality set. Right two: TIL performance for two different orders of tasks on the single modality set.

While Strategy I was adopted for kernel training in the above evaluations, similar findings were obtained using the other two kernel training strategies as shown in Fig. 5, i.e., our method outperforms all baselines and can preserve knowledge of old tasks over TIL. Here the pre-trained feature extractor was trained using a subset (randomly selected 200 classes) of ImageNet dataset. Figure 5 also shows Strategy I and Strategy III ('Ours-I' and 'Ours-III') for kernel training resulted in similar TIL performance, and Strategy II ('Ours-II') is better than the other two probably because the pre-trained feature extractor is more transferable as observed in previous studies.

3.4   Generalizability Assessment

Our method is not limited to specific classifier backbone and medical image datasets. As shown in Table 3, our method consistently performs best with all three representative CNN backbones (ResNet18, VGG19 [31], AlexNet [24]) compared to the strong baselines. On the other hand, considering that all the baselines were evaluated only on natural image datasets in previous studies, our method was also evaluated on CIFAR100 [23]. As expected, our method showed the best performance when incrementally learning 10 or 20 tasks (correspondingly 10 and 5 classes within each task), regardless of the feature train-

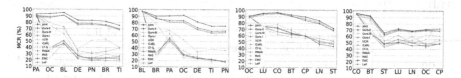

Fig. 5. Effect of kernel training strategy on task-incremental learning. Left two: TIL performance for two different orders of tasks on the multi-modality set. Right two: TIL performance for two different orders of tasks on the single-modality set. All baselines here used the same pre-trained feature extractor for the first task.

Table 3. Performance comparison between baselines and ours with different CNN backbones on the multi-modality set. The average and standard deviation of mean class recall (%) at the end of incremental learning over five runs were reported.

ResNet18			VGG19			AlexNet		
LwF	UCIR	Ours	LwF	UCIR	Ours	LwF	UCIR	Ours
$23.57_{\pm 1.3}$	$40.11_{\pm 0.5}$	$\mathbf{66.48_{\pm 2.1}}$	$24.8_{\pm 3.0}$	$44.3_{\pm 2.4}$	$\mathbf{61.84_{\pm 1.5}}$	$22.15_{\pm 3.5}$	$36.2_{\pm 1.7}$	$\mathbf{49.3_{\pm 2.7}}$

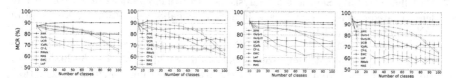

Fig. 6. Task-incremental learning performance on the CIFAR100 dataset. First two subfigures: incremental learning of total 10 and 20 tasks respectively, using Strategy I for kernel training. Last two subfigures: TIL performance using Strategies II and III for kernel training, where the same pre-trained feature extractor was used for baselines.

ing strategies (Fig. 6). UCIR often performs better than the other baselines, probably because it additionally used a subset of old data when learning new tasks. Note that when Strategies II and III were used, the 200 ImageNet classes used for feature extractor pre-training are not overlapped with the classes in CIFAR100. These results clearly suggest our method is generalizable to various image domains.

4   Conclusions

In this study, we propose a task-specific batch normalization approach for task-incremental classification. Extensive evaluations mainly on medical image datasets support our method is effective in preserving knowledge of old tasks, even if multiple imaging modalities appear across tasks over the course of incremental learning. This opens a door for development of a unified intelligent system which can be incrementally trained for multiple tasks of disease diagnosis. Its potential in lesion detection and segmentation can be investigated as well.

References

1. Acevedo, A., Merino, A., Alférez, S., Molina, Á., Boldú, L., Rodellar, J.: A dataset of microscopic peripheral blood cell images for development of automatic recognition systems. Data in Brief 30 (2020). https://doi.org/10.1016/j.dib.2020.105474
2. Al-Dhabyani, W., Gomaa, M., Khaled, H., Fahmy, A.: Dataset of breast ultrasound images. Data in Brief 28, 104863 (2020). https://doi.org/10.1016/j.dib.2019.104863
3. Aljundi, R., Babiloni, F., Elhoseiny, M., Rohrbach, M., Tuytelaars, T.: Memory aware synapses: learning what (not) to forget. In: ECCV, pp. 139–154 (2018)
4. Bejnordi, B.E., et al.: Diagnostic assessment of deep learning algorithms for detection of lymph node metastases in women with breast cancer. JAMA 318(22), 2199–2210 (2017)
5. Borkowski, A.A., Bui, M.M., Thomas, L.B., Wilson, C.P., DeLand, L.A., Mastorides, S.M.: Lung and colon cancer histopathological image dataset (LC25000). arXiv preprint arXiv:1912.12142 (2019)
6. Chaudhry, A., Dokania, P.K., Ajanthan, T., Torr, P.H.: Riemannian walk for incremental learning: understanding forgetting and intransigence. In: ECCV, pp. 532–547 (2018)
7. Codella, N., et al.: Skin lesion analysis toward melanoma detection 2018: a challenge hosted by the international skin imaging collaboration (ISIC). arXiv preprint arXiv:1902.03368 (2019)
8. Ghamsarian, N., et al.: LensID: a CNN-RNN-based framework towards lens irregularity detection in cataract surgery videos. In: MICCAI, pp. 76–86 (2021)
9. He, K., Zhang, X., Ren, S., Sun, J.: Deep residual learning for image recognition. In: CVPR, pp. 770–778 (2016)
10. Hou, S., Pan, X., Loy, C.C., Wang, Z., Lin, D.: Learning a unified classifier incrementally via rebalancing. In: CVPR, pp. 831–839 (2019)
11. Institute, N.C.: TCGA dataset (2006). https://www.cancer.gov/about-nci/organization/ccg/research/structural-genomics/tcga

12. Janowczyk, A., Madabhushi, A.: Deep learning for digital pathology image analysis: a comprehensive tutorial with selected use cases. JPI **7**(1), 29 (2016)
13. Karras, T., et al.: Alias-free generative adversarial networks. In: NeurIPS (2021)
14. Karras, T., Laine, S., Aila, T.: A style-based generator architecture for generative adversarial networks. In: CVPR, pp. 4401–4410 (2019)
15. Kather, J.N., Halama, N., Marx, A.: 100,000 histological images of human colorectal cancer and healthy tissue (2018). https://doi.org/10.5281/zenodo.1214456
16. Kather, J.N., et al.: Predicting survival from colorectal cancer histology slides using deep learning: a retrospective multicenter study. PLOS Med. **16**(1), 1–22 (2019). https://doi.org/10.1371/journal.pmed.1002730
17. Ke, Z., Liu, B., Ma, N., Xu, H., Shu, L.: Achieving forgetting prevention and knowledge transfer in continual learning. In: NeurIPS (2021)
18. Kebede, A.F.: Oral cancer dataset, version 1 (2021). https://www.kaggle.com/datasets/ashenafifasilkebede/dataset
19. Kermany, D.S., et al.: Identifying medical diagnoses and treatable diseases by image-based deep learning. Cell **172**(5), 1122–1131.e9 (2018). https://doi.org/10.1016/j.cell.2018.02.010
20. Kermany, D.S., Zhang, K., Goldbaum, M.H.: Large dataset of labeled optical coherence tomography (Oct) and chest X-RAY images (2018). https://doi.org/10.17632/rscbjbr9sj.3
21. Kingma, D.P., Ba, J.: Adam: a method for stochastic optimization. ICLR (2015)
22. Kirkpatrick, J., et al.: Overcoming catastrophic forgetting in neural networks. PNAS **114**(13), 3521–3526 (2017)
23. Krizhevsky, A., Hinton, G.: Learning multiple layers of features from tiny images. Tech. Rep. **4**(7) (2009)
24. Krizhevsky, A., Sutskever, I., Hinton, G.E.: ImageNet classification with deep convolutional neural networks. Commun. ACM **60**(6), 84–90 (2017)
25. Li, Z., Hoiem, D.: Learning without forgetting. PAMI **40**(12), 2935–2947 (2017)
26. Li, Z., Zhong, C., Wang, R., Zheng, W.S.: Continual learning of new diseases with dual distillation and ensemble strategy. In: MICCAI, pp. 169–178 (2020)
27. Liu, Y., Schiele, B., Sun, Q.: Adaptive aggregation networks for class-incremental learning. In: CVPR, pp. 2544–2553 (2021)
28. Ljosa, V., Sokolnicki, K.L., Carpenter, A.E.: Annotated high-throughput microscopy image sets for validation. Nat. Methods **9**(7), 637–637 (2012)
29. PourKeshavarzi, M., Zhao, G., Sabokrou, M.: Looking back on learned experiences for class/task incremental learning. In: ICLR (2022)
30. Rebuffi, S.A., Kolesnikov, A., Sperl, G., Lampert, C.H.: ICARL: incremental classifier and representation learning. In: CVPR, pp. 2001–2010 (2017)
31. Simonyan, K., Zisserman, A.: Very deep convolutional networks for large-scale image recognition. In: ICLR (2015)
32. Tschandl, P.: The ham10000 dataset, a large collection of multi-source dermatoscopic images of common pigmented skin lesions (2018). https://doi.org/10.7910/DVN/DBW86T
33. Veeling, B.S., Linmans, J., Winkens, J., Cohen, T., Welling, M.: Rotation equivariant CNNs for digital pathology. In: MICCAI, pp. 210–218 (2018)
34. Wei, J., et al.: A petri dish for histopathology image analysis. In: Artificial Intelligence in Medicine, pp. 11–24 (2021)
35. Xie, C., Tan, M., Gong, B., Wang, J., Yuille, A.L., Le, Q.V.: Adversarial examples improve image recognition. In: CVPR, pp. 819–828 (2020)

36. Yang, J., et al.: MedMNIST v2: a large-scale lightweight benchmark for 2D and 3D biomedical image classification. arXiv preprint arXiv:2110.14795 (2021)

37. Yang, Y., Cui, Z., Xu, J., Zhong, C., Wang, R., Zheng, W.S.: Continual learning with Bayesian model based on a fixed pre-trained feature extractor. In: MICCAI, pp. 397–406 (2021)

Hybrid Encoded Attention Networks for Accurate Pulmonary Artery-Vein Segmentation in Noncontrast CT Images

Ming Wu[1], Hao Qi[1], Hui-Qing Zeng[3], Xiangxing Chen[3(✉)], Xinhui Su[4], Sunkui Ke[3], Yinran Chen[1], and Xiongbiao Luo[1,2(✉)]

[1] Department of Computer Science and Technology,
Xiamen University, Xiamen, China
[2] National Institute for Data Science in Health and Medicine,
Xiamen University, Xiamen, China
[3] Zhongshan Hospital, Xiamen University, Xiamen, China
1207148830qq.com
[4] The First Affiliated Hospital, Zhejiang University School of Medicine,
Hangzhou, China

Abstract. Pulmonary artery-vein segmentation in computed tomography image is essential to lung disease diagnosis. It is still a challenge to segment small distal vessels, crossover and adhesion of arterioles due to complicated arteriovenous structures and limited computed tomography resolution. This work proposes a new U-shaped architecture of hybrid encoded attention networks that employ stacked hybrid units and 3D resample-free attention gates for automatic pulmonary artery-vein segmentation. Specifically, the hybrid unit uses normal operations, 3D hybrid easy operation, and channel shuffle to extract diverse features while the 3D resample-free attention gate can detect regions of interest such as pulmonary arteries and veins and suppress task-independent responses. We validated our method on 50 computed tomography volumes from LIDC-IDRI, with the experimental results demonstrating that it works more stable and effective than currently available approaches, improving the average dice similarity coefficients of the arteries and veins to (84.32%, 86.41%), respectively.

Keywords: Pulmonary Artery-Vein Segmentation · Classification · Attention Gates · Channel Shuffle · Noncontrast CT

1 Introduction

Lung diseases such as cancers, pulmonary embolism and pneumonia are one of high-morbidity diseases [16]. Computed tomography (CT) is widely used for diagnosis and assessment of these diseases. Automatic CT pulmonary artery-vein segmentation facilitates clinical diagnosis, surgical planning and treatment with precise localization of lesions or tumors [5]. The challenge of pulmonary artery-vein segmentation mainly lies in the complex structures of pulmonary vascular

© The Author(s), under exclusive license to Springer Nature Singapore Pte Ltd. 2024
Q. Liu et al. (Eds.): PRCV 2023, LNCS 14437, pp. 321–332, 2024.
https://doi.org/10.1007/978-981-99-8558-6_27

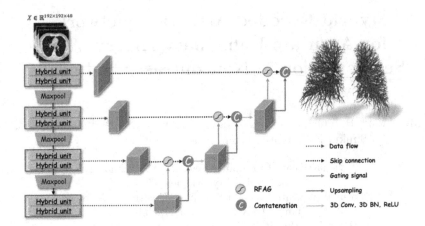

Fig. 1. Architecture of HEAN with hybrid units and 3D resample-free attention gates

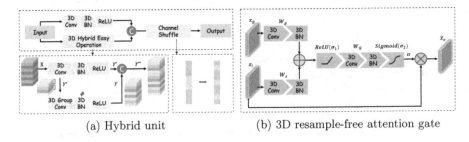

(a) Hybrid unit   (b) 3D resample-free attention gate

Fig. 2. Details of the hybrid unit and 3D resample-free attention gate

trees. Moreover, CT scanners have partial volume effect as well as CT images have limited resolution and noise artifacts. Additionally, pulmonary arteriovenous structures have intensity similar with tumors, nodules, and trachea walls.

Many researchers have worked on automatic segmentation of pulmonary arteries and veins in CT images. Nardelli et al. [12] first segmented pulmonary vessels by scale-space particles and then used 3-D convolutional neural networks (CNN) to classify arterial and venous vessels, followed by a graph-cut method to refine the results. Jimenez-Carretero et al. [10] still discussed a graph-cut approach to separate pulmonary arteries and veins in noncontrast CT images. Yu et al. [19] first extracted entire vascular structures and separated the adhesion points, and then employed anatomical knowledge to classify pulmonary arteries and veins. Pan et al. [14] built a Twin-Pipe network to classify arteries and veins in a complete topological tree and obtained the final classification results by topologically optimizing the interbranch and intrabranch relationships to maintain spatial consistency. Qin et al. [17] trained tubule-sensitive CNN with well-designed modules such as feature recalibration module and attention distillation module to perform an end-to-end pulmonary artery-vein segmentation in CT images. More recently, Pu et al. [15] combined CNN with differential geometry to separate pulmonary arteries and veins in noncontrast CT images. However, the above methods have little ability to discriminate arteriovenous adhesions and the results of small distal vessel segmentation are too coarse.

This work aims to automatically and accurately segment or separate pulmonary arteries and veins in noncontrast CT images. The contribution of this work is twofold. Technically, we redesign a new U-shaped architecture called hybrid encoded attention networks (HEAN) in an end-to-end mode. Specifically, its encoder consists of stacked hybrid units. The hybrid unit performs normal operations, a 3D hybrid easy operation with easy linear operations, and channel shuffle to extract diverse feature maps. 3D hybrid easy operation enriches feature maps with vascular topological structure to separate arteriovenous adhesions. While its decoder places 3D resample-free attention gates (RFAG) on skip connections to highlight vessel areas and suppress irrelevant regions. With application to pulmonary vessel separation, our method outperforms other models.

2 Approaches

Figure 1 shows the structure of HEAN with four encoders and decoders. Each encoder uses stacked hybrid units to extract feature maps with different resolutions, and doubles the channel of feature maps. Each decoder contains upsampling and two $3 \times 3 \times 3$ 3D convolutions, followed by ReLU to half the channel of deep feature maps, to generate high-resolution feature maps. Moreover, inspired by attention gates [13], we employ 3D resample-free attention gates to highlight salient regions and suppress task-independent responses.

2.1 Hybrid Unit

Since convolution (Conv) layers generate feature maps that usually contain much redundancy in which some feature maps are similar to the others, we design a hybrid unit that consists of three operations, normal convolution, 3D hybrid easy operation, and channel shuffle to generate richer feature maps and reduce the parameters and floating point operations. Figure 2 (a) shows the structure of the hybrid unit with various operations.

Inspired by ghost module [6] and vascular 3D structure, we designed a 3D hybrid easy operation (HEO), which extends the original 2D linear operation into the 3D space and enriches the feature map. We employ 3D convolution and 3D group convolution for generating m and s 3D feature maps, respectively. Specifically, we set a parameter V, called easy rate. The parameter V controls the channels of the feature map of HEO with the following relationship: $c = v \times m$ and $s = (v - 1) \times m$.

Given an input $X \in \mathbb{R}^{c \times h \times w \times d}$ (where c, h, w, and d are the channel number, height, width, and depth of feature maps), our module uses 3D convolution to generate m feature maps $Y' \in \mathbb{R}^{m \times h \times w \times d}$ and then uses 3D group convolution to perform easy linear operations on each feature map in Y' to produce s 3D feature maps:

$$Y' = ReLU(BN(3DConv(X)))$$

(1)

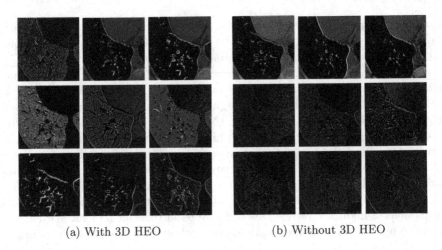

(a) With 3D HEO (b) Without 3D HEO

Fig. 3. Feature maps of HEAN (left) and feature maps of the network without 3D HEO(right)

$$Y = \Phi(Y') = ReLU(BN(3DGroupConv(Y'))) \tag{2}$$

$$y_{ij} = \phi_{i,j}(y_i'), \forall i = 1, \dots m, j = 1, \dots s, \tag{3}$$

where y_i is the i-th feature map in Y' and y_{ij} is generated by the j-th corresponding 3D linear operation ϕ_{ij}. After producing $m \times s$ 3D feature maps, we concatenate the two tensors generated in different ways at channel dimensions to obtain the output $Y'' \in \mathbb{R}^{n \times h \times w \times d}$ of the 3D HEO:

$$Y'' = Y' \oplus Y \tag{4}$$

At the parallel path in the hybrid unit, we generate feature maps $S \in \mathbb{R}^{n \times h \times w \times d}$ by normal operations which consist of a $3 \times 3 \times 3$ 3D convolution with stride 1, padding 1, batch normalization (BN) and ReLU activation. Then, we concatenate the two tensors $Y'' \in \mathbb{R}^{n \times h \times w \times d}$ and $S \in \mathbb{R}^{n \times h \times w \times d}$. Finally, we use a channel shuffle operation to strengthen the interaction among channels [21] and obtain the output feature maps.

Figure 3 shows the feature maps obtained by 3D HEO of HEAN. Apparently, more diverse feature maps are obtained by our method with 3D HEO. More diverse feature maps can deliver more robust information on arteriovenous boundaries, which reduce the incidence of mis-segmentation of arteriovenous adhesions. Furthermore, more diverse feature maps significantly assist in the accurate segmentation of small distal vessels. With 3D convolution, HEAN learns the topology of the up-and-down layers of vessels, making it extremely beneficial for solving this challenge at arteriovenous adhesions as well.

2.2 3D Resample-Free Attention Gate

Current U-shaped architectures with cascaded CNN such as U-Net [18] generally have a good representation ability. These multi-stage cascaded architectures

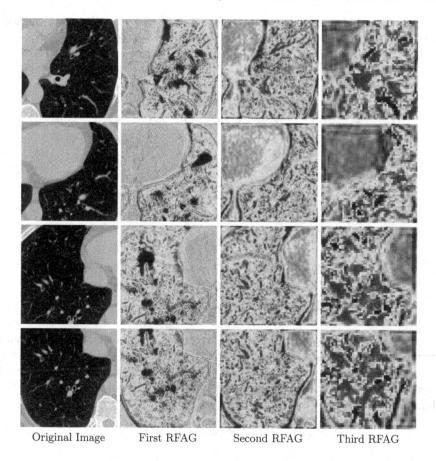

Original Image   First RFAG   Second RFAG   Third RFAG

Fig. 4. Input CT images (left) and their attention coefficient maps (right)

first locate regions of interest and then generate dense prediction in specific areas. However, they introduce a redundant use of model parameters and computational resources [13]. Particularly, these architectures fail to extract small pulmonary arteries and veins that usually show large shape variations in various patients. Also, it is reasonable to assume that the features obtained by stacked hybrid units are rich enough without resampling operations again. In this respect, inspired by attention gate [13], we employ 3D resample-free attention gates placed on skip connections to highlight small pulmonary arteries and veins and suppress the irrelevant regions like pulmonary tissues outside the lung.

Figure 2 (b) illustrates the details of the 3D resample-free attention gate module. The inputs of the 3D resample-free attention gate are the feature maps x_i from the encoding path and the feature maps x_g upsampled by 2 times from the deeper low-resolution feature maps. The 3D resample-free attention gate performs $1 \times 1 \times 1$ 3D convolution on x_g and x_i to squeeze the channel dimensions and add nonlinear activation, followed by an addition operation. After a series

Table 1. Quantitative evaluation of average DSC, Precision and Recall of the pulmonary artery-vein test data segmented by the different approaches.

Methods	DSC%		Precision%		Recall%	
	Artery	Vein	Artery	Vein	Artery	Vein
3D U-Net [3]	83.26	85.34	83.98	**88.34**	83.50	83.19
DenseVoxNet [20]	78.79	81.25	76.67	81.24	82.01	81.97
V-Net [11]	79.20	81.13	78.79	82.30	80.62	80.59
VoResNet [2]	82.48	84.57	81.51	85.61	84.54	84.20
UNETR [8]	66.75	67.16	67.39	67.53	67.22	67.36
UNet++ [22]	82.98	85.25	82.87	86.94	84.12	84.39
Attention U-Net [13]	83.82	86.14	81.84	86.03	84.89	85.80
nnU-Net [9]	80.76	83.84	**85.28**	87.52	77.70	81.28
MNet [4]	81.16	83.34	81.39	82.87	82.00	84.57
Swin UNETR [7]	81.34	83.47	81.60	86.11	82.46	81.98
HEAN (ours)	**84.32**	**86.41**	83.90	87.48	**85.60**	**86.03**

of operations including ReLU activation, $1 \times 1 \times 1$ 3D convolution, and sigmoid function, we can generate attention coefficients $\alpha \in [0, 1]$. Finally, by multiplying α and feature maps x_i, feature responses of irrelevant regions are suppressed and salient features in target regions are highlighted. The output coefficient α of the additive attention can be formulated as:

$$W(x) = BN(3DConv(x)), x \in x_g, x_i \tag{5}$$

$$\alpha = \sigma_2 \left(W_\psi(\sigma_1(gW_g + x_c W_x + b) + b_\phi) \right), \tag{6}$$

where W_g, W_x and W_ψ denote $1 \times 1 \times 1$ 3D convolution operation W, σ_1 represents ReLU activation function, σ_2 represents sigmoid function, b and b_ψ are bias terms.

Figure 4 displays the output of triple RFAG. The low-level RFAG (see the first RFAG of Fig. 4) effectively highlights the arteriovenous boundaries, helping to distinguish the arteriovenous adhesions and highlight small distal vessel features. In contrast, the high-level RFAG (see the second and third RFAG of Fig. 4) can effectively suppress the irrelevant region, preventing missegmentation of the irrelevant region. As a result of this approach, the segmentation accuracy of small distal vessels was dramatically improved and overall segmentation errors were reduced.

Table 2. Ablation experiment of HEAN. 3D denotes using 3D convolutioncusing depthwise convolution to replace 3D HEO, V_n denotes value n of the parameter V in 3D HEO, RFAG denotes resample-free attention gate.

3D	DW	V3	V4	V5	RFAG	DSC%		Precision%		Recall%	
						Artery	Vein	Artery	Vein	Artery	Vein
✔	✘	✘	✘	✘	✔	83.51	85.61	82.37	86.10	85.70	85.77
✘	✔	✘	✘	✘	✔	83.53	85.57	83.54	87.10	84.42	84.76
✘	✘	✔	✘	✘	✔	83.87	85.92	82.83	87.01	85.90	85.55
✘	✘	✘	✔	✘	✔	84.01	86.28	83.03	86.76	85.88	**86.44**
✘	✘	✘	✘	✔	✘	83.73	85.81	82.20	86.34	**86.28**	86.01
✘	✘	✘	✘	✔	✔	**84.32**	**86.41**	**83.9**	**87.48**	85.60	86.03

2.3 Hybrid DSC Loss Function

We use hybrid dice similarity coefficient (DSC) loss to measure the similarity of prediction and ground truth regions:

$$L_{ArteryDSC} = 0.5 - \frac{|G_A \cap P_A|}{|G_A| + |P_A|}, \tag{7}$$

$$L_{VeinDSC} = 0.5 - \frac{|G_V \cap P_V|}{|G_V| + |P_V|}, \tag{8}$$

$$L_{HybridDSC} = L_{ArteryDSC} + L_{VeinDSC}, \tag{9}$$

where G_A, G_V is the ground truth of artery and vein and P_A, P_V denotes the prediction of artery and vein. Separate calculation of DSC for arteries and veins improves recognition of arteriovenous adhesions (Table 2).

3 Experiments

We selected 50 cases of chest CT volumes from LIDC-IDRI [1] and separated them into 40 cases for training and 10 cases for testing. The CT spacing parameters were $0.488{\sim}0.943{\times}0.488{\sim}0.943{\times}1.0{\sim}2.5mm^3$. Ground truth data were initially generated by a region growing algorithm in the mask of pulmonary vessels. Then, three experts manually refined these automatically generated data. We adopt windows of [-800, 600] HU for intensity normalization. All data were cropped by a HU value and then normalized to [0, 1]. Due to a small amount of CT images, we implement data augmentation like randomly shifting intensity uniformly for CT images to enrich our data. We randomly sample the volumetric input at a size of [192, 192, 48] for training and use the AdamW optimizer with a weight decay of 1e-5 in training. While the initial learning rate was 1e-4, we introduce a stepped learning rate schedule by multiplying a specific value every several epochs. The model training was performed on a workstation with a CPU of Intel(R) Xeon(R) Silver 4210 CPU @ 2.20GHz and an Nvidia GeForce GTX 3090 GPU with 24GB of memory.

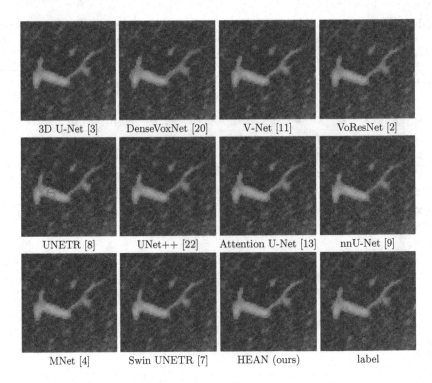

3D U-Net [3]	DenseVoxNet [20]	V-Net [11]	VoResNet [2]
UNETR [8]	UNet++ [22]	Attention U-Net [13]	nnU-Net [9]
MNet [4]	Swin UNETR [7]	HEAN (ours)	label

Fig. 5. 2D detail visual comparison of pulmonary artery-vein segmentation results of the eleven compared methods. At the sites of arteriovenous adhesions, all methods tend to mis-classify veins as arteries, but our method exhibits the smallest mis-segmentation and the highest similarity to the ground truth labels in small distal veins.

We use a five-fold cross-validation and compare our model with 3D U-Net [3], DenseVoxNet [20], V-Net [11], VoResNet [2], UNETR [8], UNet++ [22], Attention U-Net [13], nnU-Net [9], MNet [4] and Swin UNETR [7] to demonstrate the effectiveness of our approach. We employ three measures of the dice similarity coefficient (DSC), Precision and Recall to quantitatively evaluate the performance of these methods.

4 Results and Discussion

Table 1 shows the average DSC, Precision and Recall values for different models tested on 10 CT volumes. The experimental results demonstrate the effectiveness of our proposed method. We achieve an average artery DSC of 84.32% and an average vein DSC of 86.41%. Our method improves artery Recall to 85.60% and vein Recall to 86.03%, and has a low probability of over-segmentation outside the lung while it can maintain the accuracy in artery-vein segmentation.

Figure 5 intuitively compares the 2D artery-vein segmentation results. Our method can significantly reduce missed segmentation between peripheral arteries

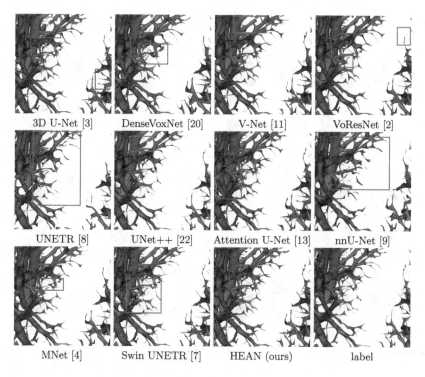

Fig. 6. 3D detail visual comparison of pulmonary artery-vein segmentation results of the eleven compared methods. Pink squares highlight the mis-segmentation results of other methods. (Color figure online)

and veins which are essential to clinical diagnosis. Unfortunately, all the methods mis-classification to some extent. Particularly, DenseVoxNet [20] and UNETR [8] obtain totally wrong segmentation results(see the first row of Fig. 5), while our method can keep better performance in integrity and consistency of arteries and veins than other compared networks. Figure 6 and Fig. 7 intuitively compare the 3D artery-vein segmentation results. In Fig. 6, unlike other methods, our method is free from problems such as vessel breakage, mis-classification, and inconsistent shape and labeling. Also, from the overall segmentation results in Fig. 7, our method exhibited no mis-segmentation compared to other methods for the outside region of the vessel.

The effectiveness of our method lies in several aspects. First, the hybrid unit uses dual-path operations rather than single-path operations to extract more diverse features, resulting in accurate pulmonary artery-vein segmentation particularly at adjacent and crossed regions. Next, 3D resample-free attention gates are introduced in skip connections in decoding. This 3D reinforced attention can extract thin and peripheral vessels and enable our method to keep the consistency of a same vessel branch, while other networks such as 3D U-Net [3] and UNETR [8] classify the same vessel branch into two different arterial and venous

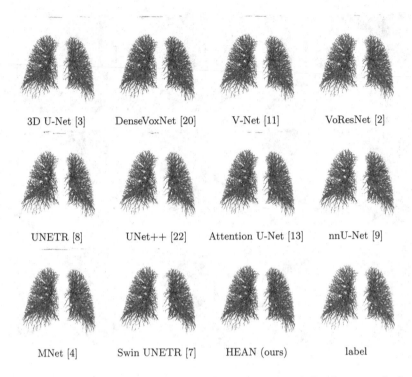

Fig. 7. 3D pulmonary artery-vein segmentation results of the eleven methods

vessels. Unfortunately, HEAN still fails to segment arteries or veins with low resolution, resulting in insufficient segmentation results at the end of the arteriovenous trees. Furthermore, the CT voxel imbalance between small foreground (arteries and veins) and large background critically degenerates the segmentation results, while it makes our method converge slowly in the training procedure.

5 Conclusion

This work proposes a new U-shaped architecture named hybrid encoded attention networks with hybrid units and attention gates in an end-to-end manner without any postprocessing. Specifically, HEAN uses stacked hybrid units as encoders to capture feature maps with 3D hybrid easy linear operations, while 3D resample-free attention gates enable HEAN to highlight target areas and suppress irrelevant regions. The experimental results show that our model outperforms other existing methods, achieving better or comparable DSC and Recall, as well as improving the accuracy and consistency of automatic CT pulmonary artery-vein segmentation. Furthermore, it has made major progress in the two challenges of arteriovenous adhesions and distal small vessels compared to other existing methods.

Acknowledgements. This work was supported in part by the National Natural Science Foundation of China under Grants 61971367, 82272133, and 62001403, in part by the Natural Science Foundation of Fujian Province of China under Grants 2020J01004 and 2020J05003, and in part by the Fujian Provincial Technology Innovation Joint Funds under Grant 2019Y9091.

References

1. Armato, S.G., III., et al.: The lung image database consortium (LIDC) and image database resource initiative (IDRI): a completed reference database of lung nodules on CT scans. Med. Phys. **38**(2), 915–931 (2011)
2. Chen, H., Dou, Q., Yu, L., Qin, J., Heng, P.A.: VoxResNet: deep voxelwise residual networks for brain segmentation from 3D MR images. Neuroimage **170**, 446–455 (2018)
3. Çiçek, Ö., Abdulkadir, A., Lienkamp, S.S., Brox, T., Ronneberger, O.: 3D U-Net: learning dense volumetric segmentation from sparse annotation. In: Ourselin, S., Joskowicz, L., Sabuncu, M.R., Unal, G., Wells, W. (eds.) MICCAI 2016. LNCS, vol. 9901, pp. 424–432. Springer, Cham (2016). https://doi.org/10.1007/978-3-319-46723-8_49
4. Dong, Z., et al.: MNet: rethinking 2D/3D networks for anisotropic medical image segmentation (2022). arXiv preprint arXiv:2205.04846
5. El-Baz, A., Suri, J.S.: Lung Imaging and Computer Aided Diagnosis. CRC Press (2011)
6. Han, K., Wang, Y., Tian, Q., Guo, J., Xu, C., Xu, C.: GhostNet: more features from cheap operations. In: Proceedings of the IEEE/CVF Conference on Computer Vision and Pattern Recognition, pp. 1580–1589 (2020)
7. Hatamizadeh, A., Nath, V., Tang, Y., Yang, D., Roth, H.R., Xu, D.: Swin UNETR: swin transformers for semantic segmentation of brain tumors in MRI images. In: Crimi, A., Bakas, S. (eds.) Brainlesion: Glioma, Multiple Sclerosis, Stroke and Traumatic Brain Injuries. BrainLes 2021. LNCS, vol. 12962. Springer, Cham (2022). https://doi.org/10.1007/978-3-031-08999-2_22
8. Hatamizadeh, A., et al.: UNETR: transformers for 3D medical image segmentation. In: Proceedings of the IEEE/CVF Winter Conference on Applications of Computer Vision, pp. 574–584 (2022)
9. Isensee, F., et al.: nnU-Net: Self-adapting framework for U-Net-based medical image segmentation (2018). arXiv preprint arXiv:1809.10486
10. Jimenez-Carretero, D., Bermejo-Pelaez, D., Nardelli, P., et al.: A graph-cut approach for pulmonary artery-vein segmentation in noncontrast CT images. Med. Image Anal. **52**, 144–159 (2019)
11. Milletari, F., Navab, N., Ahmadi, S.A.: V-Net: fully convolutional neural networks for volumetric medical image segmentation. In: 2016 Fourth International Conference on 3D Vision (3DV), pp. 565–571. IEEE (2016)
12. Nardelli, P., et al.: Pulmonary artery-vein classification in CT images using deep learning. IEEE Trans. Med. Imaging **37**(11), 2428–2440 (2018)
13. Oktay, O., et al.: Attention U-Net: Learning where to look for the pancreas (2018). arXiv preprint arXiv:1804.03999
14. Pan, L., et al.: Automatic pulmonary artery-vein separation in CT images using twin-pipe network and topology reconstruction (2021). arXiv:2103.11736

15. Pu, J., Leader, J.K., Sechrist, J., et al.: Automated identification of pulmonary arteries and veins depicted in non-contrast chest CT scans. Med. Image Anal. **77**, 102367 (2022)
16. Pulagam, A.R., Kande, G.B., Ede, V.K.R., Inampudi, R.B.: Automated lung segmentation from HRCT scans with diffuse parenchymal lung diseases. J. Digit. Imaging **29**(4), 507–519 (2016)
17. Qin, Y., et al.: Learning tubule-sensitive CNNs for pulmonary airway and artery-vein segmentation in CT. IEEE Trans. Med. Imaging **40**(6), 1603–1617 (2021)
18. Ronneberger, O., Fischer, P., Brox, T.: U-Net: convolutional networks for biomedical image segmentation. In: Navab, N., Hornegger, J., Wells, W.M., Frangi, A.F. (eds.) MICCAI 2015. LNCS, vol. 9351, pp. 234–241. Springer, Cham (2015). https://doi.org/10.1007/978-3-319-24574-4_28
19. Yu, K., Zhang, Z., Li, X., Liu, P., Zhou, Q., Tan, W.: A pulmonary artery-vein separation algorithm based on the relationship between subtrees information. J. Healthc. Eng. **2021**, 5550379 (2021)
20. Yu, L., et al.: Automatic 3D cardiovascular MR segmentation with densely-connected volumetric ConvNets. In: Descoteaux, M., Maier-Hein, L., Franz, A., Jannin, P., Collins, D.L., Duchesne, S. (eds.) MICCAI 2017. LNCS, vol. 10434, pp. 287–295. Springer, Cham (2017). https://doi.org/10.1007/978-3-319-66185-8_33
21. Zhang, X., Zhou, X., Lin, M., Sun, J.: ShuffleNet: an extremely efficient convolutional neural network for mobile devices. In: Proceedings of the IEEE Conference on Computer Vision and Pattern Recognition, pp. 6848–6856 (2018)
22. Zhou, Z., Rahman Siddiquee, M.M., Tajbakhsh, N., Liang, J.: UNet++: a nested U-Net architecture for medical image segmentation. In: Stoyanov, D., et al. (eds.) DLMIA/ML-CDS -2018. LNCS, vol. 11045, pp. 3–11. Springer, Cham (2018). https://doi.org/10.1007/978-3-030-00889-5_1

Multi-modality Fusion Based Lung Cancer Survival Analysis with Self-supervised Whole Slide Image Representation Learning

Yicheng Wang, Ye Luo[✉], Bo Li, and Xiaoang Shen

School of Software Engineering, Tongji University, Shanghai, China
{2131490,yeluo,1911030,sxa}@tongji.edu.cn

Abstract. Whole slide pathological images (WSIs) are the gold standard for lung cancer prognosis. However, due to their high resolution and limited annotations, lung cancer survival analysis based on WSIs becomes a challenging task. Some recent methods that fuse WSI and other modalities have achieved certain results. However, these methods tend to focus on integrating gene related information while overlooking relatively easily obtainable clinical variables and often rely on labor-intensive ROI annotations. In this work, we propose a novel framework for lung cancer survival analysis, which obviates the need for ROI annotations and fully exploits WSIs and clinical information by introducing a multi-modality fusion module and multi-task learning. We also utilizes self-supervised learning to eliminate the heterogeneity between WSIs and natural images. Experimental results via 5-fold cross-validation on 1,225 WSIs from 444 patients from NLST validate the state-of-the-art performance of our proposed method.

Keywords: Survival Analysis · Lung Cancer · Whole Slide Images · Self-supervised Learning · Multi-modality Fusion

1 Introduction

Lung cancer stands as the predominant cause of cancer-related fatalities worldwide. Therefore, it is of great significance to conduct survival analysis studies on lung cancer patients. Survival analysis serves as a method to predict the length of time from the initiation of follow-up until the occurrence of specific events (e.g. death, cancer recurrence). This approach enables early treatment decisions, thereby prolonging the lives of lung cancer patients.

In recent years, driven by the advent of digital pathology, the emergence of the survival prediction methods based on digital whole-slide pathological images (WSIs) has gained attention. However, relying solely on single-modal information from WSIs typically results in limited predictive performance. An increasing number of researchers have reached a consensus that integrating data from other modality could boost the performance of survival prediction. Mobadersany et al. [1] integrated genome data with pathological images by simply concatenation. Subramanian et al. [2] used canonical correlation analysis to fuse the gene expression and WSI features. Chen et al. [3] used the Kronecker product for the pathological image and genomic data fusion. However, these existing multimodality fusion methods have certain limitations. Firstly, they primarily focus on genomic data rather than easily accessible clinical data. Secondly, some simple fusion methods cannot represent complex feature interactions, while fusion strategies based on outer product often have high computational complexity and lack flexibility. Thirdly, they often require labor-intensive ROI annotations, which restrict their realistic application to large datasets. Recently, Several unimodal methods [4–6] have emerged that aggregate patch-level features to patient-level through multiple instance learning (MIL), allowing models to learn WSI features without the need for ROI annotations. However, these cluster-based MIL approaches have not been applied in multi-modality methods, and they always use pre-trained CNNs from on ImageNet [7] for WSI patch feature learning, overlooking the vast heterogeneity between WSI and natural images.

In order to address all of these aforementioned issues, we propose a Self-supervsised learning and Gated Multimodal Factorized Bilinear pooling based Survival prediction framework (**SGMFBSurv**), in which features from both WSI and clinical data are integrated by the proposed **GAMFB** module. Moreover, in order to tackle the huge resolution of WSIs without pixel-level annotations, a specified feature extractor is learned via SSL, and more attentions have been put into WSI data via considering the heterogeneity between WSI and natural images, the differences of different phenotypes on WSIs, etc. To further leverage information from other modalities, we also employed Multi-Task Learning (MTL) to predict the diagnostic outcomes of other related diseases simultaneously. At last, extensive experiments on NLST validate the superiority of our method to the state-of-the-arts. The main contributions of this paper can be summarized as follows: (1) We propose a multi-modality fusion based survival prediction framework that integrates WSIs and clinical data to achieve excellent prediction performance without ROI annotations. (2) A deep neural network based on the self-supervised and multiple instance learning for better WSI representation is particularly designed to address the heterogeneity between the pathological images and the natural images as well as the challenge of survival analysis based on WSIs with huge resolutions. (3) Extensive experiments on a large dataset NLST demonstrate the superiority of our method on lung cancer survival analysis with significant C-index improvement to its competitors.

2   The Proposed Method

2.1   Architecture Overview

The primary motivation of this study is to introduce multiple instance learning and self-supervised learning into multi-modality survival analysis on WSIs to address the challenges associated with labor-intensive pixel-level annotations and the significant differences between WSIs and natural images. Moreover, unlike most studies that focus on genetic information, our approach emphasizes the fusion of WSIs and clinical data which is more easy to obtain, and introduce the Multimodal Factorized Bilinear pooling to effectively combine information from both modalities and enhance the accuracy and robustness of our model. The proposed method is shown in Fig. 1.

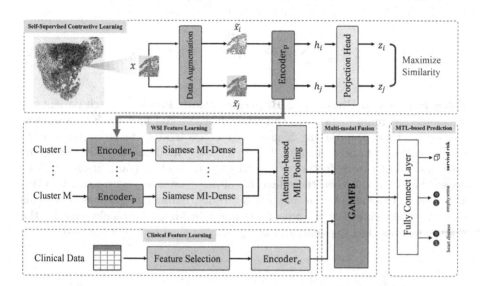

Fig. 1. Overview of the proposed framework. Firstly, a multitude of unlabeled patches are sampled from WSIs and a self-supervised learning-based encoder is trained and extract features from these patches and perform clustering. After applying the proposed Siamese MI-Dense to learn the representation of each phenotype, all of these representations are aggregated through an attention-based MIL pooling to patient-level. The clinical variables will be filtered and passed through an encoder to extract features. Finally, features from WSIs and clinical data are fused through the proposed GAMFB and predict survival risk and outcomes of two related diseases simultaneously.

2.2   Whole Slide Image Representation Learning

Patch-Level Feature Extractor by SSL. Previous methods [4–6] have over-looked the significant gaps between WSIs and natural images and directly applied pre-trained models on ImageNet for feature extraction. Hence, we train a WSI specified feature extractor named $Encoder_p$ from massive unlabeled patches by leveraging SimCLR [8], the details of which is shown on the top of Fig. 1. Specifically, given a patch x, a positive pair (x_i, x_j) is firstly obtained through two random augmentation transformations. Then $Encoder_p$ extracts feature vectors from these augmented pairs $h_i = Encoder_p(x_i)$ and $h_j = Encoder_p(x_j)$. A projection head $g(\cdot)$ then maps the two feature representations to a latent space: $z_i = g(h_i)$ and $g(h_i) = W^{(2)}\sigma(W^{(1)}h_i)$. z_j can be similarly obtained. And finally the contrastive loss is applied to the latent space to maximize agreement between differently augmented views of the same sample:

$$\ell_{i,j} = -\log \frac{\exp\left(\text{sim}\left(\boldsymbol{z}_i, \boldsymbol{z}_j\right)/\tau\right)}{\sum_{k=1}^{2N} \mathbb{1}_{[k \neq i]} \exp\left(\text{sim}\left(\boldsymbol{z}_i, \boldsymbol{z}_k\right)/\tau\right)} \tag{1}$$

where $sim(\cdot)$ is the cosine similarity, and $\mathbb{1}_{[k \neq i]} \in \{0, 1\}$ is the indicator function. And the final loss function of the contrastive learning process is:

$$\mathcal{L} = \frac{1}{2N} \sum_{k=1}^{N} [\ell(2k-1, 2k) + \ell(2k, 2k-1)]. \tag{2}$$

From Patches to Phenotype Representation. Patches of each patient are sent into the trained $Encoder_p$ to extract features and perform K-means clustering. We then propose the Siamese MI-Dense module to learn the feature representations of these clusters. Figure 2 shows the detailed structure of one MI-Dense block. Deep features of patches in each cluster are firstly fed into dense blocks [9], and then these features of one cluster are aggregated by average pooling to eliminate the impact of varying number of patches in various clusters. The Siamese structure of the multiple MI-Dense blocks enables the sharing weights thus the generalization ability of the blocks is improved.

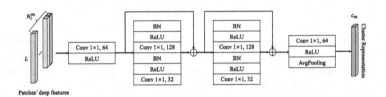

Fig. 2. Detailed structure of the proposed MI-Dense. N_i^m denotes the number of patches in the m-th cluster of the i-th patient. L is the length of the feature vector obtained by $Encoder_p$. c_m is the final output of cluster representation.

Cluster Aggregation via Attention-Based MIL Pooling. We then adapt attention-based MIL pooling [10] to aggregate cluster-level representation to patient-level, and the importance of the m_{th} cluster can be represented by attention weight a_m. The patient-level WSI representation can be calculated as:

$$\mathbf{z_{wsi}} = \sum_{m=1}^{M} a_m \mathbf{c}_m, \quad a_m = \frac{\exp\left\{\mathbf{w}^\top \tanh\left(\mathbf{V}\mathbf{c}_m^\top\right)\right\}}{\sum_{j=1}^{M} \exp\left\{\mathbf{w}^\top \tanh\left(\mathbf{V}\mathbf{c}_j^\top\right)\right\}}. \tag{3}$$

Here $\mathbf{c}_m$ denotes the m_{th} cluster representation of a patient, $\mathbf{w}$ and $\mathbf{V}$ are network parameters. The purpose of using $tanh(\cdot)$ rather than $sigmoid(\cdot)$ is to include both positive and negative values for proper gradient flow.

2.3 Clinical Feature Learning

In particular, we initially obtained 90 meaningful clinical variables from NLST with the advice of doctors. In the preprocessing stage, we applied z-score standardization for continuous variables and one-hot encoding for categorical variables. To ensure practical application and identify clinically significant variables, we used the Kaplan-Meier method [11] to estimate five-year survival curves and conducted log-rank tests to screen covariates. Variables with p-values less than 0.3 were selected, resulting in a final set of 16 variables, the categories of which including demographic, smoking, screening, work history and so on. The selected features $x_{selected\_cli}$ are then fed into the clinical feature extractor $Encoder_c$, which consists of the fully-connected layers and the dropout layers, to obtain the deep representation of clinical data: $\mathbf{z_{cli}} = Encoder_c(x_{selected\_cli})$.

We also conducted further analysis on the selected variables using the random survival forest (RSF) [12] and measured their importance by using the variable importance (VIMP), providing a comprehensive evaluation of their relative significance. Details can be referred to Sect. 3.5.

2.4 Multi-modality Fusion by GAMFB

Current multi-modality survival prediction methods [1–3] possess certain limitations. Simple fusion modules such as concatenation and bit-wise addition prove inadequate in representing complex feature interactions, while some fusion strategies based on outer product often lead to high computational complexity. Drawing inspiration from the successful application of multi-modality fusion strategy in visual question answering (VQA) task [13], we discovered that the Multimodal Factorized Bilinear pooling (MFB) [14] can effectively integrate two modalities with lower computational complexity. We are the first to extend its application to multi-modal survival analysis. Flowchart is shown in Fig. 3.

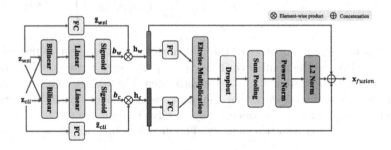

Fig. 3. The Proposed GAMFB module. Features of the two modalities are firstly fed into a gating-based attention module then perform fusion through MFB.

Gating-Based Attention Unit. We firstly use a gating-based attention unit in both clinical feature representation $\mathbf{z}_{cli}$ and the whole slide image feature representation $\mathbf{z}_{wsi}$ before fusion to control the expressiveness of each modalities:

$$\forall\, i \in \{wsi, cli\}, \mathbf{b}_i = \sigma(Bilinear(\mathbf{z}_{wsi}, \mathbf{z}_{cli})), \tilde{\mathbf{z}}_i = ReLU(\mathbf{z_i}), \mathbf{h}_i = \mathbf{b}_i * \tilde{\mathbf{z}}_i \quad (4)$$

where $\mathbf{b}_i$ is the weight for each modality, which is calculated by the bilinear transformations with softmax.

Fusion by Multimodal Factorized Bilinear Pooling. Given two feature vectors $h_{wsi} = [w_1, ..., w_m]^T$ and $h_{cli} = [c_1, ..., c_n]^T$, the simplest bilinear fusion can be denotes as $\mathbf{h}_{fuse} = \mathbf{h}_{wsi} W_i \mathbf{h}_{cli}^T$. To reduce the complexity, the projection matrix W_i can be factorized into low-rank matrices:

$$\mathbf{h}_{fuse} = \text{SumPooling}\left(U^T \mathbf{h}_{wsi} \circ V^T \mathbf{h}_{cli}, k\right). \quad (5)$$

where U and V are the projection matrices, o is the Hadmard product, and $SumPooling(x, k)$ refers to sum pooling on a one-dimensional non-overlapping window of size k over x. The fused result is later concatenated with the former two feature vectors by residual connection to prevent vanishing gradient problem. Thus, the final fusion result is $\mathbf{x}_{fusion} = Concat(\mathbf{h}_{fuse}, \mathbf{h}_{wsi}, \mathbf{h}_{cli})$.

2.5 MTL-Based Prediction and Loss Function

Some previous studies [15,16] have shown that heart disease and emphysema are associated with the survival risk of lung cancer, so we use multi-task learning (MTL) to predict the diagnosis result of these diseases simultaneously. By introducing additional supervision and the method of fully sharing parameters, the network can learn more robust representations with low risk of over-fitting.

We use the average negative log partial likelihood with L2 regularization as the survival loss function, which promotes overall agreement by penalizing any disagreement between the survival time of high-risk and low-risk patients.

$$\mathcal{L}_{surv}(\theta) = \frac{-1}{N_E} \sum_{i \in E} \left(\widehat{\varphi}_\theta\left(x_i\right) - \log \sum_{j:t_j \geq t_i} \exp(\widehat{\varphi}_\theta\left(x_j\right)) \right) + \lambda \|\theta\|_2^2, \quad (6)$$

where θ means the parameters of the network, E means the set of uncensored patients, and N_E is the number of them. $t_1 < t_2 < ... < t_n$ represent the ordered event time of these patients. $\widehat{\varphi}_\theta\left(x_i\right)$ means their hazard risk, which is one of the outputs from the network. For the other two auxiliary classification tasks, we use the weighted binary cross entropy to solve the unbalanced data problem between positive and negative samples. The final loss function is denoted as:

$$\mathcal{L}_{total} = \mathcal{L}_{surv} + \sum_{t=1}^{T} \frac{-\lambda_t}{N_t} \sum_i \left[\beta_t y_{t_i} \log\left(\hat{y}_{t_i}\right) + (1 - y_{t_i}) \log\left(1 - \hat{y}_{t_i}\right)\right], \quad (7)$$

where λ is the weight of loss between different tasks, and β is used to adjust the weight of unbalanced problems of classification tasks.

3   Experiments and Results

3.1   Dataset and Preprocessing

We selected the National Lung Screening Trial (NLST) [17] dataset to assess the effectiveness of our approach, which consists of 444 patients with 1,225 WSIs. Firstly, we use Otsu algorithm [18] to segment the lesion regions and crop patches at 20x magnification. Some studies [19] have indicated that tumor regions tend to be darker due to hematoxylin-stained nuclei. So we designed a simple sampling strategy called the "**biased sampling strategy**" to select patches. Patches are sorted in descending order of illumination, and 400 patches are sampled from the first 800 dark patches and another 400 patches from the remaining patches. In total, 681,959 patches were sampled from all WSIs. Next, all patches are used to train $Encoder_p$ (see Sect. 2.2 for details), and the trained $Encoder_p$ is later used to extract the features of these patches. To eliminate heterogeneity among WSI patches, we cluster them into different phenotypes by K-means.

3.2   Implementation Details

All of our experiments are conducted on a single NVIDIA GTX 3090. $Encoder_p$ is trained for 200 epoch with batch size of 128 and learning rate of 0.0375. Survival prediction model is trained for 400 epochs with early stopping to avoid over-fitting. We use Adam as the optimizer, with L2 regularization with $\lambda = 1e^{-5}$, task-specific loss weight $\lambda_1 = \lambda_2 = 0.1$ and unbalanced weight $\beta_1 = 6.79$ and $\beta_1 = 5.08$. We use C-index [20], the most widely used performance metric for survival analysis, which quantifies the probability of correctly ranking the predicted event times between two selected individuals. The formula is:

$$C = \frac{1}{n} \sum_{i \in \{1...N | e_i = 1\}} \sum_{t_j > t_i} I\left[\widehat{\varphi}_i > \widehat{\varphi}_j\right], \tag{8}$$

where n is the number of pairs and $I[.]$ represents the indicator function. $\widehat{\varphi}$ is the corresponding risk. All the experiments are carried out in 5-fold cross-validation.

3.3   Comparison with State-of-the-Arts

We compared our proposed model with 8 survival analysis approaches, which can be classified into three categories. Comparing our approach with existing multi-modalality based methods is not feasible as they require ROI annotations, which our method does not. Therefore, we used DeepAttnMISL [6](denoted as DAM-) as the backbone for WSI representation learning and adopted the fusion strategy of thier approaches.

The comparison results are presented in Table 1. From the results, deep learning models tend to outperform traditional methods due to their ability to automatically learn complex and non-linear features. Using only one modality may be insufficient and multi modality deep fusion models can integrate both clinical

Table 1. Performance comparison in 5-fold cross-validation.

	Method	C-index (± std)
Only Clinical	LassoCox [21]	0.6227 (± 0.0690)
	Survival SVM [22]	0.6377 (± 0.0582)
	RSF [12]	0.6465 (± 0.0741)
Only WSI	WSISA [4]	0.6450 (± 0.0910)
	DeepMISL [5]	0.6638 (± 0.0993)
	DeepAttnMISL [6]	0.6806 (± 0.0912)
WSI & Clinical	DAM-DeepCorrSurv [23]	0.6845 (± 0.0854)
	DAM-PathomicFusion [3]	0.6954 (± 0.1041)
	SGMFBSurv (ours)	**0.7304 (± 0.1096)**

and WSI information that result in more accurate prognostic analysis. Among all of these methods, our method has obvious superiority. It can be attributed to the advanced WSI representation learning method with SSL, as well as the utilization of multi-modality fusion and multi-task learning.

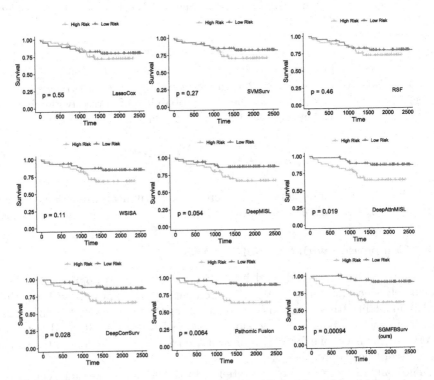

Fig. 4. Kaplan-Meier curves of different methods. High-risk and low-risk groups are marked as red and cyan lines respectively. "+" means the censored data. The wider the separation of two curves, the better the survival prediction performance of the model. (Color figure online)

To further evaluate the prognostic capability of our proposed models, we conducted the log-rank test and draw Kaplan-Meier curves [11] for each model on a test fold. Our results is illustrated in Fig. 4. The log-rank p-value serves as a measure of the survival difference between two groups, with a lower p-value indicating superior performance. The proposed model exhibited the best performance, with a p-value of 0.00094, surpassing that of all other models.

3.4   Ablation Study

Effectiveness of Multi-modality Fusion Strategy. We compared the proposed GAMFB module with other methods. Experimental results are shown in Table 2, which indicates that the proposed GAMFB module outperforms other methods. This suggests that transferring multi-modality fusion models from other fields to the medical domain for integrating WSIs and clinical features can also be effective. The result also demonstrate the effectiveness of the gating-based attention mechanism, which can improve the flexibility of the network.

Biased Sampling vs Random Sampling. Here, "✓" indicate the usage of the corresponding module in the experiments, while the absence of a checkmark suggests that an alternative base module was utilized instead. Specifically, "BiasedSamp" refers to the proposed biased sampling strategy, whereas the absence of a checkmark denotes the use of completely random sampling. The results are shown in Table 3. From the table, we can see that our biased sampling strategy can improve the C-index of the model by 1.44%, demonstrating the effectiveness of the proposed strategy.

Table 2. Comparisons of different strategies for multi-modality fusion.

Fusion Strategies	C-index (mean ± std)
No Fusion	0.6941 (± 0.0587)
Concatenation	0.7048 (± 0.0913)
Element-wise sum	0.7037 (± 0.0986)
Kronecker Product	0.6986 (± 0.0876)
Gating-based Kronecker Product	0.7113 (± 0.0736)
MFB	0.7140 (± 0.0939)
GAMFB	**0.7304 (± 0.1096)**

Self-supervised Learning vs Pre-trained Feature Extractor. Here, "SSL" represents the self-supervised learning method, with the absence of a checkmark indicating the use of ImageNet pre-trained VGG16 network as the patch feature extractor. The comparison results are shown in Table 3. We can see that the integration of SSL enhances the C-index compared to the pre-trained model from ImageNet. That means that our patch extractor learning strategy can eliminate the defects caused by heterogeneity between natural images and WSIs.

Table 3. Ablation study of the sampling strategy and SSL in preprocessing phases.

BiasedSamp	SSL	C-index (mean ± std)
		0.7038 (± 0.1150)
✓		0.7167 (± 0.1043)
	✓	0.7138 (± 0.1321)
✓	✓	**0.7304 (± 0.1096)**

Influence of the Phenotype Number in Patch Clustering. The number of phenotypes is an important hyperparameter, so we performed parametric analysis to determine the optimal value. The results, presented in Table 4, indicate that the best result is achieved when the number of phenotypes is set to 10.

Table 4. Influence of the number of phenotypes.

Phenotype Number M	4	6	8	10	12	14
C-Index	0.6844	0.7076	0.7236	**0.7304**	0.7133	0.7022

Effectiveness of MI-Dense Block. Here, "DenseBlock" refers to the dense block [9], whereas the absence of a checkmark indicates to replace the dense block with simple FCN. "Siamese" refers to the Siamese structure, whereas the absence of a checkmark indicates that the parameters of network for each cluster are not shared. "**ABMIL**" means the Attention-based MIL Pooling, whereas the absence of a checkmark indicates the simple mean pooling is used for aggregation. The experimental results are shown in Table 5. The lack of a dense block leads to a decrease in results by 1.7%, demonstrating the effectiveness of the introducing of DenseNet. The absence of Siamese led to significant performance degradation in results, as the excessive parameter volume made it difficult to train. The lack of attention-based MIL leads to a decrease of about 2.2%, as the lack of attention can lead to a loss of flexibility in the network.

Table 5. Ablation Study of MI-Dense Network.

DenseBlock	Siamese	ABMIL	C-index (± std)
	✓	✓	0.7137 (± 0.1003)
✓		✓	0.6654 (± 0.0764)
✓	✓		0.7085 (± 0.1036)
✓	✓	✓	**0.7304 (± 0.1096)**

Effectiveness of Multi-task Learning. We also conduct ablation experiments on the multi-task learning. $\text{Task}_{\text{emph}}$ refers to the emphysema prediction task, and $\text{Task}_{\text{hear}}$ refers to the heart disease prediction task. Results are shown in Table 6. It can be seen that only predicting the survival rate leads to a decrease in results comparing to conduct the predictions of two relative tasks simultaneously. That shows that MTL can better improve the survival prediction performance by simultaneously predicting the outcome of other related diseases.

Table 6. Performance comparisons of multi-task learning.

$\text{Task}_{\text{emph}}$	$\text{Task}_{\text{hear}}$	C-index ($\pm$ std)
		0.7184 ($\pm$ 0.1020)
✓		0.7230 ($\pm$ 0.0926)
	✓	0.7236 ($\pm$ 0.1199)
✓	✓	**0.7304 ($\pm$ 0.1096)**

3.5 Interpretability Analysis

Interpretability of WSI. In order to show that the proposed method can capture the important WSI regions for survival analysis, we visualize the phenotypes according to its importance via the attention-based MIL. We visualize the phenotypes ranked by their importance for survival analysis, which are shown in left of Fig. 5. Here, different colors represent different phenotypes. The results show the effectiveness and interpretability of our method on WSI.

Interpretability of Clinical Variables. We calculated the importance of each clinical variable (VIMP) by setting it to median. The degree of reduction of the result reflects the importance of the variable. We compared VIMP from our

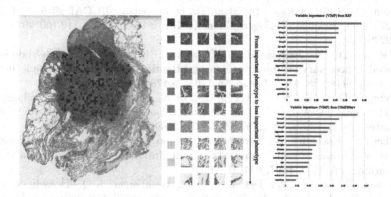

Fig. 5. Left: Visualization result of a patient. Right: VIMP from RSF and our model.

model with that from RSF, as shown in the right of Fig. 5. It can be observed that VIMP we obtained are similar to RSF, demonstrating that our model can learn deep representations and has good interpretability for clinical variables of lung cancer patients.

4 Conclusion

In this study, we propose a novel framework that integrates WSIs and clinical data for lung cancer survival analysis. Our approach overcomes limitations of previous models with improvements such as biased sampling, self-supervised contrastive learning, phenotype learning networks, multi-modality fusion, and multi-task learning. Through comparative and ablation experiments, we validate the effectiveness of each proposed modules. These results further validate the potential of deep learning models in improving prognostic analysis.

Acknowledgments. This work was partially supported by the General Program of National Natural Science Foundation of China (NSFC) under Grant 62276189, and the Fundamental Research Funds for the Central Universities No. 22120220583.

References

1. Mobadersany, P., et al.: Predicting cancer outcomes from histology and genomics using convolutional networks. Proc. Nat. Acad. Sci. **115**(13), E2970–E2979 (2018)
2. Subramanian, V., Chidester, B., Ma, J., Do, M.N.: Correlating cellular features with gene expression using CCA. In: 2018 IEEE 15th International Symposium on Biomedical Imaging (ISBI 2018), pp. 805–808. IEEE (2018)
3. Chen, R.J., et al.: Pathomic fusion: an integrated framework for fusing histopathology and genomic features for cancer diagnosis and prognosis. IEEE Trans. Med. Imaging **41**(4), 757–770 (2020)
4. Zhu, X., Yao, J., Zhu, F., Huang, J.: WSISA: making survival prediction from whole slide histopathological images. In: Proceedings of the IEEE Conference on Computer Vision and Pattern Recognition, pp. 7234–7242 (2017)
5. Yao, J., Zhu, X., Huang, J.: Deep multi-instance learning for survival prediction from whole slide images. In: Shen, D., et al. (eds.) MICCAI 2019. LNCS, vol. 11764, pp. 496–504. Springer, Cham (2019). https://doi.org/10.1007/978-3-030-32239-7_55
6. Yao, J., Zhu, X., Jonnagaddala, J., Hawkins, N., Huang, J.: Whole slide images based cancer survival prediction using attention guided deep multiple instance learning networks. Med. Image Anal. **65**, 101789 (2020)
7. Deng, J., Dong, W., Socher, R., Li, L.J., Li, K., Fei-Fei, L.: ImageNet: a large-scale hierarchical image database. In: 2009 IEEE Conference on Computer Vision and Pattern Recognition, pp. 248–255. IEEE (2009)
8. Chen, T., Kornblith, S., Norouzi, M., Hinton, G.: A simple framework for contrastive learning of visual representations. In: International Conference on Machine Learning, pp. 1597–1607. PMLR (2020)
9. Huang, G., Liu, Z., Van Der Maaten, L., Weinberger, K.Q.: Densely connected convolutional networks. In: Proceedings of the IEEE Conference on Computer Vision and Pattern Recognition, pp. 4700–4708 (2017)

10. Ilse, M., Tomczak, J., Welling, M.: Attention-based deep multiple instance learning. In: International Conference on Machine Learning, pp. 2127–2136. PMLR (2018)
11. Kaplan, E.L., Meier, P.: Nonparametric estimation from incomplete observations. J. Am. Stat. Assoc. **53**(282), 457–481 (1958)
12. Ishwaran, H., Kogalur, U.B., Blackstone, E.H., Lauer, M.S.: Random survival forests (2008)
13. Antol, S., et al.: VQA: visual question answering. In: Proceedings of the IEEE International Conference on Computer Vision, pp. 2425–2433 (2015)
14. Yu, Z., Yu, J., Fan, J., Tao, D.: Multi-modal factorized bilinear pooling with co-attention learning for visual question answering. In: Proceedings of the IEEE International Conference on Computer Vision, pp. 1821–1830 (2017)
15. de Torres, J.P., et al.: Assessing the relationship between lung cancer risk and emphysema detected on low-dose CT of the chest. Chest **132**(6), 1932–1938 (2007)
16. Kravchenko, J., Berry, M., Arbeev, K., Lyerly, H.K., Yashin, A., Akushevich, I.: Cardiovascular comorbidities and survival of lung cancer patients: medicare data based analysis. Lung Cancer **88**(1), 85–93 (2015)
17. Team, N.L.S.T.R.: The national lung screening trial: overview and study design. Radiology **258**(1), 243–253 (2011)
18. Otsu, N.: A threshold selection method from gray-level histograms. IEEE Trans. Syst. Man Cybern. **9**(1), 62–66 (1979)
19. Zubiolo, A.: Extraction de caractéristiques et apprentissage statistique pour l'imagerie biomédicale cellulaire et tissulaire. Ph.D. thesis, Université Nice Sophia Antipolis (2015)
20. Harrell, F.E., Jr., Lee, K.L., Mark, D.B.: Multivariable prognostic models: issues in developing models, evaluating assumptions and adequacy, and measuring and reducing errors. Stat. Med. **15**(4), 361–387 (1996)
21. Tibshirani, R.: The lasso method for variable selection in the cox model. Stat. Med. **16**(4), 385–395 (1997)
22. Pölsterl, S., Navab, N., Katouzian, A.: Fast training of support vector machines for survival analysis. In: Appice, A., Rodrigues, P.P., Santos Costa, V., Gama, J., Jorge, A., Soares, C. (eds.) ECML PKDD 2015. LNCS (LNAI), vol. 9285, pp. 243–259. Springer, Cham (2015). https://doi.org/10.1007/978-3-319-23525-7_15
23. Yao, J., Zhu, X., Zhu, F., Huang, J.: Deep correlational learning for survival prediction from multi-modality data. In: Descoteaux, M., Maier-Hein, L., Franz, A., Jannin, P., Collins, D.L., Duchesne, S. (eds.) MICCAI 2017. LNCS, vol. 10434, pp. 406–414. Springer, Cham (2017). https://doi.org/10.1007/978-3-319-66185-8_46

Incorporating Spiking Neural Network for Dynamic Vision Emotion Analysis

Binqiang Wang[1,2] and Xiaoqiang Liang[3(✉)]

[1] Shandong Massive Information Technology Research Institute, Jinan 250101, China
[2] Inspur (Beijing) Electronic Information Industry Co., Ltd., Beijing, China
[3] China North Vehicle Research Institute, Beijing 100072, China
liangxiaoqiang15@mails.ucas.ac.cn

Abstract. In the domain of affective computing, researchers have sought to enhance the performance of models and algorithms by leveraging the complementarity of multimodal information. However, the rapid emergence of new modalities has outpaced the development of suitable datasets, posing a challenge in keeping up with the advancements in modal sensing technology. The collection and analysis of multimodal data present intricate and substantial tasks. To address the partial missing data challenge within the research community, we have curated a novel homogeneous multimodal gesture emotion recognition dataset, augmenting existing datasets through meticulous analysis. This dataset not only fills the gaps in homogeneous multimodal data but also opens up new avenues for emotion recognition research. Additionally, we propose a pseudo dual-flow network based on this dataset, establishing its potential application in the affective computing community. Experimental findings indicate the feasibility of utilizing traditional visual information and spiking visual information derived from homogeneous multimodal data for visual emotion recognition.

Keywords: Spiking neural network · Affective Computing · Dynamic vision sensor

1 Introduction

Multimodal emotion recognition endeavors to enhance the robustness and precision of emotion identification by harnessing the complementarity inherent in multimodal information [16]. By synergistically fusing these modalities, it becomes possible to compensate for the limitations of individual modalities in certain information domains. Furthermore, this advancement in multimodal information processing provides valuable insights for other domains such as matching, tracking, and brain imaging classification, paving the way for advancements in those fields [5,6,13].

The availability of datasets has significantly propelled the advancements in multimodal emotion recognition. Based on the sources of multimodal information, existing datasets can be categorized into two main types: homogeneous

© The Author(s), under exclusive license to Springer Nature Singapore Pte Ltd. 2024
Q. Liu et al. (Eds.): PRCV 2023, LNCS 14437, pp. 346–357, 2024.
https://doi.org/10.1007/978-981-99-8558-6_29

multimodal data and heterogeneous multimodal data. Homogeneous multimodal data pertains to data derived from uniform information sources, such as diverse visual perspectives or devices [7]. On the other hand, heterogeneous multimodal data encompasses data originating from distinct information sources, such as multimodal data involving both visual and auditory components [9]. In contrast, homogeneous multimodal data typically stems from different manifestations of the same information source, such as the fusion of conventional RGB video information with depth camera data or the acquisition of data in non-traditional formats beyond single RGB video [7]. The research goal of homogeneous multimodality is to mine the task-related information contained in the data as much as possible from multiple angles under the same information source.

This study primarily focuses on information processing sourced from visual data. Conventional video data consists of sequentially arranged frames, enabling the capture of coarse-grained temporal dynamics and facilitating task prediction through contextual modeling over time. However, inherent limitations exist within the traditional fixed-position video capture, resulting in blind spots. To enable more comprehensive information capture, some researchers have proposed a novel approach that involves utilizing multiple cameras placed at different locations, subsequently modeling the acquired 3D data for emotion recognition [7]. In contrast to prioritizing spatial integrity, this paper endeavors to enhance the temporal integrity of video data, specifically aiming to progress from coarse-grained dynamic information to fine-grained dynamic information. The focus of visual transmission predominantly revolves around posture information. While preliminary research on dynamic postures has been conducted by scholars, this paper presents a more systematic exploration and analysis within the domain of emotion recognition.

To capture the nuanced dynamic information inherent in gestures, alongside conventional video frame data, we incorporate event stream data as an additional modality in our study. Event stream data possesses the distinctive capability of capturing fine-grained dynamic information with high temporal resolution and a wider dynamic range, solely recording changes in picture position and polarity [8]. Motivated by these characteristics, we aim to explore the untapped potential of homogeneous multimodal data-comprising video frames and event streams-for emotion recognition. Overcoming the challenge of scene registration when utilizing two different devices is essential, as the device locations may differ. To circumvent this complication, we employ a collection device capable of the simultaneous output of both video frames and event stream data, thereby obviating registration-related errors that may arise at the hardware level [2].

Pure vision-based emotion recognition technology finds utility in scenarios where alternative modal information is inaccessible, such as long-distance communication with substantial ambient noise where visual information alone proves effective. Sign language serves as a crucial mode of communication, particularly for individuals with hearing impairments, circumventing reliance on vocal communication [15]. Nonetheless, acquiring proficiency in formal sign language demands substantial human resources. The ability to differentiate gestures based

on common human understanding, devoid of specialized sign language training, bridges this gap by fostering emotional comprehension rather than precise
semantic interpretation.

Through an analysis of the inherent data characteristics of the two modalities,
it becomes evident that video frame data encompasses a richer set of semantic information, whereas event stream data emphasizes dynamic information.
Artificial Neural Network (ANN) architectures excel in effectively modeling the
semantic information present in video frames, while Spiking Neural Network
(SNN) architectures inherently possess advantageous properties for capturing
and modeling event-based dynamics. Consequently, the key contributions of this
paper can be summarized as follows:

- In light of the limited availability of fine-grained dynamic information in existing datasets, we introduce a novel approach that combines high-resolution
  temporal event data with concurrent camera data captured by event cameras.
  This integration proves to be effective in enhancing visual gesture emotion
  recognition.
- We present a baseline method, termed the pseudo-double-stream network,
  which leverages ANN to model video frame data and exploits the inherent
  capabilities of SNN to represent event stream data. Experimental results substantiate the efficacy of our proposed method.

2   Related Work

To acquire event data, prevalent event cameras in use include DVS (Dynamic
Vision Sensor) [8], DAVIS (Dynamic and Active-pixel VIsion Sensor) [2], ATIS
(Asynchronous Time-based Image Sensor) [14], and CeleX [4]. Among these
options, DVS and DAVIS emerge as particularly favorable choices in terms of
equipment availability. Specifically, while DVS solely captures pure event stream
data, DAVIS also enables the acquisition of corresponding video frame data that
encompasses rich semantic information. Consequently, DAVIS (more precisely,
DAVIS346) is selected as the designated acquisition device for this study.

The dynamic nature of gestures has spurred a growing body of research
investigating the utilization of event cameras. Mueggler *et al.* developed a versatile event data simulator capable of serving multiple tasks, including pose estimation, visual odometry, and simultaneous localization and mapping, thereby
providing valuable data for the advancement of these research endeavors [12].
However, distinctions between simulated and real event data arise due to factors
such as variations in noise distribution. Consequently, many researchers opt to
employ physical cameras for collecting authentic event data. Among the existing
event datasets, the majority have been captured using DVS [3,10,15], while the
datasets obtained through the use of DAVIS cameras remain relatively scarce.

The existing datasets primarily consist of single-mode event data from DVS
or ATIS cameras, without specific targeting of information or emotion transmission. This paper addresses these limitations in the field by proposing a new
dataset. Table 1 provides relevant information on the aforementioned datasets,

demonstrating that this paper exhibits a relatively larger data scale and features more homogeneous video frame data compared to other datasets.

Table 1. Descriptions of different datasets related to gestures.

Dataset Name/Contributor	Device	Category number	Sample number	Spatial resolution
ROSHAMBO17 [10]	DVS	3	5 million	64×64
DvsGesture [1]	DVS128	10+1(other)	1342	128×128
Neuro ConGD Dataset [3]	DVS	16+1(blank)	2040	128×128
Maro and Benosman [11]	ATIS	6	1621	None
SL-Animals-DVS [15]	DVS	19	1100	None
SGED	DAVIS346	9+1(other)	2500	346×260

3   Proposed Dataset

3.1   Basic Information

In order to establish a multimodal dataset that encompasses both conventional visual information and dynamic visual information, we utilized the Dynamic and Active Pixel Vision Sensor (DAVIS346) for data capture. The DAVIS346 offers a temporal resolution of 1 microsecond, enabling the retention of valuable information that may be lost by standard cameras. Notably, the DAVIS346 demonstrates enhanced performance in backlight conditions due to its high dynamic range. Ten volunteers participated in our experiments, wherein carefully designed emotional gestures were performed under various lighting conditions and body positions. There are totally nine categories: ok, hello, no, kill, victory, good, yes, love, and fighting, which include three potential emotion dispositions: positive, neutral, and negative.

3.2   Labeling Process

The process of labeling multimodal data comprising video frames and event streams differs from traditional multimodal data annotation methods. In traditional multimodal annotation, such as video and audio synchronization, timestamps from the original video can be directly used to label the corresponding audio segments. However, this direct alignment is not feasible for video frame data and event stream data due to their disparate time resolutions. The DAVIS346 output format deviates from traditional video, as it is stored as compressed and segmented video frames. Similarly, the event stream data is stored in tensor format, making direct alignment with video frames for visualization purposes challenging. The discrepancy in time resolution leads to a significant disparity in the number of frames in video frame data and the number of events in event stream data. To address this challenge and accomplish accurate labeling of the collected data, this paper introduces a binary search algorithm based on a scaling factor to align the two distinct data types. This alignment process enables the determination of the labeling serial number for the event stream data.

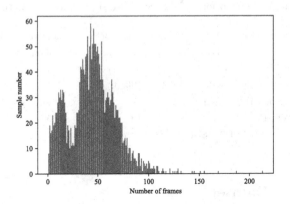

Fig. 1. Distribution of sample number and frame number of video frame data.

3.3   Statistical Analysis

A comprehensive statistical analysis was conducted on the dataset to gain a comprehensive understanding of its characteristics. The dataset consists of two main components: video frame data and event stream data. The primary parameter of interest for the video frame data is the number of frames in each sample. A histogram was generated to depict the distribution of samples based on the number of video frames they contain. Figure 1 illustrates this distribution, revealing that the majority of samples have a concentration of 0–100 frames. This information can be utilized as a priori knowledge when designing the input for the video frame module, allowing for the truncation of the input video frame sequence based on this threshold of 100 frames.

To gain a deeper understanding of the distribution law of event data, we have made statistics on positive and negative events respectively. We compiled statistics on the number of positive and negative events present in each sample across different categories and visualized the mean values and outlier information using box plots. Figures 2 and 3 illustrate the findings. Firstly, we observed variations in the number of events, both positive and negative, across different emotional gestures. Secondly, we can infer the expected number of events based on the generation process of gestures. For instance, emotional postures involving complex movements (e.g., 'love') tend to contain a higher number of events compared to those with simpler head movements (e.g., 'yes' and 'no'). Moreover, we observed a positive correlation between the distribution of positive and negative events. This implies that samples with a larger number of positive events also tend to have a higher count of negative events. This correlation can be attributed to the nature of gesture movements, which predominantly involve rigid body motion.

Consequently, event triggers exhibit positive and negative polarities simultaneously, albeit with differences in location.

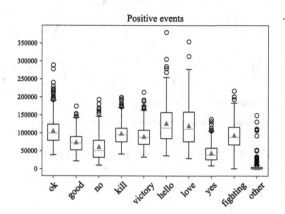

Fig. 2. Box diagram of the number of positive events.

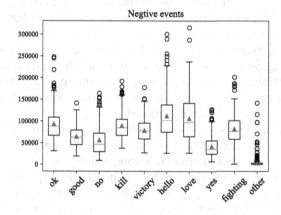

Fig. 3. Box diagram of the number of negative events.

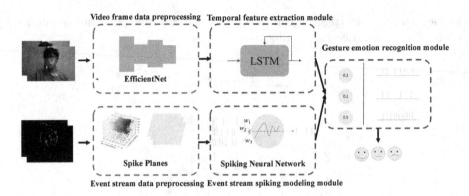

Fig. 4. The framework of the proposed method.

4 Methods

4.1 Overview

To enable the recognition of gesture emotions using a homogeneous multimodal input dataset, we have devised a model framework termed the Homogeneous Pseudo Dual Flow Network. Inspired by the concept of dual flow networks [18], we have adapted and extended this concept to the realm of homogeneous multimodal data. As shown in Fig. 4, the framework comprises several key components, including preprocessing modules, temporal feature extraction module, event stream spiking modeling module, and a gesture emotion recognition module. In the following sections, we will delve into the details and provide a comprehensive overview of each of these modules.

For the symbolic representation of homogeneous multimodal data, we use $v_i, i = 1, .., N$ to represent video frame data, where N represents the number of video frames; $s_j, i = 1, ..., L$ represents a spike train (a spike can also be seen as an event), with L represents the number of spikes. More specifically, each spike is represented by a vector, (t, x, y, p), where t represents the time at which the event occurred, (x, y) represents the spatial location at which the event occurred, and p represents the polarity at which the event occurred. The polarity can be divided into two cases, positive and negative, represented by 1 and 0. The goal of the model is to obtain output $\hat{y}$ based on the input data. Ideally, this $\hat{y}$ should be the same as the ground truth category y that represents real gesture emotions.

4.2 Details

Firstly, we introduce the preprocessing operations of data.

Video Data: For the extraction of video frame features, we choose the model, EfficientNet, from the perspective of efficiency. In this way, the video frame sequence eventually becomes a sequence composed of feature vectors and the above process can be expressed as:

$$vp_i = f_{EN}(v_i), i = 1, ..., N, \tag{1}$$

where f_{EN} represents the feature extraction of EfficientNet, vp_i is the feature vector of the video frame v_i, which is used as the input of the temporal feature extraction module.

Event Data: Existing event-based processing algorithms try to map the sequence of events back to the original two-dimensional space, which is commonly referred to as the spike plane. There are two main schemes available, one is to compress based on a fixed time interval, and the other is to compress based on the number of events. Due to the uncertainty of the occurrence of events, there may be situations where no events occur for a period of time depending on the time interval, resulting in an invalid spike plane (i.e., all data is zero). Therefore, compression based on a fixed number of events is used here.

Specifically, first, the number of all events included in a sample is counted, without distinguishing the polarity of the event. Then, according to a preset K, representing the number of dense spike planes after data preprocessing. It should be noted that due to two different polarity events, the dense spike plane is essentially a three-dimensional tensor with a third dimension of 2. To represent that the data is obtained by compressing the original spike plane sequence, it is called a dense spike plane. The numerical value in the dense spike plane is obtained by separately counting the number of corresponding positive and negative spike events. So far, the processing of the original event data stream has been completed, and dense spike plane sequence data has been obtained. The formal description of the above process is as follows:

$$sp_1, ..., sp_K = f_{dense}(s_j), j = 1, ..., L, \tag{2}$$

where f_{dense} denotes the compress from spike planes to dense spike planes, $sp_1, ..., sp_K$ are the dense spike planes. Note the for s_j, the channel number is 1 which means the positive and negative events are mixed. Secondly, we model temporal dependencies in video frame sequences. The output results obtained in the video frame data processing stage are processed separately for each frame, and the extracted feature vectors contain the semantic information of a single frame. In order to capture timing information between consecutive frames, it is necessary to model specifically. Here, the classic time series modeling model LSTM is used to provide a baseline. LSTM can control the compression of effective information from front to back through a gating mechanism, using the output data of the last step of LSTM as the temporal feature representation of the video. The timing modeling process of LSTM can be expressed as follows:

$$h_{last} = LSTM(vp_1, ..., vp_N), \tag{3}$$

where h_{last} is the output of last hidden states of LSTM.

It is important to note that LSTM is used here to verify the effectiveness of the complementarity of homogeneous multimodal data, and can be completely replaced by other temporal modeling tools, such as Transformer. Then, we focus on modeling event data, which has unique spike characteristics that make it well-suited for SNNs. This approach is commonly used for gesture recognition with event cameras. In this work, we employ a classic framework for modeling event flow data, with the key modification being the adjustment of the number of intermediate-layer neurons to match the output types [1]. The SNN topology includes standard convolution, pooling, and fully connected layers, which are used to spatially compress input data and extract meaningful features. The main difference in our approach is the configuration of the number of spiking neurons in the fully connected layer prior to the output layer.

The specific spiking neuron model uses Leaky Integrate-and-Fire (LIF) [17]. An effective event flow feature extraction process can be expressed as:

$$s_{DG} = DvsGesture\left(sp_1, ..., sp_K\right),\tag{4}$$

where $DvsGesture$ is the network structure in [1], s_{DG} is the output spike array whose size is equal to the number of category. Note that s_{DG} is after averaged with K, and direct used to compute the MSE loss. Finally, we describes how to synthesize the output of two different networks to obtain the final result. For SNNs, we average pool the outputs of final spiking neurons output by the corresponding module (i.e. event stream spiking modeling module) to reduce the dimension to the emotional category domain space. For the LSTM structure, we take the hidden state of the final step output as input and map it to the emotional output space through two fully connected layer with dropout rate 0.5. Due to the structure of the two branches involved, the output fusion method is directly used here.

$$\hat{y} = s_{DG} + f_o\left(h_{last}\right),\tag{5}$$

where f_o represents two fully connected layer with dropout rate 0.5.

In addition, in experiments, we found that SNNs need to use MSE losses to achieve good performance, while ANNs are more effective in using cross entropy losses for this type of classification problem. Therefore, here, we balance the contribution of two different outputs by setting a hyperparameter and ultimately output the recognized gesture's emotional tendencies.

5   Experiments

5.1   Experimental Setup

We will conduct experimental verification on the dataset released in this work. Because the goal of this paper is to identify the emotions involved in gestures, we ultimately abstract the task into a three-class problem. According to the introduction to the SGED in the previous section, it can be found that the number of neutral and negative samples is close, while the number of positive samples is more than the other two combined. Thus we add the class numbers to weight different classes when computing cross-entropy loss. This point reminds us that when carrying out model evaluation, we cannot simply measure the quality of the model based on accuracy, but should consider the imbalance of the sample.

The number of dense spike planes K is set to 12. Adam is used to optimize the parameters. The comparison method used is the classic gesture recognition method [1]. Because gesture emotion recognition can be seen as a classification task, the accuracy and confusion matrix are used to evaluate the model. In order to take into account sample imbalance in the evaluation of model performance, we also used weighted precision, weighted recall, and weighted F1 score as the evaluation criterion.

5.2   Experimental Results

Table 2. Evaluation on SGED.

	Accuracy	Weighted Precision	Weighted Recall	Weighted F1
DvsGesture	63.4	66.0	63.4	60.8
ours	64.5	64.1	64.5	64.1

The experimental results on the SGED are shown in the Table 2. From the perspective of indicators, our proposed method has higher accuracy, weighted recall, and weighted F1 than the compared method. Specifically, the accuracy of our method is 64.5%, which is higher than the accuracy of DvsGesture that is 63.4%. As the accuracy is computed by the overall samples without considering the imbalance distribution of SGED, more in-depth and detailed analysis is needed. For weighted precision metric, the DvsGesture is better than ours, this phenomenon may caused by the DvsGesture's tendency to predict positive categories with more samples, which can be verified again by the confusion matrix later. Our method takes into account the imbalance of samples, so the performance indicators of the method proposed in this paper are better on indicators weighted recall and weighted F1.

To more intuitively see the distribution of different categories of predictions, Fig. 5 shows the confusion matrix for the results of the two methods. From the confusion matrix, it can be seen that the advantage of the method proposed in this paper comes from the correct prediction of neutral and negative gesture emotions. For positive gesture emotions, our performance is not comparable to the compared method. This reason is attributed to the uneven distribution of gesture emotion categories. Although we considered the factors of category imbalance in the design process of the loss function, there are still models that tend to output positive gesture emotion results. We believe that this is also one of the challenging features of the dataset released in this paper.

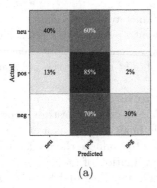 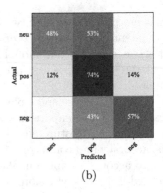

(a)   (b)

Fig. 5. (a) The confusion matrix of the results of the compared method. (b) The confusion matrix of the method proposed in this paper.

6   Conclusion

This paper introduces a novel dataset for homogeneous multimodal gesture emotion recognition, which combines event stream data and video frame data. Additionally, a pseudo dual stream structured network is proposed as a benchmark solution. The dataset bridges the gap between the fields of affective computing and gesture recognition, offering the advantage of incorporating homogeneous multimodal data to enhance data diversity and leverage the high dynamic range capabilities of event cameras.

References

1. Amir, A., et al.: A low power, fully event-based gesture recognition system. In: Proceedings of the IEEE Conference on Computer Vision and Pattern Recognition, pp. 7243–7252 (2017)
2. Brandli, C., Berner, R., Yang, M., Liu, S.C., Delbruck, T.: A 240× 180 130 db 3 μs latency global shutter spatiotemporal vision sensor. IEEE J. Solid-State Circ. **49**(10), 2333–2341 (2014)
3. Chen, G., Chen, J., Lienen, M., Conradt, J., Röhrbein, F., Knoll, A.C.: FLGR: fixed length GISTS representation learning for RNN-hmm hybrid-based neuromorphic continuous gesture recognition. Front. Neurosci. **13**, 73 (2019)
4. Guo, M., Huang, J., Chen, S.: Live demonstration: a 768× 640 pixels 200meps dynamic vision sensor. In: 2017 IEEE International Symposium on Circuits and Systems (ISCAS), pp. 1–1. IEEE (2017)
5. Jiang, J., Fares, A., Zhong, S.H.: A context-supported deep learning framework for multimodal brain imaging classification. IEEE Trans. Hum. Mach. Syst. **49**(6), 611–622 (2019)
6. Li, S., et al.: Unsupervised RGB-T object tracking with attentional multi-modal feature fusion. Multimedia Tools Appl. **82**(15), 1–19 (2023)
7. Li, X., et al.: 4DME: a spontaneous 4D micro-expression dataset with multimodalities. IEEE Trans. Affect. Comput. **14**(4), 3031–3047 (2022)

8. Lichtsteiner, P., Posch, C., Delbruck, T.: A 128× 128 120 db 15 μs latency asynchronous temporal contrast vision sensor. IEEE J. Solid-State Circ. **43**(2), 566–576 (2008)

9. Lu, X., Wang, B., Zheng, X.: Sound active attention framework for remote sensing image captioning. IEEE Trans. Geosci. Remote Sens. **58**(3), 1985–2000 (2019)

10. Lungu, I.A., Corradi, F., Delbrück, T.: Live demonstration: convolutional neural network driven by dynamic vision sensor playing RoShamBo. In: 2017 IEEE International Symposium on Circuits and Systems (ISCAS), pp. 1–1. IEEE (2017)

11. Maro, J.M., Ieng, S.H., Benosman, R.: Event-based gesture recognition with dynamic background suppression using smartphone computational capabilities. Front. Neurosci. **14**, 275 (2020)

12. Mueggler, E., Rebecq, H., Gallego, G., Delbruck, T., Scaramuzza, D.: The event-camera dataset and simulator: event-based data for pose estimation, visual odometry, and slam. Int. J. Robot. Res. **36**(2), 142–149 (2017)

13. Ning, H., Zheng, X., Lu, X., Yuan, Y.: Disentangled representation learning for cross-modal biometric matching. IEEE Trans. Multimedia **24**, 1763–1774 (2021)

14. Simon Chane, C., Ieng, S.H., Posch, C., Benosman, R.B.: Event-based tone mapping for asynchronous time-based image sensor. Front. Neurosci. **10**, 391 (2016)

15. Vasudevan, A., Negri, P., Linares-Barranco, B., Serrano-Gotarredona, T.: Introduction and analysis of an event-based sign language dataset. In: 2020 15th IEEE International Conference on Automatic Face and Gesture Recognition (FG 2020), pp. 675–682. IEEE (2020)

16. Wang, B., Dong, G., Zhao, Y., Li, R., Cao, Q., Chao, Y.: Non-uniform attention network for multi-modal sentiment analysis. In: Þór Jónsson, B., et al. (eds.) MMM 2022. LNCS, vol. 13141, pp. 612–623. Springer, Cham (2022). https://doi.org/10.1007/978-3-030-98358-1_48

17. Wang, B., et al.: Spiking emotions: Dynamic vision emotion recognition using spiking neural networks. vol. 3331, pp. 50–58. Virtual, Online, China (2022)

18. Wang, H., Chen, H., Wang, B., Jin, Y., Li, G., Kan, Y.: High-efficiency low-power microdefect detection in photovoltaic cells via a field programmable gate array-accelerated dual-flow network. Appl. Energy **318**, 119203 (2022)

PAT-Unet: Paired Attention Transformer for Efficient and Accurate Segmentation of 3D Medical Images

Qingzhi Zou[1,2,3], Jing Zhao[1,2,3]([✉]), Ming Li[4]([✉]), and Lin Yuan[1,2,3]

[1] Key Laboratory of Computing Power Network and Information Security, Ministry of Education, Shandong Computer Science Center, Qilu University of Technology (Shandong Academy of Sciences), Jinan, China
zjstudent@126.com
[2] Shandong Engineering Research Center of Big Data Applied Technology, Faculty of Computer Science and Technology, Qilu University of Technology (Shandong Academy of Sciences), Jinan, China
[3] Shandong Provincial Key Laboratory of Computer Networks, Shandong Fundamental Research Center for Computer Science, Jinan, China
[4] School of Intelligence and Information Engineering, Shandong University of Traditional Chinese Medicine, Jinan, China

Abstract. Due to the remarkable performance of Transformers in 2D medical image segmentation, recent studies have incorporated them into 3D medical segmentation tasks. Compared to convolution operations in CNNs, Transformer-based models possess self-attention, allowing them to capture long-range dependencies among pixels. To address the high computational cost of the Transformer architecture when dealing with volumetric images containing a large number of slices, we propose an efficient hybrid CNN-Transformer architecture for 3D medical image segmentation named PAT-Unet. Firstly, our proposed Paired Attention Transformer (PAT) blocks effectively reduce spatial dimensions while proficiently learning channel and spatial information in 3D feature maps. This leads to improved segmentation performance by reducing parameter count and accelerating computation speed. Secondly, our Deformable Enhanced Skip Connection (DESC) module captures detailed features in irregular lesion areas by learning volume spatial offsets. Finally, we experimentally validate the effectiveness and efficiency of our model on the Synapse and ACDC benchmark datasets. On the Synapse dataset, our model achieves a Dice similarity score of 87.17%, reducing parameters and FLOPs by 67% compared to the best existing methods reported in the literature.

Keywords: 3D Medical Image Segmentation · Model complexity · Channel and Spatial Attention · Skip Connections Enhancement

1 Introduction

Due to the richer and more detailed spatial information present in 3D medical images compared to 2D images, 3D voxel segmentation plays a crucial role in visualizing medical images, aiding in diagnosis, and facilitating treatment planning [2]. The U-Net [15] model, originally proposed by Ronneberger et al., is commonly employed for 3D biomedical image segmentation. Its encoder generates layered 3D voxel feature maps by utilizing information across all three dimensions of the input data. Meanwhile, the decoder progressively upsamples features to obtain the 3D voxel segmentation results. However, CNN-based methods [6,11,15] have limitations in capturing geometric and structural information in medical image data due to the restricted receptive field of convolutional operations. These limitations hampers their ability to learn long-range dependencies between pixels. In contrast, Transformer-based approaches address this issue by incorporating self-attention mechanisms, allowing for the learning of long-range dependencies among pixels. While this improves performance, it also increases the model complexity.

Recently, there have been efforts [4,8,18] to design hybrid architectures that combine the U-Net model with Transformers, aiming to leverage convolutions for local feature extraction while incorporating global self-attention to learn global-local contextual information. UNETR [8] utilizes Transformers as encoders and convolutional operations for upsampling as decoders. Models such as Swin-Unet [4] and nnformer [18] have been developed, which introduce hybrid modules based on Transformers and convolutions in both the encoder and decoder. However, these research works primarily focus on improving segmentation accuracy at the cost of increased model parameters and computational complexity, leading to poor model robustness and high computational resource consumption. Additionally, existing methods fail to capture the dependencies between channel and spatial features in 3D voxel feature maps and do not effectively learn complementary channel-spatial features to enhance segmentation quality. To address these issues, we propose PAT-Unet, a novel approach that effectively integrates dependency information between channels and rich spatial information using PAT blocks. This integration significantly improves segmentation performance while simultaneously reducing the model's parameter count and accelerating inference speed. Our key contributions are as follows:

- We propose an efficient hybrid U-shaped 3D medical image segmentation model, PAT-Unet, designed based on the PAT block. The PAT block reduces spatial dimensionality and effectively learns rich channel and spatial information in 3D feature maps. This achieves a balance between reducing model parameters and accelerating model computation speed while enhancing segmentation performance.
- We introduce the Deformable Enhanced Skip Connection (DESC) module to improve the model's performance in irregular lesion areas. The DESC module adjusts the 3D feature maps by learning volume spatial offsets, enabling the capture of detailed features in regions with diverse shape contours, further enhancing segmentation performance.
- We validate the performance of our model and the effectiveness of the modules on publicly available datasets, conducting visual analysis and parameter comparisons. The experimental results demonstrate that our model

outperforms the comparison models in terms of performance, while reducing the model parameters by more than 67%.

2   Related Work

2.1   CNN-Based Segmentation Methods

Several variants of the U-Net model, such as U-Net++ [19], U-Net 3+ [9], and Residual U-Net [1], have achieved excellent performance on various datasets by processing 3D voxel data as 2D slices. Çiçek et al. [6] extended the U-Net architecture by using 3D operations instead of 2D convolutions for segmentation in sparsely labeled volumetric images. Isensee et al. [11] proposed a nnU-Net model based on U-Net, which incorporates automated configuration to extract features from multi-level images. Furthermore, researchers have also explored the learning of local-global information using pure CNN architectures, including deformable convolutions [16], depthwise convolutions [17], and large kernel convolutions [14].

2.2   Transformers-Based Segmentation Methods

Recent breakthroughs in Vision Transformers [7] have addressed the challenge of capturing long-range dependencies, particularly in medical image segmentation. Transformers employ self-attention mechanisms to learn correlations among all input tokens, enabling the capture of distant relationships. While several efforts [4,10,18] have focused on improving the structure of Transformers for more accurate segmentation, there is limited research on addressing the computational complexity of self-attention mechanisms within Transformers. Swin-Unet [4] is a U-shaped encoder-decoder structure consisting of Swin Transformer blocks. The nnFormer [18] proposed by Zhou et al. continues to utilize convolutional layers for extracting local image details and adopts a hierarchical structure for modeling multi-scale features. Huang et al. [10] proposed MISSFormer, a layered encoder-decoder network with the Enhanced Transformer Context Bridge.

2.3   CNN-Transformer Hybrid Segmentation Methods

Recently, there has been a growing interest in designing hybrid architectures [5,8,13] that combine the U-Net model with Transformers. These architectures aim to leverage the strengths of both convolutional neural networks (CNNs) for local feature extraction and Transformers for capturing global-local contextual information using self-attention mechanisms. TransUNet [5] has introduced an encoder with a hybrid CNN-Transformer architecture, enhancing segmentation performance by introducing convolutional neural networks into the Transformer structure to recover local spatial information. UNETR [8] introduced a novel Transformer-based approach for semantic segmentation of medical images, redefining the task as a 1D sequence-to-sequence prediction problem. Additionally, 3D UX-NET [13] incorporates a lightweight volume ConvNet module with large-kernel depth-wise convolutions to adjust layered features and improve volume segmentation.

3 Method

3.1 Hybrid 3D Segmentation Architecture

Our PAT-Unet model architecture, depicted in Fig. 1, comprises a multi-scale encoder-decoder with deformable enhanced skip connections in between. The decoder's output undergoes 3D convolutions to generate the model's predicted output. In the encoding phase, our PAT-Unet adopts a multi-scale hierarchical design to progressively decrease the resolution of the feature maps, rather than utilizing feature maps with the same resolution.

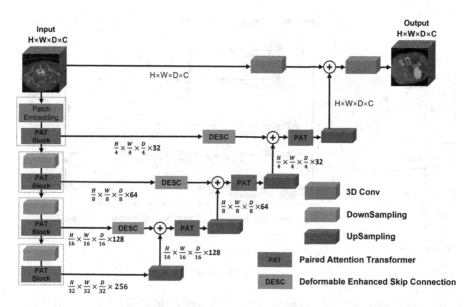

Fig. 1. Architecture of PAT-Unet. Our PAT-Unet consists of a multi-stage hierarchical encoder-decoder with deformable enhanced skip connections in between. The encoder consists of PAT blocks and downsampling, while the decoder includes PAT blocks, upsampling, and DESC aggregation of features from the decoder. The output of the decoder undergoes 3D convolutions to generate the model's predicted output.

In the PAT-Unet model, the encoder comprises four stages. The first stage begins by partitioning the volumetric data into non-overlapping 3D blocks, followed by Transformer blocks (PAT Blocks) with effective multi-head paired attention. In the remaining three encoder stages, we use downsampling layers with 3D convolutions to reduce the resolution of the feature maps by a factor of two, followed by PAT blocks.

The decoder also consists of four stages. In the first three stages, we first use an upsampling layer composed of deconvolutions to double the resolution of the feature maps, followed by concatenation with feature maps from the deformable enhanced skip connections. These concatenated features are then processed by

PAT blocks. With each decoder stage, the number of channels is halved to extract crucial feature information. In the fourth stage of the decoder, the upsampled feature maps are further enhanced by fusion with the feature maps obtained from convolutions applied to the original input image, restoring spatial information. Finally, the fused result undergoes 3D convolutions to generate the model's predicted output.

3.2   Paired Attention Transformer (PAT)

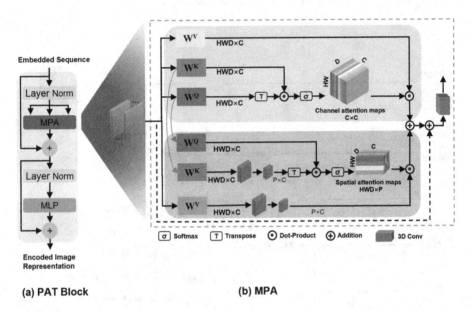

(a) PAT Block   **(b) MPA**

Fig. 2. (a) Paired Attention Transformer (PAT) Block, (b) Multi-Head Paired Attention (MPA).

Most hybrid CNN-Transformer models with self-attention mechanisms suffer from quadratic complexity in terms of the number of tokens, leading to significant computational costs for 3D medical image segmentation. To address this challenge, we propose the Paired Attention Transformer (PAT), as depicted on Fig. 2(a). PAT addresses the complexity issue by performing self-attention across channels instead of the volume dimension, reducing the computational complexity from quadratic to linear with respect to the volume dimension. Additionally, the paired attention mechanism decreases the dimension of the spatial matrix with keys and values, effectively capturing richer spatial information while reducing computational complexity. By combining inter-channel dependency information and rich spatial information, PAT enhances the contextual channel-spatial feature representation, resulting in improved prediction performance.

Figure 2(b) illustrates the specific operations of the Multi-Head Paired Attention (MPA) within our PAT. MPA consists of two attention modules that capture inter-channel dependency information in 3D voxel feature maps, reducing the dimension of the spatial matrix to lower computational costs and extract more comprehensive spatial information. Both attention modules utilize the same Q and K linear layer weights while having their own V layer weights. The formulas for the channel attention and spatial attention modules in MPA are as follows:

$$X_c = \text{CA}\left(Q_{\text{channel}}, K_{\text{channel}}, V_{\text{channel}}\right), \tag{1}$$

$$X_s = \text{SA}\left(Q_{\text{spatial}}, K_{\text{spatial}}, V_{\text{spatial}}\right), \tag{2}$$

where X_c and X_s represent the obtained channel attention map and spatial attention map, respectively. CA refers to the channel attention module, and SA refers to the spatial attention module. $Q_{channel}$, $K_{channel}$, $Q_{spatial}$, $K_{spatial}$, represent the channel query vector, channel keys vector, spatial query vector, and spatial keys vector, respectively. $Q_{spatial}$ is a copy of $Q_{channel}$, and $K_{spatial}$ is a copy of $K_{spatial}$. $V_{channel}$ and $V_{spatial}$ represent the channel value vector and spatial value vector, respectively.

Channel Attention: Let X be a normalized tensor of shape $HWD \times C$. We calculate the $Q_{channel}$, $K_{channel}$, and $V_{channel}$ tensors of size $HWD \times C$ by applying the formulas $Q_{channel} = W^Q X$, $K_{channel} = W^K X$, and $V_{channel} = W^V X$, where W^Q, W^K, and W^V are the linear transformation matrices for $Q_{channel}$, $K_{channel}$, and $V_{channel}$, respectively. In the multi-head paired attention, the channel attention first performs a scaled dot product operation between the transposed $Q_{channel}$ and $K_{channel}$, applying Softmax to measure the similarity between each feature and the rest of the channel features, resulting in channel attention maps. Then, the channel attention maps are multiplied by the $V_{channel}$ vector to capture the inter-channel dependencies in the feature maps. The channel attention is expressed as follows:

$$X_c = V_{\text{channel}} \cdot \text{Softmax}\left(\frac{Q_{\text{channel}}^{\top} K_{\text{channel}}}{\sqrt{d}}\right), \tag{3}$$

where X_c represents the output obtained through channel attention. $Q_{channel}$, $K_{channel}$, and $V_{channel}$ represent the channel query vector, channel keys, and channel values, respectively, where d denotes the size of each vector.

Spatial Attention: The spatial attention in the multi-head paired attention effectively learns richer spatial information. It reduces the dimensionality of the spatial matrix with keys and values, thereby reducing the computational complexity from $O(n^2)$ to $O(np)$, where n represents the number of tokens and p represents the dimensionality of the projected spatial matrix with keys and values, and $p << n$. The spatial attention is then computed in three steps. First, we project the $V_{spatial}$ tensor of dimensions $HWD \times C$ onto the spatial dimension of size $P \times C$. Second, we perform scaled dot product between the transposed

$K_{spatial}$ tensor of dimensions $P \times C$ and $Q_{spatial}$, followed by applying Softmax to obtain the spatial attention maps of dimensions $HWD \times P$. Finally, we multiply the spatial attention maps with the projected $V_{spatial}$ tensor to generate the spatial attention map of dimensions $HWD \times C$. The spatial attention can be defined as follows:

$$X_s = V_{\text{spatial\_proj}} \cdot \text{Softmax}\left(\frac{Q_{\text{spatial}} \, K^{\top}_{\text{spatial\_proj}}}{\sqrt{d}}\right), \tag{4}$$

where $Q_{spatial}$, $K_{spatial\_proj}$, and $V_{spatial\_proj}$ represent the spatial query vector, the projection of spatial keys, and the projection of spatial values, respectively, with d denoting the size of each vector.

Finally, we combine the outputs of channel and spatial attention, fuse them with the original 3D voxel features, and further extract deeper-level feature representations through a 3D convolution operation. The final output of the multi-head paired attention module is represented as:

$$X = \text{Conv}\left(X_s + X_c\right), \tag{5}$$

where X_c and X_s represent the output feature maps of channel attention and spatial attention, respectively, and Conv denotes a three-dimensional convolution block with a kernel size of $3 \times 3 \times 3$.

3.3   Deformable Enhanced Skip Connection Module (DESC)

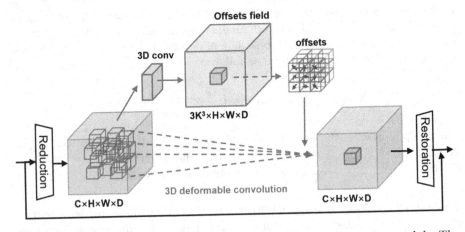

Fig. 3. The representation of the Deformable Enhanced Skip Connection module. The Reduction layer and Restoration layer are two 3D convolutional blocks utilized to decrease and restore the channel dimension of the 3D feature map, respectively. K represents the size of the 3D convolutional kernel.

Due to the diversity and irregular shapes of lesion areas, accurately delineating the lesion regions in medical image segmentation is challenging. To achieve outstanding segmentation performance, the model needs to capture local details such as fine-grained structures and shape contours. To address this issue, we propose the Deformable Enhanced Skip Connection (DESC) module, which incorporates deformable convolutions to capture detailed features in lesion areas with varying shapes.

As illustrated in Fig. 3, the DESC module first reduces the dimensionality of the input 3D voxel feature maps through the Reduction layer. Then, a 3D convolution layer is used to learn the volumetric spatial offsets corresponding to the reduced feature maps in the volume space. The learned volumetric spatial offsets have dimensions of $3K^3 \times H \times W \times D$, where K represents the kernel size of the 3D deformable convolution. While traditional 2D deformable convolutions can only learn 2D offsets in the spatial dimension, our proposed DESC module enables learning of 3D offsets in the volume dimension, allowing appropriate adjustments in both slice and spatial dimensions during feature extraction.

Specifically, our Deformable Enhanced Skip Connection module can be computed as follows:

$$z\left(p_0\right) = \sum_{p_n \in \Re} \omega\left(p_n\right) \cdot x\left(p_0 + p_n + \triangle p_n\right), \tag{6}$$

where z represents the output feature map, R is a regular grid, x denotes the input 3D voxel feature maps, and ω represents the convolution kernel. p_n and p_0 correspond to the enumerated regular grid R and positions in the output feature map, respectively. $\triangle p_n$ represents the learned offset values of the volumetric space. $p_n + \triangle p_n$ indicates an irregular offset added to the original sampling position p_0.

4   Experiments

4.1   Implementation Details

Our model was developed using PyTorch v1.12.1 and leveraged the MONAI library. The training was performed on 4*RTX 2080 Ti (11 GB) GPUs for 1000 epochs. The initial learning rate was set to 0.01, and the initial weight decay was set to 3e−5. For the Synapse dataset [12], the model input size was set to 128 × 128 × 64. In the results section, we provide the Dice Similarity Score (DSC) and the 95% Hausdorff distance (HD95) of the model for eight abdominal organs. For the ACDC dataset [3], the model input size was 160 × 160 × 16 for the three-dimensional volumetric data. All input images were cropped to the predetermined dimensions and underwent contrast adjustments, scaling, rotations, and other augmentation operations.

4.2   Results

Experimental Results on Synapse Dataset: Table 1 presents a comprehensive comparison between our model and other models on the Synapse dataset, focusing on model parameters, FLOPs, HD95, and Dice similarity coefficient (DSC). The U-Net model, utilizing a pure convolutional network, achieves a DSC of 76.85% on the Synapse dataset. The TransUNet and Swin-UNETR models, which utilize a hybrid CNN-Transformers architecture, achieve DSC scores of 77.49% and 83.48%, respectively. Notably, among the latest literature on this dataset, the nnFormer model attains the highest DSC score of 86.57%. In comparison to nnFormer, our PAT-Unet demonstrates a decrease of 1.9% in the HD95 metric while achieving a DSC of 87.17%. Furthermore, PAT-Unet improves both the efficiency and accuracy of segmentation, achieving excellent DSC results while reducing model parameters and FLOPs by more than 67% in comparison to nnformer.

Table 1. Comparative performance analysis of our PAT-Unet with other existing methods on Synapse dataset. Best results are in bold.

Methods	Params (M)	FLOPs (G)	Average DSC ↑	HD95 ↓	Spl	RKid	LKid	Gal	Liv	Sto	Aor	Pan
U-Net	–	–	76.85	–	86.67	68.60	77.77	69.72	93.43	75.58	89.07	53.98
TransUNet	96.07	88.91	77.49	31.69	85.08	77.02	81.87	63.16	94.08	75.62	87.23	55.86
Swin-Unet	–	–	79.13	21.55	90.66	79.61	83.28	66.53	94.29	76.60	85.47	56.58
UNETR	92.49	75.76	78.35	18.59	85.00	84.52	85.60	56.30	94.57	70.46	89.80	60.47
MISSFormer	–	–	81.96	18.20	91.92	82.00	85.21	68.65	94.41	80.81	86.99	65.67
Swin-UNETR	62.83	384.2	83.48	10.55	95.37	86.26	86.99	66.54	95.72	77.01	91.12	68.80
nnFormer	150.5	213.4	86.57	10.63	90.51	86.25	86.57	70.17	**96.84**	**86.83**	92.04	**83.35**
PAT-Unet (Ours)	**46.02**	**59.52**	**87.17**	**8.73**	**95.60**	**87.03**	**87.49**	**71.11**	96.58	85.21	**92.55**	81.79

Experimental Results on ACDC Dataset: Table 2 present a comparative analysis of our model's experimental results with other models on the ACDC dataset. Models such as TransUNet and UNETR, which employ a hybrid CNN-Transformers architecture, achieve average DSC scores of 89.71% and 88.61%

Table 2. Comparison results on the ACDC dataset.

Methods	Average DSC (%)	RV	Myo	LV
U-Net	87.55	87.10	80.63	94.92
TransUNet	89.71	88.86	84.54	95.73
Swin-UNet	90.00	88.55	85.62	95.83
UNETR	88.61	85.29	86.52	94.02
MISSFormer	87.90	86.36	85.75	91.59
nnFormer	92.06	90.94	89.58	95.65
PAT-Unet (Ours)	**92.77**	**91.67**	**90.59**	**96.05**

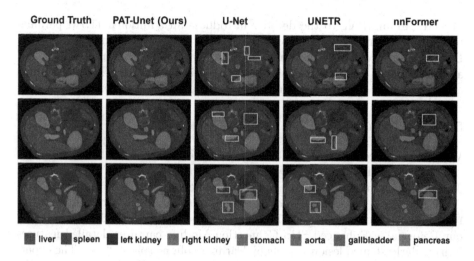

Ground Truth   PAT-Unet (Ours)   U-Net   UNETR   nnFormer

■ liver ■ spleen ■ left kidney ■ right kidney ■ stomach ■ aorta ■ gallbladder ■ pancreas

Fig. 4. Comparison results of visualization on the Synapse dataset.

Table 3. Ablation experiments on the Synapse dataset. We present the performance of incorporating the proposed modules into the baseline.

PAT in Encoder	PAT in Decoder	DESC module	Params (M)	FLOPs (G)	DSC (%)
✔	✘	✘	31.84	39.26	84.44
✔	✔	✘	43.54	54.34	86.89
✔	✔	✔	46.02	59.52	87.17

respectively. The nnFormer model achieves an average DSC of 92.06%. However, our PAT-Unet surpasses them, demonstrating superior segmentation results with an average DSC of 92.77%.

Figures 4 provide visual analysis results comparing our model with other models on the Synapse dataset. From the qualitative comparison, it can be observed that while nnFormer provides clear and accurate segmentation results, our model achieves better segmentation performance with smoother and more precise contours.

4.3   Ablation Experiment

Table 3 presents the performance of incorporating the proposed modules into the baseline U-Net on the Synapse dataset. We have also documented the changes in model parameters, FLOPs, and Dice similarity scores after integrating the enhancement modules. As previously mentioned, our PAT-Unet follows a hierarchical design in both the encoding and decoding stages, progressively reducing the size of feature maps in the encoder and performing upsampling operations in the decoder. Experimental results demonstrate that incorporating the PAT block into the encoder of the baseline model increases the parameter count to

31.84M. Adding PAT to the decoder introduces an additional 11.7M parameters. By incorporating PAT in both the encoder and decoder, and including the DESC module in skip connections, the model parameter count reaches 46.02M, accompanied by an increase in the DSC to 87.17%. The proposed PAT module and hierarchical architecture significantly reduce model complexity.

5   Conclusions

We have proposed an efficient and accurate U-shaped 3D medical image segmentation model called PAT-Unet, which combines the strengths of both CNN and Transformer architectures. Our model achieves remarkable performance by effectively leveraging the channel and spatial information in 3D feature maps while reducing the model parameters through the innovative PAT block. Furthermore, our Deformable Enhanced Skip Connection Module captures detailed features in irregularly shaped lesion regions by learning volume space offsets. The experimental results demonstrate that our PAT-Unet outperforms existing methods by achieving more accurate segmentation results and improving segmentation efficiency.

Acknowledgements. This work is supported in part by The Key R&D Program of Shandong Province (2021SFGC0101), The 20 Planned Projects in Jinan (202228120), National Key Research and Development Plan under Grant No. 2019YFB1404700.

References

1. Alom, M.Z., Yakopcic, C., Hasan, M., Taha, T.M., Asari, V.K.: Recurrent residual U-Net for medical image segmentation. J. Med. Imaging **6**(1), 014006 (2019)
2. Azad, R., et al.: Medical image segmentation review: the success of U-Net. arXiv preprint arXiv:2211.14830 (2022)
3. Bernard, O., et al.: Deep learning techniques for automatic MRI cardiac multi-structures segmentation and diagnosis: is the problem solved? IEEE Trans. Med. Imaging **37**(11), 2514–2525 (2018)
4. Cao, H., et al.: Swin-Unet: Unet-like pure transformer for medical image segmentation. In: Karlinsky, L., Michaeli, T., Nishino, K. (eds.) ECCV 2022, Part III. LNCS, vol. 13803, pp. 205–218. Springer, Cham (2023). https://doi.org/10.1007/978-3-031-25066-8_9
5. Chen, J., et al.: TransUnet: transformers make strong encoders for medical image segmentation. arXiv preprint arXiv:2102.04306 (2021)
6. Çiçek, Ö., Abdulkadir, A., Lienkamp, S.S., Brox, T., Ronneberger, O.: 3D U-Net: learning dense volumetric segmentation from sparse annotation. In: Ourselin, S., Joskowicz, L., Sabuncu, M.R., Unal, G., Wells, W. (eds.) MICCAI 2016, Part II. LNCS, vol. 9901, pp. 424–432. Springer, Cham (2016). https://doi.org/10.1007/978-3-319-46723-8_49
7. Dosovitskiy, A., et al.: An image is worth 16x16 words: transformers for image recognition at scale. arXiv preprint arXiv:2010.11929 (2020)
8. Hatamizadeh, A., et al.: UNETR: transformers for 3d medical image segmentation. In: Proceedings of the IEEE/CVF Winter Conference on Applications of Computer Vision, pp. 574–584 (2022)

9. Huang, H., et al.: Unet 3+: a full-scale connected Unet for medical image segmentation. In: ICASSP, pp. 1055–1059. IEEE (2020)
10. Huang, X., Deng, Z., Li, D., Yuan, X.: MISSFormer: an effective medical image segmentation transformer. CoRR abs/2109.07162 (2021)
11. Isensee, F., Jaeger, P.F., Kohl, S.A., Petersen, J., Maier-Hein, K.H.: NNU-Net: a self-configuring method for deep learning-based biomedical image segmentation. Nat. Methods 18(2), 203–211 (2021)
12. Landman, B., Xu, Z., Igelsias, J., Styner, M., Langerak, T., Klein, A.: Miccai multi-atlas labeling beyond the cranial vault-workshop and challenge. In: Proceedings of the MICCAI Multi-Atlas Labeling Beyond Cranial Vault-Workshop Challenge, vol. 5, p. 12 (2015)
13. Lee, H.H., Bao, S., Huo, Y., Landman, B.A.: 3D UX-Net: a large kernel volumetric convnet modernizing hierarchical transformer for medical image segmentation. arXiv preprint arXiv:2209.15076 (2022)
14. Li, H., Nan, Y., Yang, G.: LKAU-Net: 3D large-kernel attention-based U-Net for automatic MRI brain tumor segmentation. In: Yang, G., Aviles-Rivero, A., Roberts, M., Schönlieb, C.B. (eds.) MIUA 2022. LNCS, vol. 13413, pp. 313–327. Springer, Cham (2022). https://doi.org/10.1007/978-3-031-12053-4_24
15. Ronneberger, O., Fischer, P., Brox, T.: U-Net: convolutional networks for biomedical image segmentation. In: Navab, N., Hornegger, J., Wells, W.M., Frangi, A.F. (eds.) MICCAI 2015, Part III. LNCS, vol. 9351, pp. 234–241. Springer, Cham (2015). https://doi.org/10.1007/978-3-319-24574-4_28
16. Yang, X., Li, Z., Guo, Y., Zhou, D.: DCU-Net: a deformable convolutional neural network based on cascade U-Net for retinal vessel segmentation. Multimedia Tools Appl. 81(11), 15593–15607 (2022)
17. Zeng, N., et al.: Factoring 3d convolutions for medical images by depth-wise dependencies-induced adaptive attention. In: 2022 IEEE International Conference on Bioinformatics and Biomedicine (BIBM), pp. 883–886. IEEE (2022)
18. Zhou, H.Y., Guo, J., Zhang, Y., Yu, L., Wang, L., Yu, Y.: NNFormer: interleaved transformer for volumetric segmentation. arXiv preprint arXiv:2109.03201 (2021)
19. Zhou, Z., Siddiquee, M.M.R., Tajbakhsh, N., Liang, J.: Unet++: redesigning skip connections to exploit multiscale features in image segmentation. IEEE Trans. Med. Imaging 39(6), 1856–1867 (2019)

Cell-CAEW: Cell Instance Segmentation Based on ConvAttention and Enhanced Watershed

Liang Zeng[✉]

Zhejiang University of Technology, Hangzhou, China
1935984404@qq.com

Abstract. Cell instance segmentation in microscopy images is a challenging task. The morphological differences between different types of cells are significant, it is difficult to distinguish the boundaries between adjacent or overlapping cells. To address these issues, we improved Cellpose's framework and proposed Cell-CoaT. Cell-CoaT adopts CoaT as the encoder and designs a decoder that can integrate features from different scales, and predicts the center region and gradient fields of cells. In the post-processing stage, we utilized a Marker-Controlled Watershed Segmentation with center point labels predicted by the network to alleviate under-segmentation and over-segmentation. Cell-CAEW obtains an F1 score of 0.7724 on the tuning set. The code will be released soon.

Keywords: Cell Instance Segmentation · Attention · Enhanced Watershed

1 Introduction

Instance segmentation is a common task in biomedical applications, which involves detecting individual object instances and performing segmentation [3]. Many biological applications require the segmentation of cell bodies, membranes, and nuclei from microscopy images. However, for the network model to learn better, we often need to collect a large amount of labeled data. The labeling process requires professional domain knowledge and can be prohibitively expensive. Hence much effort has been devoted to semi-supervised. Our paper was generated from a Cell Segmentation competition, and the dataset used in our paper was also from the competition.

Although general instance segmentation has achieved good results in natural images [3,4], there are some differences between the universal instance segmentation of natural images and cell instance segmentation. We summarize this task's key points and difficulties: (1) Universal Instance Segmentation always has multiple class instances in each image, but cell segmentation is different. Most of the time we study only a few or a single class of cells under the same imaging instrument. (2) The target cell sizes are relatively balanced over the whole dataset, but the cell size is stable in the same image during cell segmentation. (3) The density

Q. Liu et al. (Eds.): PRCV 2023, LNCS 14437, pp. 370–381, 2024.
https://doi.org/10.1007/978-981-99-8558-6_31

of image instances is different, and the task of dense segmentation of the same type of instances is performed during cell segmentation. (4) Boundary separation is more difficult. In cell segmentation, the area close to the cell membrane can hardly be distinguished by significant edge features. Shape information in a wide field of view is required, which requires a large receptive field.

Cellpose [11] generated topological maps through a process of simulated diffusion that used ground-truth masks. A neural network was then trained to predict the horizontal and vertical gradients of the topology map and a binary prediction mask to indicate whether a given pixel was inside or outside the ROI. On the test image, a vector field is also obtained by predicting the binary mask as well as the horizontal gradient and vertical gradient by the network, and the exact shape of individual cells is recovered by this vector field to obtain the final instance segmentation results.

In this study, we followed Cellpose's strategy and constructed a model to predict the central regions of individual cells indirectly, and then to obtain the final instance segmentation results by the watershed algorithm to enhance the concept of the "cell's center" compared with vanilla Cellpose. For the architecture part, we used CoaT-Lite [13] as the encoder of the network and designed a decoder to fuse the outputs of different layers of the encoder. In addition, to obtain a model with better generalization capability, we use the mean-teacher approach to perform semi-supervised training using the provided unlabeled data.

2   Related Work

2.1   Cell Instance Segmentation

Many instance segmentation algorithms have been proposed in recent years, including the detection-base method [18] and the segmentation-base method [19]. As the detection-based models always achieve high accuracy in instance segmentation, we focused on this line in this study. Fully Convolutional Networks (FCN) [1] improves the accuracy of semantic segmentation but its segmentation results often have poor details because contextual information is not considered. Based on FCN, Ronneberger, and colleagues propose the U-Net [2], which uses a U-shaped encoder-decoder architecture that combines low-resolution and high-resolution information through skip connections. U-Net has achieved good segmentation performance on medical images and has become a widely adopted baseline model. Still, its architecture is not designed to handle instance segmentation of single cells. Mask R-CNN [3] performs classification and coordinate regression on the proposals. It can handle segmentation in the presence of overlap between image objects but tends to produce incorrect boundary predictions in regions of tight cell contacts and incomplete segmentation masks in regions of weak signals [20].

2.2   Vision Transformer

As a pioneered work, ViT [6] validated the feasibility of pure transformer architectures for computer vision tasks. The results of ViT on image classification are encouraging, but its architecture is unsuitable for use as a general-purpose backbone network on dense vision tasks or when the input image resolution is high, due to its low-resolution feature maps and the quadratic increase in complexity with image size. Also, the self-attention mechanism in Vit often ignores local feature details. To address these two issues, CoaT [13] proposes the co-scale and Conv-Attention mechanism, which have achieved significant results.

3   Method

In this chapter, we will detail the strategy and process of the adopted method in three parts: pre-processing, model architecture, and post-processing.

3.1   Preprocessing

To follow Cellpose's instance segmentation strategy, we process the provided instance labels to obtain the corresponding gradient vector fields and binary classification masks. Figure 1 shows an example of pre-processing the instance labels.

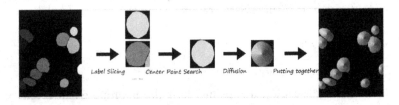

Fig. 1. Instance of splitting labels to generate gradient fields.

3.2   Model Architecture

Overall Structure: The overall structure of Cell-CoaT is show in Fig. 2. Cell-Coat uses Coat as an encoder, aggregates feature maps of different scales using mixupsample and concat and merges information from different scales through a fusion block. Finally, it predicts the result and calculates the Loss.

Encoder: The encoder part on the left is CoaT-Lite, through which we can get four different scales of feature maps. The structure of the CoaT Serial Block is shown in Fig. 3. The module first downsamples the input features through a patch embedding layer and then flattens them into N patch tokens and 1 class token obtains $X \in R^{(N+1) \times C}$. Then add position embedding by Convolutional Position Encoding:

$$X = reshape2(DConv(reshape1(X))) + X, \tag{1}$$

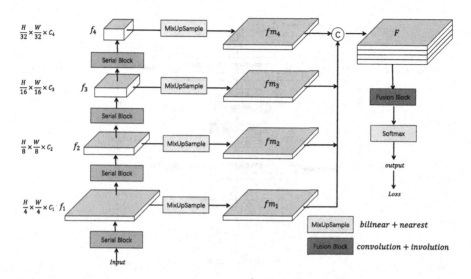

Fig. 2. The overall structure of Cell-CoaT, where the encoder part on the left is CoaT-Lite, through which we can get four different scales of feature maps $f_i, i = 1, 2, 3, 4$. The MixUpSample Block samples these different scales of feature maps to get a set of features of the same size $fm_i \in R^{\frac{H}{4} \times \frac{W}{4} \times C_1}, i = 1, 2, 3, 4$ and concat in the channel dimension obtain F. We design a Fusion Block to fully fuse these features, and finally, F passes a Softmax layer to obtain the final gradient field and binary prediction results.

where $reshape1$ is used to reshape the X into shape of $(\sqrt{N}, \sqrt{N}, C)$, $DConv$ means Depthwise Conv and $reshape2$ recover the X into shape of $(N + 1, C)$.

The attention mechanism of this module adopts a combination of the Factorized Attention Mechanism and Convolutional Relative Position Encoding:

$$ConvAtt(X) = \frac{Q}{\sqrt{C}}(softmax(K)^{\mathrm{T}}V) + E(V), \tag{2}$$

$$X = X + ConvAtt(X), \tag{3}$$

where Q, K, V are obtained by applying linear transformations $W^Q, W^K, W^V \in R^{C \times C}$ to X, E presents convolutional relative position encoding.

Finally, x goes through MLP and reshapes to obtain a feature map:

$$f = MLP(reshape2(X)) + X, \tag{4}$$

MixUpSample Block: The MixUpSample Block uses a mixture of bilinear interpolation and nearest neighbor interpolation to upsample feature maps of different sizes to the same size for the subsequent fusion of these feature maps:

$$fm_i = (1 - \lambda)bilinear(f_i) + \lambda nearest(f_i), i = 1, 2, 3, 4, \tag{5}$$

where λ is a learnable parameter, $bilinear$ is bilinear interpolation and $nearest$ is nearest neighbor interpolation.

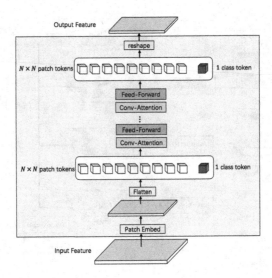

Fig. 3. Schematic illustration of the serial block in CoaT [13].

Fusion Block: Consider that convolutions are limited by spatial-agnostic and channel-specific, while Involution has the opposite property [14]. We use both convolution and involution in Fusion Block so that the feature maps after concat on the channels can be fully integrated into both spatial and channel dimensions to enhance the performance of the model, which shows in Fig. 4.

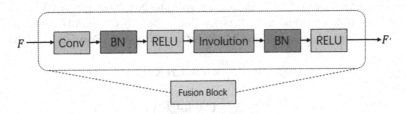

Fig. 4. Structure of fusion block.

Loss Function: We use the summation between Dice loss and cross entropy loss because compound loss functions have been proven to be robust in various medical image segmentation tasks [9]. In addition, we used MSE loss on the predicted resulting gradient fields:

$$DiceLoss = 1 - \frac{2}{NC} \sum_{n=1}^{N} \sum_{c=1}^{C} \frac{yl_{n,c} \sum_{i=1}^{H} \sum_{j=1}^{W} exp(pl_{n,c}^{i,j})}{\sum_{i=1}^{H} \sum_{j=1}^{W} exp(pl_{n,c}^{i,j})}, \tag{6}$$

$$CELoss = -\frac{1}{N} \sum_{n=1}^{N} \sum_{c=1}^{C} log \frac{exp(pl_{n,c})}{\sum_{c=1}^{C} exp(pl_{n,i})} yl_{n,c}, \tag{7}$$

$$MSELoss = \frac{1}{N}\sum_{n=1}^{N}(yg_n - pg_n)^2, \tag{8}$$

$$Loss = \alpha DiceLoss + \alpha CELoss + \beta MSELoss, \tag{9}$$

where N is the batch size, C is the class number, H and W are the width and height of the image, pl represents the predicted mask by the model, yl is the ground truth, pg is the predicted gradient fields by the model, and yg is the gradient fields of the ground truth obtained by preprocessing. And α and β are hyperparameters.

To make full use of a large amount of unlabeled data and to make the model more generalizable, we conducted semi-supervised experiments by the mean teacher [12] approach.

3.3 Post-processing

First, the gradient field and binary mask obtained by the backbone network are used to generate a prediction map containing the center pixels' set of each cell instance. It's different from Cellpose which marks all pixels inside the cells' masks because we find that some center points with similar features are easy to connect into one center point set in the dynamic programming of these center points. To overcome this, we only collect the center region sets using dynamic programming with only 3 iterations (that means only the points within 3 pixels near the center point will participate in the generation of the center point), and then use the Marker-Controlled Watershed Segmentation to finish the instance segmentation.

4   Experiments

4.1   Dataset and Evaluation Metric

We use the official dataset provided by the competition for training and validation, and some rare cell shapes were automatically synthesized offline. The training set consists of 1000 labeled images and over 2200 unlabeled images from various microscopic types, tissue types, and stain types. And there are four microscopy modalities in the training set, including Brightfield, Fluorescent, Phase-contrast, and Differential interference contrast. Noting that some of the cell types in the validation set were fewer in the test set, we used the taichi physics engine [15] to simulate some simply shaped soft cells and generate the cell images automatically. Some of the generated samples are shown in Fig. 5. To measure the performance of our method, we use the F1 as the evaluation metric.

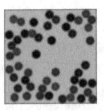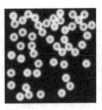

Fig. 5. Samples generated by taichi.

4.2   Implementation Details

Data Augmentation: The online data augmentation includes Flip, Rotate, Random Crop, Gaussian Smooth, Gaussian Noise, Adjust Contrast, Histogram Shift, and Random Zoom. Additionally,.Color Jitter, Scale Intensity and Cutout are adopted. Moreover, we use the Taichi Physical Engine and mpm18 (Material Point Method) [15] to simulate cell distribution.

Training Protocols: Patch sampling strategy during training (e.g., randomly sample 512 × 512 size) and inference (slide window with an image size 768 × 768).Our encoder is loaded with the weights of the CoaT-Lite Medium version pre-trained under the ImageNet dataset [13]. Other details are shown in Table 1.

Table 1. Training protocols.

Training Parameter	Value
Network initialization	he normal initialization [17]
Crop size	512 × 512
Total epochs/Max epochs	1000/186
Optimizer	AdamW
Initial learning rate (lr)	6e−5
α	1
β	1
Training time	<48 h
Loss function	CE Loss + Dice Loss + MSE Loss
Number of model parameters	50.58M[1]
Number of flops	404.90G[2]
GPU (number and type)	4 * NVIDIA 2080Ti 11G
CPU	Intel(R) Core(TM) i9-10900X CPU@3.30 GHz
Batch Size	6

[1] https://github.com/sksq96/pytorch-summary
[2] https://github.com/facebookresearch/fvcore

5   Results and Discussion

We try to train our model under the framework of mean-teacher and some improved methods of it. But all the methods we tried failed (see F1 score below).

By the local validation set and tuning set, we compare the quantitative of all the cells' segmentation results by F1 score or observation.

We check the validation set output segment results, small cells whose area is less than $16 \times N$ ($N > 16$), and fluorescent cells always get low F1 scores.

Our method follows Cellpose, so the total test time is amortized between the model inference stage and the post-processing generation mask stage. The inference time is almost based on the model's FLOPs, so the input size affects much, we tried sizes 256, 512, 768, and 1024, the size 768 get the best score and cost nearly the least time on large images. Besides, a valid way to reduce the inference time is to transfer the model parameters and structure to ONNX/tflite and then use other high-performance inference engines, but no one in our team is good at model deployment (due to the use of a special operator called involution, we failed many times), we give it up. The whole-slide image in our method costs about 1 min within our vision, therefore we did not try to optimize inference time for such images.

5.1   Quantitative Results on Tuning Set

We spend a lot of time exploring the effect of different cell representations and network model structures on segmentation results. The SDF-based (Signed Distance Field) cell representation is equivalent to the gradient field based cell representation in theory. In other word, the SDF-based method is the cumulative function of a gradient field based cell in the spatial dimension. But this may cause a lower fault tolerance rate for the probability distribution of the center points of the cell with the SDF-based method. We compare the different representations at Table 2. From the table, it can be seen that using both gradient field and watershed gives the best F1 score in the tuning set.

Table 2. Quantitative results on tuning set. Where 3class means the same way as the official baseline provided to get the output of 3 class from the network, SwinUNetr presents Swin Transformer + U-Net, * present without fine-tune at the train dataset, DFC present DeFormable sparse Convolution [8].

Method	F1 Score
MaskRCNN	74.72
UNet+3class	54.72
SwinUNetr+3class	54.72
SwinUNetr+3class+DFC	56.62
CoaT+3class	72.30
Cellpose*	51.42
CoaT+SDF	58.32
CoaT+grad	75.45
CoaT+grad+Watershed	**77.24**

In addition, we performed partial ablation experiments on the fusion module, including using only 3 × 3 convolution, using only Involution, replacing 3 × 3 convolution with an oversized convolution kernel of 31 × 31 size [10]. The results are visible in Table 3. In particular, in the experiment with the 31×31 convolution kernel size, we used large convolution kernels in multiple places of the decoder instead of just in the fusion block.

Table 3. Different implementation schemes for fusion blocks based on Coat+3class.

Fusion Method	F1 Score
3 × 3 Conv	74.2
31 × 31 Conv [10]	68.42
Involution [14]	**77.24**

We also tested the effect of the size of the input feature maps under the model of CoaT+3class, and the results are shown in Table 4. Based on the results of this experiment, we choose 768 × 768 resolution as the input feature map size for all other models.

Table 4. Different size of the input feature maps under the model of CoaT+3class.

Input Size	F1 Score
256 × 256	67.86
512 × 512	69.72
768 × 768	**72.30**
1024 × 1024	71.90

We performed semi-supervised training using the unlabeled dataset using the methods [7,12] under the mean teacher framework. Perhaps because the distribution of the combination of unlabeled images and labeled images is shifted compared with the train/validation dataset's distribution, the results on the tuning set are not satisfactory. Some of the results are shown in Table 5.

Table 5. Performances under the mean teacher framework.

Semi-supervisory strategy	F1 Score
w/o mean teacher [7]	**77.24**
mean teacher finetune based	77.06
mean teacher + confidence loss	74.16
mean teacher + confidence loss + cutout augmentation	74.17

5.2   Qualitative Results on Validation Set

Figure 6 presents some easy and hard examples of validation set. Where the first row is the original graph, the second row is the predicted result after full-supervised training, and the third row is the predicted result after semi-supervised training. As you can see from the figure, the semi-supervised approach does not work well.

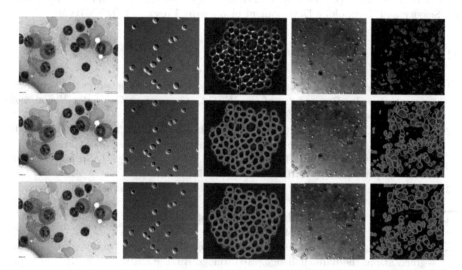

Fig. 6. Qualitative results of the full-supervised and semi-supervised model on easy (Three columns on the left) and hard examples. Each prediction graph is the result of ground truth and predicted label overlap, where the red outline is the ground truth and the yellow outline is the predicted label obtained. (Color figure online)

5.3   Segmentation Efficiency Results on Validation Set

The average running time is 2.76 s per case in the inference phase (a total of 127 images in our augment data set), and the average used GPU memory is 799.52 MB. The area under GPU memory-time curve is 247845 and the under CPU utilization-time curve is about 1395.

6   Conclusion

The main challenge of this task is to separate the adherent cells from each other. Compared to the methods in baseline and Cellpose, the method proposed in this paper is significantly improved and better able to accomplish this task.

The results of the semi-supervised experiments we have conducted so far are not satisfactory, so we will follow up by learning more about semi-supervision and conducting experiments. We note that omnipose [16] is a proposed method based on Cellpose and was recently published in Nature Methods, and we may subsequently refer to these changes to improve the performance of our model. In addition, integrating multiple models and using ONNX to accelerate inference might be a good way to improve the results.

Ackonwledgments. This work is partially supported by the National Natural Science Foundation of China (Grant No. U20A20171), Zhejiang Provincial Natural Science Foundation of China (Grant Nos. LY21F020027, LY23F020023), and Key Programs for Science and Technology Development of Zhejiang Province (2022C03113).

References

1. Long, J., Shelhamer, E., Darrell, T.: Fully convolutional networks for semantic segmentation. In: Proceedings of the IEEE Conference on Computer Vision and Pattern Recognition, pp. 3431–3440 (2015)
2. Ronneberger, O., Fischer, P., Brox, T.: U-Net: convolutional networks for biomedical image segmentation. In: Navab, N., Hornegger, J., Wells, W.M., Frangi, A.F. (eds.) MICCAI 2015, Part III 18. LNCS, vol. 9351, pp. 234–241. Springer, Cham (2015). https://doi.org/10.1007/978-3-319-24574-4_28
3. He, K., et al.: Mask R-CNN. In: Proceedings of the IEEE International Conference on ICCV (2017)
4. Girshick, R.: Fast R-CNN. In: Proceedings of the IEEE International Conference on Computer Vision, pp. 1440–1448 (2015)
5. Ma, J., Zhang, Y., Gu, S., et al.: Fast and low-GPU-memory abdomen CT organ segmentation: the flare challenge. Med. Image Anal. **82**, 102616 (2022)
6. Dosovitskiy, A., Beyer, L., Kolesnikov, A., et al.: An image is worth 16×16 words: transformers for image recognition at scale. arXiv preprint arXiv:2010.11929 (2020)
7. Liu, Y., Tian, Y., Chen, Y., et al.: Perturbed and strict mean teachers for semi-supervised semantic segmentation. In: Proceedings of the IEEE/CVF Conference on Computer Vision and Pattern Recognition, pp. 4258–4267 (2022)
8. Zhu, X., Hu, H., Lin, S., et al.: Deformable ConvNets v2: more deformable, better results. In: Proceedings of the IEEE/CVF Conference on Computer Vision and Pattern Recognition, pp. 9308–9316 (2019)
9. Ma, J., Chen, J., Ng, M., et al.: Loss odyssey in medical image segmentation. Med. Image Anal. **71**, 102035 (2021)
10. Ding, X., Zhang, X., Han, J., et al.: Scaling up your kernels to 31×31: revisiting large kernel design in CNNs. In: Proceedings of the IEEE/CVF Conference on Computer Vision and Pattern Recognition, pp. 11963–11975 (2022)
11. Stringer, C., Wang, T., Michaelos, M., et al.: Cellpose: a generalist algorithm for cellular segmentation. Nat. Meth. **18**(1), 100–106 (2021)
12. Tarvainen, A., Valpola, H.: Mean teachers are better role models: weight-averaged consistency targets improve semi-supervised deep learning results. In: Advances in Neural Information Processing Systems, vol. 30 (2017)
13. Xu, W., et al.: Co-scale conv-attentional image transformers. In: Proceedings of the IEEE/CVF International Conference on Computer Vision (2021)
14. Li, D., et al.: Involution: inverting the inherence of convolution for visual recognition. In: Proceedings of the IEEE/CVF Conference on Computer Vision and Pattern Recognition (2021)

15. Hu, Y.: Taichi: an open-source computer graphics library. arXiv preprint arXiv:1804.09293 (2018)
16. Cutler, K.J., Stringer, C., Lo, T.W., et al.: Omnipose: a high-precision morphology-independent solution for bacterial cell segmentation. Nat. Meth. **19**(11), 1438–1448 (2022)
17. He, K., Zhang, X., Ren, S., et al.: Delving deep into rectifiers: surpassing human-level performance on ImageNet classification. In: Proceedings of the IEEE International Conference on Computer Vision, pp. 1026–1034 (2015)
18. Chen, H., et al.: BlendMask: top-down meets bottom-up for instance segmentation. In: IEEE Conference on Computer Vision and Pattern Recognition, pp. 8573–8581 (2020)
19. Oda, H., et al.: BESNet: boundary-enhanced segmentation of cells in histopathological images. In: Frangi, A.F., Schnabel, J.A., Davatzikos, C., Alberola-López, C., Fichtinger, G. (eds.) MICCAI 2018, Part II. LNCS, vol. 11071, pp. 228–236. Springer, Cham (2018). https://doi.org/10.1007/978-3-030-00934-2_26
20. Zhou, Y., Li, W., Yang, G.: SCTS: instance segmentation of single cells using a transformer-based semantic-aware model and space-filling augmentation. In: Proceedings of the IEEE/CVF Winter Conference on Applications of Computer Vision (2023)

A Comprehensive Multi-modal Domain Adaptive Aid Framework for Brain Tumor Diagnosis

Wenxiu Chu[1], Yudan Zhou[1], Shuhui Cai[1,2], Zhong Chen[1,2], and Congbo Cai[1,2(✉)]

[1] Institute of Artificial Intelligence, Xiamen University, Xiamen, China
cbcai@xmu.edu.cn
[2] Department of Electronic Science, Xiamen University, Xiamen, China

Abstract. Accurate segmentation and grading of brain tumors from multi-modal magnetic resonance imaging (MRI) play a vital role in the diagnosis and treatment of brain tumors. The gene expression in glioma also influences the selection of treatment strategies and assessment of patient survival, such as the gene mutation status of isocitrate dehydrogenase (IDH), the co-deletion status of 1p/19q, and the value of Ki67. However, obtaining medical image annotations is both time-consuming and expensive, and it is challenging to perform tasks such as brain tumor segmentation, grading, and genotype prediction directly using label-deprived multi-modal MRI. We proposed a comprehensive multi-modal domain adaptive aid (CMDA) framework building on hospital datasets from multiple centers to address this issue, which can effectively relieve distributional differences between labeled source datasets and unlabeled target datasets. Specifically, a comprehensive diagnostic module is proposed to simultaneously accomplish the tasks of brain tumor segmentation, grading, genotyping, and glioma subtype classification. Furthermore, to learn the data distribution between labeled public datasets and unlabeled local hospital datasets, we consider the semantic segmentation results as the output capturing the similarity between different data sources, and we employ adversarial learning to facilitate the network in learning domain knowledge. Experimental results show that our end-to-end CMDA framework outperforms other methods based on direct transfer learning and other state-of-the-art unsupervised methods.

Keywords: Unsupervised Domain Adaptation · IDH Mutation · Ki67 Genotype · Brain Tumor Segmentation · Grading · Glioma Subtype

1 Introduction

Brain tumors have seriously threatened people's life and health, and glioma is one of the most common brain diseases. Accurate segmentation and grading of tumor lesions using MRI play a crucial role in tumor diagnosis and subsequent treatment.

Some gene states can be used as markers for the diagnosis and prognosis of glioma [1, 2]. IDH is one of the most clinically valuable genetic markers for glioma, and Ki67

© The Author(s), under exclusive license to Springer Nature Singapore Pte Ltd. 2024
Q. Liu et al. (Eds.): PRCV 2023, LNCS 14437, pp. 382–394, 2024.
https://doi.org/10.1007/978-981-99-8558-6_32

is an indicator of tumor cell proliferation activity. Detection of 1p/19q co-deletion is of great significance in the diagnosis of oligodendroglioma and the prognosis of patients. The relationship between these genotype statuses, glioma grade, glioma subtype, and patient survival is shown in Fig. 1.

Glioma Risk Stratification Scheme					
WHO Grade	IDH	1p/19q	Glioma Subtype	Ki-67	Survival
II, III	Mutation	Codeletion	Oligodendroglioma	low	long
			Astrocytoma		
	Wildtype	Non-Codeletion	Glioblastoma		
IV	Mutation		Astrocytoma		
	Wildtype		Glioblastoma	high	short

Fig. 1. Glioma risk stratification scheme [3]. The genetic status of IDH, Ki67, and 1p/19q, as well as the grade and subtype of brain tumors, are closely related to the survival rate of patients. The genetic status of IDH and 1p/19q determines the brain glioma subtype. The higher the grade of brain glioma, the shorter the survival time of patients, and the higher the Ki67 value, the higher the malignant degree of the tumor.

Glioma intelligent diagnosis tasks are usually trained with a fully supervised strategy. However, it is an expensive and time-consuming process to manually annotate glioma images since it requires the knowledge of experienced clinical experts. In addition, clinical data from different hospitals, different scanning instruments, and scanning parameters often have a large distribution gap. Fortunately, the public dataset BraTS2019 has many high-quality multi-modal data and labels[4], so the realization of comprehensive intelligent diagnosis of glioma clinical data with the help of knowledge transfer of public data is now urgently needed.

Motivated by this issue, we propose an end-to-end CMDA framework, which consists of two parts: a comprehensive diagnostic module and a data domain discriminator module. For the segmentation task, the comprehensive diagnosis module uses an adversarial learning strategy to trick the discriminator into generating similar distributions on the segmented output of source and target domain images. Although classification labels are more easily obtained by medical instruments than segmentation labels, tumor classification tasks are based on the tumor ROI extracted by segmentation labels. For the classification task, genotyping and subtyping of gliomas is more difficult since gliomas are often located in arbitrarily small areas in the MRI. Inspired by this, we implement a new feature training strategy, using the predicted segmentation map of the network as a mask to obtain the region of interest (ROI) of the image, which is helpful for feature extraction and improves the effects of genotyping and glioma subtype classification.

The main contributions of our work in this paper can be summarized as follows:

– A new unsupervised domain adaptive intelligent diagnosis framework is proposed to fully mine high-dimensional tumor features from public data and multi-center hospital data in an end-to-end training manner, greatly alleviating the dependence on labels for intelligent diagnosis of clinical brain tumor data.

- We have developed a general diagnostic algorithm, and to the best of our knowledge, this is the first time that the tasks of grading, classification, and segmentation have been integrated into a U-shaped network. The algorithm can simultaneously solve the challenging tasks of multi-modal brain tumor segmentation, grading, genotyping, and glioma subtype classification.
- Comprehensive experiments were conducted on two local hospital datasets, and the results demonstrate that this method significantly improves the diagnostic accuracy of multiple tasks simultaneously. Furthermore, it exhibits good robustness on different target domain data.

2 Related Work

We briefly review two types of work most relevant to our approach: 1) deep learning-based methods for brain tumor segmentation and 2) unsupervised domain adaptation methods and their applications in medical images.

2.1 Deep Learning-based Brain Tumor Segmentation

Brain image segmentation is one of the most time-consuming and challenging processes in clinical Settings. Deep learning-based methods have attracted extensive research interest due to their excellent performance and ability to extract features [5]. In the past few years, more people have studied deep learning-based brain tumor segmentation methods [6, 7]. Dense prediction methods use a fully convolutional network (FCN) [8]. For example, Dong et al. [9] use the famous FCN architecture in medical image segmentation, i.e., U-Net, to segment brain tumors in MR images. Cui et al. [10] reported on a hybridized cascade of a deep convolution neural network (DCNN) architecture that can segment 2D brain images automatically in two major steps. These methods show the effectiveness of U-Net in a fully supervised environment, while the segmentation performance would be poor in a weakly supervised environment. In this case, unsupervised domain adaptation needs to be applied to solve the brain tumor segmentation problem.

2.2 Domain Adaptation

The adaptive approach is able to solve the problem of missing labels. It is common to align the feature distributions between two domains by minimizing their distances [11, 12]. For example, Tsai et al. [13] proposed a multi-level adversarial architecture to adapt the structured output from the source domain to the target domain for semantic segmentation. Some people have also tried to address the Unsupervised Domain Adaptive (UDA) problem in medical image applications [14–16]. For example, He et al. [17] proposed a novel bidirectional Global-to-local (BiGL) adaptation framework to segment brain tumors for intermediate data distributions generated from the two domains.

3 Methods

3.1 Overall Architecture

The proposed fully automatic CMDA framework has two steps: 1) The comprehensive diagnosis module is used to obtain segmentation prediction, grading prediction, and classification prediction. 2) Adversarial learning establishes distribution consistency between images from different domains to reduce domain distribution differences. For simplicity, we take the public dataset BraTs2019 with segmentation and grading labels (source domain) and the dataset from hospitals with only classification labels of IDH genotype and Ki67 genotype (target domain) as examples to show the CMDA model, as shown Fig. 2.

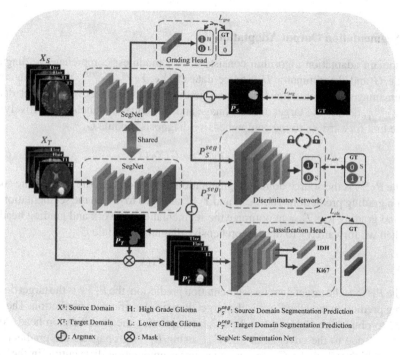

Fig. 2. The framework of CMDA. The segmentation network, grading head, and classification head of the comprehensive diagnosis module and the discriminator are shown. "GT" is the ground truth manually annotated by pathologists.

Comprehensive Diagnosis Module. The comprehensive diagnostic module of our proposed CMDA model consists of three parts: 1) A segmentation network (SegNet) based on the classic UNet++ [18]. The input has four channels corresponding to the four modalities of MR Images, and the output has two channels corresponding to the background and tumor region of brain tumor segmentation. 2) A grading head, which consists of a global average pooling block and a convolutional block, takes the output of the encoder of the segmentation network as the input, and the output is the brain tumor grading. 3) A classification head, which is composed of a convolutional block, four residual convolutional blocks, and a certain number of pooling convolutional blocks, takes the brain tumor segmentation output argmax as the mask to extract the brain tumor ROI as the input of the classification head, and the output is the target task classification. In this example, the classification task for the local hospital dataset is the prediction of the IDH and ki67 genotypes, so the number of pooling convolutional blocks is two.

Discriminator. The discriminator has a similar structure to CycleGAN [19]. The two input channels of identification take the segmentation result of the integrated diagnosis module as the input, and the output is compressed into a single-channel patch for discrimination. In this example, the output of the discriminator is a domain prediction for the input for source domain label 0 and target domain label 1.

3.2   Segmentation Output Adaptation

Our domain adaptation algorithm consists of two modules: comprehensive diagnosis module G and discriminator D. Source dataset $\{X_S, Y_S^{seg}, Y_S^{gra}\}$, denote the source domain image, segmentation labels, and grading labels, respectively. Target dataset $\{X_T, Y_T^{cls}\}$, denote the target domain image and classification labels, respectively.

We first forward X_S to the comprehensive diagnosis module G:

$$P_S^{seg}, P_S^{gra} = G(X_S) \tag{1}$$

The P_S^{seg} is the source domain segmentation prediction, and the P_S^{gra} is the source domain grading prediction. The Y_S^{seg} and Y_S^{gra} are used to obtain the segmentation loss $\mathcal{L}_{seg}$ and grading loss $\mathcal{L}_{seg}$ to optimize the segmentation network and grading head.

Then we forward X_T to the comprehensive diagnosis module G:

$$P_T^{seg}, P_T^{gra}, P_T^{cls} = G(X_T) \tag{2}$$

The P_T^{seg} is the target domain segmentation prediction, the P_T^{gra} is the target domain grading prediction, and the P_T^{cls} is the target domain classification prediction. The Y_S^{cls} is used to obtain the classification loss $\mathcal{L}_{cls}$ to optimize the classification head. At the same time, due to the adversarial loss $\mathcal{L}_{adv}$ using P_T^{seg} propagates the gradient from D to G, which will make G generate a similar segmentation distribution in the target domain as the source domain. We aim to make P_T^{seg} and P_S^{seg} close to each other, so we use these two predictions as inputs to D to distinguish whether the input is from the source or target domain.

Therefore, the training objective $\mathcal{L}_{CDM}$ of the comprehensive diagnosis module G used can be denoted as:

$$\mathcal{L}_{CDM} = \lambda_0 \mathcal{L}_{seg} + \lambda_1 \mathcal{L}_{gra} + \lambda_2 \mathcal{L}_{cls} + \lambda_3 \mathcal{L}_{adv} \tag{3}$$

Among them, $\vec{\lambda}$ denotes the weight of each class of loss, mainly based on the difficulty and importance of the tasks. In the experiment, we set $\vec{\lambda} = [\lambda_0, \lambda_1, \lambda_2, \lambda_3] = [3, 0.1, 0.5, 0.5]$, optimization principle for $\underset{D}{max}\underset{G}{min}L_{CDM}$. The ultimate goal is to minimize the segmentation and classification of the loss of G in the source image, minimize the classification loss of the target image, and maximize the probability that the target image segmentation output is regarded as the source image segmentation output simultaneously. Our optimization strategy is described in Alg. 1.

Algorithm 1 *A CDMA Framework for Brain Tumor Diagnosis.*

1 : **Input** :*Batch pair of source and target*$(\{X_S, Y_S^{seg}, Y_S^{gra}\}_i, \{X_T, Y_T^{cls}\}_i)_{i=1}^{i_{max}}$

2 : **Input** :*Comprehensive diagnosis module* **G** ← *random parameters,*
      Discriminator **D** ← *random parameters*

3 : **Output** :**G**, **D**

4 : **Hyperparameters** :*source_label* ← *tensor*(1). *fill_*(0), *target_label* ← *tensor*(1). *fill_*(1),

$$i_{max} \leftarrow 10000, i_{delay} \leftarrow 1000, \vec{\lambda} = [\lambda_0, \lambda_1, \lambda_2, \lambda_3] \leftarrow [3, 0.1, 0.5, 0.5]$$

5 : **for** $i = 0$ *to* i_{max} **do**

6 :   *make requires_grad of* **D**.*parameters*() *False*

7 : $P_{Si}^{seg}, P_{Si}^{gra} = \mathbf{G}(X_{Si})$

8 : $\mathcal{L}_{seg} = CrossEntropyLoss(P_{Si}^{seg}, Y_{Si}^{seg})$

9 : $\mathcal{L}_{gra} = CrossEntropyLoss(P_{Si}^{gra}, Y_{Si}^{gra})$

10 : $P_{Ti}^{seg}, P_{Ti}^{gra}, P_{Ti}^{cls} = \mathbf{G}(X_{Ti})$

11 : $\mathcal{L}_{adv} = MSELoss(\mathbf{D}(P_{Ti}^{seg}), source\_label. exapand\_as(\mathbf{D}(P_{Ti}^{seg})))$

12 :   **if** $i \geq i_{delay}$ **then**

13 : $\mathcal{L}_{cls} = CrossEntropyLoss(P_{Ti}^{cls}, Y_{Ti}^{cls})$

14 : $\mathcal{L}_{CDM} = \lambda_0\mathcal{L}_{seg} + \lambda_1\mathcal{L}_{gra} + \lambda_2\mathcal{L}_{cls} + \lambda_3\mathcal{L}_{adv}$

15 :   **else**

16 : $\mathcal{L}_{CDM} = \lambda_0\mathcal{L}_{seg} + \lambda_1\mathcal{L}_{gra} + \lambda_3\mathcal{L}_{adv}$

17 :   **end if**

18 : $\mathcal{L}_{CDM}.backward()$

19 : $\mathbf{G} \leftarrow \mathbf{G}'$

20 :   *make requires_grad of* **D**.*parameters*() *True*

21 : $\mathcal{L}_D = 0.5 * MSELoss(\mathbf{D}(P_{Si}^{seg}. detach()), source\_label. exapand\_as(\mathbf{D}(P_{Si}^{seg}. detach())))$
         $+0.5 * MSELoss(\mathbf{D}(P_{Ti}^{seg}. detach()), target\_label. exapand\_as(\mathbf{D}(P_{Ti}^{seg}. detach())))$

22 : $\mathcal{L}_D.backward()$

23 : $\mathbf{D} \leftarrow \mathbf{D}'$

24 : **end for**

Discriminator Loss. For given segmentation prediction $P_{(C)}^{seg}$, where C represents the image from the source or target domain, and we forward $P_{(C)}^{seg}$ to D and use the loss of mean-square error to obtain the discriminator training objective $\mathcal{L}_D$, where source domain label 0, target domain label 1, and the total number of pixels n:

$$\mathcal{L}_D(P_T^{seg}, P_S^{seg}) = 0.5 * \frac{1}{n}\sum_{i=0}^{n}(D(P_T^{seg})_i - 1)^2 + 0.5 * \frac{1}{n}\sum_{i=0}^{n}(D(P_S^{seg})_i - 0)^2$$

(4)

Comprehensive Diagnosis Module Objective Function. First, we use the cross entropy loss to obtain the losses of three tasks: the segmentation loss $\mathcal{L}_{seg}$ between P_S^{seg} and Y_S^{seg}, the grading loss $\mathcal{L}_{gra}$ between P_S^{gra} and Y_S^{gra}, and the classification loss $\mathcal{L}_{cls}$ between $P_T{}^{cls}$ and $Y_T{}^{cls}$. The loss functions for $\mathcal{L}_{seg}$, $\mathcal{L}_{gra}$, and $\mathcal{L}_{cls}$ are as follows:

$$\mathcal{L}(P, Y) = -\frac{1}{n}\sum_{i=0}^{n}\left[Y_i log(P_i) + (1 - Y_i)log(1 - P_i)\right] \tag{5}$$

Finally, in order to make the distribution of $P_T{}^{seg}$ closer to $P_S{}^{seg}$, we forward $P_T{}^{seg}$ to D and obtain the adversarial loss $\mathcal{L}_{adv}$ using the loss of mean-square error, where source domain label 0:

$$\mathcal{L}_{adv}\left(P_T{}^{seg}\right) = \frac{1}{n}\sum_{i=0}^{n}\left(D\left(P_T{}^{seg}\right)_i - 0\right)^2 \tag{6}$$

4   Experiments

We used a public dataset (source domain) and two local hospital datasets (target domain) to demonstrate the superiority of the proposed CMDA method on multiple diagnostic tasks of brain tumor segmentation, grading, genotyping, and glioma subtype classification. Finally, ablation experiments show that adversarial learning in CMDA can effectively relieve the data distribution gap between the two domains.

4.1   Experimental Setup

Datasets. Three datasets are used for the experiments: the public dataset BraTS2019, the hospital glioma dataset from Zhengzhou (BrainZ), and the hospital glioma dataset from Fuzhou (BrainF). **BraTS2019** contains 335 subjects in total, and each subject consists of four different modalities of MR scans, T1, T2, T1-enhanced (T1ce), T2 with FluidAttenuated Inversion Recovery (Flair), the segmentation mask segmented manually, and the graded labels(high-grade or low-grade). There are a total of 200 cases of the same multi-modal glioma MRI in the **BrainZ** dataset from Zhengzhou with IDH genotype classification labels and Ki67 genotype classification labels (89 cases of IDH mutant and 101 cases of wild type; 109 cases with high Ki67 positive rate and 91 cases with low Ki67 positive rate). There are a total of 100 cases of the same multi-modal glioma MRI in the **BrainF** dataset from Fuzhou with glioma subtype classification labels (8 cases of oligodendrogliomas, 70 cases of glioblastomas, and 22 cases of astrocytomas). The images of the four modalities are stitched in a cross-sectional manner to convert the 3D images into 4-channel 2D images. Due to the large slice gap of hospital data, in this experiment, the multi-modal data is randomly divided according to the ratio of 8:1:1 for training, validation, and testing in 2D. In this work, two local datasets were collected in collaboration with two hospitals. We've performed some processing on the data, including registration, skull stripping, abnormal data removal, etc.

Implementation Details. Our approach is implemented using the open-source framework PyTorch. The experiments were accelerated by an Nvidia GeForce RTX 3090 GPU with 24 GB memory for six hours of training and trained end-to-end. During fine-tuning, the comprehensive diagnosis module is fine-tuned with the target domain validation set, and all layers except the output layer, grading head, and classification head of the comprehensive diagnosis module are frozen. During training, the network parameters are randomly initialized, the Adam optimizer is used, the initial learning rate is set to 2e-3, the batch size is set to 64, and the polynomial decay with the power of 0.99 mentioned in [20] is used. The model is trained over 10000 iters. The training of the classification head starts after 1000 iters. We use standard evaluation metrics to evaluate the outcome, including Dice coefficient score, accuracy (ACC), positive predictive value (PPV), true positive rate (TPR), and F1-score.

4.2 Experimental Results

Result of BrainZ Dataset. We test the performance of the proposed CMDA method by using BraTS2019 dataset with expert segmentation and grading labels (source domain) and using the BrainZ dataset with IDH genotype and Ki67 genotype classification labels (target domain). The best performance of segmentation and grading is obtained by supervised training in the comprehensive diagnostic module with expert labels. The performance of segmentation and grading is compared with the following networks, such as Attention UNet [21] as a segmentation network with grading head (AUNetGraHd) network, UNet [22] as a segmentation network with grading head (UNetGraHd) network. AdaptSegNet [13], ADVENT [11], and FADA [12]. The results are shown in Table 1.

Table 1. Comparison of average results of different models on BrainZ dataset.

Methods	Segmentation			Grading	
	Dice(%)	PPV(%)	TPR(%)	ACC(%)	F1-score(%)
Expert Supervised	**87.73**	**88.22**	**89.39**	88.26	93.62
CMDA	**85.42**	**87.17**	**86.27**	**90.89**	**94.73**
AUNetGraHd	74.15	75.50	82.22	87.11	93.11
UNetGraHd	81.95	83.23	84.28	78.59	87.79
AdaptSegNet	80.20	81.69	83.32	–	–
ADVENT	80.17	79.28	85.98	–	–
FADA	79.96	82.32	82.35	–	–

Compared with the AUNetGraHd network and UNetGraHd network, our method improves Dice by 11.27% and 3.47%, and the ACC of grading reaches 90.89% while that of the UNetGraHd is 78.59%. This indicates that the proposed CMDA method significantly relieves the distribution gap between the two domains. At the same time, it is observed that the Dice of segmentation of the proposed CMDA method is the closest to the expert labels supervised training. In addition, the ACC and F1-score of grading are increased by 2.63% and 1.11%. This is mainly attributed to our CMDA method using adversarial learning to extract features of source domain images, which enhances the robustness of network transfer learning and thus improves the classification effect on the target domain. Compared with AdaptSegNet, ADVENT, and FADA, the proposed CMDA method improves Dice by 5.22%, 5.25%, and 5.46%. As seen from the segmentation results in Fig. 3, our proposed method can accurately segment the tumor ROI and preserve the shape of the tumor ROI intact.

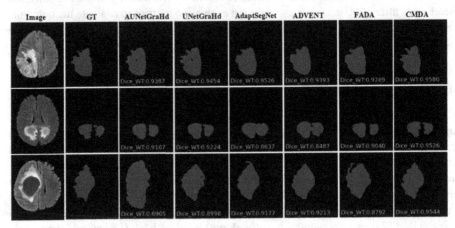

Fig. 3. Visualization of segmentation results of different networks

We take the classification head trained on the tumor ROI generated by the expert segmentation as the best performance baseline and compare the performance of our method with the tumor ROI generated by the AUNetGraHd and the UNetGraHd as the input of the classification head. The results are shown in Table 2.

Table 2. Classification performance evaluation of different methods on the BrainZ dataset.

Methods	IDH		Ki67	
	ACC(%)	F1-score(%)	ACC(%)	F1-score(%)
Expert ROI	**99.56**	**99.50**	**98.22**	**98.54**
CMDA	**95.22**	**94.68**	**94.19**	**95.24**
AUNetGraHd ROI	90.33	89.86	92.48	93.63
UNetGraHd ROI	93.48	92.72	92.15	93.38

The ACC of IDH and Ki67 genotype classification of our CMDA model is 95.22% and 94.19%, significantly improved compared with the results of the other two comparison methods. Compared with other methods, our CMDA method also leads in the F1 metric. In addition, all metrics of our CMDA approach on both IDH and Ki67 genotype classification tasks were closest to expert ROI baselines.

Result of BrainF Dataset. We test the performance of the proposed CMDA method by using BraTS2019 dataset with expert segmentation and grading labels (source domain)

and using the BrainF dataset with glioma subtype classification labels (target domain). The best performance of segmentation and grading is obtained by supervised training in the comprehensive diagnostic module with expert labels, and the best performance of classification is obtained by training the classification head with the tumor ROI generated by the expert segmentation. The performance of segmentation and grading is compared with the following networks: the UNetGraHd network, the AdaptSegNet [13], and FADA [12]. The classification performance is compared with training the classification head with the tumor ROI generated by these networks. The results are shown in Table 3.

Table 3. Comparison of average results of different methods on the BrainF dataset.

Methods	Segmentation		
	Dice(%)	PPV(%)	TPR(%)
Expert supervised	**81.38**	**82.68**	**82.97**
CMDA	**78.75**	**83.25**	79.06
UNetGraHd	69.37	69.19	76.39
AdaptSegNet	72.73	75.02	77.78
FADA	73.43	72.36	**81.63**
Methods	Grading		Classification
	ACC(%)	F1-score(%)	ACC(%)
Expert supervised	**94.68**	**97.08**	80.85
CMDA	**84.04**	**90.32**	**82.98**
UNetGraHd	75.53	85.71	75.53

Compared with UNetGraHd, AdaptSegNet, and FADA, the segmentation Dice of our CMDA model is improved by 9.38%, 6.02%, and 5.32%. At the same time, we observe that the segmentation TPR is decreased by 2.57% compared with FADA. Compared with the baseline of expert ROI supervised training, our Dice and TPR indicators of the segmentation results, as well as the grading ACC and F1-score indicators, are the closest to the expert ROI method, and our method also improves the ACC of the classification results by 2.13%, which is mainly due to the adversarial learning of the proposed CMDA method to extract the features of the source domain image and enhance the robustness of the network, so the classification effect on the target domain is improved. Our CMDA model has greatly improved with the UNetGraHd network in all segmentation, grading, and classification metrics.

4.3 Ablation Study

The ablation experiments were completed using the BraTS2019 dataset (source domain) and BrainZ dataset (target domain). A further comparison is made in our CMDA model. The two methods of changing the position of the classification head of CMDA at the end of the segmentation output (CMDA*) and removing the discriminator of CMDA (CMDA**) are compared with the proposed CMDA method, and the results are shown in Table 4.

Table 4. Impact of using different training structures on BrainZ.

Methods	Segmentation			Grading	
	Dice(%)	PPV(%)	TPR(%)	ACC(%)	F1-score(%)
CMDA	**85.42**	**87.17**	**86.27**	**90.89**	**94.73**
CMDA*	83.63	85.27	86.08	88.63	93.24
CMDA**	69.99	65.50	85.69	87.00	93.05
Methods	IDH Genotyping			Ki67 Genotyping	
	ACC(%)	F1-score(%)		ACC(%)	F1-score(%)
CMDA	**95.22**	**94.68**		**94.19**	**95.24**
CMDA*	92.89	91.93		92.30	93.67

Compared with CMDA*, CMDA performs better in segmentation, grading, and genotype classification. The grading head located at the end of the encoder is better than that of the segmentation output, which can help the segmentation network extract features better to improve the effect of the three tasks. At the same time, compared with CMDA**, the proposed CMDA method improves the Dice of segmentation and the ACC of graded by 15.43% and 3.89%. This shows that the data distribution gap between the two domains exists and that adversarial learning can effectively relieve it, so unsupervised segmentation and grading in the target domain can perform better.

5 Conclusion

In this paper, we propose a novel multi-modal domain adaptive comprehensive aid brain tumor diagnosis framework for the tasks of segmentation, grading, and classification of genotype or glioma subtype without expert segmentation and grading labels. Our approach employs adversarial learning to narrow the distribution between different datasets. The experimental results show that our proposed framework outperforms the methods of direct transfer learning and other unsupervised methods.

The unsupervised domain adaptation framework provides a feasible way to predict the prognosis of brain tumors, which is of great clinical interest for datasets from hospitals that lack annotated labels. Follow-up work will further expand our diagnostic model in the heart, abdomen, and other organs to achieve systemic diagnosis.

Acknowledgments. This work was supported in part by the National Key R&D Program of China under Grant 2022YFC2402102, and in part by the National Natural Science Foundation of China under grant numbers 82071913 and 22161142024.

References

1. Parsons, D.W.: An integrated genomic analysis of human glioblastoma multiforme. Science **321**(5897), 1807–1812 (2008)
2. Oronsky, B.: A review of newly diagnosed glioblastoma. Front. Oncol. **10**, 574012 (2021)
3. Komori, T.: Grading of adult diffuse gliomas according to the 2021 WHO Classification of Tumors of the Central Nervous System. Lab. Invest. **102**(2), 126–133 (2022)
4. Menze, B.H.: The multimodal brain tumor image segmentation benchmark (BRATS). IEEE Trans. Med. Imaging **34**(10), 1993–2024 (2014)
5. Fu, J.: Dual attention network for scene segmentation. In: Proceedings of the IEEE/CVF Conference on Computer Vision and Pattern Recognition, pp. 3146–3154 (2019)
6. Menze, B.H.: A generative model for brain tumor segmentation in multi-modal images. In: Jiang, T., Navab, N., Pluim, J.P.W., Viergever, M.A. (eds.) Medical Image Computing and Computer-Assisted Intervention – MICCAI 2010. MICCAI 2010. LNCS, vol. 6362. Springer, Berlin, Heidelberg (2010). https://doi.org/10.1007/978-3-642-15745-5_19
7. Havaei, M.: Brain tumor segmentation with deep neural networks. Med. Image Anal. **35**, 18–31 (2017)
8. Long, J.: Fully convolutional networks for semantic segmentation. In: Proceedings of the IEEE Conference on Computer Vision and Pattern Recognition, pp. 3431–3440. (2015)
9. Dong, H.: Automatic brain tumor detection and segmentation using U-Net based fully convolutional networks. In: Hernández, M.V., González-Castro, V. (eds.) MIUA 2017. CCIS, vol. 723, pp. 506–517. Springer, Cham (2017). https://doi.org/10.1007/978-3-319-60964-5_44
10. Cui, S.: Automatic semantic segmentation of brain gliomas from MRI images using a deep cascaded neural network. J. Healthcare Eng. **2018** (2018)
11. Vu, T.H.: Advent: adversarial entropy minimization for domain adaptation in semantic segmentation. In: Proceedings of the IEEE/CVF Conference on Computer Vision and Pattern Recognition, pp. 2517–2526. (2019)
12. Wang, H.: Classes matter: a fine-grained adversarial approach to cross-domain semantic segmentation. In: Vedaldi, A., Bischof, H., Brox, T., Frahm, J.-M. (eds.) ECCV 2020. LNCS, vol. 12359, pp. 642–659. Springer, Cham (2020). https://doi.org/10.1007/978-3-030-58568-6_38
13. Tsai, Y.H.: Learning to adapt structured output space for semantic segmentation. In: Proceedings of the IEEE Conference on Computer Vision and Pattern Recognition, pp. 7472–7481. (2018)
14. Zhang, Y.: Task driven generative modeling for unsupervised domain adaptation: application to X-ray image segmentation. In: Frangi, A.F., Schnabel, J.A., Davatzikos, C., Alberola-López, C., Fichtinger, G. (eds.) MICCAI 2018. LNCS, vol. 11071, pp. 599–607. Springer, Cham (2018). https://doi.org/10.1007/978-3-030-00934-2_67
15. Chen, C.: Deep learning for cardiac image segmentation: a review. Front. Cardiovasc. Med. **7**, 25 (2020)
16. Chen, C.: Synergistic image and feature adaptation: towards cross-modality domain adaptation for medical image segmentation. In: Proceedings of the AAAI Conference on Artificial Intelligence, pp. 865–872 (2019)
17. He, K.: Cross-Modality Brain Tumor Segmentation via Bidirectional Global-to-Local Unsupervised Domain Adaptation. arXiv preprint arXiv:2105.07715 (2021)
18. Zhou, Z.: Unet++: redesigning skip connections to exploit multiscale features in image segmentation. IEEE Trans. Med. Imaging **39**(6), 1856–1867 (2019)
19. Zhu, J.Y.: Unpaired image-to-image translation using cycle-consistent adversarial networks. In: Proceedings of the IEEE International Conference on Computer Vision, pp. 2223–2232 (2017)

20. Chen, L.C.: Deeplab: semantic image segmentation with deep convolutional nets, atrous convolution, and fully connected crfs. IEEE Trans. Pattern Anal. Mach. Intell. **40**(4), 834–848 (2017)
21. Oktay, O.: Attention U-Net: Learning Where to Look for the Pancreas. arXiv preprint arXiv: 1804.03999 (2018)
22. Ronneberger, O.: U-Net: convolutional networks for biomedical image segmentation. In: Navab, N., Hornegger, J., Wells, W., Frangi, A. (eds.) Medical Image Computing and Computer-Assisted Intervention – MICCAI 2015. MICCAI 2015. LNCS, vol. 9351. Springer, Cham (2015). https://doi.org/10.1007/978-3-319-24574-4_28

Joint Boundary-Enhanced and Topology-Preserving Dual-Path Network for Retinal Layer Segmentation in OCT Images with Pigment Epithelial Detachment

Xiaoming Liu$^{(\boxtimes)}$ (ID) and Xiao Li (ID)

School of Computer Science and Technology, Wuhan University of Science and Technology, Wuhan 430065, China
lxmspace@gmail.com

Abstract. Automatic retinal layer segmentation methods are currently successful in normal Optical Coherence Tomography (OCT) images, but they face great challenges for eyes with Pigment Epithelial Detachment (PED), where the morphology and structure of the retina change dramatically. Therefore, we propose a novel dual-path network that uses residual blocks as the encoders, and semantic path and boundary paths for layer segmentation and boundary regression, respectively. Specifically, to capture the shape and boundary information of objects accurately, we design a Boundary-Enhanced Global Attention (BEGA) module for semantic path to provide global guidance for high-level features. Additionally, we use a Gradient-Guided Spatial Attention (GGSA) module in the boundary path to acquire boundary features, alleviating interference in non-boundary regions. Finally, the segmentation probability map and the distance map output by the network are fed into the Topology Correction (TC) module to obtain the topology-guaranteed results. We investigate the proposed method on the OCTA-500 dataset and the experimental results prove that the proposed method outperformed the other four existing methods.

Keywords: OCT · retinal layer segmentation · global attention · Distance map · topology-guarantee

1 Introduction

Changes in retinal layer thickness are a key indicator for ophthalmologists to assess the severity and progression of retinal diseases [1], such as glaucoma, Diabetic Retinopathy (DR) and Age-related Macular Degeneration (AMD). OCT is a novel technique for high-resolution, non-invasive laminar and biomicroscopic imaging that allows visualization of three-dimensional cross-sections of human tissue at micron resolution, providing clear images of the structure of the retinal layers [2]. To quantify retinal OCT images, ophthalmologists need to segment retinal layer tissue and then diagnose disease progression based on its thickness

© The Author(s), under exclusive license to Springer Nature Singapore Pte Ltd. 2024
Q. Liu et al. (Eds.): PRCV 2023, LNCS 14437, pp. 395–406, 2024.
https://doi.org/10.1007/978-981-99-8558-6_33

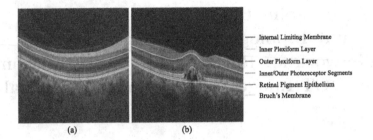

(a)   (b)

Fig. 1. (a) is a healthy OCT B-scan and its annotation, (b) is an OCT B-scan image with PED and its annotation.

changes. However, manually annotating retinal layers is tedious and subjective. Therefore, the development of an objective and accurate retinal layer segmentation tool is essential.

Research on automatic layer segmentation of retinal OCT images emerged early. Early studies mainly included traditional learning methods such as level sets [3], Markov Random Fields (MRF) [4] and graph theory [5] as well as machine learning methods [6,7]. While these methods have greatly advanced the field, traditional methods, which mostly rely on prior knowledge or manual features obtained by complex processing, are less robust and may be difficult to generalize. Machine learning methods are more robust than traditional methods. However, these methods usually rely on pixel-by-pixel classifiers, which are very inefficient. In addition, there is a need for hand-crafted features or parameter tuning, which greatly limits the generalization to other datasets. Recently, deep learning has shown excellent performance in medical images [8,9]. Various studies have implemented deep learning-based retinal layer segmentation techniques [10–12] and have achieved better results than classical methods. However, although these methods achieve superior performance in some cases, most of them exhibit degraded performance when segmenting images in the presence of severe pathological abnormalities. As shown in Fig. 1, retinas with large PED pose the following two challenges to these methods compared to healthy retinas:

- The PED is accompanied by massive fluid accumulation in the retinal layers, which results in huge morphological and structural changes in the retinal layers.
- The retina is a strictly layered biological tissue [13], and traditional methods segment it pixel-wise without considering its topology.

To tackle these challenges, we present a new layer segmentation network that uses two parallel decoders for simultaneous segmentation and regression. Specifically, we design a series of Boundary-Enhanced Global Attention (BEGA) modules in the semantic path to gradually integrate features of different scales and the boundary path. This module not only integrates neighborhood features more precisely, but also strengthens the boundary shape awareness of the semantic path. In addition, the boundary path encodes the distance information of layer boundaries through a set of Gradient-Guided Spatial Attention (GGSA) modules, which use the gradient of the semantic output as a spatial attention guide to highlight the boundary information while suppressing responses from non-boundary regions. Finally, we have designed a topology correction module (TC) to obtain segmentation results with topology guarantees. The experimental results show the effectiveness of the proposed method.

2 Method

Our method mainly consists of three parts: (1) ResBlock as the encoder, (2) BEGA module and GGSA module for layer segmentation and boundary regression respectively, (3) TC module that corrects the topological errors of the predicted map output by the semantic path. The specific process of the method is as follows: we first input a B-scan image into the network and generate the probability map $\hat{P}$ and the boundary distance map $\hat{D}$, respectively. Considering the morphological and structural changes produced by lesions, we design the BEGA modules and the GGSA modules for the two decoding paths respectively to enhance the boundary shape perception of the semantic path. Finally, we feed the output $\hat{P}$ and $\hat{D}$ of the network together into the TC module to generate the final prediction map. It is worth noting that the TC module is only used in the inference phase.

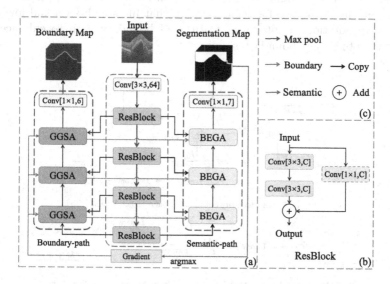

Fig. 2. Framework of the proposed method (a) and (b) Detailed structure of the network and the ResBlock, Where Conv [3 × 3, C] represents a 3 × 3 convolution with C output channels.

2.1   Network Structure

As shown in Fig. 2(a), our network consists of a common encoding path and two decoding paths, defined as the semantic path and the boundary path, respectively. The encoding path extracts high-level features from the input OCT B-scan through four consecutive residual blocks (see Fig. 2(b)) and applies max-pooling to reduce the feature size. Then, we feed these representative features into two decoding paths to obtain task-relevant features. Specifically, the semantic path utilizes channel and spatial attention mechanisms to achieve feature complementation and enhancement at each level. It also fuses the features of the boundary path to enhance the perception of boundary shapes, thus improving the segmentation performance. The boundary path gradually refines the boundary with the GGSA module. It introduces the gradient of semantic segmentation as attention guidance, effectively enhances object boundary features and suppresses non-boundary features.

2.2   Boundary Enhanced Global Attention (BEGA) Module

Recently, attention mechanisms have been widely used in semantic segmentation [12,14], which can be designed in the decoder or embedded in the backbone. For retinal layer segmentation aiming at recognizing layer structure, it's not only necessary to extract rich semantic and spatial information from retinal images but also to keep the boundaries clear and continuous. To this end, we design a global attention module for semantic path that fuses high-level and low-level features. In addition, it integrates features of boundary path to enhance the perception of network boundary shapes.

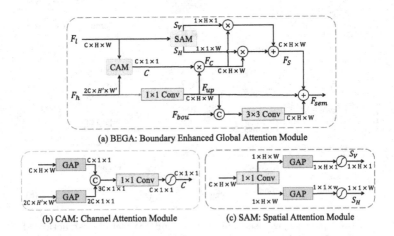

(a) BEGA: Boundary Enhanced Global Attention Module

(b) CAM: Channel Attention Module   (c) SAM: Spatial Attention Module

Fig. 3. (a) Detailed structure of the BEGA, where F_l and F_h represent high-resolution features from the encoder and low-resolution features from the previous level of decoder respectively, and the red arrows represent up-sampling. (b) and (c) respectively represent the CAM and SAM used in (a).

As shown in Fig. 3, the BEGA module consists of a Spatial Attention Module (SAM) and a Channel Attention Module (CAM). SAM recodes the weight of each pixel for low-level features and utilizes the pooling to catch the distance dependence between pixels in two different directions. In addition, to better utilize the channel information at each level, CAM weights the channels of different levels of features to strengthen important feature channels.

CAM: As shown in Fig. 3(b), we first use a global average pooling (GAP) for low-level features $F_l \in \mathbb{R}^{C \times H \times W}$ and high-level features $F_h \in \mathbb{R}^{2C \times H' \times W'}$ respectively, to extract global information. Then, we concatenate these two features together and fuse features through a 1×1 convolution. Finally, we use a sigmoid function $\sigma(\cdot)$ to generate a channel attention map $C \in \mathbb{R}^{C \times 1 \times 1}$:

$$C = \sigma \left(Conv \left(concat \left[pool \left(F_l \right), pool \left(F_h \right) \right] \right) \right) \tag{1}$$

where $Conv(\cdot)$ denotes a convolution operation, $concat[\cdot]$ denotes concatenation in the channel dimension, and $pool(\cdot)$ denotes global average pooling.

SAM: Low-level features can offer finer details for better target recognition. Therefore, as shown in Fig. 3(c), the low-level feature F_l is convolved by 1×1 to produce the attention map. Then, two GAPs are used to encode the pixel-wise long-range dependencies of the attention map from the horizontal and vertical directions, respectively. Finally, two spatial attention maps $S_V \in \mathbb{R}^{1 \times H \times 1}$ and $S_H \in \mathbb{R}^{1 \times 1 \times W}$ are generated by the sigmoid function:

$$S_V = \sigma \left(pool \left(Conv \left(F_l \right) \right) \right), S_H = \sigma \left(pool \left(Conv \left(F_l \right) \right) \right) \tag{2}$$

Feature Integration: We first upsample F_h to match the size of F_l, and then use a 1×1 convolution to reduce its dimension to obtain a feature map $F_{up} \in \mathbb{R}^{C \times H \times W}$. Finally, F_{up} is reweighted by C to obtain a channel attention enhancement feature map F_C:

$$F_C = C \otimes F_{up} \tag{3}$$

where $\otimes$ denotes element-wise manipulation.

Next, F_C is reweighted by S_V and S_H respectively, and then added to obtain the spatial attention enhanced feature $F_S \in \mathbb{R}^{C \times H \times W}$:

$$F_S = \left(S_V \odot F_C \right) \oplus \left(S_H \odot F_C \right) \tag{4}$$

where $\oplus$ and $\odot$ denotes pixel-wise addition and row or column-wise multiplication.

Furthermore, the boundary and shape information of the distance map can well compensate for the shape information lost by down-sampling [15]. Therefore, we use a residual mechanism for F_{up} to learn attention features by integrating the features F_{bou} of the boundary path. Finally, perform residual connection with F_S to obtain F_{sem}:

$$F_{sem} = Conv\left(concat\left[F_{up}, F_{bou}\right]\right) \oplus F_{up} \oplus F_S \tag{5}$$

where F_{sem} denotes output of BEGA module.

2.3   Gradient-Guided Spatial Attention (GGSA) Module

Although previous work [12,16] proposes to utilize boundary information to assist layer segmentation, these methods are usually disturbed by non-boundary parts, which may not be conducive to identify object boundaries accurately. In addition, the encoder produces many redundant features, while object edges can be identified with few channels. Based on this, as shown in Fig. 4, we design a set of light-weight GGSA modules for the boundary path to deal with the above problems.

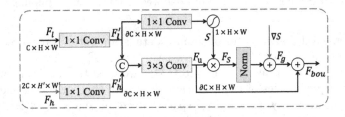

Fig. 4. The detailed structure of the GGSA module, where the red arrows represent up-sampling. (Color figure online)

As shown in Fig. 4, a 1×1 convolution is first used for $F_l \in \mathbb{R}^{C \times H \times W}$ and $F_h \in \mathbb{R}^{2C \times H' \times W'}$ respectively, to obtain two related features $F_l' \in \mathbb{R}^{\partial C \times H \times W}$ and $F_h' \in \mathbb{R}^{\partial C \times H \times W}$ with same channel number, where the scaling factor $\partial \in (0, 1]$ is used to control the feature channel (in this paper, $\partial = \frac{1}{16}$). Like the BEGA module, we use F_l' to generate the spatial attention map. Specifically, we input F_l' into a 1×1 convolution and a sigmoid function to generate the spatial attention map $S \in \mathbb{R}^{1 \times H \times W}$. Then, we concatenate F_l' and F_h' together and fuse by a 3×3 convolution to generate an updated feature $F_u \in \mathbb{R}^{\partial C \times H \times W}$. Finally, we use S to weight F_u to obtain the spatial attention enhanced feature F_s:

$$F_S = S \otimes F_u \tag{6}$$

To highlight boundary pixels while suppressing non-boundary regions, we use the segmentation results of the semantic path as an auxiliary. Specifically, we compute the gradient map $\nabla S \in \mathbb{R}^{1 \times H \times W}$ of the segmentation result, and then add it to each channel of F_S. It is worth noting that since the eigenvalues of

F_S have no fixed range, it may not be appropriate to add them directly to ∇S. Therefore, we normalize F_S to range from 0 to 1, consistent with ∇S:

$$F_g = Norm\,(F_S) \oplus \nabla S \tag{7}$$

Finally, we add F_u and F_g to help end-to-end training:

$$F_{bou} = F_u \oplus F_g \tag{8}$$

where F_{bou} represents the output of the GGSA module.

2.4 Topology Correction (TC) Module

During the inference phase, there are two outputs produced by the proposed network: the segmentation probability map and the normalized boundary distance map. To guarantee the topological order in the segmentation prediction map, we use a topology correction algorithm from [12] to iteratively correct the topological errors on each layer. The following briefly describes the steps of the algorithm: first, the initial prediction result $\hat{L}$ is obtained according to the segmentation probability map $\hat{P}$. Then, the distance map $\hat{D}$ corresponding to each boundary is calculated according to Eq. (10). Next, for each boundary, two masks are generated according to its corresponding distance map $\hat{D}$ and the threshold ϵ, which represent the areas above and below that boundary in $\hat{L}$ that need to be corrected, respectively. Finally, the category value of the corresponding region in $\hat{L}$ is corrected according to the mask to obtain the corrected prediction result L_c. Where the threshold ϵ is used to control the influence of the distance map $\hat{L}$ on $\hat{P}$. When $\epsilon = 0$, the final prediction structure is completely determined by the distance map, and when $\epsilon = 1$, $\hat{D}$ has no influence on $\hat{P}$. In our method, the threshold ϵ is set to 0.2.

2.5 Loss Function

To train the semantic path of the segmentation network, we defined a joint segmentation loss for the probability map $\hat{P}$, which includes a Cross-Entropy (CE) loss and a Dice loss. The loss is defined as follows:

$$\mathcal{L}_{seg} = -\frac{1}{H \times W} \sum_{i=1}^{H} \sum_{j=1}^{W} \sum_{c=1}^{C} \left(P_{i,j,c} log \hat{P}_{i,j,c} + \frac{2 P_{i,j,c} \hat{P}_{i,j,c}}{P_{i,j,c}{}^2 + \hat{P}_{i,j,c}{}^2} \right) \tag{9}$$

where $\hat{P}_{i,j,c}$ and $P_{i,j,c}$ represent the probability value and the ground truth, respectively.

For the boundary path, we consider multiple boundaries in a B-scan as target objects, where the ground truth distance map for each boundary is constructed from the corresponding boundary set:

$$D_C(x) = \begin{cases} \min\left(\inf_{y \in S_C} \|x - y\|_2 \right), & \text{if } x \in \Omega_U \\[2mm] -\min\left(\inf_{y \in S_C} \|x - y\|_2 \right), & \text{if } x \in \Omega_L \end{cases} \tag{10}$$

where Ω_U and Ω_L denote the pixels above and below the cth boundary S_c, respectively. $||x - y||_2$ denotes the Euclidean distance between x and y. We normalize the distance map using the tanh function so that pixels close to the border get more attention.

Then we use the Mean Square Error (MSE) loss to train the boundary path. The formula for the MSE loss is defined as follows:

$$L_{reg} = \frac{1}{H \times W} \sum_{i=1}^{H} \sum_{j=1}^{W} \sum_{b=1}^{B} \left(D_{i,j,b} - \hat{D}_{i,j,b} \right)^2 \tag{11}$$

where $D_{i,j,b}$ and $\hat{D}_{i,j,b}$ represent the ground truth and the predicted distance value of the pixel with coordinates (i,j) in b^{th} distance map, respectively.

Finally, the overall loss of our method is:

$$\mathcal{L}_{All} = \mathcal{L}_{seg} + \mathcal{L}_{reg} \tag{12}$$

3 Experiment

3.1 Dataset

We use the OCTA-500 [17] dataset to evaluate our method, which consists of retinal OCT volume from 30 PED patients. Each patient has 400 macular OCT B-scan images with a resolution of 640 × 400. We selected 10 severely distorted B-scan for each patient and annotated them with 5 retinal layer regions by an expert ophthalmologist. For data preprocessing, we crop each B-scan to 400×400 to eliminate most of the background and resize it to 512 × 512 for training convenience. Following [17], we randomly split the dataset into training (18 B-scans), validation (20 B-scans), and test (100 B-scans) sets.

3.2 Comparison Methods and Evaluation Metric

We compare our method with four existing methods on the OCTA-500 dataset to demonstrate its effectiveness: BRU-Net [18], Relay-Net [10], U-Net++ [19] and BAU-Net [20]. We use the Dice coefficient to measure the accuracy of our region segmentation. Additionally, we also use the mean absolute deviation (MAD) and the standard deviation (Std) to evaluate the precision and stability of the layer boundaries.

3.3 Result

The results of our method and four other methods on the OCTA-500 dataset are shown in Fig. 5. Due to the sharp gradient changes on the ILM and less affected by lesions, all methods can achieve relatively robust results. However, the BM is severely damaged by the PED and is hard to segment. The deformation also makes other layers thinner than normal subjects, reducing other methods' performance. To overcome this limitation and enhance the effective receptive field, BRU-Net utilizes dilated convolutions with different dilation rates, but it ignores the layer boundary information, so it cannot ensure boundary precision and continuity. Relay-Net uses an unpooling layer to preserve the layer boundary's spatial information but has difficulty in segmenting the BM. U-Net++ is a widely used segmentation network, but it does not consider retina topology. BAU-Net improves boundary segmentation with some modules, and a topology loss is designed to supervise the topology. However, BAU-Net does not handle images with severe lesion deformation well because it does not consider the lesion features. As shown in Fig. 5, our method achieves the best results on the layer boundary details, because we use the BEGA module and GGSA module to alleviate the posed by lesions challenge and use distance map through the TC module to obtain topologically guaranteed results.

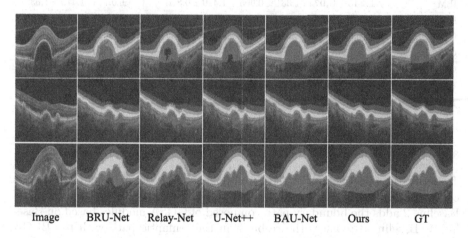

Image   BRU-Net   Relay-Net   U-Net++   BAU-Net   Ours   GT

Fig. 5. Segmentation results on three different B-scans. The regions of ILM~IPL, INL~OPL, ONL, IS~OS, and RPE~BM are labelled in red, orange, yellow, green, and blue, respectively. (Color figure online)

Table 1 lists the Dice value of all methods in each layer area and the average Dice value. Our method achieves the highest Dice values in most of the layer regions (except ILM~IPL). Especially in the most difficult segment RPE~BM, our method outperforms the suboptimal BAU-Net by about 1.5%, which shows that our method has advantages in segmenting images with PED. Table 2 shows the accuracy and stability of each boundary segmentation quantified by MAD and Std. All methods have lower MAD and Std at the ILM compared to the others due to its sharp gradient variation. For other boundaries, both MAD and Std of our method are lower than other methods, which indicates that our method obtains more accurate and stable segmentation results. The results in Table 1 and Table 2 show that our proposed method outperforms other methods in retinal layer segmentation.

Table 1. The Dice value of comparison methods.

Layer region	BRU-Net	Relay-Net	U-Net	BAU-Net	Ours
ILM~IPL	0.9726	**0.9746**	0.9636	0.9692	0.9716
INL~OPL	0.9261	0.9237	0.9077	0.9358	**0.9364**
ONL	0.9249	0.9218	0.9185	0.9227	**0.9257**
IS~OS	0.9045	0.9097	0.9046	0.9119	**0.9147**
RPE~BM	0.8806	0.8695	0.8835	0.9070	**0.9207**
Average	0.9217	0.9199	0.9156	0.9293	**0.9338**

Table 2. The MAD (μm) $\pm$ Std of comparison methods.

Layer boundary	BRU-Net	Relay-Net	U-Net++	BAU-Net	Ours
ILM	**1.246 $\pm$ 0.977**	1.331 $\pm$ 0.905	1.419 $\pm$ 0.876	1.656 $\pm$ 0.774	1.479 $\pm$ 1.032
IPL/INL	**2.097 $\pm$ 7.764**	3.952 $\pm$ 10.355	4.721 $\pm$ 10.195	3.729 $\pm$ 6.707	3.325 $\pm$ 3.347
OPL/ONL	4.163 $\pm$ 8.672	3.910 $\pm$ 9.771	4.468 $\pm$ 6.563	3.126 $\pm$ 5.681	**2.953 $\pm$ 3.539**
ONL/IS	4.312 $\pm$ 10.115	4.432 $\pm$ 11.662	5.223 $\pm$ 9.898	3.940 $\pm$ 9.302	**3.492 $\pm$ 3.215**
OS/RPE	3.548 $\pm$ 6.308	3.693 $\pm$ 6.335	4.576 $\pm$ 8.819	3.370 $\pm$ 6.087	**3.037 $\pm$ 3.673**
BM	4.956 $\pm$ 9.839	6.856 $\pm$ 8.449	5.743 $\pm$ 11.249	4.347 $\pm$ 8.418	**3.745 $\pm$ 4.337**
Average	3.387 $\pm$ 7.279	4.029 $\pm$ 7.913	4.358 $\pm$ 7.933	3.361 $\pm$ 6.162	**3.005 $\pm$ 3.191**

3.4 Ablation Experiment

To verify the effectiveness of these components. We set up four baselines for comparison: Baseline 1 is the Resnet that only includes layer segmentation tasks, Baseline 2 adds the boundary path composed of GGSA modules based on Baseline 1, Baseline 3 replaces the resblock in the semantic path with the BEGA module, and Baseline 4 is Baseline 3 with the topology correction module, which is our proposed method.

Table 3. The Dice value of comparison methods.

Baseline	Components			Layer regions					Average
	BEGA	GGSA	TC	ILM~IPL	INL~OPL	ONL	IS~OS	RPE~BM	
Baseline-1				0.9686	0.9137	0.9161	0.9037	0.8461	0.9096
Baseline-2	✓			0.9705	0.9278	0.9185	0.9078	0.8815	0.9212
Baseline-3	✓	✓		**0.9723**	0.9319	0.9214	0.9104	0.9153	0.9303
Baseline-4	✓	✓	✓	0.9716	**0.9364**	**0.9257**	**0.9147**	**0.9207**	**0.9338**

Table 3 shows the Dice similarity for each baseline. Baseline 2 integrates the boundary path based on Baseline 1, which improves the feature extraction ability of the network. In addition, we design a GGSA module for boundary path, suppressing interferences in non-boundary regions to better recognize boundaries. Compared with Baseline 1, Baseline 2 significantly improves in each layer area, especially in the RPE~BM area, which is about 4.2% higher than Baseline 1. Based on Baseline 2, Baseline 3 replaces the resblock of the semantic path with the BEGA module to utilize the global information and the boundary shape information of the boundary path. From the table, it is evident that Baseline 3 has varying degrees of performance improvement in each layer, and the overall Dice value is improved to 0.9303. To further modify the topological errors in the segmentation prediction map, we add a topology correction module to Baseline 3, which iteratively modifies the topology error in the segmentation prediction map using the distance map, resulting in the highest Dice score across all baselines.

4   Conclusion

In this work, we propose a new two-path network for layer segmentation in OCT images with PED. Specifically, we improved Resnet by adding a boundary path to supervise the network to learn the retinal layer topology. The semantic path uses a BEGA module to accurately locate and distinguish different layer structures in retinal images. Moreover, we design a set of GGSA modules in the boundary path to better recognize layer boundaries. The information of these two paths is fused and interacted with each other to solve the challenges brought about by PED deformation. Finally, we correct topological errors in segmentation predictions using distance maps via a fusion module to obtain topologically guaranteed segmentation results. The method is tested on the severely deformed PED dataset, and the results show that the performance of the method is better than the other four existing methods.

Acknowledgements. This work was supported in part by the National Natural Science Foundation of China under Grant 62176190.

References

1. Fleckenstein, M., Issa, P.C., Helb, M., et al.: High-resolution spectral domain-OCT imaging in geographic atrophy associated with age-related macular degeneration. Invest. Ophthalmol. Vis. Sci. **49**(9), 4137–4144 (2008)
2. Podoleanu, A.G.: Optical coherence tomography. J. Microsc. **247**(3), 209–219 (2012)
3. Novosel, J., Vermeer, K.A., De Jong, J.H., Wang, Z., Van Vliet, L.J.: Joint segmentation of retinal layers and focal lesions in 3-D OCT data of topologically disrupted retinas. IEEE Trans. Med. Imaging **36**(6), 1276–1286 (2017)
4. Koozekanani, D., Boyer, K., Roberts, C.: Retinal thickness measurements from optical coherence tomography using a Markov boundary model. IEEE Trans. Med. Imaging **20**(9), 900–916 (2001)
5. Chiu, S.J., Li, X.T., Nicholas, P., Toth, C.A., Izatt, J.A., Farsiu, S.: Automatic segmentation of seven retinal layers in SDOCT images congruent with expert manual segmentation. Opt. Exp. **18**(18), 19413–19428 (2010)

6. Xiang, D., et al.: Automatic retinal layer segmentation of OCT images with central serous retinopathy. IEEE J. Biomed. Health Inform. **23**(1), 283–295 (2018)

7. Lang, A., et al.: Retinal layer segmentation of macular OCT images using boundary classification. Biomed. Opt. Exp. **4**(7), 1133–1152 (2013)

8. Liu, X., Liu, Q., Zhang, Y., Wang, M., Tang, J.: TSSK-Net: weakly supervised biomarker localization and segmentation with image-level annotation in retinal OCT images. Comput. Biol. Med. **153**, 106467 (2023)

9. Liu, X., Zhang, D., Yao, J., Tang, J.: Transformer and convolutional based dual branch network for retinal vessel segmentation in OCTA images. Biomed. Sig. Process. Control **83**, 104604 (2023)

10. Roy, A.G., et al.: ReLayNet: retinal layer and fluid segmentation of macular optical coherence tomography using fully convolutional networks. Biomed. Opt. Exp. **8**(8), 3627–3642 (2017)

11. Liu, X., et al.: Semi-supervised automatic segmentation of layer and fluid region in retinal optical coherence tomography images using adversarial learning. IEEE Access **7**, 3046–3061 (2018)

12. Liu, X., Cao, J., Wang, S., Zhang, Y., Wang, M.: Confidence-guided topology-preserving layer segmentation for optical coherence tomography images with focus-column module. IEEE Trans. Instrum. Meas. **70**, 1–12 (2020)

13. Waldstein, S.M., Wright, J., Warburton, J., Margaron, P., Simader, C., Schmidt-Erfurth, U.: Predictive value of retinal morphology for visual acuity outcomes of different ranibizumab treatment regimens for neovascular AMD. Ophthalmology **123**(1), 60–69 (2016)

14. Fu, J., et al.: Dual attention network for scene segmentation. In: Proceedings of the IEEE/CVF Conference on Computer Vision and Pattern Recognition, pp. 3146–3154 (2019)

15. Ma, J., et al.: How distance transform maps boost segmentation CNNs: an empirical study. In: Medical Imaging with Deep Learning, pp. 479–492. PMLR (2020)

16. Lu, Y., Shen, Y., Xing, X., Ye, C., Meng, M.Q.H.: Boundary-enhanced semi-supervised retinal layer segmentation in optical coherence tomography images using fewer labels. Comput. Med. Imaging Graph. **105**, 102199 (2023)

17. Li, M., et al.: Image projection network: 3D to 2D image segmentation in OCTA images. IEEE Trans. Med. Imaging **39**(11), 3343–3354 (2020)

18. Apostolopoulos, S., De Zanet, S., Ciller, C., Wolf, S., Sznitman, R.: Pathological OCT retinal layer segmentation using branch residual U-shape networks. In: Descoteaux, M., Maier-Hein, L., Franz, A., Jannin, P., Collins, D.L., Duchesne, S. (eds.) MICCAI 2017. LNCS, vol. 10435, pp. 294–301. Springer, Cham (2017). https://doi.org/10.1007/978-3-319-66179-7_34

19. Zhou, Z., Siddiquee, M.M.R., Tajbakhsh, N., Liang, J.: UNet++: redesigning skip connections to exploit multiscale features in image segmentation. IEEE Trans. Med. Imaging **39**(6), 1856–1867 (2019)

20. Wang, B., Wei, W., Qiu, S., Wang, S., Li, D., He, H.: Boundary aware U-net for retinal layers segmentation in optical coherence tomography images. IEEE J. Biomed. Health Inform. **25**(8), 3029–3040 (2021)

Spatial Feature Regularization and Label Decoupling Based Cross-Subject Motor Imagery EEG Decoding

Yifan Zhou[1], Tian-jian Luo[1(✉)], Xiaochen Zhang[1], and Te Han[2]

[1] College of Computer and Cyber Security,
Fujian Normal University, Fuzhou 350108, China
{createopenbci,zhangxch2008}@fjnu.edu.cn
[2] School of Management and Economics,
Beijing Institute of Technology, Beijing 100081, China
hant15@bit.edu.cn

Abstract. Motor imagery (MI) serves as a vital approach to constructing brain-computer interfaces (BCIs) based on electroencephalogram (EEG) signals. However, the time-variant and label-coupling characteristics of EEG signals, combined with the limited sample sizes, often necessitate MI-EEG decoding across subjects. Unfortunately, existing methods encounter challenges related to interference from out-of-distribution features and feature-label coupling, resulting in the deterioration of decoding performance. To address these issues, this paper proposes a novel MI-EEG feature learning framework that focuses on decoupling features from labels and regularizing the feature representation. The proposed framework leverages aligned MI-EEG samples to extract Gaussian weighting regularized spatial features. Subsequently, a domain adaptation method is employed to decouple the extracted features from labels across different subjects' domains, thereby facilitating cross-subject MI-EEG decoding. To evaluate the effectiveness and efficiency of the proposed method, we conducted experiments using three benchmark MI-EEG datasets, consisting of four distinct groups of experiments. The experimental results demonstrate the effectiveness, efficiency, and parameter insensitivity of the proposed method, indicating its significant application value in the field of MI-EEG decoding.

Keywords: Motor Imagery · EEG Decoding · Pattern Recognition · Spatial Feature Regularization · Label Decoupling

1 Introduction

Brain-computer interfaces (BCIs) [10] have emerged as a promising technology for facilitating direct communication and control between the brain and external devices, circumventing conventional neuro-muscular pathways. BCIs hold

Supplementary Information The online version contains supplementary material available at https://doi.org/10.1007/978-981-99-8558-6_34.

immense potential in offering novel communication and control alternatives for individuals with profound motor disabilities, including stroke patients, spinal cord injury patients, and individuals afflicted with neurodegenerative disorders. Among various BCI paradigms, motor imagery-based BCIs (MI-BCIs) [8] have garnered substantial interest due to their non-invasive nature and their potential for broad clinical applicability. MI-BCIs rely on the modulation of brain activity during motor imagery tasks, making them an attractive avenue for research and development in the field.

The MI-BCIs include electroencephalogram (EEG) signals recording, feature extraction, and pattern recognition for MI-EEG decoding [1]. With the development, existing methods of decoding MI-EEG signals also have limitations, which can be summarized in two main points: Firstly, Inadequate sample size: there is a lack of large-scale datasets that can effectively extract EEG signals. Secondly, EEG signals vary from subject to subject and even within the same subject at different time sessions, leading to reduced performance of the classifier and decreased computational efficiency. Therefore, it is necessary to improve the performance of MI-EEG decoding by utilizing a cross-subject approach. Based on transfer learning methods, most MI-EEG samples across subjects can be used for constructing pattern recognition models.

To construct the cross-subject MI-EEG decoding method, domain adaptation (DA), as a subfield of transfer learning, presents a promising approach for leveraging knowledge acquired from different subjects' domains. Within the realm of cross-subject MI-EEG decoding, the objective of the DA method is to adapt feature distributions across subjects. This adaptation process assists in addressing challenges associated with inter-subject variability, signal quality variations, and category imbalance. Spatial features, such as common spatial patterns (CSP) and Riemannian tangent space (RTS) [2], are widely employed in MI-EEG decoding for pattern recognition tasks. Recently, researchers have begun utilizing these spatial features for distribution adaptation across subjects, thereby enhancing the performance and robustness of cross-subject MI-EEG decoding methodologies.

Despite existing cutting-edge research on cross-subject MI-EEG decoding, two major challenges still remain: Firstly, no matter CSP or RTS features, the time-variant and spatial coupling across subjects should be prevented, which needed to be regularized to improve the performance of MI-EEG decoding. Secondly, MI tasks-related labels across subjects have been coupled within the feature representations, which should be decoupled during DA. To solve these issues, this paper innovates a novel method, Spatial Feature Regularization and Label Decoupling (SFRLD), to meet such two challenges simultaneously, and the overall framework is divided into three parts: Firstly, aligning of raw MI-EEG samples across subjects using the centroid alignment of covariance. Secondly, extracting Gaussian weighting regularized spatial features from aligned MI-EEG samples to prevent time-variant and spatial coupling characteristics. Lastly, decoupling labels from the regularized features during domain adaptation and pattern recognition.

Figure 1 presents the overall flow chart illustrating the proposed method. Our approach encompasses two main aspects as follows. (I) Thorough extraction of spatial features. We aim to extract spatial features comprehensively, enabling access to a Gaussian weighting distribution. (II) Label decoupling domain adaptation. To enhance discriminability among different classes, we utilize label decoupling domain adaptation to dissociate labels from the regularized features. By adopting this framework, we strive to improve the representation learning process and facilitate effective adaptation between domains.

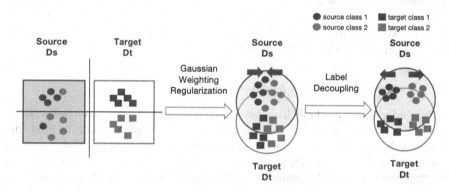

Fig. 1. The overall framework of our SFRLD. The underlying concept of our SFRLD framework revolves around the use of circles and squares to denote samples from the source domain and target domain, respectively. Furthermore, the colors blue and orange are employed to distinguish between different features. (Color figure online)

The main contributions of this paper are therefore summarized as follows:

– Gaussian weighting regularization is introduced to extract features from MI-EEG samples, effectively addressing temporal variations and overfitting spatial features.
– A joint distribution alignment method with label decoupling is employed for MI-EEG decoding based on the regularized features, reducing coupling characteristics between features and labels.
– Extensive experiments on various MI-EEG benchmark datasets validate the effectiveness and the efficiency of the proposed method, meanwhile ensuring the interpretability and parameter insensitivity of the proposed method.

The remainder of this paper is organized as follows: Sect. 2 reviews the MI-EEG domain adaptation methods. Section 3 introduces the proposed SFRLD method. Section 4 introduces the experiments and results. Section 5 conducts the ablations study and analysis of sensitivity. Finally, Sect. 6 concludes and points out some promising future research works.

2   Related Works

2.1   Sample Adaptation Methods

Since the MI-EEG samples are multi-variate time series, the covariance matrix can be used to measure the distributions of different subjects. Therefore, the alignment of the covariance matrix has been proposed as a sample adaptation method. These methods align the covariance of MI-EEG samples to the identity matrix to make the distributions similar across subjects. Such as Riemannian alignment (RA) [17] which transforms MI-EEG samples into Riemannian manifold space for data streaming alignment, and Euclidean alignment (EA) [9] which maps MI-EEG samples from different subjects into a computationally efficient Euclidean space for the alignment of the identity matrix.

2.2   Feature Adaptation Methods

Feature adaptation methods have been derived from cross-domain images classification, such as the joint distribution alignment (JDA) method [12] by adapting marginal distribution and conditional distribution, and transfer joint matching (TJM) method [13] provides sparse feature selection during JDA. By adopting such methods on spatial features of MI-EEG samples, researchers have proposed sub-band target alignment CSP (SB-TA-CSP) method [19] to utilize sub-band target alignment and extracts CSP features for MI-EEG decoding, and manifold embedded knowledge transfer (MEKT) method [18] to construct alignment on a manifold embedded tangent space features, and manifold embedded transfer learning (METL) method [5] to extract features from the Grassmann manifolds and symmetric positive definite manifold, and performs JDA on such two manifolds.

2.3   Deep Adversarial Methods

Cross-domain deep adversarial learning has greatly advanced domain adaptation applications. These methods, based on the principles of Generative Adversarial Networks (GAN) [6], blur the boundaries between source and target domains. The Domain-Adversarial Neural Network (DANN) [7] confounds discriminators to hinder domain differentiation. Building on DANN, the Domain Separation Network (DSN) [4] divides samples into public and private parts. The public part learns shared features, while the private part retains distinct domain characteristics. To incorporate adversarial learning into deep networks with conditional data, Conditional Domain Adversarial Networks (CDAN) [11] have been formulated for domain adaptation tasks. The deep learning models require high computational resources, and the conventional domain adaptation methods lack of considering label coupling problem during domain adaptation. Therefore, we propose the SFRLD method to solve such issues.

3 Methods

3.1 Notations and Problem Definition

The formal mathematical definition of cross-subject MI-EEG decoding using DA is as follows: Given a set of EEG signals containing a total number of subjects of S, and each subject had N MI-EEG samples. Our task is to select a subject as the target domain $D_t = \{x_j\}_{j=1}^N, x_j \in \mathbb{R}^S$ with labels unknown, the remaining subjects are used as source domain $D_s = \{(x_i, y_i)\}_{i=1}^{(S-1)*N}, x_i \in \mathbb{R}^S$ with labels are known and perform a DA method to train a decoder on the source domain to predict labels on the target domain with the minimum overall loss. The optimization objection can be defined as follows:

$$\min \frac{1}{N} \sum_{i=1}^N L\left(\mathcal{M}(X_i, \theta), Y_i\right) \tag{1}$$

Table 1. Frequently used notations in this paper.

Symbol	Description
Ds/Dt	Source/target domain space
Xs/Xt	Source/target domain samples
Ys/Yt	Source/target domain labels
C	Number of classes for each domain
n_s/n_t	Number of samples for source/target domain
$\mathcal{X}s/\mathcal{X}t$	Source/target domain feature space
Fs/Ft	Source/target domain features
F	Feature matrix combining source and target

where L is the overall loss function, $\mathcal{M}$ is the decoder model and θ are the trainable parameters. For convenience, all the frequently used notations in this paper are summarized in Table 1.

3.2 Centroid Alignment of Covariance

Generally, MI-EEG samples recorded from different subjects are distributed differently. To eliminate the distribution differences across subjects at the sample level, followed by the thought of EA [9] and RA [17], we need to align the centroid of the covariance matrix for each subject as the identity matrix. For the samples from each subject $\{x_j\}_{j=1}^N, x_j \in \mathbb{R}^{ch*t}$, where ch is the number of channels, and t is the number of sampling points. The first step of performing the EA method is computing the reference matrix $\overline{R} = \frac{1}{N} \sum_{i=1}^N x_i x_i^T$, where R is the arithmetic average of all covariance matrices. MI-EEG samples from each subject can be aligned with the help of reference matrix $\overline{R}$:

$$x_i' = \overline{R}^{-\frac{1}{2}} x_i \tag{2}$$

After alignment, the average covariance of each subject can be equal to the identity matrix, namely $\overline{R} = I$, which is commonly used to measure the distribution. Now, the sample distribution of each subject could be similar.

3.3 Gaussian Weighting Regularized Spatial Features

After alignment, spatial features should be extracted for classification. For a binary decoding task of MI-EEG signals, the commonly used CSP [2] features want to learn an optimal spatial vector w to maximize the covariance of one class and minimize the covariance of the other class, which can be defined as:

$$\max J\left(w\right) = \frac{w^T R_1 w}{w^T \left(R_1 + R_2\right) w} \tag{3}$$

where J represents the objective function to be solved, R_1 and R_2 represent the covariance of such two classes. Solving the above Eq. (3) is equivalent to solving the eigenvector matrix of the following matrix:

$$R_1 w = \lambda \left(R_1 + R_2\right) w \tag{4}$$

where λ and w are the generalized eigenvalues and the corresponding eigenvector. However, CSP features will cause overfitting across subjects. Researchers used the regularization strategy on CSP features (RCSP) [14] of two classes, which can be defined as:$J_{r1}\left(w\right) = \frac{w^T R_1 w}{w^T (R_2 + \delta K) w}$, $J_{r2}\left(w\right) = \frac{w^T R_2 w}{w^T (R_1 + \delta K) w}$, where δ is the regularization parameter and K is a predefined matrix. Here, $K = I$ is the identity matrix.

However, original RCSP features neglect the weights of different features and only select m features to construct spatial vector w, resulting in the loss of a large number of transferable valid features. To solve this problem, the feature weighting regularization strategy [15] is adopted for CSP features in this paper and the optimal spatial vector w^* is constructed after obtaining the feature values and the corresponding feature vectors as $w^* \in \mathbb{R}^{2m*ch}$, which is consisted of the corresponding eigenvectors of the first m largest eigenvalues and last m largest eigenvalues solved by Eq. (4).

$$f_i^k = diag(w^* \times x_i'^T \times x_i' \times w^{*T}), k = 1, 2, ..., 2m \tag{5}$$

where $f_i^k \in \mathbb{R}^{2m*N}$ is the extracted regularized CSP features, and $k = 1, 2, ..., N$ is the number of samples. By using all spatial features, the dimension $2m$ will be the same as the number of channels: $2m = ch$. To give the weight of spatial features, the Gaussian distribution α with the same number of channels suggested in [15] is computed:$\alpha_k = e^{-\frac{(k-1)^2}{2\delta^2}}$, where $k = 1, 2, ..., ch$, then project the feature f_i^k to the Gaussian distribution $f_i^\alpha = \alpha_k \times f_i^k$

Finally, the feature weighting CSP features extracted in this paper can be concatenated:

$$F = [f_1^\alpha, f_2^\alpha, ..., f_N^\alpha] \tag{6}$$

where $F \in \mathbb{R}^{ch*N}$ retains all regularized weighting spatial features from the source/target domain and will be used in the later domain adaptation methods.

3.4 Regularized Spatial Feature Adaptation and Label Decoupling

Overall Model. To adapt the regularized spatial features across subjects, the efficient JDA [12] or TJM [13] methods can be selected. However, different from image samples, MI-EEG samples suffer from nonlinear and non-stationary characteristics of different classes, which couples regularized features from labels. Inspired by the label disentangled analysis method [16], during joint distribution alignment, the labels of different classes can be decoupled during iterations. Therefore, we construct the regularized spatial feature adaptation and label decoupling method for the domain adaptation. The overall model can be described as:

$$L = L_{DIS} + L_{FMMD} + F_{LMMD} - F_{COUP} \tag{7}$$

where L_{DIS} is the label decoupling loss, L_{FMMD} and F_{LMMD} are the joint distribution alignment loss, and F_{COUP} is the label reconstruction loss. During regularized spatial feature adaptation and label decoupling, we want to find the optimal projection matrix M to further minimize the difference of feature distributions between the source domain and target domain. To project regularized spatial features and labels, we first define the optimal projection for both features and labels: $Q_s = \frac{1}{n_s} \sum_{i=1}^{n_s} M y_i^s$, $Q_t = \frac{1}{n_s} \sum_{i=1}^{n_s} M \hat{y}_i^t$, $P_s = \frac{1}{n_t} \sum_{j=n_s+1}^{n_s+n_t} M f_j^{\alpha,s}$, $P_t = \frac{1}{n_t} \sum_{j=n_s+1}^{n_s+n_t} M f_j^{\alpha,t}$.

The four parts of Eq. (7) work together to constitute an effective and efficient objective loss function. By minimizing this objective function, a new distribution can be learned to solve the label coupling problem more thoroughly. To construct label decoupling during joint distribution alignment, there are three steps: Label Decoupling, Joint Distribution Adaption, and Label Reconstruction as shown in the following sections.

Label Decoupling. First, the labels are decoupled and independent from the regularized spatial features by minimizing the association between label-reorganized feature centers $P(y)$ and original feature centers $P(x)$, so we gave the label decoupled loss $minL_{DIS}$:

$$\min_{P_x,P_y} L_{DIS}(P_x, P_y) = Q_s \odot Q_t + P_s \odot P_t \tag{8}$$

where $\odot$ is the dot product operation between two vectors. For two vectors in space, if their dot product is smaller, then the association between them is smaller. Accordingly, the purpose of decoupling labels from data are achieved.

Joint Distribution Adaption. To minimize the features distribution M_x and labels distribution M_y between the source domain and target domain, instructed by the JDA [12] method, the maximum mean discrepancy (MMD) [3] is used as the metric to minimize such distributions for both features and labels:

$$\min_{M_x} L_{FMMD} (M_x) = \|Q_t - P_t\|^2 \tag{9}$$

$$\min_{M_y} L_{LMMD} (M_y) = \|Q_s - P_s\|^2 \tag{10}$$

Label Reconstruction. Although the decoupled regularized spatial features and labels have been adapted, before decoding and recognition it is also necessary to reorganize the decoupled labels and regularized spatial features, which can be defined to maximize the coupling loss L_{COUP}:

$$\max_{M_x, M_y} L_{COUP} (M_x, M_y) = P_s \odot Q_t + P_t \odot P_s \tag{11}$$

Combining the above Eq. (8)–Eq. (11), we can yield the final objective loss function L as follows: By introducing the l_2 norm regularization of M, the final label decoupling domain adaptation loss function can be obtained:

$$L = \min_M tr \left(M^T FAF^T M \right) + \lambda \|M\|_F^2 \text{ s.t. } M^T FHF^T M = I \tag{12}$$

where $A = A_0 + A_c$ is the MMD matrix, let $Z = Concat[f^\alpha, \mu y]$ be the concatenation of features and labels, where $z_i = Concat[f_i^\alpha, \mu y_i]$, $z_j = Concat[f_j^\alpha, \mu y_j]$, $\hat{z}_j = Concat[f_j^\alpha, \mu \hat{y}_j]$. μ and λ are the trade-off parameters.

To solve Eq. (12), the MMD matrix A between z_i and z_j can be defined as:

$$(\mathbf{A}_0)_{ij} = \begin{cases} \frac{1}{n_s n_s}, & z_i, z_j \in \mathcal{D}_s \\ \frac{1}{n_t n_t}, & z_i, z_j \in \mathcal{D}_t \\ \frac{-1}{n_s n_t}, & \text{otherwise} \end{cases} \tag{13}$$

During iteration, the MMD of A_c between z_i and $\hat{z}_j$ can be defined as:

$$(\mathbf{A}_c)_{ij} = \begin{cases} \frac{1}{n_s^{(c)} n_s^{(c)}}, & z_i, \hat{z}_j \in D_s^{(c)} \\ \frac{1}{n_t^{(c)} n_t^{(c)}}, & z_i, \hat{z}_j \in D_t^{(c)} \\ \frac{1}{n_s^{(c)} n_t^{(c)}}, & \begin{cases} z_i \in D_s^{(c)}, \hat{z}_j \in D_t^{(c)} \\ \hat{z}_j \in D_s^{(c)}, z_i \in D_t^{(c)} \end{cases} \\ 0, & \text{otherwise} \end{cases} \tag{14}$$

where c is the different MI classes, we all have C classes.

In addition, the centering matrix H is defined as $H = \mathbf{I}_n - \frac{1}{n}\mathbf{1}$, where $\mathbf{I}_n$ is the identity matrix of n, and $\mathbf{1}$ is the all-1 matrix.

Model Solving. Further, since L in Eq. (12) is a convex function, the solution can be solved using Lagrange multipliers. Let $\Phi = diag\,(\Phi_1, \Phi_2, ..., \Phi_{ch}) \in \mathbb{R}^{(ch+C)*(ch+C)}$ as the Lagrange multiplier and construct the Lagrange function for the problem as:

$$\min_M L = \lambda tr\left(M^T M\right) + tr\left(M^T Z A Z^T M\right) + tr\left(\left(I - M^T Z H Z^T M\right)\Phi\right) \quad (15)$$

By setting $\frac{\partial L}{\partial M} = 0$, Eq. (15) is converted to the generalized eigenvalue decomposition as follows:

$$\left(Z A Z^T + \lambda I\right) M = Z H Z^T M \Phi \quad (16)$$

After solving the above equation, the optimal domain adaptation matrix M^* is finally obtained, which is used to map the feature matrix of source and target domain $F = [F_s; F_t]$ into the subspace by a linear transformation to obtain the adapted features for decoding and classification:

$$F_M = M^* \times F, M^* \in \mathbb{R}^{k*(2m+C)} \quad (17)$$

where F_M is the adapted features, and k is the dimension of subspace. For the final cross-subject MI-EEG decoding, the eigenfeature regularization, and extraction (ERE) classifier [15] is adopted, both in terms of accuracy and efficiency.

At the end of this section, the flow chart of the proposed SFRLD method is shown below:

Algorithm 1: SFRLD for MI-BCI

---

input : Raw EEG matrix of source and target $X = [Xs; Xt]$; μ, λ, k: parameter.

output: Iterative decoding and classification results.

1 Align EEG samples by using Eq. (2);
2 Obtain spatial filter w^* and spatial features by solving Eq. (4) and Eq. (5);
3 Construct a Gaussian distribution α;
4 Calculate Gaussian weighting regularized spatial feature f^α;
5 Construct spatial feature matrix F by using Eq. (6);
6 **repeat**
7 $\quad$ Initialize optimal projection matrix M;
8 $\quad$ Generate pseudo labels of the target domain;
9 $\quad$ Construct MMD of A_0 and A_c by Eq. (13) and Eq. (14), and centering matrix H;
10 $\quad$ Construct features and labels concatenation of Z by μ;
11 $\quad$ Construct final loss function L by Eq. (15) using λ and above matrices;
12 $\quad$ Solve optimal adaption matrix M^* by Eq. (16) under k;
13 $\quad$ Use M^* to project F into F_M by Eq. (17);
14 $\quad$ Train an ERE classifier on $\{F_M, \hat{y}_i\}_{i=1}^{n_s}$;
15 **until** *Convergence or Maximum iteration*;

---

4   Experiment

4.1   Datasets

To validate the feasibility and effectiveness of the proposed SFRLD method. Three publicly available datasets were selected, BCI Competition IV[1] 2a & 2b, and BCI Competition III[2] 3a. The three datasets share a similar motor imagery process: subjects sit comfortably in a chair in a quiet room and each motor imagery task begins with a cue tone, followed by a cross in the center of the screen to focus the subject's attention. 2 s later, the subject performs motor imagery based on the proposed arrow, and the motor imagery lasts for 4 s, with a 1.25-s break after the motor imagery is completed. The motor imagery generally consisted of the left hand (L), right hand (R), foot (F), and tongue (T) tasks. Table 2 gives the basic information of the three data sets.

Table 2. Basic information of the three datasets.

Datasets	Subjects	Samples	Channels	Sampling points	Classes
BCI IV 2a	9	288	22	750	4
BCI IV 2b	9	160	3	750	2
BCI III 3a	5	280	118	300	2

All MI-EEG samples were preprocessed with a 50-order [8, 30] Hz band-pass filter according to the procedure in the literature [18]. The BCI IV 2a dataset selected samples from the training session and divided the L, R, F, and T motor imagery tasks into two sets of classification tasks, L vs. R and F vs. T. The BCI IV 2b dataset selects samples from the third training session and contains the classification task of L vs. R. The BCI III 3a dataset selects the classification task of R vs. F.

4.2   Results

To extensively evaluate the proposed method, we compare the SFRLD method[3] with the state-of-the-art methods: **MDM** (Minimum Distance to Mean) [17], **JDA** (Joint Distribution Adaptation) [12], **TJM** (Transfer Joint Matching) [13], **MEKT** (Manifold Embedded Knowledge Transfer) [18], **SB-TA-CSP** (Sub-Band Target Alignment CSP) [19], **METL** (Manifold Embedded Transfer Learning) [5]. All experiments in this paper were run in Matlab_R2022B on Mac OS with Intel Core i7. Table 3 illustrates the cross-subject MI-EEG decoding performance comparisons on four groups of experiments between the state-of-the-arts.

---

[1] BCI III: https://www.bbci.de/competition/iii/.

[2] BCI IV: https://www.bbci.de/competition/iv/.

[3] The code is available at: https://github.com/Geniusyingmanji/SFRLD.git.

Table 3. Cross-subject MI-EEG decoding performance comparisons.

Datasets	MDM	JDA	TJM	MEKT	SB-TA-CSP	METL	ours
BCI IV 2a($Lvs.R$)	73.46	74.00	75.23	76.00	75.15	76.00	**76.39**
BCI IV 2a($Fvs.T$)	64.74	67.44	69.06	70.29	-	-	**71.22**
BCI IV 2b	73.82	71.18	75.56	73.75	-	-	**78.68**
BCI III 3a	76.93	78.43	**80.50**	80.29	-	-	78.93
Average	72.24	72.76	75.09	75.08	-	-	**76.15**

where '-' indicates no codes provided to work on a particular dataset

From Table 3, it can be seen that our proposed SFRLD method achieved a classification accuracy of 76.15% on four cross-subject MI-EEG classification tasks, surpassing the RA-based MDM method, CSP-based JDA, TJM, and SB-TA-CSP methods, as well as the tangent space feature-based MEKT and METL methods. Specifically, the best classification performance was achieved on the BCI IV 2a and 2b datasets, while the accuracy on the BCI III 3a dataset was lower than that of the TJM method. In fact, the TJM method selects effective features through $l_{2,1}$ regularization during the DA process, which is why it performs better on the higher-dimensional 3a dataset. However, the SFRLD method does not include feature selection, so it performs poorly on datasets with a larger number of channels. In future research, we plan to add $l_{2,1}$ regularization to select effective features in label decoupling domain adaptation, thus enabling the SFRLD method to adapt to more diverse EEG datasets.

4.3 Efficiency Comparison

Efficiency is another important index for measuring the cross-subject MI-EEG decoding method. For the compared five methods, we have measured and compared the time consumption of each decoding task, and the results have illustrated in Table 4. For the datasets of BCI IV 2a & 2b, our method shows a large advantage in time efficiency, benefiting from the regularized spatial features and label decoupling. For the dataset of BCI III 3a, which has a large number of channels, the RTS feature-based MDM and METK required a large number of running time to extract features. However, for the CSP feature-based JDA and METK methods, a fast speed is shown for recognition. Our method needs to construct the Gaussian weighting regularization, which achieves a significant improvement in classification accuracy within an acceptable time overhead.

Table 4. Cross-subject MI-EEG decoding time cost (unit: s) comparisons.

Datasets	MDM	JDA	TJM	MEKT	ours
BCI IV 2a($Lvs.R$)	1.11	3.63	2.67	1.03	**0.32**
BCI IV 2b	0.32	3.98	3.35	**0.27**	0.35
BCI III 3a	494.85	5.05	**3.31**	365.052	43.37

4.4   Feature Visualization

To apparently demonstrate the indispensable meanings of each module in the proposed SFRLD method, the MI-EEG features of Subject 8 from the BCI IV 2a($Lvs.R$) dataset have been selected for the feature visualization experiment. To visualize the features, the t-SNE tool is used to reduce the dimension of the feature to 2, and Fig. 2 shows the results. Figure 2(a), (b), and (c) exhibited the features without (w/o) centroid alignment, Gaussian weighting regularization, and label decoupling, respectively. Compared with Fig. 2(d), the lack of centroid alignment will deteriorate the discriminativity of different MI classes, the lack of feature regularization leads to mis-classification of key samples, and the lack of label decoupling reduces the boundary of different classes when performing the recognition. Based on the comparison of feature visualization, any module in the proposed SFRLD method is indispensable and effective for cross-subject MI-EEG decoding.

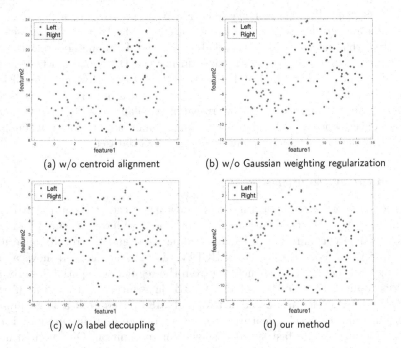

(a) w/o centroid alignment   (b) w/o Gaussian weighting regularization

(c) w/o label decoupling   (d) our method

Fig. 2. Feature visualization of ablation experiments. The blue and orange represent the Left MI class and the Right MI class from the BCI IV 2a (L $vs.$ R) task respectively. (Color figure online)

5   Discussion

To discuss the proposed method, we gave the ablation study of each component among SFRLD method, the parameter sensitivity, and the convergence.

5.1   Ablation Study

To explore the most effective of centroid alignment measurements, this paper compares four measurements of covariance centroid: **non** (no alignment), **RA** (alignment in Riemannian space) [17], **logEA** (alignment in logarithmic Euclidean space) [18], and **EA** (alignment in Euclidean space) [9]. The classification results of different data alignment methods have been given in Fig. 3(a). As depicted in Fig. 3(a), the EA exhibits a notable advantage in three out of the four classification tasks. It is worth mentioning that although the accuracy of EA on the BCI IV 2a (L vs. R) task falls short of that achieved by the RA, it is important to highlight that RA's utilization of the Riemannian centroid comes at a prohibitively computational time cost, rendering it far less efficient compared to EA.

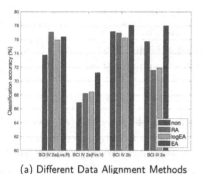 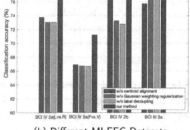

(a) Different Data Alignment Methods   (b) Different MI-EEG Datasets

Fig. 3. Ablation study experiments.

To evaluate the efficacy of each component in the SFRLD method, we conducted the ablation study with the following configurations: (1) **method 1**: w/o centroid alignment; (2) **method 2**: w/o Gaussian weighting regularization; (3) **method 3**: w/o label decoupling. Figure 3(b) presents the classification results obtained from the three ablation study methods, in comparison with the proposed SFRLD. Based on the findings depicted in Fig. 3(b), it is evident that the components comprising the SFRLD method contribute positively to the performance of cross-subject MI-EEG decoding. The collective performance achieved by all three components surpasses the performance attainable by removing any one of them individually. When comparing the three components, it is notable that spatial feature weighting regularization proves to be more effective in enhancing performance compared to Euclidean alignment and label decoupling. Furthermore, the proposed SFRLD method, benefiting from the synergistic integration of all three components, exhibits a significant improvement in decoding performance for the challenging BCI IV 2a (F vs. T) decoding task.

5.2   Parameters Sensitivity

To analyze the sensitivity of the parameters, we conclude the optimal parameters of dim, μ, and λ for each dataset by using a trial-and-error strategy. Table 5 illustrates the optimal parameters selection for four groups of experiments from three datasets. As can be seen in Table 5, the three parameters will be varied for different classification tasks, so we conduct the parameter sensitivity experiments by selecting one parameter to measure the sensitivity while keeping the rest two parameters from Table 5.

Table 5. Optimal parameters setting on each experiment from different datasets.

Datasets	dim	μ	λ	acc
BCI IV 2a (L vs. R)	14	0.48	1.38	76.39
BCI IV 2a (F vs. T)	10	0.35	0.38	71.22
BCI IV 2b	2	0.5	1.19	78.06
BCI III 3a	10	0.4	0.46	78.93

Due to a fast convergence, the maximum iteration is set to $T = 5$

The experimental results are illustrated in Fig. 4. For Fig. 4(a), due to the limited number of channels in the BCI IV 2b dataset, only one dimension can be selected, resulting in the unavailability of sensitivity results. As depicted in Fig. 4(a), the classification accuracy initially increases and then decreases with increasing dimensions, indicating the presence of the phenomenon of underfitting followed by overfitting.

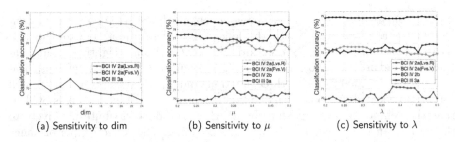

(a) Sensitivity to dim   (b) Sensitivity to μ (c) Sensitivity to λ

Fig. 4. Parameters sensitivity on each experiment from different datasets. The value of dim in BCI IV 2b is only 2 or 4, so the analysis of parameter dim in BCI IV 2b is not considered.

Optimal values for the dimension (dim) can be obtained for each classification task within a suitable range. In contrast, for Fig. 4(b) and (c), the parameters of μ and λ exhibit minimal influence on the classification accuracy across the four groups of experiments, highlighting the high robustness of the SFRLD method in cross-subject MI-EEG decoding. In general, the proposed SFRLD method demonstrates insensitivity to the parameters μ and λ, allowing for the selection of an appropriate value for dim based on the dataset to achieve the highest classification accuracy.

5.3 Convergence

To measure the convergence of the proposed SFRLD method, by setting a suitable iteration of $T = 5$, we gave the convergence results of nine subjects from the BCI IV 2a dataset and five subjects from the BCI III 3a dataset, as can be seen in Fig. 5(a), (b), and (c). Notably, regardless of whether a single subject was selected as the target domain, the SFRLD method consistently demonstrated rapid convergence, typically requiring only two iterations. This favorable convergence behavior can be attributed to the regularization of spatial features and the decoupling of labels, thereby presenting potential benefits for online motor imagery-based brain-computer interface (MI-BCI) applications.

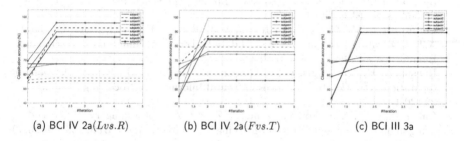

(a) BCI IV 2a($Lvs.R$)   (b) BCI IV 2a($Fvs.T$)   (c) BCI III 3a

Fig. 5. Iterations convergence of different subjects from different datasets. There are nine subjects in BCI IV 2a (L vs. R) and (F vs. T) tasks, while five in BCI III 3a task.

6 Conclusions

In the field of cross-subject MI-EEG signal decoding, the time-varying and label-coupled characteristics of EEG signals are often overlooked. State-of-the-art methods rely solely on marginal or conditional distribution adaptation, which leads to performance bottlenecks in classification. This paper proposes the SFRLD method, which combines Gaussian weighting regularized spatial features and label decoupling strategy to ensure time-variant and label decoupling characteristics during domain adaptation, overcoming the shortcomings of classic DA methods for cross-subject MI-BCI in feature distribution adaptation. Results from four groups of experiments on three representative public datasets demonstrate that the proposed method has higher cross-subject MI-EEG decoding performance and faster convergence. Future research will focus on two aspects:

422   Y. Zhou et al.

first, adding more regularization during the DA, such as sparse feature selection and graph Laplacian, to further enhance the effectiveness of cross-domain features. Second, incorporating large-scale data and integrating knowledge distillation after DA to create a lighter model that enables transfer learning on larger datasets.

Acknowledgments. This research was supported by National Natural Science Foundation of China (grant number 62106049); Natural Science Foundation of Fujian Province of China (grant number 2022J01655); Education and Research Project for Middle and Young Teachers in Fujian Province (grant number JAT210050).

References

1. Arpaia, P., Esposito, A., Natalizio, A., Parvis, M.: How to successfully classify EEG in motor imagery BCI: a metrological analysis of the state of the art. J. Neural Eng. **19**(3), 031002 (2022)
2. Biesmans, W., Bertrand, A., Wouters, J., Moonen, M.: Optimal spatial filtering for auditory steady-state response detection using high-density EEG. In: 2015 IEEE International Conference on Acoustics, Speech and Signal Processing (ICASSP), pp. 857–861. IEEE (2015)
3. Borgwardt, K.M., Gretton, A., Rasch, M.J., Kriegel, H.P., Schölkopf, B., Smola, A.J.: Integrating structured biological data by kernel maximum mean discrepancy. Bioinformatics **22**(14), e49–e57 (2006)
4. Bousmalis, K., Trigeorgis, G., Silberman, N., Krishnan, D., Erhan, D.: Domain separation networks. In: Advances in Neural Information Processing Systems, vol. 29 (2016)
5. Cai, Y., She, Q., Ji, J., Ma, Y., Zhang, J., Zhang, Y.: Motor imagery EEG decoding using manifold embedded transfer learning. J. Neurosci. Meth. **370**, 109489 (2022)
6. Courville, A., Bengio, Y.: Generative adversarial nets. In: Advances in Neural Information Processing Systems (2014)
7. Ganin, Y., Lempitsky, V.: Unsupervised domain adaptation by backpropagation. In: International Conference on Machine Learning, pp. 1180–1189. PMLR (2015)
8. Gao, X., Wang, Y., Chen, X., Gao, S.: Interface, interaction, and intelligence in generalized brain-computer interfaces. Trends Cogn. Sci. **25**(8), 671–684 (2021)
9. He, H., Wu, D.: Transfer learning for brain-computer interfaces: a Euclidean space data alignment approach. IEEE Trans. Biomed. Eng. **67**(2), 399–410 (2019)
10. Khademi, Z., Ebrahimi, F., Kordy, H.M.: A review of critical challenges in MI-BCI: from conventional to deep learning methods. J. Neurosci. Meth. **383**, 109736 (2023)
11. Long, M., Cao, Z., Wang, J., Jordan, M.I.: Conditional adversarial domain adaptation. In: Advances in Neural Information Processing Systems, vol. 31 (2018)
12. Long, M., Wang, J., Ding, G., Sun, J., Yu, P.S.: Transfer feature learning with joint distribution adaptation. In: Proceedings of the IEEE International Conference on Computer Vision, pp. 2200–2207. IEEE (2013)
13. Long, M., Wang, J., Ding, G., Sun, J., Yu, P.S.: Transfer joint matching for unsupervised domain adaptation. In: Proceedings of the IEEE Conference on Computer Vision and Pattern Recognition, pp. 1410–1417. IEEE (2014)
14. Lotte, F., Guan, C.: Regularizing common spatial patterns to improve BCI designs: unified theory and new algorithms. IEEE Trans. Biomed. Eng. **58**(2), 355–362 (2010)

15. Mishuhina, V., Jiang, X.: Feature weighting and regularization of common spatial patterns in EEG-based motor imagery BCI. IEEE Sig. Process. Lett. **25**(6), 783–787 (2018)
16. Xiao, N., Zhang, L., Xu, X., Guo, T., Ma, H.: Label disentangled analysis for unsupervised visual domain adaptation. Knowl. Based Syst. **229**, 107309 (2021)
17. Zanini, P., Congedo, M., Jutten, C., Said, S., Berthoumieu, Y.: Transfer learning: a Riemannian geometry framework with applications to brain-computer interfaces. IEEE Trans. Biomed. Eng. **65**(5), 1107–1116 (2017)
18. Zhang, W., Wu, D.: Manifold embedded knowledge transfer for brain-computer interfaces. IEEE Trans. Neural Syst. Rehabil. Eng. **28**(5), 1117–1127 (2020)
19. Zhang, X., She, Q., Chen, Y., Kong, W., Mei, C.: Sub-band target alignment common spatial pattern in brain-computer interface. Comput. Meth. Programs Biomed. **207**, 106150 (2021)

Autism Spectrum Disorder Diagnosis Using Graph Neural Network Based on Graph Pooling and Self-adjust Filter

Aimei Dong[1,2,3(✉)], Xuening Zhang[1,2,3], Guohua Lv[1,2,3], Guixin Zhao[1,2,3], and Yi Zhai[1,2,3]

[1] Key Laboratory of Computing Power Network and Information Security, Ministry of Education, Shandong Computer Science Center (National Supercomputer Center in Jinan), Qilu University of Technology (Shandong Academy of Sciences), Jinan, China
amdong@qlu.edu.cn
[2] Shandong Provincial Key Laboratory of Computer Networks, Shandong Fundamental Research Center for Computer Science, Jinan, China
[3] Faculty of Computer Science and Technology, Qilu University of Technology (Shandong Academy of Sciences), Jinan 250353, China

Abstract. Currently, accurately identifying autism spectrum disorders (ASD) still have challenges. However, graph neural network used for ASD diagnosis only focus on the part of abnormal brain functional connectivity, ignoring the effects between brain regions and the help of phenotypic information, and the kernel is also pre-defined. To solve the above problems, a graph neural network based on self-attention graph pooling and self-adjust filter is proposed for ASD diagnosis. Specifically, first, self-attention graph pooling is used for feature extraction of the fMRI to account for the influence of nodes with abnormal brain functional connectivity and activity between brain regions in fMRI data. Then, image features were taken as graph nodes and phenotypic information as edges to form a population graph. Finally, to focus on both high and low frequency information in the graph, a graph neural network based on self-adjust filter is used to learn node embedding. Experimental results on ABIDE I showed the effectiveness of the proposed method.

Keywords: Autism prediction · graph neural network · self-attention · graph pooling · self-adjust filter

1 Introduction

Autism spectrum disorder (ASD) is a neurodevelopmental disorder [1,2], often occurring in childhood. It is clinically characterized by social difficulties and repetitive behaviors [3], which can only be relieved by training or medication.

This work was supported by the Natural Science Foundation of Shandong Province, China (ZR2022MF237 and ZR2020MF041).

However, because patients in this period cannot communicate with others, they are easily ignored and cannot receive timely diagnosis and treatment. Due to the lack of pathological markers, doctors only rely on the evaluation scale for the diagnosis of ASD, which brings difficulties for early intervention and treatment of ASD. In recent years, with the development of technology, researchers have used machine learning and other means to achieve an auxiliary diagnosis of diseases, especially in psychiatric disorders, which have gained wide application [25].

With advancements in neuroimaging and functional brain imaging technologies, magnetic resonance imaging (MRI) has become a valuable auxiliary tool for diagnosing brain diseases, particularly psychiatric disorders such as autism, Alzheimer, and schizophrenia [10,17,18]. MRI encompasses two main modalities: structural MRI (sMRI) and functional MRI (fMRI). Structural MRI focuses on capturing images of the brain's structural changes. It provides detailed information about the anatomy and morphology of the brain, allowing clinicians to identify abnormalities or structural alterations associated with various brain disorders. Functional MRI is employed to image active brain regions. By measuring changes in blood flow and oxygenation, fMRI enables the mapping of brain activity and the identification of regions that are engaged during specific tasks or at rest. This functional imaging technique has become indispensable in psychology and research related to brain dysfunction [9,22].

Graph neural network (GNN) is a neural network for feature extraction of graph structure data, which is effective in graph learning tasks such as classification and prediction. Since the topology structure of the brain network can be regarded as graph structure data, most researchers use graph neural networks to predict ASD [14,15]. For example, Mhiri et al. [19] proposed a brain network atlas-guided feature selection (NAG-FS) method to generate atlases by similarity network fusion. Ktena et al. [11] propose a siamese graph convolutional neural network (s-GCN) to learn the similarity measurement of the brain network and then use spectral graph convolutions to extract the features. Jiang et al. [7] proposed a hierarchical GCN (hi-GCN) considering the network topology and applied it to the classification of ASD. Parisot et al. [21] combined non-image information, such as phenotype information to form a sparse graph and used GCN to classify nodes. Li et al. [16] innovate the pooling method, highlight the area of interest (ROI), and design the pooling regularized-GNN (PR-GNN). Wen et al. [20] combined graph structure learning and multi-task graph embedding to carry out innovation in the recognition function network. Silverstein et al. [8] proposed a model that adds a self-focused layer based on graph convolution and considers the uniqueness of multimodal data elements. Cao et al. [4] combined scale information and imaging information into graph structure data and added the ResNet unit and DropEdge strategy to the DeepGCN model. These studies indicate that GNN can achieve excellent results in the auxiliary diagnosis of ASD.

However, the above methods focus on the abnormal brain regions when extracting features from brain regions, while the mutual influence of brain regions and the impact of phenotypic information is not paid attention to. The kernel of the graph neural network used for training is also pre-defined. The flexibility between different graphs is weak, and the mismatch between graph and kernel is easy to occur, thus affecting the training performance. To solve these problems, we propose a self-attention graph pooling method, which not only considers abnormal brain regions but also preserves the information between brain regions through node selection and edge prediction. The extracted node characteristics are combined with phenotypic information to generate the population graph. Input to the graph neural network based on self-adjust filter for training. The graph neural network based on self-adjust filter learns the graph's high and low frequency signals to retain as much essential information about the graph as possible. In summary, the contribution of this paper can be divided into three points:

(1) Based on fMRI data, features were extracted using a self-attention graph pooling, which retained vital information such as functional connections of brain regions and signals between brain regions.
(2) The nodes of the population graph are the fMRI image features of each subject, and the edge connections are generated by considering the different effects of each phenotype information through the attention mechanism.
(3) The graph neural network based on self-adjust filter is used to achieve the balance between filters by adjusting the maximum eigenvalue.

2   Method

2.1   Overview

As shown in Fig. 1, the whole framework is divided into three parts. (1) Self-attention Graph Pooling, (2) Population Graph Construction, (3) Graph neural network based on self-adjust filter. We first separated the fMRI data using the Harvard-Oxford Atlas, using brain regions as nodes and functional connections between brain regions as edges to form a brain graph for each subject. Then, the brain graph is put into the Self-attention Graph Pooling for node selection and edge prediction to generate subgraphs and extract features. Then, the brain graph features are used as node features and the phenotype information is used as edge connections to generate population maps. Finally, the population graph is put into a graph neural network based on self-adjust filter, and nodes are classified by learning the best balance between high and low frequency information. is input into the network for training.

2.2   Self-attention Graph Pooling

In this subsection, we introduce the feature extraction method with self-attention graph pooling. This method can extract features from both functional connections and graph structures. The graph pooling method adopted in this paper preserves the connectivity between brain regions and abnormal brain regions while reducing information redundancy.

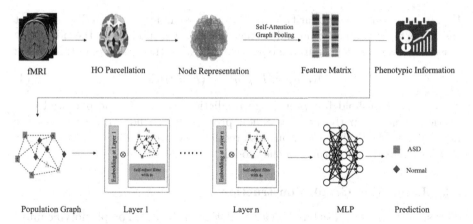

fMRI   HO Parcellation   Node Representation   Feature Matrix   Phenotypic Information

Population Graph   Layer 1   Layer n   MLP   Prediction

Fig. 1. Overview of the diagnosis of autism by graph neural network based on graph pooling and self-adjust filter.

We developed the self-attention graph pooling inspired by [12] and [26], which used node information score to select nodes and attention mechanism to predict edges. Each subject's brain could be segmented into 110 regions based on the Harvard-Oxford atlas, which converted the fMRI data into a graph structure. The labeled regions represented nodes, with functional connections between brain regions as edges.

Node Selection. Node selection is to calculate the self-attention score of each node. This score serves as an indicator of the node's significance and determines whether it will be retained or not during the feature extraction process. The larger the score, the more significant the difference between nodes and the more likely to be retained. L1 norm between the defined node itself and its neighbors:

$$S = \|I - D^{-1}AN\|_1 \tag{1}$$

Where I is the identity matrix and D is the degree matrix of the adjacency matrix A. N is the node feature matrix. The vector S holds the self-attention score for each node. Sort nodes by scores and select the top n nodes with high scores:

$$idx = top(S, \lceil r * n \rceil) \tag{2}$$

Where r is the pooling rate, the function top(·) returns the index of the top n nodes with a high score. The node feature matrix N and adjacency matrix A is updated according to the returned index subscript.

Edge Prediction. edge prediction is to predict the potential connections between subgraphs separated after node selection. We have designed a self-attention method:

$$F(p,q) = \frac{N(p,:) \cdot N(q,:)}{\|N(p,:)\| \|N(q,:)\|} + A(p,q) \tag{3}$$

F(p, q) represents the similarity score between two nodes, and N and A represent the node feature matrix and adjacency matrix. The update of the adjacency matrix A can be converted into the problem of quadratic constraint optimization:

$$A(p,q) = [F(p,q) - \tau(F(p,:))]_+ \qquad (4)$$

τ (·) is the threshold function. Edge prediction can retain the potential connections ignored in the node selection process, which not only preserves the attributes of the brain region, but also reduces the overhead of storage capacity and improves the robustness of the model.

2.3   Population Graph Construction

This subsection describes building a population graph based on phenotypic information. Each subject represents a node, node features are obtained by self-attention graph pooling, and the edge connections between nodes are determined by phenotypic information. The attention mechanism is used to quantify the different effects of phenotypic information, such as gender, age, collection site, and Full-Scale Intelligence Quotient(FIQ), which can adapt to calculate edge weights and retain more effective information. The corresponding edge weights on the population graph are calculated as:

$$A_{ij} = a_0 + a_g * r_g(g_i, g_j) + a_e * r_e(e_i, e_j)$$
$$+ a_c * r_c(c_i, c_j) + a_f * r_f(f_i, f_j), \qquad (5)$$
$$a_0 + a_g + a_e + a_c + a_f = 1$$

Where A_{ij} represents the edge weight between subjects i and j, a is the amplification coefficients, and r_g, r_e, r_c, r_f represent gender, age, collection site, and FIQ distance in phenotypic information of subjects i and j respectively. r_g, r_e, r_c, r_f are defined as:

$$r_g(g_i, g_j) = \begin{cases} 1, g_i = g_j \\ 0, g_i \neq g_j \end{cases}, \quad r_e(e_i, e_j) = \begin{cases} 1, |e_i - e_j| \leq 2 \\ 0, |e_i - e_j| > 2 \end{cases}$$

$$r_c(c_i, c_j) = \begin{cases} 1, c_i = c_i \\ 0, c_i \neq c_i \end{cases}, \quad r_f(f_i, f_j) = \begin{cases} 1, |f_i - f_i| \leq 4 \\ 0, |f_i - f_i| > 4 \end{cases} \qquad (6)$$

The choice of setting the age difference to 2 in this study is based on the understanding that the majority of individuals with ASD are children. When the sum of r_g, r_e, r_c, r_f is greater than or equal to 2, there is an edge between the default nodes. The result is a population graph with 871 nodes. Each node represents the fMRI features of a subject. These fMRI features capture the functional characteristics and patterns of brain activity for each individual. To establish connections between the nodes in the graph, we calculate the similarity distance of the phenotypic information. This connectivity structure enables our model to leverage both the fMRI features and the phenotypic information, facilitating the integration of multi-modal data for more accurate and comprehensive analysis.

2.4   Graph Neural Network Based on Self-adjust Filter

Inspired by [27], we provide a comprehensive explanation of the technical details underlying the graph neural network based on the self-adjust filter. This innovative method introduces the capability to automatically adjust the weights of both the all-pass and low-pass filters, leading to improved performance. To achieve this, we modified the Cheby-Filter applied to the graph signal f. The modified equation is as follows:

$$f' = \frac{2\lambda_{max} - 2}{\lambda_{max}}\theta I f + \frac{2}{\lambda_{max}}\theta D^{-\frac{1}{2}} A D^{-\frac{1}{2}} f \tag{7}$$

Where f' is the modulating signal, λ_{max} represents the maximum value in the diagonal eigenvalue matrix, and θ represents the learnable filter. $\frac{2\lambda_{max}-2}{\lambda_{max}}$ is the weight of the all-pass filter, and $\frac{2}{\lambda_{max}}$ is the weight of the low-pass filter. We achieve the balance between the filters by adjusting λ_{max}. The weight of the all-pass filter increases with the increase of λ_{max}, and the weight of the high-frequency signal in the figure is also higher. The weight of the low-pass filter increases with the reduction of λ_{max}, and the weight of the low-frequency signal in the figure also increases. To balance high and low frequency signals, λ_{max} at the kth layer is taken as a learnable parameter:

$$\lambda_{max}^k = 1 + relu(\phi_k) \tag{8}$$

ϕ_k is a learnable scalar, $relu(\cdot)$ is an activation function. When the initial equilibrium point of the two filters is the same, $\phi_k = 1$, $\lambda_{max} = 2$. Node embedding $H^{(k)}$ in the kth layer can be expressed as:

$$H^{(k)} = (\frac{2relu(\phi_k)}{1+relu(\phi_k)}I + \frac{2}{1+relu(\phi_k)}D^{-\frac{1}{2}}AD^{-\frac{1}{2}} + \alpha H^{(k-1)})H^{(k-1)}W_k \tag{9}$$

Where W_k represents the parameter matrix of the k-layer filter, αRepresents a learnable parameter used to control the contribution of the previous layer. $\alpha H^{(k-1)}$ can be thought of as a residual connection that allows the model to learn the difference between the output of the current layer and the input of the layer. This helps the model propagate information more efficiently through the network and reduces the problem of disappearing gradients. Importantly, this model effectively learns the optimal balance between low-frequency and high-frequency information without the need for manual hyperparameter tuning.

3   Experiments

In this particular section, our primary objective is to thoroughly examine and evaluate the effectiveness of the proposed method, with the intention of showcasing its inherent advantages when compared to alternative approaches. To achieve this, we conducted a comprehensive evaluation of our method on the ABIDE I dataset. Through this evaluation, we aimed to establish the superiority and distinctiveness of our proposed method.

3.1   ABIDE I Dataset

The ABIDE I is a public multi-site data repository that gathers information on fMRI and phenotypic data from 17 sites [6]. The demographics are shown in Table 1.

Table 1. Demographic characteristics of the ABIDE I.

Site	Group		Age		Male/Female	FIQ	
	ASD	NC	Min	Max		Min	Max
CALTECH	5	10	17.0	56.2	2/1	96	128
CMU	6	5	19.0	40.0	7/4	99	125
KKI	12	21	8.0	12.8	8/3	69	131
LEUVEN	26	30	12.1	32.0	7/1	89	146
MAX_MUM	19	27	7.0	58.0	21/2	79	129
NYU	74	98	6.5	39.1	34/9	76	148
OHSU	12	13	8.0	15.2	25/0	67	132
OLIN	14	14	10.0	24.0	23/5	71	135
PITT	24	26	9.3	35.2	43/7	81	131
SBL	12	14	20.0	64.0	26/0	95	125
SDSU	8	19	8.7	17.2	7/2	88	141
STANFORD	12	13	7.5	12.9	18/7	80	141
TRINITY	19	25	12.0	25.7	44/0	72	135
UCLA	48	37	8.4	17.9	74//11	75	132
UM	47	73	8.2	28.8	31/9	77	148
USM	43	24	8.8	50.2	67/0	65	144
YALE	22	19	7.0	17.8	25/16	41	141

Due to the differences in scanners and parameters at different sites, the Configurable Pipeline for the Analysis of Connectomes (C-PAC) was used to preprocess data, including time slice, motion correction, skull dissection, and other operations. This ensures that the data is standardized and comparable across different sites. This study used data from 403 ASD subjects and 468 normal control subjects from 17 sites [5]. To obtain node characteristics, the fMRI data after C-PAC pretreatment were divided into 110 brain regions using the Harvard-Oxford Atlas, and the average time series of each brain region was calculated. After standardization, the Pearson's correlation coefficient between each pair of time series was calculated to obtain the functional correlation matrix of each subject. This matrix captures the functional connectivity patterns between different brain regions, revealing how they interact and work together.

As a widely used atlas of the brain, the Harvard-Oxford Atlas provides a detailed anatomical reference of the brain. The map includes both cortical regions and subcortical regions and also provides the possibility that brain regions are present in specific locations. The probabilistic maps of cortical and subcortical regions have helped study the differences in brain structure and function in autism.

3.2 Experimental Setting

In the experiment, we employed a 10-fold cross-validation strategy to ensure robustness and minimize bias. To assess the performance of the proposed method, we utilized five key indicators as evaluation criteria. These indicators include: (1) accuracy (ACC), (2) sensitivity (SEN), (3) specificity (SPE), (4) balance accuracy (BAC), and (4) Area under ROC curve (AUC). Throughout the experimentation process, we trained the entire framework on an RTX 2080Ti computer. The parameters were set as follows: a learning rate of 0.01, a total of 200 epochs for training, a discard rate of 0.1, and a weight decay rate of 5e-4. These parameter settings were chosen to optimize the performance and ensure effective training of the model.

3.3 Results

In Table 2, we compare our experiments with other models on the ABIDE I dataset preprocessed by C-PAC. The comparison includes four traditional methods and five methods proposed in recent years. We evaluate the performance of these different methods, and the best results are highlighted in bold.

Table 2. Performance comparison of various methods.

	ACC	SEN	SPE	BAC	AUC
SVM	63.46 ± 3.09	71.43 ± 4.27	79.43 ± 5.13	61.21 ± 5.26	71.35 ± 5.63
LSTM [13]	70.57 ± 5.81	67.80 ± 5.61	70.86 ± 3.57	67.59 ± 4.77	72.62 ± 2.94
GCN [20]	77.14 ± 3.23	78.47 ± 2.87	74.31 ± 4.60	72.75 ± 2.53	77.93 ± 4.52
GAT [23]	70.98 ± 6.67	77.28 ± 3.16	72.34 ± 4.66	73.95 ± 2.05	75.70 ± 3.59
MVS-GCN [24]	76.51 ± 6.29	77.76 ± 4.48	74.56 ± 5.23	72.01 ± 3.81	78.63 ± 2.72
POP [21]	73.14 ± 4.71	74.75 ± 5.57	71.19 ± 5.58	72.15 ± 4.44	74.45 ± 6.84
Multi-GCN [8]	76.27 ± 6.87	71.02 ± 5.12	75.67 ± 4.22	71.56 ± 1.96	72.78 ± 3.97
DeepGCN [4]	75.98 ± 3.41	73.70 ± 4.18	72.74 ± 2.55	72.40 ± 3.62	76.35 ± 5.25
NAG-FS [19]	78.16 ± 5.65	76.67 ± 3.08	80.32 ± 5.85	74.91 ± 6.56	81.76 ± 3.95
Ours	**85.27 ± 4.04**	**81.39 ± 3.18**	**82.45 ± 3.12**	**82.50 ± 3.27**	**84.78 ± 5.05**

The results obtained demonstrate that our proposed method outperforms other approaches in terms of several evaluation metrics. Specifically, our method achieves the highest ACC at 85.27%, SEN at 81.39%, SPE at 82.45%, BAC at 82.50%, and AUC at 84.78%. Compared to traditional classifiers like SVM, which focus solely on individual features without considering the interconnections between individuals, our method showcases superior performance. This emphasizes the importance of capturing the relationships and interactions between individuals in the dataset. Moreover, our method demonstrates adeptness in handling high and low-frequency information during the training phase of the graph neural network. This capability is particularly crucial in the context of medical imaging, where different frequency components can offer valuable insights into the underlying physiological and pathological processes. By effectively incorporating and leveraging frequency information, our method enhances its ability to capture the diverse and complex characteristics of brain function, resulting in more accurate and reliable diagnosis. Overall, our proposed method surpasses existing neural network models by addressing their limitations, preserving essential information, emphasizing abnormal brain regions and their interactions, and effectively handling frequency information. These advancements contribute to the improved performance of our method in ASD diagnosis using medical imaging data.

3.4   Ablation Analysis

To evaluate the contributions of different modules in our approach, we conducted ablation experiments. These experiments aimed to assess the effectiveness of specific components and their impact on the overall performance of the model. One aspect we investigated was the superiority of self-attention graph pooling for node selection in brain regions. To compare its effectiveness, we designed a model without self-attention graph pooling, as depicted in Table 3. The results indicated that the inclusion of self-attention graph pooling improved the extraction fMRI features and led to higher classification accuracy. This suggests that self-attention graph pooling plays a crucial role in capturing relevant information from brain regions, enhancing the model's performance.

Table 3. Results of the quantitative assessment of ablation studies on the ABIDE I.

self-attention graph pooling	ACC	SEN	SPE	BAC	AUC
✓	80.32	78.24	79.58	77.31	82.23
	85.27	**81.39**	**82.45**	**82.50**	**84.78**

Additionally, we explored the impact of graph neural network layers based on self-adjust filter on model performance. Figure 2 illustrates the classification performance as the number of layers in the network increases. It can be observed that the model's accuracy improves with an increase in the number of layers, enhancing the expressive and learning abilities of the model. The highest accuracy is typically achieved around the 5th layer, where the model captures the maximum features and relationships in the graph data while maintaining stability. This demonstrates that our model has an advantage in addressing the oversmoothing problem often encountered in graph neural networks. Furthermore, we did not experiment with the self-tuning filter-based graph neural network beyond the 10th layer. Going beyond this threshold would make it challenging to distinguish the target nodes, impacting the model's discriminative power.

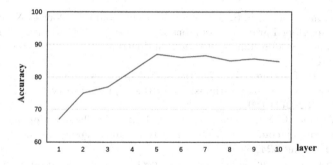

Fig. 2. Impact of the number of layers on accuracy.

4   Conclusion

In this paper, we propose an ASD diagnosis method using graph neural networks based on self-attention graph pooling and self-adjust filter. Self-attention pooling can detect brain region abnormalities and the interaction between brain regions The extracted features and phenotypic information are combined into a population graph, and input to the graph neural network based on self-adjust filter for training, which is balanced the filter by modifying the maximum eigenvalue of the graph Laplacian. The effectiveness of the proposed method is validated by comparing it with other methods on the ABIDE I dataset.

References

1. Amaral, D.G., Schumann, C.M., Nordahl, C.W.: Neuroanatomy of autism. Trends Neurosci. **31**(3), 137–145 (2008)
2. American Psychiatric Association, D., Association, A.P., et al.: Diagnostic and statistical manual of mental disorders: DSM-5, vol. 5. American psychiatric association Washington, DC (2013)
3. Baio, J., et al.: Prevalence of autism spectrum disorder among children aged 8 years-autism and developmental disabilities monitoring network, 11 sites, united states, 2014. MMWR Surveill. Summ. **67**(6), 1 (2018)
4. Cao, M.: Using deepgcn to identify the autism spectrum disorder from multi-site resting-state data. Biomed. Signal Process. Control **70**, 103015 (2021)
5. Craddock, C., et al.: The neuro bureau preprocessing initiative: open sharing of preprocessed neuroimaging data and derivatives. Front. Neuroinform. **7**, 27 (2013)
6. Di Martino, A., et al.: The autism brain imaging data exchange: towards a large-scale evaluation of the intrinsic brain architecture in autism. Mol. Psychiatry **19**(6), 659–667 (2014)
7. Jiang, H., Cao, P., Xu, M., Yang, J., Zaiane, O.: HI-GCN: a hierarchical graph convolution network for graph embedding learning of brain network and brain disorders prediction. Comput. Biol. Med. **127**, 104096 (2020)

8. Kazi, A., Shekarforoush, S., Kortuem, K., Albarqouni, S., Navab, N., et al.: Self-attention equipped graph convolutions for disease prediction. In: 2019 IEEE 16th International Symposium on Biomedical Imaging (ISBI 2019), pp. 1896–1899. IEEE (2019)

9. Khazaee, A., Ebrahimzadeh, A., Babajani-Feremi, A.: Identifying patients with Alzheimer's disease using resting-state fMRI and graph theory. Clin. Neurophysiol. **126**(11), 2132–2141 (2015)

10. Kong, Y., Gao, J., Xu, Y., Pan, Y., Wang, J., Liu, J.: Classification of autism spectrum disorder by combining brain connectivity and deep neural network classifier. Neurocomputing **324**, 63–68 (2019)

11. Ktena, S.I., et al.: Metric learning with spectral graph convolutions on brain connectivity networks. Neuroimage **169**, 431–442 (2018)

12. Lee, J., Lee, I., Kang, J.: Self-attention graph pooling. In: International Conference on Machine Learning, pp. 3734–3743. PMLR (2019)

13. Li, J., Zhong, Y., Han, J., Ouyang, G., Li, X., Liu, H.: Classifying ASD children with LSTM based on raw videos. Neurocomputing **390**, 226–238 (2020)

14. Li, X., Dvornek, N.C., Zhou, Y., Zhuang, J., Ventola, P., Duncan, J.S.: Graph neural network for interpreting task-fMRI biomarkers. In: Shen, D., et al. (eds.) MICCAI 2019. LNCS, vol. 11768, pp. 485–493. Springer, Cham (2019). https://doi.org/10.1007/978-3-030-32254-0_54

15. Li, X.: Braingnn: interpretable brain graph neural network for fMRI analysis. Med. Image Anal. **74**, 102233 (2021)

16. Li, X., et al.: Pooling regularized graph neural network for fMRI biomarker analysis. In: Martel, A.L., et al. (eds.) MICCAI 2020. LNCS, vol. 12267, pp. 625–635. Springer, Cham (2020). https://doi.org/10.1007/978-3-030-59728-3_61

17. Liu, J., Li, M., Lan, W., Wu, F.X., Pan, Y., Wang, J.: Classification of alzheimer's disease using whole brain hierarchical network. IEEE/ACM Trans. Comput. Biol. Bioinf. **15**(2), 624–632 (2016)

18. Liu, J., Wang, X., Zhang, X., Pan, Y., Wang, X., Wang, J.: Mmm: classification of schizophrenia using multi-modality multi-atlas feature representation and multi-kernel learning. Multimedia Tools Appl. **77**, 29651–29667 (2018)

19. Mhiri, I., Rekik, I.: Joint functional brain network atlas estimation and feature selection for neurological disorder diagnosis with application to autism. Med. Image Anal. **60**, 101596 (2020)

20. Parisot, S., et al.: Disease prediction using graph convolutional networks: application to autism spectrum disorder and alzheimer's disease. Med. Image Anal. **48**, 117–130 (2018)

21. Parisot, S., et al.: Spectral graph convolutions for population-based disease prediction. In: Descoteaux, M., Maier-Hein, L., Franz, A., Jannin, P., Collins, D.L., Duchesne, S. (eds.) MICCAI 2017. LNCS, vol. 10435, pp. 177–185. Springer, Cham (2017). https://doi.org/10.1007/978-3-319-66179-7_21

22. Silverstein, B.H., Bressler, S.L., Diwadkar, V.A.: Inferring the dysconnection syndrome in schizophrenia: interpretational considerations on methods for the network analyses of fmri data. Front. Psych. **7**, 132 (2016)

23. Veličković, P., Cucurull, G., Casanova, A., Romero, A., Lio, P., Bengio, Y.: Graph attention networks. arXiv preprint arXiv:1710.10903 (2017)

24. Wen, G., Cao, P., Bao, H., Yang, W., Zheng, T., Zaiane, O.: MVS-GCN: a prior brain structure learning-guided multi-view graph convolution network for autism spectrum disorder diagnosis. Comput. Biol. Med. **142**, 105239 (2022)

25. Yan, Y., Zhu, J., Duda, M., Solarz, E., Sripada, C., Koutra, D.: Groupinn: Grouping-based interpretable neural network for classification of limited, noisy brain data. In: Proceedings of the 25th ACM SIGKDD International Conference on Knowledge Discovery & Data Mining, pp. 772–782 (2019)
26. Zhang, Z., et al.: Hierarchical graph pooling with structure learning. arXiv preprint arXiv:1911.05954 (2019)
27. Zhu, M., Wang, X., Shi, C., Ji, H., Cui, P.: Interpreting and unifying graph neural networks with an optimization framework. In: Proceedings of the Web Conference 2021, pp. 1215–1226 (2021)

CDBIFusion: A Cross-Domain Bidirectional Interaction Fusion Network for PET and MRI Images

Jie Zhang[1], Bicao Li[1(✉)], Bei Wang[2], Zhuhong Shao[3], Jie Huang[1], and Jiaxi Lu[1]

[1] School of Electronic and Information Engineering, Zhongyuan University of Technology, Zhengzhou 450007, China
lbc@zut.edu.cn

[2] University Infirmary, Zhongyuan University of Technology, Zhengzhou 450007, China

[3] College of Information Engineering, Capital Normal University, Beijing 100048, China

Abstract. The existing methodologies for image fusion in deep learning predominantly focus on convolutional neural networks (CNN). However, recent investigations have explored the utilization of Transformer models to enhance fusion performance. Hence, we propose a novel generative adversarial fusion framework that combines both CNN and Transformer, referred to as the Cross-Domain Bidirectional Interaction Fusion Network (CDBIFusion). Specifically, we devise three distinct pathways for the generator, each serving a unique purpose. The two CNN pathways are employed to capture local information from MRI and PET images, and the other pathway adopts a transformer architecture that cascades both source images as input, enabling the exploitation of global correlations. Moreover, we present a cross-domain bidirectional interaction (CDBI) module that facilitates the retention and interaction of information deactivated by the ReLU activation function between two CNN paths. The interaction operates by cross-cascading ReLU activation features and deactivation features from separate paths by two ReLU rectifiers and then delivering them to the other path, thus reducing the valuable information lost through deactivation. Extensive experiments have demonstrated that our CDBIFusion surpasses other current methods in terms of subjective perception and objective evaluation.

Keywords: Cross domain bidirectional interaction · Image fusion · Convolutional neural network · Transformer · PET and MRI images

1 Introduction

According to various imaging technologies, medical images of different modalities can capture distinct information regarding human tissues or organs. For example, magnetic resonance imaging (MRI) provides detailed soft tissue information and offers high resolution. Positron emission tomography (PET) images display cell activity and biomolecular metabolism. The combination of multi-modal medical pictures combines essential biological function information with precise anatomic structure, resulting in more comprehensive and trustworthy clinical data [1, 2].

© The Author(s), under exclusive license to Springer Nature Singapore Pte Ltd. 2024
Q. Liu et al. (Eds.): PRCV 2023, LNCS 14437, pp. 436–447, 2024.
https://doi.org/10.1007/978-981-99-8558-6_36

Numerous traditional techniques have been proposed to realize that the fused image retains clear and comprehensive information from the source images. Typically, Multi-Scale Transform (MST) fusion algorithms [3–6] are commonly employed to obtain a representation of the source images at multiple scales. These methods derive fused multi-scale coefficients based on specific rules and ultimately generate the fused image through a multi-scale inverse transformation. Additionally, Sparse Representation (SR) is also a frequently utilized traditional approach for image fusion. For instance, Liu et al. [7] introduced an adaptive sparse representation algorithm for medical image fusion, which enhanced the computational efficiency of decomposing image matrices in the sparse transform domain. However, these approaches exhibit limited feature extraction capabilities, and their fusion rules are manually crafted and progressively intricate.

Subsequently, to capitalize on the robust feature extraction capabilities of convolutional operations, image fusion techniques [8–11] based on convolutional neural networks (CNN) emerged. These fusion methods typically employ pre-trained CNN to assess the activity level of pixels in medical images and generate fusion weight maps, which are implemented with classic decomposition and reconstruction procedures. However, the manual design of fusion strategies imposes significant limitations. To address this limitation, approaches based on generative adversarial networks (GAN) [12–18] were developed. They employ cyclic optimization of the network's generator during adversarial training, aiming to align the distribution of the fusion result with that of the source image. Notably, Ma et al. presented FusionGAN [12], in which both the visible image and the resulting image are continuously optimized and learned by the discriminator, gradually incorporating more details from the visible image into the generated image until the discriminator can no longer differentiate them.

While the above methods have achieved commendable performance, there are still certain challenges that require further attention. Firstly, the aforementioned approaches primarily depend on convolutional procedures to gather local information, overlooking the crucial aspects of global correlation and the modeling of long-term dependencies. Secondly, they fail to address the issue of intermediate information loss during the process of fusion, which will have an influence on the final fusion consequence.

To solve these limitations, we propose a GAN fusion network based on a union of CNN and Transformer networks for PET and MRI image fusion. In particular, our CDB-IFusion consists of three paths for the generator, two of which are CNN paths to extract the local features of source images, and the other is a Swin Transformer path with the concatenation of two source images as input, fully mining the depth features with global correlation and generating deep features containing high-level semantic information. In addition, we construct the incorporation of a cross-domain bidirectional interaction (CDBI) module, which serves to enhance the preservation and interaction of information that is deactivated by the ReLU activation function across two distinct CNN paths. This module operates by performing cross-cascading of the ReLU activation and deactivation features from different paths using two ReLU rectifiers. The resulting combined features are then shared and provided to the other path, effectively mitigating the loss of beneficial information caused by deactivation. By implementing this mechanism, we aim to reduce the detrimental impact of information loss and enable the retention and integration of valuable details between the paths in a bidirectional manner. In addition, we use a dual discriminator to learn the differences and complementarities between MRI and PET images to better preserve the structural and functional information of the original images and produce more informative and complete fused images.

The following are the primary contributions of our CDBIFusion fusion framework:

1. We propose a GAN fusion framework for PET and MRI images based on the union CNN-Transformer structure that can fully mine local and global information to provide more proficient complimentary feature integration.
2. We construct a CDBI module to store the comprehensive information of the source images. The deactivation information is transmitted from one stream to another by the ReLU rectifiers, alleviating the problems of information loss and dead ReLU.
3. In multi-modal medical image fusion techniques, large-scale trials show that our CDBIFusion outperforms existing sophisticated fusion technologies in subjective vision assessment and objective evaluation.

2 Method

2.1 Network Overview

Generator. The CDBIFusion network framework is depicted in Fig. 1. Within the generator component, we implemented two separate CNN pathways for processing MRI and PET images, along with a Swin Transformer pathway that considers the combination of the two source images as input. Specifically, the two CNN pathways initially employ two convolutional layers to extract additional localized and low-level information, such as edges, colors, textures, and other related features. The shallow features F and F are defined as:

$$\left\{ F_{SF}^1, F_{SF}^2 \right\} = \{Convs(I_{MRI}), Convs(I_{PET})\} \tag{1}$$

where Convs $(\cdot)$ denotes two consecutive convolution operations for which the kernel size is 3×3, and the stride is 1. I_{MRI} and I_{PET} are raw MRI and PET images.

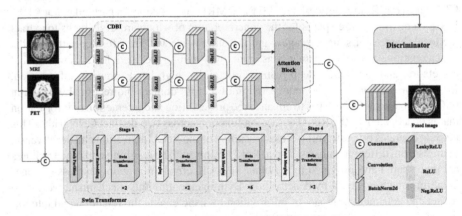

Fig. 1. The network architecture of CDBIFusion

The extracted shallow features are then fed into the CDBI module, which can motivate complementary information between branches, and deactivated information can also be passed to another branch through interaction to provide some contextual information, thus enhancing the robustness and diversity of features. The deep features F1 DF and F2 DF gained by going through the CDBI module are denoted by Eq. 2.

$$\left\{ F^1_{DF}, F^2_{DF} \right\} = \left\{ CDBI(F^1_{SF}), CDBI(F^2_{SF}) \right\} \tag{2}$$

where CDBI ($\cdot$) denotes the CDBI module with two ReLU rectifiers to avoid data loss. It is going to be described in complete detail in Sect. 2.2.

Meanwhile, the Swin Transformer path exploits a local self-attention mechanism to capture short-range dependencies in the input sequence and a global self-attention mechanism to preserve long-range dependencies. This hierarchical attention mechanism allows the model to efficiently reflect local and global features in the image while maintaining efficiency. This path takes advantage of the cascaded input of PET and MRI images to capture the global features F shown in Eq. 3.

$$F^3_{GF} = H_{ST}(I_{MRI} \oplus I_{PET}) \tag{3}$$

where $H_{ST}(\cdot)$ denotes the Swin Transformer feature extraction layer and $\oplus$ represents the cascade operation.

The features derived from the above three paths are subsequently merged to acquire a composite feature F that combines comprehensive local and global information. Finally, these features are processed through the reconstruction module in order to derive the final generated image I_F as defined by Eq. 4.

$$\begin{aligned} F &= F^1_{DF} \oplus F^2_{DF} \oplus F^3_{GF}, \\ I_F &= H_{RE}(F) \end{aligned} \tag{4}$$

where H_{RE} ($\cdot$) denotes the reconstruction unit, which consists of three convolution operations and three LeakyReLU activation functions.

Discriminator. The structure of the discriminator is shown in Fig. 2(a). To maintain the sophisticated texture structure of the fused images, the fused images and MRI images are fed into the discriminator architecture for learning within the discriminator. The discriminator consists of four convolutional blocks, the first three using a convolutional layer with step size 2, followed by a normalization layer, then a convolutional layer with step size 1, and then a LeakyReLU activation layer. The fourth convolutional block is composed of a convolutional layer with a step size of 1 and a sigmoid function. The size of all convolution kernels is 3×3. Throughout the training process, the discriminator forces the fused image to gradually retain higher-level texture details from the MRI image. Achieving a balanced interaction between the generator and the discriminator means that the former successfully tricks the latter and obtains the desired results.

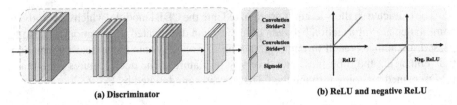

(a) Discriminator   (b) ReLU and negative ReLU

Fig. 2. Structure of (a) discriminator and functional behaviors of (b) ReLU and negative ReLU

2.2 CDBI Module

The architecture of CDBI module is shown in Fig. 1. The ReLU activation function is widely employed in neural networks to enhance the nonlinear properties of the model and effectively address the vanishing gradient issue, thereby speeding up the training process. However, ReLU has concerns that may cause valuable information to be lost. When the input falls below zero, the gradient of ReLU becomes zero, causing the neurons in that region to remain inactive and preventing the corresponding weights from being updated. This can render certain neurons ineffective and result in information loss within the ReLU. Despite the introduction of various ReLU variants, they have not adequately resolved the problem of information loss. Inspired by YTMT [19], we can construct an interaction path by transferring the deactivated information from one pathway to another through the ReLU rectifier rather than discarding it outright. It is important to note that the deactivated features of one pathway are not treated as garbage but rather delivered to another pathway as compensation or valuable information. The CDBI module shows two key advantages: 1) The application of two ReLU rectifiers allows for the exchange of deactivated information between the two branches. This interaction not only prevents feature drop-out but also promotes information complementarity between the branches, thus improving feature representation and model performance. 2) Mitigates the problem that, for some neurons, the weights are updated in such a way that they may not be activated.

We begin by defining the negative ReLU function below:

$$\text{ReLU}^-(\theta) = \theta - \text{ReLU}(\theta) = \min(\theta, 0) \tag{5}$$

where $\text{ReLU}(\theta) = \max(\theta, 0)$. The deactivated features can be readily kept thanks to the negative ReLU. Figure 2(b) depicts the ReLU and negative ReLU behaviors.

The retrieved shallow features F and F are sent into the CDBI module. Then, active features are derived from ReLU activation, while deactivated features are obtained from negative ReLU activation using Eq. 6.

$$\text{ReLU}^-(F_{SF}^1) = F_{SF}^1 - \text{ReLU}(F_{SF}^1),$$
$$\text{ReLU}^-(F_{SF}^2) = F_{SF}^2 - \text{ReLU}(F_{SF}^2) \tag{6}$$

The two paths receive the ReLU information and the negative ReLU information cross-cascaded as the input features Y and Y of the next convolution block, and after the above operation for a total of z (z = 1, 2, 3) times, the output features Y and Y of the z_{th} ReLU rectifier are obtained, after which they are input to the attention module to integrate these features to obtain the final extracted features of the two CNN paths. The depth features F and F are extracted by the two CNN paths, respectively. The ReLU rectifiers and attention module are shown as:

$$Y_1^1 = \text{ReLU}(F_{SF}^1) \oplus \text{ReLU}^-(F_{SF}^2),$$
$$Y_2^1 = \text{ReLU}(F_{SF}^2) \oplus \text{ReLU}^-(F_{SF}^1) \tag{7}$$

$$Y_1^{z+1} = \text{ReLU}(Y_1^z) \oplus \text{ReLU}^-(Y_2^z),$$
$$Y_2^{z+1} = \text{ReLU}(Y_2^z) \oplus \text{ReLU}^-(Y_1^z), (z = 1, 2, 3) \tag{8}$$

$$\left\{F_{DF}^1, F_{DF}^2\right\} = \left\{Attention(Y_1^3), Attention(Y_2^3)\right\} \tag{9}$$

where the attention block is CBAM attention. As can be seen from Eqs. 6 and 7, the amount of information in Y and Y is equal to that in F and F. The interaction process ensures that no information flows out, which largely avoids the problems of vanishing or bursting gradients and dead ReLU.

2.3 Swin Transformer

Swin Transformer [20] introduced the Swin Block, which is the basic building block of Swin Transformer and consists of two sub-blocks: the Local Window and the Global Window, as shown in Fig. 3(a). The local module allows the local self-attention mechanism to capture short-range dependencies in the input sequence, while the global module enables the global self-attention mechanism to detect long-range dependencies. This hierarchical attention mechanism permits the model to efficiently collect both local and global features in the image while maintaining high efficiency.

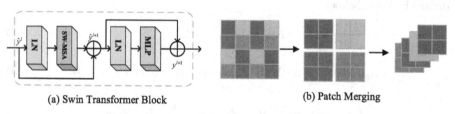

(a) Swin Transformer Block

(b) Patch Merging

Fig. 3. The structure of Swin Transformer Block (a) and the principle of Patch Merging (b)

Specifically, the input image is divided into a series of patches, each of which is regarded as a sequence. The initial input sequence is subjected to a Patch Merging operation, shown in Fig. 3(b), before being fed into a set of Swin Blocks for feature extraction. It operates similarly to pooling in that it takes the maximum or average value of the identical position for every small window to create an additional patch and then concatenates all the patches. Patch Merging is used to decrease the resolution of every stage in Swin Transformer. The local and global modules in each Swin Block process the input sequence through a self-attention mechanism to obtain the feature representation.

After multiple Swin Blocks to compute local attention in a non-overlapping local window, the global attention is implemented through a shift window mechanism. The final feature representation is sent to a global pooling layer for dimensionality compression. The compressed feature representations are delivered to a fully connected layer and a Softmax layer for classification prediction. The computational equation for layer l + 1 is shown below:

$$\hat{y}^{l+1} = SW - MSA(LN(y^l)) + y^l,$$
$$y^{l+1} = MLP(LN(\hat{y}^{l+1})) + \hat{y}^{l+1} \tag{10}$$

where y^l denote the output features of the MLP module for block l (l = 1, 2, 3) and SW-MSA indicates shifted window partitioning setups.

2.4　Loss Founction

In the training phase, the total loss (Loss$_{total}$) function of our network consists of two parts, including the content loss (L_{con}) and the GAN loss (L_{adv}). The loss function formula is defined as:

$$Loss_{total} = L_{con} + L_{adv} \tag{11}$$

Content loss includes structural similarity loss and L1 norm. Synthetic images usually need to maintain a consistent structure with the original image to ensure that the fused results appear realistic and natural. By using structure loss, the generated image can be forced to be more structurally similar to the real image, thus maintaining the overall structural consistency of the image. L1 norm is a common pixel-level difference metric that calculates the absolute difference per pixel between the generated image and the real image and can reduce noise and artifacts in the generated image. The calculation formula is shown below:

$$L_{con} = L_{SSIM} + \alpha L_{l1} \tag{12}$$

$$L_{SSIM} = \lambda_1(1 - SSIM(I_F, I_{MRI})) + \lambda_2(1 - SSIM(I_F, I_{PET})) \tag{13}$$

$$L_{l1} = \frac{1}{HW}(\gamma_1 \sum |I_F - I_{MRI}| + \gamma_2 \sum |I_F - I_{PET}|) \tag{14}$$

where SSIM (·) and |·| denote the structural similarity operation and the absolute value operation, respectively. We set the parameter $\lambda_1 = \lambda_2 = \gamma_1 = \gamma_2 = 0.5$. α is the hyperparameter that controls the tradeoff between the two loss terms.

Applying GAN loss to both MRI and PET images can drive the generated images to be consistent with the original images in terms of structure and features. The fused images will better match the anatomical structure and functional information of the original images, enhancing the correspondence of the fusion results. Adversarial loss and discriminator loss are calculated as follows:

$$L_{adv}(I_F, I_i) = L_G = \frac{1}{N} \sum_{i=1}^{2} \sum_{n=1}^{N} (D(G(I_i^n)) - 1))^2,$$

$$L_{D_i} = \frac{1}{N} \sum_{n=1}^{N} [(D(I_F^n) - 1)^2 + (D(I_i^n) - 0)^2](i = 1, 2)$$

(15)

where $n \in \mathbb{N}_N$, N is the number of images. Note that we set I_1 and I_2 to denote the initial images I_{MRI} and I_{PET}, respectively.

3 Experiments

3.1 Experimental Setup

During the training phase, we utilized the Harvard Medical Dataset, which consists of 58 pairs of PET and MRI images, to train our network. Due to the limited size of the dataset, we employed augmentation techniques such as flipping and panning to expand it. Specifically, each image was flipped twice, both vertically and horizontally, resulting in a total of 174 images. Additionally, we applied shifting operations to each image, including upward, downward, leftward, and rightward shifts, resulting in four separate images for each original image. This augmentation process yielded a total of 696 image pairs. To facilitate network training, we resized all images to a size of 224 × 224. The Adam optimizer was utilized, with a learning rate of 1×10^{-3}, a batch size of 4, and training conducted for 100 epochs. In the testing phase, we selected an additional 20 pairs of MRI and PET images to be fed into our trained generator to retrieve our fusion results. Our proposed CDBIFsion is implemented in the PyTorch framework and trained with a NVIDIA-GTX1080GPU with 12G of memory.

3.2 Qualitative Assessments

The outcomes of CDBIFusion and the other nine approaches are depicted in Fig. 4. There are two original MRI and PET images and the results of CBF [21], TIF [22], FPDE [23], LatLRR [24], U2Fusion [25], EMF [26], SeAFusion [27], CUDF [28], MSDRA [29], and our CDBIFusion, respectively. Clearly observable is the fact that while TIF, EMF, and SeAFusion retain the color information of PET images in their fusion results, TIF sacrifices the intricate details of brain edges, and both EMF and SeAFusion produce excessively blurred structural information within the brain tissue. The brain texture information in MRI images of EMF and SeAFusion is virtually eradicated. On the contrary, CBF primarily preserves the structural intricacies of the complete MRI images but struggles to integrate the brightness information from the PET image. The fusion image generated from CUFD and MSDRA exhibits diminished brightness compared to the original PET image, with the detailed information within the skull appearing significantly blurred. Moreover, the fusion results of LatLRR are overexposed, while FPDE retains the weakened source image information. Although U2Fusion successfully integrated most of the information of the original image, it can be clearly seen from the local magnification of the temporal area that our CDBIFusion model outperforms it by more fully retaining the high-resolution texture details of the brain structure and the edge information near the eyes. In addition, our method effectively integrates the luminance information of PET images to produce more remarkable visual effects.

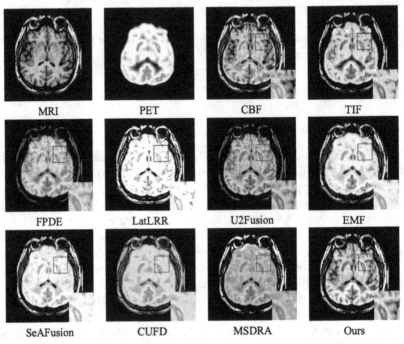

Fig. 4. Qualitative comparisons of CDBIFusion with nine state-of-the-art methods

3.3 Quantitative Comparisons

To validate the effectiveness of our approach, we employed six evaluation metrics: entropy (EN), image noise ($N^{ab/f}$), mutual information (MI), normalized mutual information (QMI), nonlinear correlation information entropy (NCIE), and mutual information between images ($MI^{ab/f}$). According to Table 1, it is apparent that our CDBIFusion outperformed all others in terms of EN, MI, QNCIE, and $MI^{ab/f}$, with the best and second-best values highlighted in bold and underlined, respectively. The EN metric indicates that the fused image successfully preserves the intricate texture and detailed information of the original image. This accomplishment is due to our suggested CDBI module, which minimizes information loss during the stages of fusion. The higher MI and $MI^{ab/f}$ values imply a strong correlation between the fused image and the source image, indicating that our method effectively retains rich and valuable information from the source. The largest NCIE value shows the significant relationship between the fused result and the original image. This is due to our method's utilization of discriminators to monitor and optimize the generator using a specific loss function, resulting in well-balanced fusion outcomes. The noise value is not optimal because our fusion result has relatively comprehensive information about the source image that can be interpreted. According to the statistical results, our method definitely beats others in terms of objective evaluation measures, suggesting better performance.

Table 1. The quantitative evaluations results on the PET and MRI dataset

	EN	$N^{ab/f}$	MI	QMI	NCIE	$MI^{ab/f}$
CBF	5.138	0.012	10.270	0.673	0.8069	2.918
TIF	4.975	0.017	9.950	0.601	0.8053	2.502
FPDE	5.081	0.002	10.16	**0.696**	0.8066	2.950
LatLRR	3.827	0.093	7.655	0.595	0.8046	2.142
U2fusion	5.034	**0.001**	10.070	0.656	0.8061	2.769
EMF	4.984	0.007	9.969	0.696	0.8064	2.910
SeAFusion	5.243	0.008	10.486	0.663	0.8063	2.869
CUFD	5.071	0.003	10.142	0.646	0.8057	2.709
MSDRA	4.064	0.032	8.128	0.696	0.8054	2.575
Ours	**5.390**	0.004	**10.780**	0.676	**0.8079**	**3.233**

3.4 Ablation Study

In our network, the proposed CDBI module utilizes the deactivation information of the ReLU activation function to fully keep the information in the source images. In addition, the Swin Transformer module is exploited to draw global information and establish long-range dependencies. We evaluated the effectiveness of two modules through ablation experiments. NO_Neg.ReLU indicates that the CDBI module contains only normal ReLU operations (removal of negative ReLU operations), and NO_Swin indicates a network without a Swin Transformer. Table 2 displays the outcomes of the ablation trials. The optimal average value of the evaluation metrics is marked in bold. Our network obviously exceeds the other two architectures in four metrics: EN, MI, QMI, and NCIE. This indicates that the CDBI and Swin Transformer modules are able to promote information exchange and complementarity between different branches, enhance the representation of important features, prevent feature loss while extracting global information and establishing long-term dependencies, give full play to their respective advantages, and strengthen the network's feature representation capability.

Table 2. The quantitative results of ablation comparisons

Fusion tasks	Negative ReLU	Swin Tranfomer	Metrics					
			EN	$N^{ab/f}$	MI	QMI	NCIE	$MI^{ab/f}$
No_Neg.ReLU		✓	5.189	0.0048	10.378	0.677	**0.8080**	3.259
No_Swin	✓		5.181	**0.0039**	10.361	0.688	0.8078	3.203
CDBIFusion	✓	✓	**5.390**	0.0049	**10.78**	**0.699**	0.8079	**3.311**

4 Conclusion

In this paper, we present a fusion network combining GAN and a joint CNN and transformer for PET and MRI images. Our network effectively harnesses the strengths of CNN and transformer models, enabling comprehensive extraction of both local and global information from the primary images. Moreover, we introduce a CDBI module that employs two ReLU rectifiers to minimize information loss while simultaneously preserving and integrating valuable details bidirectionally across two CNN paths. The results of extensive experiments indicate that CDBIFusion surpasses other leading fusion algorithms, validating its effectiveness in image fusion tasks. We will continue to refine our network's development in the future to optimize its performance and capabilities.

References

1. Zhao, Z., et al.: Cddfuse: correlation-driven dual-branch feature decomposition for multi-modality image fusion. In: Proceedings of the IEEE/CVF Conference on Computer Vision and Pattern Recognition, pp. 5906–5916 (2023)
2. Zhao, Z., et al.: Equivariant Multi-Modality Image Fusion. ArXiv abs/2305.11443 (2023)
3. Al-Mualla, M.: Ebrahim, bull, david, hill, paul: perceptual image fusion using wavelets. IEEE Trans. Image Process. **26**, 1076–1088 (2017)
4. Du, J., Li, W., Xiao, B., Nawaz, Q.: Union Laplacian pyramid with multiple features for medical image fusion. Neurocomputing **194**, 326–339 (2016)
5. Singh, S., Gupta, D., Anand, R.S., Kumar, V.: Nonsubsampled shearlet based CT and MR medical image fusion using biologically inspired spiking neural network. Biomed. Signal Process. Control (2015)
6. Bhatnagar, G., Wu, Q., Zheng, L.: Directive contrast based multimodal medical image fusion in NSCT domain. IEEE Trans. Multimedia **9**, 1014–1024 (2014)
7. Liu, Y., Wang, Z.: Simultaneous image fusion and denoising with adaptive sparse representation. Image Process. Iet **9**, 347–357 (2014)
8. Yu, L., Xun, C., Cheng, J., Hu, P.: A medical image fusion method based on convolutional neural networks. In: 2017 20th International Conference on Information Fusion (2017)
9. Wang, K., Zheng, M., Wei, H., Qi, G., Li, Y.: Multi-modality medical image fusion using convolutional neural network and contrast pyramid. Sensors **20**, 2169 (2020)
10. Lahoud, F., Süsstrunk, S.: Zero-learning fast medical image fusion. In: 2019 22th International Conference on Information Fusion (2019)
11. Song, S., Wang, J., Wang, Z., Su, J., Ding, X., Dang, K.: Bilateral-Fuser: A Novel Multi-cue Fusion Architecture with Anatomical-aware Tokens for Fovea Localization. arXiv preprint arXiv:2302.06961 (2023)

12. Ma, J., Yu, W., Liang, P., Li, C., Jiang, J.: FusionGAN: a generative adversarial network for infrared and visible image fusion. Inform. Fus. **48**, 11–26 (2019)
13. Ma, J., Xu, H., Jiang, J., Mei, X., Zhang, X.P.: DDcGAN: a dual-discriminator conditional generative adversarial network for multi-resolution image fusion. IEEE Trans. Image Process. (2020)
14. Guo, X., Nie, R., Cao, J., Zhou, D., Mei, L., He, K.: FuseGAN: learning to fuse multi-focus image via conditional generative adversarial network. IEEE Trans. Multimedia 1982–1996 (2019)
15. Ma, J., Zhang, H., Shao, Z., Liang, P., Xu, H.: GANMcC: a generative adversarial network with multiclassification constraints for infrared and visible image fusion. IEEE Trans. Instrument. Measure. 1 (2020)
16. Hao, Z.A., Zl, A., Zs, B., Han, X.A., Jm, A.: MFF-GAN: an unsupervised generative adversarial network with adaptive and gradient joint constraints for multi-focus image fusion. Inform. Fusion **66**, 40–53 (2021)
17. Wang, Z., Shao, W., Chen, Y., Xu, J., Zhang, X.: Infrared and Visible Image Fusion via Interactive Compensatory Attention Adversarial Learning (2022)
18. Ma, T., Li, B., Liu, W., Hua, M., Dong, J., Tan, T.: CFFT-GAN: cross-domain Feature Fusion Transformer for Exemplar-based Image Translation. ArXiv abs/2302.01608, (2023)
19. Hu, Q., Guo, X.: Trash or Treasure? An Interactive Dual-Stream Strategy for Single Image Reflection Separation (2021)
20. Liu, Z., et al.: Swin transformer: hierarchical vision transformer using shifted windows. In: Proceedings of the IEEE/CVF international conference on computer vision, pp. 10012–10022 (2021)
21. Kumar, B.S.: Image fusion based on pixel significance using cross bilateral filter. Signal, image and video processing (2015)
22. Dhuli, Ravindra, Bavirisetti, Prasad, D.: Two-scale image fusion of visible and infrared images using saliency detection. Infrared Phys. Technol. (2016)
23. Bavirisetti, D.P.: Multi-sensor image fusion based on fourth order partial differential equations. In: 20th International Conference on Information Fusion (2017)
24. Li, H., Wu, X.J.: Infrared and visible image fusion using Latent Low-Rank Representation (2018)
25. Xu, H., Ma, J., Jiang, J., Guo, X., Ling, H.: U2Fusion: a unified unsupervised image fusion network. IEEE Trans. Pattern Anal. Mach. Intell. 1–1 (2020)
26. Xu, H., Ma, J.: EMFusion: an unsupervised enhanced medical image fusion network. Inform. Fusion **76**, 177–186 (2021)
27. Tang, L., Yuan, J., Ma, J.: Image fusion in the loop of high-level vision tasks: a semantic-aware real-time infrared and visible image fusion network. Inform. Fusion **82**, 28–42 (2022)
28. Xu, H., Gong, M., Tian, X., Huang, J., Ma, J.: CUFD: an encoder–decoder network for visible and infrared image fusion based on common and unique feature decomposition. Comput. Vis. Image Understand. 218 (2022)
29. Li, W., Peng, X., Fu, J., Wang, G., Huang, Y., Chao, F.: A multiscale double-branch residual attention network for anatomical–functional medical image fusion. Comput. Biol. Med. **141**, 105005 (2022)

LF-LVS: Label-Free Left Ventricular Segmentation for Transthoracic Echocardiogram

Qing Kang[1,2], Wenxiao Tang[1], Zheng Liu[1], and Wenxiong Kang[1,2(✉)]

[1] South China University of Technology, Guangzhou 510640, China
auwxkang@scut.edu.cn
[2] Pazhou Lab, Guangzhou 510335, China

Abstract. Left ventricular segmentation for transthoracic echocardiographic (TTE) is crucial for advanced diagnosis of cardiovascular disease and measurement of cardiac function parameters. Recently, some TTE ventricular segmentation methods achieved satisfactory performance with large amounts of labeled data. However, the process of labeling medical segmentation data requires specialist surgeons and is highly time-consuming. To reduce reliance on segmentation annotations, we propose a label-free approach for left ventricular segmentation for TTE named LF-LVS. Specifically, we design multiple sets of templates and employ three common data enhancement strategies to generate pseudo-ultrasound masks and their corresponding pseudo-ground truths (pseudo-GTs). Then, we utilize CycleGAN with real-world TTE images to construct a synthetic transthoracic echocardiographic left ventricular segmentation dataset (STE-LVS), which will play an important role in the research of TTE left ventricular segmentation. Finally, we feed both the synthetic and real-world TTE data into a weight-shared segmentation network, and devise a domain adaptation discriminator to ensure their similarity in the output space of the segmentation network. Extensive experiments demonstrate the effectiveness of our proposed LF-LVS, which achieves satisfactory performance on the EchoNet-Dynamic dataset without any annotation. Our STE-LVS dataset and code are available at https://github.com/SCUT-BIP-Lab/LF-LVS.

Keywords: Transthoracic echocardiogram · Synthetic datasets · Label-free semantic segmentation · Domain adaptation

1 Introduction

The heart plays a crucial role in the circulation of human blood. Cardiac muscle dysfunction can result in various conditions, such as heart failure, cardiomyopathy, and pericardial effusion. In the field of surgery, transthoracic echocardiography (TTE) is regarded as one of the most effective diagnostic tools for heart diseases due to its non-invasive nature and cost-effectiveness. In addition, TTE data provides valuable information about cardiac structures (e.g., cardiac chamber size, cardiac valves) and measure various cardiac function parameters. The

Q. Liu et al. (Eds.): PRCV 2023, LNCS 14437, pp. 448–459, 2024.
https://doi.org/10.1007/978-981-99-8558-6_37

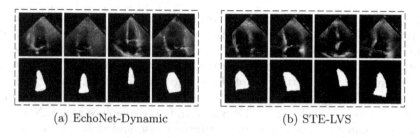

(a) EchoNet-Dynamic   (b) STE-LVS

Fig. 1. Some TTE images from the EchoNet-Dynamic dataset and the STE-LVS dataset, and their corresponding left ventricular segmentation annotations.

left ventricular ejection fraction (LVEF), a crucial metric in TTE for assessing cardiac function, is typically calculated by end-systolic volume (ESV) and end-diastolic volume (EDV) of the left ventricle. Professional surgeons analyze TTE videos of multiple cardiac cycles and manually label the left ventricular structures to estimate LVEF, which is complex and time-consuming. Therefore, it is important to automatically segment the left ventricle in the TTE data, which is essential for the real-time estimation of LVEF.

Recently, some works [10,12,14,16] had successfully estimated the LVEF for TTE. Jafari et al. [10] trained segmentation networks on 2,562 left ventricular segmentation annotated datasets and implemented LVEF estimation on mobile devices. Ouyang et al. [12] created a large-scale EchoNet-Dynamic dataset to estimate the ejection fraction by a video classification method. Zhang et al. [16] constructed a dataset of 14,035 echocardiograms over ten years, and they estimated ejection fraction by implementing ventricular segmentation for quantifying ventricular volume and left ventricular mass. However, these methods required large-scale annotated data, which took much time of the surgeon. Therefore, a label-free left ventricular segmentation method to estimate the LVEF is urgently required.

In this paper, we propose a label-free left ventricular segmentation (LF-LVS) method for TTE, which avoids the large demand for annotated segmentation data. Specifically, we draw 89 cardiac templates and generate pseudo-ultrasound masks and their corresponding pseudo-GTs with three commonly used data augmentation strategies. We utilize CycleGAN [21] to generate pseudo-ultrasound images from the augmented templates with similar distribution to the real-world TTE data. Meanwhile, we treat the mask in the template as the pseudo-GT of its corresponding image to construct the synthetic transthoracic echocardiographic left ventricular segmentation dataset (STE-LVS) with 9,960 synthetic TTE images. Note that the construction of our STE dataset requires only the real-world TTE images without their annotation. For more detailed information, please refer to Fig. 1, where we can find that our synthetic data is similar to the real-world data.

Furthermore, we propose a domain adaptation method to narrow the output distribution of synthetic and real-world TTE data. We feed them to a weight-sharing segmentation network and devise a discriminator to close their output distribution. In this stage, our synthetic data is regarded as the source domain, while the real-world TTE data serves as the target domain. We train the segmentation network with synthetic data, while using discriminator to constrain the similarity of the distribution of the source and target domains in the output space of the segmentation network. Therefore, our segmentation network is able to work on the real-world TTE data during the inference phase. Note that the discriminator is removed in the inference phase and thus causes no additional computational overhead. Experiments on various segmentation networks demonstrate the effectiveness of our method.

The main contributions are summarized as follows:

- We draw some templates and utilize CycleGAN to construct the STE-LVS dataset with 9,960 synthetic images, which will significantly contribute to the study of left ventricular segmentation in the medical field.
- We propose a domain adaptation method consists of a segmentation network and a discriminator network, and the latter strictly constrains the real-world TTE data and the synthetic data to share a similar distribution in the output space of the segmentation network.
- Extensive experiments demonstrate that our proposed LF-LVS achieves superior segmentation performance on real-world TTE data without sacrificing the inference speed.

2 Related Work

Automatic left ventricular segmentation of transthoracic echocardiograms is challenging due to heartbeat during the cardiac cycle and artifacts in ultrasound imaging [19]. Meanwhile, the annotation of medical image datasets usually requires experts, which is very expensive and results in few datasets for left ventricle segmentation. The deep learning-based segmentation algorithm requires larger numbers of labeled data and higher quality, which also significantly limits its performance.

Recently, deep learning has achieved satisfactory performance in medical image segmentation tasks. The full convolutional network (FCN) [11] was the first exploration of deep learning for image semantic segmentation tasks. Subsequently, U-Net [14] designed the U-shaped network to generate pixel-level predictions for the input images. Considering the characteristics of pooling and multi-scale convolution, many current semantic segmentation works (SPPNet [8], DeepLab series [3–5]) have been proposed. U-Net++ [20] improved the weakness of U-Net [14] by designing an architecture with dense skip connections to promote feature fusion effectively. To improve the complementarity between features at different levels, MSNet [18] uses a multi-scale subtractive network to segment polyps in colonoscopy images. Based on MSNet [18], MS2Net [17] proposes a general medical image segmentation method through subtractive aggregation and modified single-scale subtractive units. FLANet [15] proposes a fully attentional network for semantic segmentation based on channel and spatial self-attention.

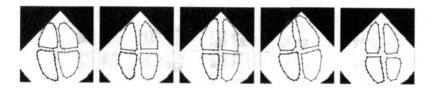

Fig. 2. Visualization of some four-chamber heart templates with hand-drawn contours in transthoracic echocardiography.

Recently, many networks based on transformer have been proposed for medical image segmentation as they learn global semantic information of images, which is helpful for segmentation. TransUNet [2] applies the Vit structure [6] to U-Net [14] for 2D image segmentation. Other transformer-based networks like MedT [2] and UTNet [7] has also been proposed for medical image segmentation. Migrated and modified computer vision segmentation algorithms perform well on echocardiographic left ventricle segmentation. Zhang et al. [16] designed a left ventricular segmentation model based on the U-Net [14] network structure to interpret echocardiographic studies. Similar to this method, Jafari et al. [10] proposed a modified U-Net [14] to improve model performance by adversarial training for LV segmentation. To deeply study the automatic segmentation of transthoracic echocardiography, Ouyang et al. [12,13] construct a large transthoracic echocardiography dataset EchoNet-Dynamic and use DeepLabv3 [4] framework to achieve left ventricle segmentation. Chen et al. [1] proposed a weakly supervised algorithm that pays more attention to the edge of the left ventricle based on DeepLabv3 [4].

3   Proposed Method

In this section, we elaborate on our proposed LF-LVS. It first constructs a synthetic STE-LVS dataset without the involvement of any experienced surgeons. Then, we propose a domain adaptation method, taking the synthetic data as the source domain and the real-world TTE data as the target domain, which effectively ensures that the segmentation output distribution of the real-world TTE data could share the similarity with that of the synthetic data.

3.1   The Construction of STE-LVS Dataset

In computer vision tasks, the data and its annotation play a crucial role in model performance. However, the annotation of TTE left ventricle segmentation requires specialist surgeons with considerable time and effort to complete, which makes it challenging to construct large-scale segmentation datasets. To avoid annotation costs, we constructed the STE-LVS dataset for TTE left ventricular segmentation. Specifically, we invited six non-medical data annotators to view the annotation of four-chambered hearts of some TTE data, which mainly

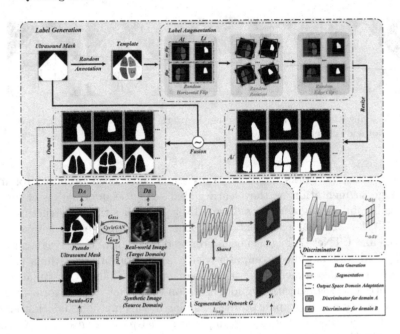

Fig. 3. The pipeline of our proposed label-free approach for left ventricular segmentation for TTE (LF-LVS). We first draw the templates and perform three data augmentation strategies on them. Then, we utilize CycleGAN with real-world TTE data to transform the pseudo ultrasound masks into the synthetic TTE images. Finally, the segmentation network and discriminator are alternately trained to ensure that the output distribution of the real TTE data and the synthetic data are consistent. In the inference phase, we keep only the segmentation network.

reflected the learning of segmentation contours. After that, they draw 15 contours on the ultrasound mask based on their impressions and personal subjective will, without any TTE data and annotations as a reference. Finally, we gathered these contours and calculated the Intersection over Union (IoU) between them. we found a pair with a high degree of overlap and we dropped one of them. Thus, we drew 89 templates in total, and some of the drawn contours are shown in Fig. 2.

To fully simulate the collection of TTE data in a realistic scenario, we performed data augmentation of these templates, including random horizontal flip $\mathcal{F}$, random rotation $\mathcal{R}$, and random edge crop $\mathcal{C}$. The horizontal flip enables expansion of the template image, the random rotation with angle $\theta \in (-5°, 5°)$ is performed to simulate the changing position of the probe during the ultrasound examination, and the random cropping is used to simulate the systolic and diastolic situation of the heart at different moments. Assume that $\{A_t, L_t\} \in \mathbb{R}^{H \times W \times 3}$ is the data pair of the template and its GT, the data augmentation can be formulated as follows:

$$\{A_i', L_i'\}_{i \in \{1,2,...,N\}} = \mathcal{C}(\mathcal{R}(\mathcal{F}(\{A_t, L_t\})), \theta), l, b) \tag{3.1}$$

where $N = 120$ represents the number of augmented templates from each hand-drawn template, $l \in [0, 40]$ and $b \in [0, 50]$ represent the range of left and bottom edge cropping, respectively. Note that the left ventricle is located in the upper right corner of the template. We generate a new template after the flip operation, but the position of our four-chambered heart remains the same, with the left ventricle still in the upper right corner. Meanwhile, we only crop the left and bottom edges of the template to ensure the integrity of the left ventricle in the image.

After cropping, we resize the image from $(H - b) \times (W - l) \times 3$ to $H \times W \times 3$, which ensures that the input images share the same resolution. Finally, we utilize the ultrasound mask to generate a pseudo-ultrasound mask with its corresponding pseudo-GT.

We utilize CycleGAN [21] to generate synthetic TTE images from pseudo-ultrasonic masks, as shown in Fig. 3. The CycleGAN [21] includes two generators $G_{A2B}: A \to B$ and $G_{B2A}: B \to A$, where A and B denote the pseudo-ultrasonic four-chamber mask and the real-world TTE four-chamber image, respectively. Their data distributions can be represented as $x \sim P_{data}(A)$ and $y \sim P_{data}(B)$. We selected 6 hand-drawn ($6 \times 120 = 720$ augmented) templates to train the CycleGAN network, while the remaining 83 hand-drawn ($83 \times 120 = 9960$ augmented) templates were used to construct the STE-LVS dataset. Furthermore, we design two adversarial discriminators D_A and D_B, where D_A aims to constrain the similarity of the image $\{x\}$ and the translated image $\{G_{B2A}(y)\}$, while D_B aims to constrain that of image $\{y\}$ and $\{G_{A2B}(x)\}$.

In the training phase, G_{A2B} ensures that the distribution of the images generated from the pseudo-ultrasound mask is similar to that of the real-world TTE data. In the inference phase, we fix the parameters of G_{A2B} and generate 9,960 synthetic TTE images from the augmented templates. The STE-LVS dataset we constructed will significantly contribute to the research of TTE segmentation.

3.2   Domain Adaptation for Real-World TTE Data

Undeniably, even though the distribution of our synthetic data is similar to that of the real-world TTE data, it still remains differences between them. Note that our synthetic data is given with pseudo-GT, while the real-world TTE data remains unlabeled, and we need to achieve satisfactory performance on the real-world TTE test set. A direct way is to train the segmentation model on the STE-LVS dataset and apply it to the real-world TTE test set. It is evident that this approach drastically degrades the model's performance and fails to achieve satisfactory performance, please refer to Sect. 4.4. To this end, we propose a simple but effective domain adaptive approach, which incorporates a segmentation network and a discriminator to constrain the consistency of the output distribution of the real-world TTE and synthetic data.

Segmentation Network. Our segmentation network can employ the current mainstream segmentation algorithms, such as FCN [11], U-Net [14], and DeepLabv3 [14]. We feed the synthetic and real-world TTE data into the segmentation network simultaneously, with their outputs denoted as Y_s and Y_t, respectively. Note that our method is label-free, where the real-world TTE data is unlabelled and we only have the synthetic data with the segmentation label. Assuming that the ground truth of the synthetic data is Y_{gt} and the output of the segmentation network (G) is $Y_s = G(I_s) \in \mathbb{R}^{H \times W \times C}$, where C is the number of categories. We utilize the cross-entropy loss to train the segmentation network, and the L_{seg} can be written as follows:

$$L_{seg} = -\sum_{h,w}\sum_{c \in C} Y_{gt}^{h,w,c} log(Y_s^{h,w,c}) \qquad (3.2)$$

To make the data distribution of Y_t closer to Y_s, we utilize an adversarial loss L_{adv} for the discriminator (D) as:

$$L_{adv} = -\sum_{h,w} log(D(Y_t)^{(h,w,1)}) \qquad (3.3)$$

The L_{adv} loss is designed to train the domain-independent segmentation network and fool the discriminator by maximizing the probability of the target prediction being considered as the source prediction. Therefore, We optimize the parameters of the segmentation network by the following optimization guidelines $\max_{D} \min_{G} L$, which can be written as follows:

$$L = L_{seg} + \alpha L_{adv} \qquad (3.4)$$

where α represents the hyper-parameters to balance segmentation and adversarial loss.

Discriminator. The design of the discriminator D is inspired by PatchGAN [9] and its output is a 4×4 feature map. We feed the synthetic data and the real-world TTE data into the segmentation network and the outputs are noted as Y_s and Y_t, respectively. We take $Y_{i \in \{s,t\}}$ as the input of the fully-convolutional discriminator using a cross-entropy loss L_{dis} for two classes (i.e., synthetic and real-world TTE data). The loss can be written as:

$$L_{dis} = -\sum_{h,w}(1-z)log(D(Y_s)^{(h,w,0)}) + zlog(D(Y_t)^{(h,w,1)}) \qquad (3.5)$$

where $z = 0$ and $z = 1$ denote the output of segmentation network is draw from synthetic and real-world TTE data domain, respectively. The core aim of the discriminator We designed is to make the output distribution of the real-world TTE data as similar as possible to that of our synthesised data.

4   Experiments

4.1   Dataset

EchoNet-Dynamic Dataset. EchoNet-Dynamic [12] is a public TTE dataset containing 10,030 four-chamber echocardiographic videos with a resolution of

112×112, which is released by Stanford University Hospital. The dataset consists of 7,465 training videos, 1,277 validation videos, and 1,288 test videos, with pixel-level binary labels of left ventricular end-systolic (ES) and end-diastolic (ED). In our experiments, the ES and ED images of each video from the training set without their segmented label are utilized to ensure a consistent style of the synthetic image. Moreover, we evaluate the performance of different methods on the test set.

4.2 Evaluation Metrics

In our experiments, we mainly use the following three evaluation metrics for performance evaluation:

- *Dice Similarity Coefficient* (DSC) computes the overlay between the surface manually annotated by the cardiologist and the surface automatically segmented by the algorithm. The higher DSC indicates better segmentation performance.
- *Jaccard Coefficient* (JC) is also known as the IoU coefficient. It is used for comparing the similarity and diversity of the segmented region.
- *Hausdorff Distance* (HD) measures the maximum distance between the contour of the surface manually delineated and the automated contour. HD_{95} is a commonly used medical image segmentation indicator. The lower HD_{95} reflects the better performance.

4.3 Training Settings

We implement our proposed LF-LVS in Pytorch with an NVIDIA A5000. In our experiment, we compare our proposed method with seven segmentation networks. Following their vanilla papers [4,11,14,15,17,18,20], we choose ResNet-50 or Res2Net-50 as their backbone. For a fair comparison, all the segmentation networks are trained from scratch by an Adam optimizer with a batch size of 32, and the random seed is set to 2. The learning rate of the generator and the discriminator network is set to 1×10^{-4} and 2.5×10^{-4}, respectively. As for the CycleGAN network training, the learning rate of the generator and discriminator is set to 0.0002, and the batch size is set to 1.

4.4 Experimental Analysis

In this section, we validate our LF-LVS on some classical segmentation networks and the experimental results are shown in Table 1. The segmentation networks we applied include, traditional segmentation algorithms (FCN [11], U-Net [14], DeepLabv3 [4], U-Net++ [20]) and medical image segmentation algorithms (MSNet [18], FLANet [15], MS2Net [17]).

The baseline in Table 1 refers to the experiments where we trained different segmentation networks on synthetic datasets and directly made inference on real-world TTE test set. We argue that if the two datasets are not similar, the

Table 1. Comparisons between baseline methods and our proposed method on the test set of EchoNet-Dynamic dataset.

Method		Overall			ES			ED		
		DSC(%)	JC(%)	HD$_{95}$(mm)	DSC(%)	JC(%)	HD$_{95}$(mm)	DSC(%)	JC(%)	HD$_{95}$(mm)
FCN [11]	Baseline	80.23	69.33	11.62	83.15	72.92	10.07	77.32	65.74	13.16
(2015)	Ours	85.63↑	75.67↑	6.56↓	87.42↑	78.33↑	6.48↓	83.84↑	73.01↑	6.65↓
U-Net [14]	Baseline	68.40	54.39	30.96	73.25	59.70	27.75	63.54	49.09	34.17
(2015)	Ours	81.70↑	70.85↑	8.55↓	83.94↑	73.89↑	8.29↓	79.46↑	67.81↑	8.80↓
DeepLabv3 [4]	Baseline	80.10	69.17	11.23	83.08	72.82	10.39	77.12	65.52	12.07
(2017)	Ours	85.06↑	74.80↑	7.29↓	87.26↑	77.87↑	6.83↓	82.86↑	71.72↑	7.74↓
U-Net++ [20]	Baseline	78.37	66.30	14.15	81.84	70.57	12.14	74.89	62.02	16.15
(2018)	Ours	83.76↑	72.90↑	7.73↓	86.20↑	76.32↑	7.21↓	81.32↑	69.47↑	8.26↓
MSNet [18]	Baseline	77.59	65.88	13.28	79.76	68.64	11.63	75.41	63.11	14.93
(2021)	Ours	86.53↑	76.75↑	6.25↓	88.11↑	79.12↑	6.25↓	84.96↑	74.38↑	6.24↓
FLANet [15]	Baseline	74.28	62.23	16.03	78.07	66.61	13.14	70.50	57.85	18.95
(2022)	Ours	84.11↑	73.68↑	7.84↓	86.19↑	76.43↑	7.26↓	82.04↑	70.94↑	8.42↓
M2SNet [17]	Baseline	78.44	66.67	12.43	79.90	68.60	11.53	76.98	64.74	13.32
(2023)	Ours	85.59↑	75.60↑	6.41↓	87.01↑	77.71↑	6.50↓	84.16↑	73.49↑	6.31↓

Table 2. The DSC performance of our method based on different model and impact of α on EchoNet-Dynamic test set.

Based on	Year	$\alpha(\times 10^{-5})$								
		1	2	3	4	5	6	7	8	9
FCN [11]	2015	85.28	85.24	**85.63**	85.28	84.48	84.89	85.22	84.98	84.16
U-Net [14]	2015	79.90	78.92	80.27	80.61	**81.70**	80.57	81.07	79.71	80.39
DeepLabv3 [4]	2017	84.64	84.54	84.30	84.40	**85.06**	84.41	84.33	83.06	83.12
U-Net++ [20]	2021	83.37	82.21	82.99	**83.76**	82.68	82.37	82.57	82.76	81.61
MSNet [18]	2021	85.82	85.87	86.03	86.31	86.29	86.48	**86.53**	86.43	86.22
FLANet [15]	2022	83.48	83.35	**84.11**	83.01	82.58	81.44	83.47	83.30	82.68
MS2Net [17]	2023	85.53	**85.59**	85.11	85.03	85.22	85.11	85.44	85.40	85.50

generalization performance of the segmentation network will drop dramatically. We can clearly find from Table 1 that our baseline algorithm achieves moderate results on the real-world TTE dataset without any auxiliary operations, which indicates that the distribution of our synthetic dataset is somewhat similar to that of the real-world TTE dataset.

Compared to the baseline method, our LF-LVS achieves a performance improvement of more than 5% on the overall DSC evaluation metric in different segmentation networks. Meanwhile, our method also achieves significant improvements in the JC and HD$_{95}$ metrics. Our LF-LVS achieves optimal performance when MSNet [18] is utilized as the segmentation network. To analyse the reasons behind this, we counted the number of samples in the test set at different metrics, as shown in Fig. 4(a). It is obvious that our method has increased the number of samples for better performance at different metrics. We also visualised the segmentation output of MSNet [18] and our LF-LVS on the real TTE test set,

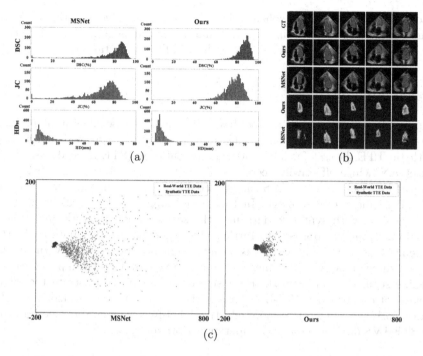

Fig. 4. (a) DSC, JC, and HD_{95} show the statistical information of the performance of MSNet and our method based on MSNet, overall frames. (b) Visual comparison of output results and class activation map of the last layer based on grad-cam on the real-world TTE data between our method and baseline based on MSNet. (c) The PCA dimensionality reduction results of the discriminator's response to the segmentation network's output.

as shown in Fig. 4(b). Our method shares a high similarity with the GT annotations in terms of segmentation completeness and structural regularity. Besides, to demonstrate the effectiveness of our domain adaptation algorithm, we feed the predicted output of MSNet [18] and LF-LVS into the discriminator D. We flatten the output heatmap (4×4) of the discriminator into a one-dimensional vector and utilize PCA to visualise the distribution of the synthetic and real-world TTE datasets, as shown in Fig. 4(c). It can be found that our domain adaptation enables to close the segmented output distribution of the synthetic and real-world TTE data, which demonstrates the effectiveness of our domain adaptation. Thus, our method naturally achieves better performance.

Furthermore, we conducted sufficient ablation experiments on the loss weights α during training of the segmentation network, and the experimental results are shown in Table 2. It can be found that, with all parameters fixed, the variation of α could make an impact on the segmentation performance of different algorithms. As the α gradually increases, the overall trend in segmentation performance remains first increasing and then decreasing. Moreover, there is some variation in the α when different algorithms achieve optimal performance, which proves that our STE-LVS dataset is challenging. We believe that our dataset will contribute to the research on TTE segmentation.

5 Conclusion

In this paper, we propose a label-free left ventricular segmentation method (LF-LVS) for transthoracic echocardiogram. In the first stage, we build a large synthetic TTE dataset for left ventricle segmentation (STE-LVS) dataset with CycleGAN, which effectively decreases the cost of segmentation annotations in medical images. Our STE-LVS dataset could make a significant contribution to TTE segmentation research. In the second stage, we feed real-world TTE data together with synthetic data into the segmentation network, and design domain adaptation approach to further close their output distribution. In the inference phase, we only keep the trained segmentation network without the domain discriminator, which increases no inference overhead compared to the baseline segmentation network. Extensive experiments demonstrate the effectiveness of our proposed LF-LVS on a variety of segmentation algorithms. In future work, we will explore more effective label-free segmentation methods on the STE-LVS dataset to achieve superior performance.

Acknowledgement. This work was supported by the Fundamental Research Funds for the Central Universities under Grant 2022ZYGXZR099, the National Natural Science Foundation of China under Grant 61976095 and the Natural Science Foundation of Guangdong Province, China, under Grant 2022A1515010114.

References

1. Chen, E., Cai, Z., Lai, J.h.: Weakly supervised semantic segmentation of echocardiography videos via multi-level features selection. In: Pattern Recognition and Computer Vision: 5th Chinese Conference, PRCV 2022, Shenzhen, China, 4–7 November 2022, Proceedings, Part II, pp. 388–400. Springer, Heidelberg (2022). https://doi.org/10.1007/978-3-031-18910-4_32
2. Chen, J., et al.: Transunet: transformers make strong encoders for medical image segmentation. arXiv preprint arXiv:2102.04306 (2021)
3. Chen, L.C., Papandreou, G., Kokkinos, I., Murphy, K., Yuille, A.L.: Deeplab: semantic image segmentation with deep convolutional nets, atrous convolution, and fully connected crfs. IEEE Trans. Pattern Anal. Mach. Intell. 40(4), 834–848 (2017)
4. Chen, L.C., Papandreou, G., Schroff, F., Adam, H.: Rethinking atrous convolution for semantic image segmentation. arXiv preprint arXiv:1706.05587 (2017)
5. Chen, L.C., Zhu, Y., Papandreou, G., Schroff, F., Adam, H.: Encoder-decoder with atrous separable convolution for semantic image segmentation. In: Proceedings of the European conference on computer vision (ECCV), pp. 801–818 (2018)
6. Dosovitskiy, A., et al.: An image is worth 16×16 words: transformers for image recognition at scale. arXiv preprint arXiv:2010.11929 (2020)

7. Gao, Y., Zhou, M., Metaxas, D.N.: UTNet: a hybrid transformer architecture for medical image segmentation. In: de Bruijne, M., et al. (eds.) MICCAI 2021. LNCS, vol. 12903, pp. 61–71. Springer, Cham (2021). https://doi.org/10.1007/978-3-030-87199-4_6

8. He, K., Zhang, X., Ren, S., Sun, J.: Spatial pyramid pooling in deep convolutional networks for visual recognition. IEEE Trans. Pattern Anal. Mach. Intell. **37**(9), 1904–1916 (2015)

9. Isola, P., Zhu, J.Y., Zhou, T., Efros, A.A.: Image-to-image translation with conditional adversarial networks. In: Proceedings of the IEEE Conference on Computer Vision and Pattern Recognition, pp. 1125–1134 (2017)

10. Jafari, M.H., et al.: Automatic biplane left ventricular ejection fraction estimation with mobile point-of-care ultrasound using multi-task learning and adversarial training. Int. J. Comput. Assist. Radiol. Surg. **14**, 1027–1037 (2019)

11. Long, J., Shelhamer, E., Darrell, T.: Fully convolutional networks for semantic segmentation. In: Proceedings of the IEEE Conference on Computer Vision and Pattern Recognition, pp. 3431–3440 (2015)

12. Ouyang, D., et al.: Echonet-dynamic: a large new cardiac motion video data resource for medical machine learning. In: NeurIPS ML4H Workshop: Vancouver, BC, Canada (2019)

13. Ouyang, D., et al.: Video-based AI for beat-to-beat assessment of cardiac function. Nature **580**(7802), 252–256 (2020)

14. Ronneberger, O., Fischer, P., Brox, T.: U-Net: convolutional networks for biomedical image segmentation. In: Navab, N., Hornegger, J., Wells, W.M., Frangi, A.F. (eds.) MICCAI 2015. LNCS, vol. 9351, pp. 234–241. Springer, Cham (2015). https://doi.org/10.1007/978-3-319-24574-4_28

15. Song, Q., Li, J., Li, C., Guo, H., Huang, R.: Fully attentional network for semantic segmentation. In: Proceedings of the AAAI Conference on Artificial Intelligence, vol. 36, pp. 2280–2288 (2022)

16. Zhang, J., et al.: Fully automated echocardiogram interpretation in clinical practice: feasibility and diagnostic accuracy. Circulation **138**(16), 1623–1635 (2018)

17. Zhao, X., et al.: M^2 snet: multi-scale in multi-scale subtraction network for medical image segmentation. arXiv preprint arXiv:2303.10894 (2023)

18. Zhao, X., Zhang, L., Lu, H.: Automatic polyp segmentation via multi-scale subtraction network. In: de Bruijne, M., et al. (eds.) MICCAI 2021. LNCS, vol. 12901, pp. 120–130. Springer, Cham (2021). https://doi.org/10.1007/978-3-030-87193-2_12

19. Zhou, J., Du, M., Chang, S., Chen, Z.: Artificial intelligence in echocardiography: detection, functional evaluation, and disease diagnosis. Cardiovasc. Ultrasound **19**(1), 1–11 (2021)

20. Zhou, Z., Rahman Siddiquee, M.M., Tajbakhsh, N., Liang, J.: UNet++: a nested u-net architecture for medical image segmentation. In: Stoyanov, D., et al. (eds.) DLMIA/ML-CDS -2018. LNCS, vol. 11045, pp. 3–11. Springer, Cham (2018). https://doi.org/10.1007/978-3-030-00889-5_1

21. Zhu, J.Y., Park, T., Isola, P., Efros, A.A.: Unpaired image-to-image translation using cycle-consistent adversarial networks. In: Proceedings of the IEEE International Conference on Computer Vision, pp. 2223–2232 (2017)

Multi-atlas Representations Based on Graph Convolutional Networks for Autism Spectrum Disorder Diagnosis

Jin Liu[1], Jianchun Zhu[1], Xu Tian[1], Junbin Mao[1], and Yi Pan[2(✉)]

[1] Hunan Province Key Lab on Bioinformatics, School of Computer Science and Engineering, Central South University, Changsha 410083, China
[2] Shenzhen Key Laboratory of Intelligent Bioinformatics, Shenzhen Institute of Advanced Technology, Shenzhen 518055, China
yi.pan@siat.ac.cn

Abstract. Constructing functional connectivity (FC) based on brain atlas is a common approach to autism spectrum disorder (ASD) diagnosis, which is a challenging task due to the heterogeneity of the data. Utilizing graph convolutional network (GCN) to capture the topology of FC is an effective method for ASD diagnosis. However, current GCN-based methods focus more on the relationships between brain regions and ignore the potential population relationships among subjects. Meanwhile, they limit the analysis to a single atlas, ignoring the more abundant information that multi-atlas can provide. Therefore, we propose a multi-atlas representation based ASD diagnosis. First, we propose a dense local triplet GCN considering the relationship between the regions of interests. Then, further considering the population relationship of subjects, a subject network global GCN is proposed. Finally, to utilize multi-atlas representations, we propose multi-atlas mutual learning for ASD diagnosis. Our proposed method is evaluated on 949 subjects from the Autism Brain Imaging Data Exchange. The experimental results show that the accuracy and an area under the receiver operating characteristic curve (AUC) of our method reach 78.78% and 0.7810, respectively. Compared with other methods, the proposed method is more advantages. In conclusion, our proposed method guides further research on the objective diagnosis of ASD.

Keywords: Autism Spectrum Disorder Diagnosis · Multi-atlas Representations · Graph Convolutional Network

1 Introduction

Autism spectrum disorder (ASD) is a developmental brain disorder characterized by stereotypic behavior, verbal communication deficits, and social difficulties [1]. The daily life behaviors of patients differ from those of normally developing individuals, and they may have some difficulties in adapting to daily life [12]. ASD has a variety of complex symptoms and the pathological mechanisms are unclear [14]. Due to the heterogeneity among patients [9], the diagnosis of ASD

© The Author(s), under exclusive license to Springer Nature Singapore Pte Ltd. 2024
Q. Liu et al. (Eds.): PRCV 2023, LNCS 14437, pp. 460–471, 2024.
https://doi.org/10.1007/978-981-99-8558-6_38

depends on the physician's subjective judgment of the patient's clinical behavioral criteria [6]. The diagnosis of ASD is a challenge.

Functional magnetic resonance imaging (fMRI), with its high temporal and spatial resolution and noninvasiveness [23], is an important technical tool for the diagnosis of ASD. A brain atlas consists of consecutive slices of the brain of a healthy or diseased developing along different anatomical planes where is a method of systematically describing and calibrating the structure and function of the brain. A functional brain atlas is made up of N regions of interest (ROI), where these regions are typically defined as spatially contiguous and functionally coherent patches of gray matter [17]. Yin et al. [24] use functional connectivity (FC) as the original input feature. Kunda et al. [10] propose a approach to second-order FC to extract better features for classification. Inspired by existing functional connectivity analysis models, Byeon et al. [2]propose a model that utilizes spatial and temporal feature extraction combined with phenotypic data for classifying ASDs. In general, brain disorders are likely to be caused by abnormalities in one or several regions of the brain, or by abnormal connections among brain regions. For these reasons, many studies have explored correlations between brain regions based on fMRI data to analyze the pathological mechanisms of ASD [15].

With the development of deep learning, more and more researchers are using deep learning to automate the diagnosis of ASD . Research has shown that functional connectivity in the brains of patients may differ from that of normally developing individuals [13]. Graph convolutional neural networks (GCNs) are used for autism recognition by capturing the topology of the network and extracting features in the brain network that are beneficial for identifying autism. Wang et al. [22] proposed a method for autism diagnosis based on multi-atlas deep feature representation and ensemble learning. You et al. [25] developed an identity-aware graph neural network by considering node information in the message delivery process. Wang et al. [20] proposed a Multi-Omics Graph cOnvolutional NETworks for classification and diagnosis of diseases. Wang et al. [21] proposed an autism diagnosis method based on Multi-Atlas Graph Convolutional Networks and an ensemble learning. The Multi-Atlas Graph Convolutional Networks method is used for feature extraction, and then an ensemble learning is proposed for the final disease diagnosis. However, existing GCNs based analysis of FC has two major shortcomings: First, several studies on ASD diagnosis with graph convolutional networks concern the relationship between paired brain regions but ignore the potential population relationship among subjects. Second, some studies only utilize a single atlas, while they ignore the association of multi-atlas.

To respond to the above challenges, we propose Multi-atlas Representations based on Graph Convolutional Networks for Autism Spectrum Disorder Diagnosis. Firstly, we calculate the FC of each subject by fMRI of multi-atlas and use the FC to calculate the local graph of ROIs. Secondly, we propose the DLTGCN based on ROIs, further build the subject network global graph and obtain new representations using the subject network GGCN. Finally, multi-atlas mutual learning is proposed to learn the representations of different atlases. Our proposed method is evaluated on the ABIDE dataset.

2   Related Work

2.1   Atlas-Based ASD Diagnosis

Diagnosis of ASD based on the atlas has been making good progress. Variability of a single atlas data (e.g., caused by multi-site acquisition, different pre-processing methods) leads to low performance of existing disease identification methods, Wang et al. [18] proposed Connectome Landscape Modeling (CLM) that can mine cross-site consistent connectome landscapes and extract representations of functional connectivity networks for the identification of brain diseases. Since there are problems of heterogeneity, dimensional catastrophe, and interpretability among atlas-based data, Wang et al. [19] proposed Multi-site Clustering and Nested Feature Extraction. They clustered heterogeneous brain functional connectivity networks. Then they designed nested singular value decomposition method to learn functional connectivity features. Finally, a linear support vector machine was used to diagnose ASD. Since previous studies used data from a single atlas from an imaging center, Huang et al. [7] proposed Self-weighted Adaptive Structure Learning (SASL) based on a multi-atlas multi-center integration scheme. The method is based on different predefined atlases to construct multiple Functional Connectivity brain networks for each subject. The final diagnostic results are obtained using SASL learning features using an integrated strategy of multi-atlas multi-center representation. However, current atlas-based studies more often utilize only a single atlas and do not make better use of the relationship among multi-atlas.

2.2   GCN-Based ASD Diagnosis

GCNs have more powerful ability to aggregate information of graphs and capture topological information. There is a growing number of studies related to GCNs. For example, Li et al. [11] proposed a method to analyze fMRI and discover neurobiological markers. In order to solve the problem that existing methods ignore the non-imaging information related to subjects and the relationship between subjects, and fail to analyze the local brain regions and biomarkers related to the disease, Zhang et al. [26] proposed a local-to-global graph neural network for autism diagnosis. A local network was designed to learn the features of local brain regions, and a global network was built to encode the non-imaging information and further learn the relationship between multiple subjects. Since it is challenging to complementarily integrate image and non-image data into a unified model when studying inter-subject relationships, Huang et al. [8] proposed a framework that can automatically learn to construct population graphs with variational edges using multimodal data. However, current studies related to the GCN have focused more on the relationship between paired brain regions and ignored the potential population relationships among subjects.

3   Materials and Methods

The flowchart of our proposed method for ASD diagnosis is shown in Fig 1. As can be seen from the Fig 1, First, we use the time series to calculate the FC of the different atlases. Second, we propose a multi-atlas graph representation based on triplet GCNs. Finally, we propose multi-atlas mutual learning, where the representations of different atlases are mutually learned and the diagnosis results are obtained after a weighted fusion of features. We describe these parts in detail in the following sections.

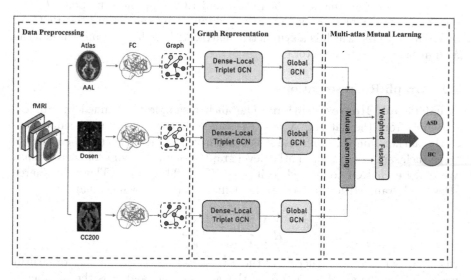

Fig. 1. An overall flowchart of our proposed method for ASD diagnosis.

3.1   Data Acquisition and Preprocessing

We use fMRI from the ABIDE dataset for the diagnosis of ASD. In this study, we used preprocessed fMRI from the Configurable Pipeline for the Analysis of Connectomes (CPAC) [3] of the Preprocessed Connectomes Project (PCP). Time series are acquired from three different atlases, Automatic Anatomical Labeling (AAL) [16], Craddock200 (CC200) [4], and dosenbach160 (DOSEN) [5]. CPAC functional pre-processing procedures include slice time correction, motion correction, skull stripping, global average intensity normalization, nuisance signal regression and alignment of functional images to anatomical space. Since there are some subjects with missing time series in the preprocessed fMRI obtained from ABIDE, we removed subjects with missing time series by quality control to obtain time series for a total of 949 subjects (419 ASD and 530 HC). The pre-processed fMRI is divided into M ROIs according to a specific atlas, and the mean value of voxels in each ROI over its time series is calculated to obtain the time series of M ROIs.

In this study, the pre-processed time series are used to calculate the FC between pairs of ROIs. The Pearson correlation coefficient (PCC) of the time series between each pair of ROIs is calculated as FC. The formula was as follows:

$$\rho_{a,b} = \frac{\sum_{t=1}^{T} (\gamma_a(t) - \overline{\gamma_a})(\gamma_b(t) - \overline{\gamma_b})}{\sqrt{\sum_{t=1}^{T} (\gamma_a(t) - \overline{\gamma_a})^2} \cdot \sqrt{\sum_{t=1}^{T} (\gamma_b(t) - \overline{\gamma_b})^2}}, \tag{1}$$

where $\overline{\gamma_a}$ and $\overline{\gamma_b}$ are the mean values of the time series of ROI a and ROI b. $\gamma_a(t)$, $\gamma_b(t)$ are the time series of ROI a and ROI b at the time point t. T is the total number of time points. FC is obtained by calculating the PCC for each pair of ROIs, which is a symmetric matrix with each row represented as an M-dimensional feature vector.

3.2   Graph Representation

Local Graph Representation: The anchor sample is obtained by random sampling from the dataset. A positive sample is the one with the same class as the anchor sample, otherwise it is a negative sample. The triplet consists of three such samples. A shared atlas-level graph structure is built for each sample. We make each ROI connected to its first K neighbors only. Thus, the shared atlas-level graph structure ($\overline{A}$) can be defined as the following equation:

$$\overline{A} = \frac{1}{H} \sum_{h=1}^{H} A_h + \frac{1}{P} \sum_{p=1}^{P} A_p, \tag{2}$$

where A_h is the graph structure of the h-th sample, H is the number of healthy people, A_p is the graph structure of the p-th sample, and P is the number of patients.

Considering the pairwise relationships between samples, DLTGCN is used to measure similarity between samples. It narrows the distance between similarities and widens the distance between dissimilarities, as shown in Fig 2. To extract the triplet relationships among samples, DLTGCN in the proposed method contains three parallel GCNs with shared parameters for each atlas. To better utilize the learned knowledge, the proposed DLTGCNs are connected by means of dense connections based on the thought of DenseNet.

As seen in Fig 1, the input of DLTGCN is a triplet. The triplet consists of an anchor sample, a sample of the same category as the anchor sample (positive sample), and a sample of a different class from the anchor sample (negative sample).Each DLTGCN calculates the similarity between positive sample and negative sample pairs respectively after learning the representation of the triplet. The equation for the loss function is described as follows:

$$M_i = \sum_{k=1}^{b} \max\left(\|f_i(X_{a,k}) - f_i(X_{p,k})\|_2^2 - \|f_i(X_{a,k}) - f_i(X_{n,k})\|_2^2 + \lambda_i, 0 \right),$$

$$\tag{3}$$

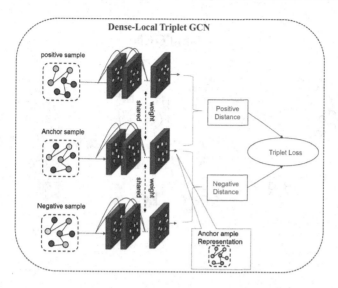

Fig. 2. Dense Local Triplet GCN.

where b is the number of triplets. $f_i(\cdot)$ is the DLTGCN graph learning representation function for the i-th atlas. λ_i is the distance hyperparameter parameter. The output of DLTGCN is the similarity of the anchor samples to different classes of sample pairs and the new graph representation of the anchor samples.

Global Graph Representation: As shown in Fig 3, the sample network graph is built by the representations of anchor samples. The similarity between the samples' representations is used to calculate the adjacency matrix of the sample network, which is calculated as follows:

$$Similar_{x,y} = 1 - distance_{x,y},$$
$$distance_{x,y} = \|x - y + \epsilon e\|_p \tag{4}$$

where x and y are the feature vectors of the two samples. ϵ is a very small number and e is a unit vector. $\|\cdot\|_p$ denotes the $p - norm$. The formula is as follows:

$$\|Z\|_p = \left(\sum_{m=1}^{n} |z_m|^p \right)^{\frac{1}{p}} \tag{5}$$

In this study, $p = 2$. We obtain the new representations of samples by sample network GGCN.

3.3 Multi-atlas Mutual Learning

Previous studies using ensemble learning cannot directly utilize the knowledge of other atlases. We propose multi-atlases mutual learning to help these DLT-GCN and GGCN (DLTGGCN) learn representations. The learning process is restricted by Kullback-Leibler (KL).

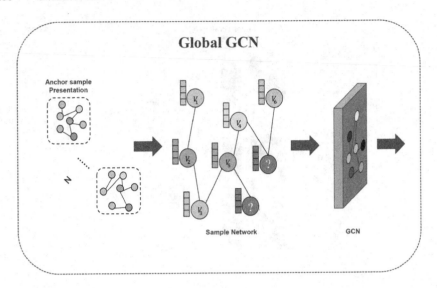

Fig. 3. Global GCN.

KL is used to measure the difference in the predicted distribution of DLTG-GCNs. the KL loss formula is as follows:

$$D_{KL}(S_i \| V_v) = \sum_{i=1, i \neq v}^{V+1} \sum_{n=1}^{N} S_i(x_n) \log \frac{S_i(x_n)}{V_v(x_n)} \tag{6}$$

where S_i is the sample distribution obtained by DLTGGCN, which is the source distribution, and V_v is the sample distribution obtained by other DLTGGCN, which is the target distribution. $S_i(x_n)$ is denoted as the i-th atlas of the probability distribution of the n-th sample, $V_v(x_n)$ is denoted as the probability distribution of the n-th sample of the other v-th atlas, and x is the sample feature vector. We build a central distribution V_{V+1} of multi-atlas fusion, and its formula is shown as follows:

$$V_{V+1}(x_n) = \frac{1}{V} \sum_{i=1}^{V} S_i(x_n) \tag{7}$$

where $S_i(x_n)$ is denoted as the category prediction probability distribution of the i-th DLTGGCN for sample x_n. The prediction scores are calculated as probabilities by $Softmax$ with the following equation:

$$S_i(x_n) = \frac{e^{x_m}}{\sum_{c=1}^{C} e^{x_c}} \tag{8}$$

where x_m is denoted as the value at the m-th feature position of sample x_n, $m \in C$, and C is the number of categories, thus obtaining the distribution of each sample under the i-th atlas.

To measure the difference between the truth and predicted labels, a cross-entropy loss function is used in the objective function, defined as follows:

$$CE_i = -\frac{1}{N} \sum_{n=1}^{N} [y_n \cdot \log s_i(x_n) + (1 - y_n) \cdot \log(1 - s_i(x_n))] \qquad (9)$$

where y_n denotes the label of sample n, positive class label is 1 and negative class label is 0; $s_i(x_n)$ denotes the probability that sample n is predicted to be positive class, and CE_i denotes the cross-entropy loss function under the current ith atlas.

The total loss function used to optimize the DLTGGCN under the ith atlas is defined as:

$$L_i = M_i + \frac{1}{V} \sum_{i=1, i \neq v}^{V+1} D_{KL}(S_i \| V_v) + CE_i \qquad (10)$$

where M_i is the triplet loss, $\frac{1}{V} \sum_{i=1, i \neq v}^{V+1} D_{KL}(S_i \| V_v)$ is the mutual learning loss, and the last is the cross-entropy loss.

Finally a weighted fusion is used to fuse the representations, with the following equation:

$$Pred_{Fusion} = \sum_{i=1}^{V} \varepsilon_i Pred_i \qquad (11)$$

where ε_i is the weight, i is the i-th atlas. The formula is as follows:

$$\varepsilon_i = \frac{\alpha_i}{\sum_{i=1}^{V} \alpha_i} \qquad (12)$$

where α_i is the accuracy of the i-th DLTGGCN.

4 Experiment and Result

4.1 Implementation Details

We conduct experiments using 5-fold cross validation on 949 subjects in the ABIDE dataset. To quantify the model performance, we evaluate it using three metrics: Accuracy (ACC), Sensitivity (SEN) and Specificity (SPE). The higher values of these metrics, the better the performance of the model. In addition, we also calculate the area under the curve (AUC) to estimate the performance of this method.

The experimental implementation is based on Pytorch and is trained and tested on a GTX 1080TI 11G server. The DLTGCN consists of three 3-layer GCNs with shared weights, followed by a GGCN. Optimization is performed by Adam optimizer with learning rate set to 0.0003, epoch of 200, and each epoch divided into 8 batches.

4.2   Comparison with Existing Methods

In order to validate the advantages of the proposed method, we compare the proposed method with some currently existing methods. The results are shown in Fig 4.

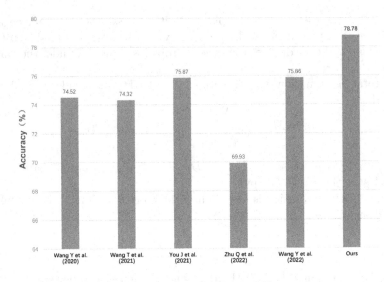

Fig. 4. Comparison with existing methods.

To make a fair comparison, we train and test the different methods on the same data. The comparison results are shown in Fig 4, from which we can see the phenomenon that the proposed method has higher accuracy compared with other methods, and the graph convolution network based multi-atlas mutual learning ASD diagnosis method can obtain a more effective topology for ASD diagnosis. From the experimental results, the proposed method is more advantages. This may be due to the representations from different atlases which benefit the ASD diagnosis.

4.3   Ablation Experiments

Impact of Multi-atlas. To evaluate the utility of the proposed method, we conduct experiments with the proposed method using three single atlas features as well as multi-atlas features. The results are shown in Table 1.

As can be seen from Table 1, the method has improved in most metrics as the number of atlases increases. This may be due to the fact that the multi-atlas mutual learning method can better learn the topologies that are beneficial for classification in different atlaes. Further, redundant information is removed, thus improving the accuracy. The results of this method with multi-atlas outperform the results of three single atlases, which indicates that multi-atlas can provide

Table 1. The performance of different numbers of atlases.

Atlas	ACC (%)	SEN (%)	SPE (%)	AUC
DOSEN	69.44	54.00	81.80	0.6790
AAL	72.67	64.75	79.00	0.7188
CC200	74.22	66.50	80.40	0.7345
AAL & *DOSEN*	75.44	66.75	82.40	0.7458
CC200 & *DOSEN*	75.67	**73.25**	77.60	0.7543
AAL & *CC*200	77.00	72.25	80.80	0.7653
AAL & *CC*200 & *DOSEN*	**78.78**	72.00	**84.20**	**0.7810**

richer topology and thus improved performance. The results further demonstrate the advantages of the proposed method.

Impact of Different Modules. First, the proposed method considers the relationship between ROIs and DLTGCN is proposed. Furthermore, the population relationship among samples is considered and GGCN is proposed. The results are shown in Table 2.

Table 2. The performance of different modules.

DLTGCN	GGCN	ACC (%)	SEN (%)	SPE (%)	AUC
✓		75.33	69.75	79.80	0.7478
	✓	77.67	68.25	**85.20**	0.7673
✓	✓	**78.78**	72.00	84.20	**0.7810**

From the experimental results, it can be concluded that the accuracy of the proposed method is not optimal when considering brain network local graphs and subject network global graphs separately, where the accuracy of DLTGCN is 75.33% and the accuracy of GGCN is improved by 2.34%. The accuracy of DLTGGCN is 78.78%, which is 3.45% and 1.11% higher compared to the DLTGCN or GGCN considering only a single one, respectively. It may be that the proposed method is able to extract the topology that is beneficial for ASD diagnosis in the local graph of ROIs and the higher-order global graph of the sample network.

5   Conclusion

In this study, we propose Multi-atlas Representations based on Graph Convolutional Networks for ASD diagnosis. First, we calculate the FC of different atlases and then build a local graph of ROIs. Triplets are formed by taking positive samples, negative samples, and anchor samples. The dense local triplet graph convolutional neural network is used for learning. Second, a network of samples is built and the global graph convolutional network is used for learning. Finally, a mutual learning method is proposed to obtain the final classification results. We evaluate and validate the method on the ABIDE dataset with good results. The comparison with other existing methods validates the advantages of our proposed method. In conclusion, our method provides guidance for further research in ASD diagnosis.

Acknowledgements. This work is supported in part by the Natural Science Foundation of Hunan Province under Grant (No.2022JJ30753), in part by the Shenzhen Science and Technology Program (No.KQTD20200820113106007), in part by Shenzhen Key Laboratory of Intelligent Bioinformatics (ZDSYS20220422103800001), in part by the Central South University Innovation-Driven Research Programme under Grant 2023CXQD018, and in part by the High Performance Computing Center of Central South University.

References

1. Abraham, A., et al.: Deriving reproducible biomarkers from multi-site resting-state data: an autism-based example. Neuroimage **147**, 736–745 (2017)
2. Byeon, K., Kwon, J., Hong, J., Park, H.: Artificial neural network inspired by neuroimaging connectivity: application in autism spectrum disorder. In: 2020 IEEE International Conference on Big Data and Smart Computing (BigComp), pp. 575–578. IEEE (2020)
3. Craddock, C., et al.: Towards automated analysis of connectomes: the configurable pipeline for the analysis of connectomes (c-pac). Front. Neuroinf. **42**, 10–3389 (2013)
4. Craddock, R.C., James, G.A., Holtzheimer, P.E., III., Hu, X.P., Mayberg, H.S.: A whole brain fMRI atlas generated via spatially constrained spectral clustering. Hum. Brain Mapp. **33**(8), 1914–1928 (2012)
5. Dosenbach, N.U., et al.: Prediction of individual brain maturity using fMRI. Science **329**(5997), 1358–1361 (2010)
6. Edition, F., et al.: Diagnostic and statistical manual of mental disorders. Am. Psychiatric Assoc. **21**(21), 591–643 (2013)
7. Huang, F., et al.: Self-weighted adaptive structure learning for ASD diagnosis via multi-template multi-center representation. Med. Image Anal. **63**, 101662 (2020)
8. Huang, Y., Chung, A.C.: Disease prediction with edge-variational graph convolutional networks. Med. Image Anal. **77**, 102375 (2022)
9. Jahedi, A., Nasamran, C.A., Faires, B., Fan, J., Müller, R.A.: Distributed intrinsic functional connectivity patterns predict diagnostic status in large autism cohort. Brain Connect. **7**(8), 515–525 (2017)
10. Kunda, M., Zhou, S., Gong, G., Lu, H.: Improving multi-site autism classification via site-dependence minimization and second-order functional connectivity. IEEE Trans. Med. Imaging **42**(1), 55–65 (2022)
11. Li, X., et al.: Braingnn: interpretable brain graph neural network for fMRI analysis. Med. Image Anal. **74**, 102233 (2021)

12. Rapin, I., Tuchman, R.F.: Autism: definition, neurobiology, screening, diagnosis. Pediatr. Clin. North Am. **55**(5), 1129–1146 (2008)
13. Sato, W., Uono, S.: The atypical social brain network in autism: advances in structural and functional MRI studies. Curr. Opin. Neurol. **32**(4), 617–621 (2019)
14. Sharma, S.R., Gonda, X., Tarazi, F.I.: Autism spectrum disorder: classification, diagnosis and therapy. Pharmacol. Therapeut. **190**, 91–104 (2018)
15. Sun, J.W., Fan, R., Wang, Q., Wang, Q.Q., Jia, X.Z., Ma, H.B.: Identify abnormal functional connectivity of resting state networks in autism spectrum disorder and apply to machine learning-based classification. Brain Res. **1757**, 147299 (2021)
16. Tzourio-Mazoyer, N., et al.: Automated anatomical labeling of activations in SPM using a macroscopic anatomical parcellation of the MNI MRI single-subject brain. Neuroimage **15**(1), 273–289 (2002)
17. Varoquaux, G., Craddock, R.C.: Learning and comparing functional connectomes across subjects. Neuroimage **80**, 405–415 (2013)
18. Wang, M., Zhang, D., Huang, J., Liu, M., Liu, Q.: Consistent connectome landscape mining for cross-site brain disease identification using functional MRI. Med. Image Anal. **82**, 102591 (2022)
19. Wang, N., Yao, D., Ma, L., Liu, M.: Multi-site clustering and nested feature extraction for identifying autism spectrum disorder with resting-state fMRI. Med. Image Anal. **75**, 102279 (2022)
20. Wang, T., et al.: Mogonet integrates multi-omics data using graph convolutional networks allowing patient classification and biomarker identification. Nat. Commun. **12**(1), 3445 (2021)
21. Wang, Y., Liu, J., Xiang, Y., Wang, J., Chen, Q., Chong, J.: Mage: automatic diagnosis of autism spectrum disorders using multi-atlas graph convolutional networks and ensemble learning. Neurocomputing **469**, 346–353 (2022)
22. Wang, Y., Wang, J., Wu, F.X., Hayrat, R., Liu, J.: Aimafe: autism spectrum disorder identification with multi-atlas deep feature representation and ensemble learning. J. Neurosci. Methods **343**, 108840 (2020)
23. Yi, P., Jin, L., Xu, T., Wei, L., Rui, G.: Hippocampal segmentation in brain MRI images using machine learning methods: a survey. Chin. J. Electron. **30**(5), 793–814 (2021)
24. Yin, W., Mostafa, S., Wu, F.X.: Diagnosis of autism spectrum disorder based on functional brain networks with deep learning. J. Comput. Biol. **28**(2), 146–165 (2021)
25. You, J., Gomes-Selman, J.M., Ying, R., Leskovec, J.: Identity-aware graph neural networks. In: Proceedings of the AAAI Conference on Artificial Intelligence, vol. 35, pp. 10737–10745 (2021)
26. Zhang, H., et al.: Classification of brain disorders in RS-fMRI via local-to-global graph neural networks. IEEE Trans. Med. Imaging (2022)

MS-UNet: Swin Transformer U-Net with Multi-scale Nested Decoder for Medical Image Segmentation with Small Training Data

Haoyuan Chen, Yufei Han, Yanyi Li, Pin Xu, Kuan Li$^{(\boxtimes)}$, and Jianping Yin

Dongguan University of Technology, Dongguan, China
likuan@dgut.edu.cn

Abstract. We propose a novel U-Net model named MS-UNet for the medical image segmentation task in this study. Instead of the single-layer U-Net decoder structure used in Swin-UNet and TransUnet, we specifically design a multi-scale nested decoder based on the Swin Transformer for U-Net. The new framework is proposed based on the observation that the single-layer decoder structure of U-Net is too "thin" to exploit enough information, resulting in large semantic differences between the encoder and decoder parts. Things get worse if the number of training sets of data is not sufficiently large, which is common in medical image processing tasks where annotated data are more difficult to obtain than other tasks. Overall, the proposed multi-scale nested decoder structure allows the feature mapping between the decoder and encoder to be semantically closer, thus enabling the network to learn more detailed features. Experiment results show that MS-UNet could effectively improve the network performance with more efficient feature learning capability and exhibit more advanced performance, especially in the extreme case with a small amount of training data. The code is publicly available at: https://github.com/HH446/MS-UNet.

Keywords: Medical Image Segmentation · U-Net · Swin Transformer · Multi-scale Nested Decoder

1 Introduction

Since U-Net [15] was proposed, the CNN network and its variants have gradually become the dominating methods in the field of medical image segmentation. However, CNN family networks [9,18,20] gradually suffer from the limitation, i.e. the inherent inductive bias of the CNN network, which makes each receptive field of the network stay only in a fixed-size window each time and could not establish long-range pixel dependencies. Recently, a novel network named Vision Transformer [5] has revolutionized most computer vision tasks, main thanks to the Multi-head Self Attention mechanism [16] which could effectively establish global connections between tokens of sequences. Subsequently, a hierarchical Swin Transformer proposed in [13] not only effectively reduces the computational complexity of [5], but also introduces the locality of CNN. The latter

Q. Liu et al. (Eds.): PRCV 2023, LNCS 14437, pp. 472–483, 2024.
https://doi.org/10.1007/978-981-99-8558-6_39

makes Transformer perform pixel-level prediction tasks more efficiently. Motivated by Swin Transformer, Swin-UNet [1] was proposed to leverage the power of Transformer for 2D medical image segmentation.

To further improve the performance on segmentation tasks, for CNN-based models, the proposal of DLA [19] and UNet++ [21] have also successfully demonstrated that deep layer aggregation could effectively improve the recognition and segmentation performance with sufficient datasets. For Transformer-based models, from TransUnet [3] to Swin-UNet [1], the researchers focus on combining CNN and Transformer, with the overall structures still following the traditional U-Net structure. However, the single-layer structure of U-Net is too simply designed (connecting only by skip-connection) to accommodate the complexity of the transformer structure, resulting in large semantic differences between the encoder and decoder parts. Things get worse if the number of training sets of data is not sufficiently large, which is common in medical image processing tasks where annotated data are more difficult to obtain than other tasks.

In this paper, we propose MS-UNet, a simple but more effective 2D medical image segmentation architecture based on Swin Transformer. MS-UNet replaces the single-layer decoder structure of Swin-UNet with a hierarchical multi-scale decoder structure inspired by [10, 12, 21]. The multi-scale nested decoder structure enables the proposed MS-UNet to learn the semantic information of the feature map from a more multi-dimensional perspective. In other words, MS-UNet could obtain a tighter semantic hierarchy of the feature map between the encoder and decoder parts, thus effectively improving its stability and generalization. Furthermore, these will somehow reduce the degree of requiring large-scale training data. This is important for medical image datasets, whose labeled data are more difficult to obtain than other datasets. Experimental results show that our proposed MS-UNet could effectively improve the performance of the network with more efficient feature learning ability in the extreme case of a very small amount of training data, and exhibits outstanding performance in the medical image segmentation field.

Our contributions are threefold:

1. A new semantic segmentation framework, MS-UNet, is proposed to effectively improve the performance of the network with more efficient feature learning capability in few-shot tasks.
2. A multi-scale nested decoder based on Swin Transformer is proposed to justify the deep layer aggregation could also effectively improve the segmentation performance of Transformer-based models.
3. MS-UNet surpasses other state-of-the-art methods in medical image segmentation tasks. More importantly, in the extreme case where only a very small amount of training data is available, MS-UNet achieves a huge improvement in segmentation results than other models in the U-Net family.

The remainder of the paper is organized as follows. Section 2 briefly reviews the related work. In Sect. 3, we describe the details of our proposed MS-UNet model. Section 4 reports the results of MS-UNet on medical image segmentation tasks, as well as a series of ablation experiments. We summarise our work in Sect. 5.

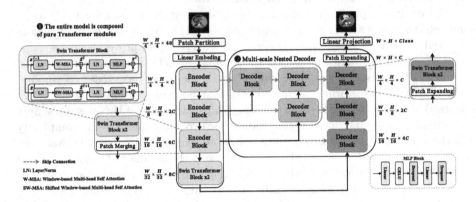

Fig. 1. The architecture of MS-UNet. Our contributions: ❶ The entire model is composed of pure Transformer modules; ❷ Multi-scale Nested Decoder.

2   Related Works

Medical Image Segmentation Based on CNNs. The variant architectures based on UNet [15] have demonstrated excellent performance in medical image segmentation. UNet++ [21] reduces the semantic difference gap through a series of nested and dense connections. Res-UNet [18] combines U-Net [15] with the classic ResNet and Attention ideas to effectively improve the performance of retinal vessel segmentation. CNN-based methods have achieved great success in medical image segmentation at present.

Vision Transformer. Vision Transformer (ViT) [5] successfully applied the transformer to CV. A series of works based on ViT [2] have been proposed since then. In [13], a hierarchical vision Transformer named Swin Transformer is proposed, which combined the advantages of Transformer and CNN. Using shifted windows, Swin Transformer not only reduces the computational complexity but also achieves state-of-the-art performances on various tasks in the field of computer vision. Subsequently, based on Swin Transformer [13], the first pure transformer network with U-shaped Encoder-Decoder architecture for 2D medical image segmentation tasks is proposed. At the same time, a novel deep medical image segmentation framework called Dual Swin Transformer U-Net (DS-TransUNet) [12], which is the first attempt to incorporate the advantages of hierarchical Swin Transformer into a standard U-shaped Encoder-Decoder architecture to improve the performance of medical image segmentation. In our work, inspired by [21] and [12], we attempt to use U-Net++ as architecture to reduce the semantic difference between encoder and decoder in [1].

Hybrid of CNN and Transformer. In recent years, people are not only exploring how to directly apply Transformer to tasks of computer vision but also trying to combine CNN and Transformer to shake the dominance of CNN in the field of computer vision. In [3], the authors propose the first Transformer-based medical image segmentation framework, which used the CNN feature map as the input of the transformer to form the encoder and achieves excellent results on medical image segmentation tasks. Inspired by [3], a U-shaped model called Hybrid Transformer U-Net (MT-UNet) is proposed. MT-UNet enables accurate medical image segmentation by using a new Hybrid Transformer Module (MTM) that is able to perform simultaneous inter and intr affinities learning [17].

3   Method

In this section, the overall structure of the proposed MS-UNet is elaborated with Fig. 1. Then we elaborate on the standard Swin-Transformer block adopted in MS-UNet. Finally, we will present the detailed designs of the encoder and decoder in MS-UNet.

3.1   Architecture Overview

The overall MS-UNet architecture is illustrated in Fig. 1. It consists of an encoder step and a multi-scale nested decoder step, connected by skip connections. First of all, We divide the medical image into non-overlapping patches by Patch Partition and get the raw-valued feature as a concatenation of the original pixel values. Subsequently, we project the raw-valued feature to some patch token with an arbitrary dimension by the Linear Embedding layer. In the encoder step, the transformed patch tokens will generate a hierarchical feature representation by a workflow consisting of several Swin Transformer blocks and Patch Merging layers. In the decoder step, we innovatively use the multi-scale nested decoder to upsample and decode the features from each encoder independently and use skip connections to concatenate the resulting features with the corresponding decoder input features in the next layer.

We believe that the new multi-scale nested decoder could learn the semantic information of the feature maps in the transformer encoder from a more multi-dimensional perspective, and solve the semantic differences between the encoder and decoder parts in the feature fusion process of the simple U-shaped network structure. This could effectively improve the stability and generalization of the network, allowing it to learn the required information better even with less labeled data. In the end, we use an upsampling Patch Expanding layer to output the same resolution as the input's resolution and perform pixel-level segmentation prediction on the upsampled features by the Linear Projection layer.

3.2   Swin Transformer Block

Combined with the characteristics of CNN, the complete construction of the Swin-Transformer block is based on a multi-head self-attention module in shifted windows instead of the traditional multi-head self-attention (MSA) module. As illustrated in Fig. 1, each Swin-Transformer block consists of two different Swin Transformer blocks, a window-based MSA (W-MSA) and a shifted window-based MSA (SW-MSA) respectively. In this case, each swin transformer block consists of LayerNorm (LN) layer, MLP module with GELU nonlinearity, and residual connection.

3.3   Encoder

The encoder branch is a hierarchical network architecture with Swin Transformer block as the backbone. In each Swin Transformer block, a Patch Merging layer downsamples the input information and increases the feature dimension to retain the valid information in the image while expanding the perceptual field. The Swin Transformer blocks perform representation learning from the input medical image.

3.4   Mutil-scale Nested Decoder and Skip Connection

The decoder branch is also composed of Swin Transformer as the backbone with upsampling modules and Patch Expanding layers, corresponding to the encoder branch. Traditional U-shaped models use simple skip-connections to connect pairs of Encoder Blocks and Decode Blocks. In this study, we come up with a multi-scale nested decoder branch to replace the decoder branch of the traditional U-shaped network. This enables the decoder network efficiently learn more effective information from the more complex Transformer encoder output features. The multi-scale nested blocks will upsample the decode features from each encoder separately, and use skip connection to concatenate the resulting features with the corresponding decoder in the next layer This would make the semantic level of the feature map in the encoder part closer to the semantic level of the feature map waiting in the decoder.

In our multi-scale nested decoder structure, skip connection is not only used to fuse the multi-scale features from the encoder with the upsampled features but also to be used for feature fusion between adjacent decoders. The resulting multi-scale nested decoder blocks allow the network to learn the required information more efficiently on datasets with fewer data.

3.5   Datasets

Synapse. Synapse [6] multi-organ segmentation dataset includes 30 abdominal CT scans and 3779 axial contrast-enhances abdominal clinical CT images. Following the literature [3,6], we randomly split the dataset into 18 training sets and 12 testing sets and evaluate our method on 8 abdominal organs (aorta,

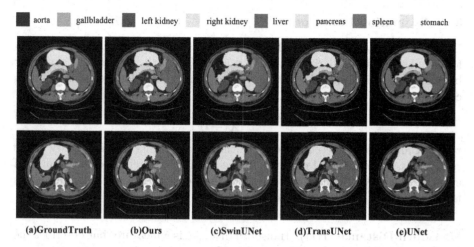

Fig. 2. Visualization of segmentation effects of different models on the Synapse multi-organ CT dataset.

gallbladder, spleen, left kidney, right kidney, liver, pancreas, spleen, stomach) using the average Dice-Similarity coefficient (DSC) and the average Hausdorff Distance (HD) as evaluation metrics.

ACDC. Automated cardiac diagnosis challenge dataset is an open-competition cardiac MRI dataset containing left ventricular (LV), right ventricular (RV), and myocardial (MYO) labels. The dataset contains 100 samples, which we divided into 70 training samples, 10 validation samples, and 20 test samples for consistency with [17].

Table 1. The average Dice-Similarity coefficient (DSC) and the average Hausdorff Distance (HD) of different methods on the complete Synapse multi-organ CT dataset.

Methods	DSC ↑	HD ↓	Aorta	Gallbladder	Kidney(L)	Kidney(R)	Liver	Pancreas	Spleen	Stomach
R50 U-Net [3]	74.68	36.87	87.74	63.66	80.60	78.19	93.74	56.90	85.87	74.16
U-Net [15]	78.23	27.48	87.91	66.77	84.97	80.33	94.05	52.44	87.20	72.14
R50 Att-UNet [3]	75.57	36.97	55.92	63.91	79.20	72.71	93.56	49.37	87.19	74.95
Att-UNet [14]	77.77	36.02	**89.55**	68.88	77.98	71.11	93.57	58.04	87.30	75.75
R50 ViT [3]	71.29	32.87	73.73	55.13	75.80	72.20	91.51	45.99	81.99	73.95
TransUnet [3]	77.93	29.31	87.03	62.19	82.80	77.76	93.98	57.36	86.38	75.93
MT-UNet [17]	78.59	26.59	87.92	64.99	81.47	77.29	93.06	**59.46**	87.75	76.81
Swin-Unet [1]	79.23	25.48	85.96	66.99	84.28	80.74	93.99	56.35	89.78	75.82
Ours	**80.44**	**18.97**	85.80	**69.40**	**85.86**	**81.66**	**94.24**	57.66	**90.53**	**78.33**

3.6   Evaluation Indicators

Dice-Similarity Coefficient (DSC). The Dice coefficient is a metric function for calculating the similarity of sets. It is usually used to calculate the similarity between two samples, and its value ranges from [0, 1]. The equation for this concept is:

$$Dice = \frac{|X \cap Y|}{|X| + |Y|}, \tag{1}$$

where X and Y are two sets. The set in the $|\cdot|$ represents the cardinality of the set, that is, the number of elements in the set. E.g. $|X|$ refers to the number of elements in set X. $\cap$ is used to represent the intersection of two sets and means the elements that are common to both sets.

Hausdorff Distance (HD). Hausdorff distance is a measure that describes the similarity between two sets of points by calculating the distance between them. Suppose there are two sets of sets: $A = \{a^1, a^2, ..., a^n\}$, $B = \{b^1, b^2, ..., b^n\}$, the Hausdorff distance between these two sets of points are defined as:

$$H(A, B) = max(h(A, B), h(B, A)), \tag{2}$$

$$h(A, B) = \max_{a \in A} \min_{b \in B} \| a - b \|, \tag{3}$$

$$h(B, A) = \max_{b \in B} \min_{a \in A} \| b - a \|, \tag{4}$$

where $\| \cdot \|$ represents the Euclidean distance between elements.

Table 2. The average Dice-Similarity coefficient (DSC) and the average Hausdorff Distance (HD) of different methods on the ACDC dataset with one-eighth of training data.

Methods	$DSC \uparrow$	$HD \downarrow$	RV	Myo	LV
U-Net [15]	86.14	3.91	82.06	83.90	92.48
TransUnet [3]	83.84	3.41	79.01	80.40	92.11
Swin-Unet [1]	85.61	2.30	81.59	82.39	92.86
Ours	**87.74**	**1.51**	**85.31**	**84.09**	**93.82**

4   Experiments

4.1   Implementation Details

In our experiments, simple data augmentation was performed on all training data with random rotations and flips. MS-UNet is achieved based on Python 3.8.8, Pytorch 1.10.0, and CUDA 11.3. The Swin Transformer backbone network in

Swin-Unet++ has been pre-trained on ImageNet [4]. The input image size will be set as 224 × 224 and the patch size will be set as 4. During training, we employ a batch size of 24, SGD optimizer with momentum 0.9 and a decay rate of $1e-4$, using Cross-Entropy and Dice-Similarity Coefficient loss functions for effective learning. All experiments are conducted using a single Nvidia RTX3090 GPU.

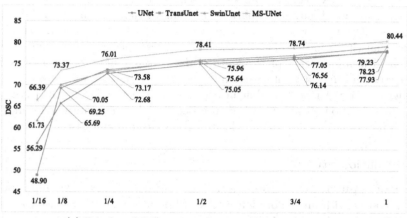

(a) Results of different methods on Synapse Dataset.

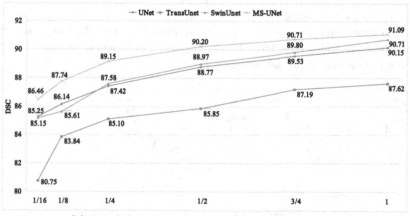

(b) Results of different methods on ACDC Dataset.

Fig. 3. Results of different methods with different ratios of training data on the Synapes and ACDC datasets.

4.2   Experiment Results

Table 1, Table 2 show the segmentation results of MS-UNet and other up-to-date methods on the Synapse multi-organ CT dataset and ACDC dataset. From the results, it could be seen that the proposed MS-UNet exhibits overall better performance than others. Fig. 2 visualizes segmentation results of different methods on the Synapse multi-organ CT dataset. On the Synapse dataset, compared to SwinUnet [1], MS-UNet achieved a 1.21% improvement in DSC and a 6.51% improvement in HD. Compared to other baseline models, MS-UNet achieved quantitative improvements in the vast majority of organ segmentation metrics.

In particular, we conduct an experiment to compare the performances of different methods if only using part of the training data. From Fig. 3, MS-UNet performs better than other models even with only half of the training data. This verifies that the proposed model with the multi-scale nested decoder could learn the semantic information of the feature maps from a more multi-dimensional perspective, forming a tighter semantic hierarchy of feature maps between the encoder and decoder parts and achieving better segmentation results even with fewer data. It also indicates that the proposed model is more suitable for medical image processing tasks, where the labeled data are rare and costly.

4.3   Ablation Study

Effect of Input Image Resolution. Since the default 224×224 input image data is converted from the original 512×512 data, we try to increase the input image resolution to 448×448 to study the impact of input image resolution on segmentation accuracy. As shown in Table. 3, high-resolution input leads to better image segmentation accuracy, but the increasing resolution will also lead to a significant increase in the computational load of the entire network. Therefore, from the perspective of model operation efficiency, we use 224×224 as the input resolution of the network.

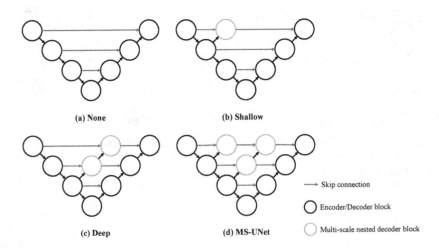

Fig. 4. The architecture of the network with different numbers and locations of the multi-scale nested blocks.

Table 3. Ablation study on the influence of input image resolution.

Resolution	DSC ↑	Aorta	Gallbladder	Kidney(L)	Kidney(R)	Liver	Pancreas	Spleen	Stomach
224 × 224	80.44	85.80	69.40	85.86	81.66	94.24	57.66	**90.53**	**78.33**
448 × 448	**81.64**	**88.87**	**70.60**	**86.93**	**84.27**	**95.09**	**60.89**	89.77	76.72

Table 4. Ablation study on the influence of multi-level nested blocks.

Structure	Parmas	DSC ↑	HD↓	Aorta	Gallbladder	Kidney(L)	Kidney(R)	Liver	Pancreas	Spleen	Stomach
None	27.17 M	79.23	25.48	**85.96**	66.99	84.28	80.74	93.99	56.35	89.78	75.82
Shallow	27.49 M	80.12	20.95	85.39	**70.82**	84.76	**82.13**	93.97	55.88	89.26	**78.82**
Deep	28.75 M	80.16	21.79	85.58	69.27	85.02	82.05	94.09	**59.83**	89.52	75.87
MS-UNet	29.06 M	**80.44**	**18.97**	85.80	69.40	**85.86**	81.66	**94.24**	57.66	**90.53**	78.33

Effect of Multi-level Nested Blocks. In this part, as shown in Fig. 4 we explore the impact of the number and location of multi-scale nested decoders on the segmentation performance of MS-UNet by changing them gradually. From the experimental results in Table. 4, even when adding only one multi-scale nested block in a shallow layer of the network, the DSC of the model improves from 79.23% to 80.12% and the DSC of the model improves from 25.48 to 20.95 when the number of parameters has increased by only 1%. More importantly, the segmentation performance of the model improves as the number of multi-scale nested blocks increases and does not fall into overfitting in the extreme case in Fig. 3. This further verifies that the structure of the network with the multi-scale nested blocks could effectively bring the feature maps between the encoder and decoder semantically closer, allowing the network to achieve better segmentation performance.

4.4   Discussion

In this work, we apply the proposed MS-UNet for medical image segmentation tasks. According to the results in Fig. 3, our model achieves overall better results relative to other models, even with a small amount/part of training data. We need to emphasize that in the extreme case where only a very small amount of training data is available, the improvement brought by the proposed method is huge. Unlike other computer vision tasks, in medical image tasks, it is usually difficult to obtain labeled medical images since it is usually costly. From this perspective, the proposed MS-UNet has a relatively smaller data scale require-ment and has great potential for different medical image processing tasks. At the same time, for better comparison with [1,3,17], we only use MS-UNet for 2D segmentation tasks. We are planning to improve MS-UNet for 3D segmentation tasks in the future. In addition, with the popularity of [7,8], We will further investigate how to pre-train the Transformer-based model on unlabelled medical images by self-supervised methods. Furthermore, drawing inspiration from [11], we consider the incorporation of boundary information of segmented objects during training, which leverages the boundary information to guide the segmentation process and improve the overall segmentation accuracy.

5   Conclusion

In this work, we propose a novel Transformer-based U-shaped medical image segmentation network named MS-UNet, which enables a tighter semantic hierarchy of the feature map between the encoder and decoder through a simple multi-scale nested decoders structure, thus effectively improving the model's stability and generalization. The experimental results show that MS-UNet has excellent performance and more efficient feature learning ability. More importantly, MS-UNet could achieve better segmentation results than other U-Net family models in the extreme case where only a very small amount of training data is available. This work has great potential for medical image processing tasks where labeled images are rare and costly.

References

1. Cao, H., et al.: Swin-Unet: Unet-like Pure Transformer for Medical Image Segmentation. arXiv:2105.05537 (May 2021). http://arxiv.org/abs/2105.05537
2. Carion, N., Massa, F., Synnaeve, G., Usunier, N., Kirillov, A., Zagoruyko, S.: End-to-end object detection with transformers. CoRR abs/ arXiv: 2005.12872 (2020)
3. Chen, J., et al.: TransUNet: Transformers Make Strong Encoders for Medical Image Segmentation. arXiv:2102.04306 (Feb 2021)
4. Deng, J., Dong, W., Socher, R., Li, L.J., Li, K., Fei-Fei, L.: ImageNet: a large-scale hierarchical image database. In: 2009 IEEE Conference on Computer Vision and Pattern Recognition. pp. 248–255 (Jun 2009). https://doi.org/10.1109/CVPR.2009.5206848, ISSN: 1063-6919
5. Dosovitskiy, A., et al.: An Image is Worth 16x16 Words: Transformers for Image Recognition at Scale. arXiv:2010.11929 (Jun 2021)
6. Fu, S., et al.: Domain adaptive relational reasoning for 3d multi-organ segmentation. CoRR abs/ arXiv: 2005.09120 (2020)
7. He, K., Chen, X., Xie, S., Li, Y., Dollár, P., Girshick, R.: Masked Autoencoders Are Scalable Vision Learners. arXiv:2111.06377 (Dec 2021)
8. He, K., Fan, H., Wu, Y., Xie, S., Girshick, R.: Momentum Contrast for Unsupervised Visual Representation Learning (Mar 2020). arXiv: 1911.05722
9. Huang, H., et al.: UNet 3+: A Full-Scale Connected UNet for Medical Image Segmentation (Apr 2020). arXiv:2004.08790
10. Jha, D., Riegler, M.A., Johansen, D., Halvorsen, P., Johansen, H.D.: DoubleU-Net: a deep convolutional neural network for medical image segmentation. In: 2020 IEEE 33rd International Symposium on Computer-Based Medical Systems (CBMS), pp. 558–564. IEEE, Rochester, MN, USA (Jul 2020). https://doi.org/10.1109/CBMS49503.2020.00111,https://ieeexplore.ieee.org/document/9183321/
11. Kuang, H., Liang, Y., Liu, N., Liu, J., Wang, J.: BEA-SegNet: body and edge aware network for medical image segmentation. In: 2021 IEEE International Conference on Bioinformatics and Biomedicine (BIBM), pp. 939–944. IEEE, Houston, TX, USA (Dec 2021). https://doi.org/10.1109/BIBM52615.2021.9669545,https://ieeexplore.ieee.org/document/9669545/
12. Lin, A., Chen, B., Xu, J., Zhang, Z., Lu, G.: DS-TransUNet: Dual Swin Transformer U-Net for Medical Image Segmentation. arXiv:2106.06716 (Jun 2021)

13. Liu, Z., et al.: Swin Transformer: Hierarchical Vision Transformer using Shifted Windows. arXiv:2103.14030 (Aug 2021)
14. Oktay, O., et al.: Attention U-Net: Learning Where to Look for the Pancreas, arXiv:1804.03999 (May 2018)
15. Ronneberger, O., Fischer, P., Brox, T.: U-Net: convolutional networks for biomedical image segmentation. In: Navab, N., Hornegger, J., Wells, W.M., Frangi, A.F. (eds.) MICCAI 2015. LNCS, vol. 9351, pp. 234–241. Springer, Cham (2015). https://doi.org/10.1007/978-3-319-24574-4_28
16. Vaswani, Aet al.: Attention Is All You Need, arXiv:1706.03762 (Dec 2017)
17. Wang, H., et al.: Mixed Transformer U-Net For Medical Image Segmentation, arXiv:2111.04734 (Nov 2021)
18. Xiao, X., Lian, S., Luo, Z., Li, S.: Weighted Res-UNet for high-quality retina vessel segmentation. In: 2018 9th International Conference on Information Technology in Medicine and Education (ITME), pp. 327–331 (Oct 2018). https://doi.org/10.1109/ITME.2018.00080, ISSN: 2474-3828
19. Yu, F., Wang, D., Shelhamer, E., Darrell, T.: Deep layer aggregation. In: Proceedings of the IEEE Conference on Computer Vision and Pattern Recognition, pp. 2403–2412 (2018)
20. Zhao, H., Shi, J., Qi, X., Wang, X., Jia, J.: Pyramid scene parsing network. In: 2017 IEEE Conference on Computer Vision and Pattern Recognition (CVPR), pp. 6230–6239 (2017). https://doi.org/10.1109/CVPR.2017.660
21. Zhou, Z., Rahman Siddiquee, M.M., Tajbakhsh, N., Liang, J.: UNet++: a nested u-net architecture for medical image segmentation. In: Stoyanov, D., et al. (eds.) DLMIA/ML-CDS -2018. LNCS, vol. 11045, pp. 3–11. Springer, Cham (2018). https://doi.org/10.1007/978-3-030-00889-5_1

GCUNET: Combining GNN and CNN for Sinogram Restoration in Low-Dose SPECT Reconstruction

Keming Chen, Zengguo Liang, and Si Li[(✉)]

School of Computer Science and Technology, Guangdong University of Technology,
Guangzhou 510006, China
sili@gdut.edu.cn

Abstract. To reduce the potential radiation risk, low-dose Single Photon Emission Computed Tomography (SPECT) is of increasing interest. Many deep learning-based methods have been developed to perform low-dose imaging while maintaining image quality. However, most of the existing methods ignore the unique inner-structure inherent in the original sinogram, limiting their restoration ability. In this paper, we propose a GNN-CNN-UNet (GCUNet) to learn the non-local and local structures of the sinogram using Graph Neural Network (GNN) and Convolutional Neural Network (CNN), respectively, for the task of low-dose SPECT sinogram restoration. In particular, we propose a sinogram-structure-based self-defined neighbors GNN (SSN-GNN) method combined with the Window-KNN-based GNN (W-KNN-GNN) module to construct the underlying graph structure. Afterwards, we employ the maximum likelihood expectation maximization (MLEM) to reconstruct the restored sinogram. The XCAT dataset is used to evaluate the performance of the proposed GCUNet. Experimental results demonstrate that, compared to several reconstruction methods, the proposed method achieves significant improvement in both noise reduction and structure preservation.

Keywords: Low-dose SPECT · Sinogram restoration · Sinogram inner-structure · GNN

1 Introduction

Single photon emission computed tomography (SPECT) is an important imaging modality that has been widely used in clinical practice and research [17]. Typically, SPECT involves a fundamental trade-off between image quality and radiation dose. Increased dose carries an increased risk of health detriment, while dose reduction inevitably leads to image degradation, which affects diagnosis accuracy [20]. As a result, under the principle of ALARA (as low as reasonably achievable) [1], there is a great deal of works devoted to the research of reducing radiation dose without compromising image quality. Over the past decade, deep

Supported by Natural Science Foundation of Guangdong under 2022A1515012379.

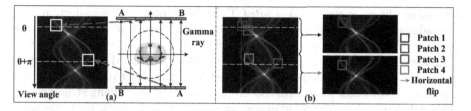

Fig. 1. The illustration of sinogram inner-structure related to parallel-beam projection geometry. (a) shows parallel-beam conjugate sampling pairs in the sinogram. (b) is the example that a patch 1 in the sinogram should have a corresponding conjugate patch 2 with a certain similarity after the horizontal flip.

neural networks (DNNs) have been widely used in low-dose image reconstruction [2,10,22]. For example, Chen et al. [2] were among the first to propose a residual encoder-decoder convolutional neural network (RED-CNN) in the image domain for low-dose CT imaging. Later, a generative adversarial network (GAN) with Wasserstein distance and perceptual similarity was introduced to image-domain denoising under low-dose scenario [22]. Li et al. [10] presented a LU-Net in the sinogram domain utilizing long short-term memory network (LSTM) to learn the sequence properties of adjacent projection angles in sparse-view sinogram. In addition, there have been innovative approaches that combine image- and sinogram-domain learning in sinogram synthesis problem [21,24]. Specifically, Zhou et al. [24] proposed a dual-domain data consistency recurrent network, dubbed DuDoDR-Net, for simultaneous sparse-view recovery and metal artifact reduction. They aimed to reconstruct artifact-free images by performing image- and sinogram-domain restoration in a recurrent manner. The critical reason behind the encouraging reconstruction performance achieved by the above CNN-based learning methods is mainly attributed to the image-specific inductive bias of CNNs when dealing with scale invariance and modeling local structures. However, CNNs have limitation in capturing global contextual information or long-range spatial dependency due to their fixed receptive fields. Recently, Transformers [3] have shown superiority for many computer vision tasks, which is bolstered by their global sensitivity and long-range dependency. Transformer-based approaches for low-dose image reconstruction [12,19,23] have been proposed and achieved promising results. For instance, Wang et al. [19] developed an Encoder-decoder Dilation network based on Token-to-Token (T2T) vision Transformer, dubbed CTformer, for low-dose CT image denoising.

Although the existing learning-based methods have made significant progress in improving image quality, they seldom consider inner-structure of sinograms, which may hinder further improvement on reconstruction quality. The parallel-beam geometry naturally generates certain special characteristics in the sinogram domain, which are also known as sinogram inner-structure. In particular, the underlying sinogram inner-structure consists of non-local and local levels, which correspond to the distant and local patches in Fig. 1(b), respectively. As shown in Fig. 1(a), a pair of conjugate sampling patches collect gamma photons along the same path during SPECT data acquisition. The projection data at conjugate patches are highly correlated rather than equivalent due to the presence of depth-dependent attenuation effect. The conjugate patches exhibit a high degree of similarity under horizontal flip, which is referred to as mirror symmetry. Remark that random noise produced during imaging process may break the above non-local-level inner-structure. Learning the underlying structure can effectively suppress noise and improve image quality. As a result, exploring sinogram inner-structure and designing a structure-aware model is significant for sinogram-domain restoration.

Graph neural networks (GNNs) are originally developed to process data with special topological structure, such as biological networks, social networks or citation networks [4,7]. Recently, Han et al. [5] proposed the Vision GNN (ViG), which stands out for its superior ability to extract non-local features, thereby achieving state-of-the-art results in high-level tasks such as classification [16] and detection [6]. Specifically, ViG partitions an image into smaller patches (nodes), which are then utilized to construct a graph. The K-nearest neighbors (KNN) algorithm is exploited to cluster and connect nodes with similar features, which facilitates effective information exchange. This mechanism exhibits substantial flexibility in handling images with repeated patterns. Building on this concept, in this study, we propose a novel network based on GNNs. The proposed network can effectively leverage the special inner-structure of sinograms, aiming to enhance the quality of images reconstructed in low-dose SPECT imaging scenarios. The main contributions are summarized as follows:

1. We propose a new GNN-based module, termed the Sinogram-structure-based Self-defined Neighbors GNN (SSN-GNN). The SSN-GNN module is designed to improve the quality of restored sinogram by modeling long-range dependencies between the image patch and its conjugate image patch.
2. We propose another GNN-based module, named Window-based K-Nearest Neighbors GNN (W-KNN-GNN). The W-KNN-GNN module is designed to model local topological feature information to enhance the representational ability of the network. In addition, the W-KNN-GNN adapts a window partition mechanism to reduce computational complexity.
3. Furthermore, we propose a novel network, GNN-CNN-UNet (GCUNet), for the task of low-dose SPECT sinogram restoration, which uses our proposed SSN-GNN and W-KNN-GNN modules. The performance of the GCUNet outperforms state-of-the-art reconstruction methods on the XCAT datasets. To the best of our knowledge, this research represents the first study to introduce GNNs into the task of low-dose SPECT reconstruction.

2   Method

2.1   Overall Architecture

Figure 2 illustrates the overall structure of the GCUNet, a U-shaped hierarchical network with skip connections between the encoder and the decoder. For

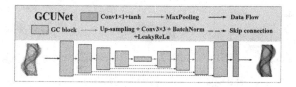

Fig. 2. The overall architecture of GCUNet includes a U-shaped hierarchical network with skip connections between the encoder and the decode; each layer consists of a Graph-Conv (GC) block.

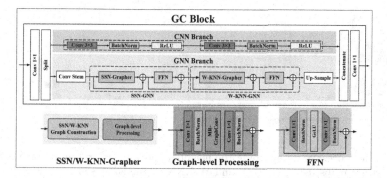

Fig. 3. The GC block, it mainly consists of a CNN branch and a parallel GNN branch.

the encoder, each layer consists of a Graph-Conv (GC) block. The feature map passes through the GC block to generate the hierarchical feature representations. Subsequently, a max pooling operation following the GC block is responsible for down-sampling. Inspired by U-Net [13], the decoder is symmetric, consisting of a GC block and an up-sampling operation. To compensate for the spatial information loss incurred during down-sampling, the extracted context features are merged with multiscale features from the corresponding encoder layers through skip connections. Figure 3 illustrates the structure of our proposed GC block, which mainly consists of a CNN branch and a parallel GNN branch. Initially, an arbitrary input feature map F is first passed through a 1×1 convolution layer, which changes the number of channels to enhance the representation ability. Subsequently, the feature map is evenly divided along the channel dimension, resulting in two feature maps, X1 and X2. Each of these feature maps is then fed into the CNN and GNN branches, respectively, to learn the structure of the different aspects of the sinogram. The outputs of the CNN and GNN branches are concatenated along the channel dimension and passed through another 1×1 convolution layer to effectively facilitate information fusion between CNN and GNN extracted features. Next, we elaborate two branches in the following parts.

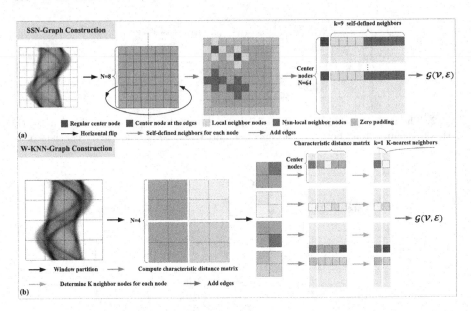

Fig. 4. The process of determining the neighbors of some central nodes on 8×8 and 4×4 feature maps using the SSN and W-KNN methods, as shown in Fig. 4(a) and Fig. 4(b), respectively. In SSN, K is a constant with a value of 9, while in W-KNN, k is a hyperparameter, and an example with k=1 is provided in the Fig.

2.2   GNN Branch

The GNN branch is specifically designed to extract both local and non-local structural features in the sinogram. As shown in Fig. 3, the GNN Branch is composed of two proposed GNN modules, including SSN-GNN and Window-KNN-based GNN (W-KNN-GNN) module. The SSN-GNN module comprises two components: the SSN-Grapher and the feed-forward network (FFN) modules. Similarly, the W-KNN-GNN module consists of W-KNN-Grapher and FFN modules. Each Grapher component is further divided into two sub-components, namely Graph Construction and Graph-level Processing. In more detail, for any given feature map X_2 of size $H \times W \times C$, we first partition the input feature map into N (the number of patches) non-overlapping patches through a shallow convolution operation (Conv Stem). Each patch is transformed into a feature vector $x_i \in \mathbb{R}^D$ with the feature dimension D, resulting in a new feature map $X_2' = H' \times W' \times D$. Here, $H' = H/P$, $W' = W/P$, where P represents the size of each patch. Next, the X_2' goes through the SSN-GNN and W-KNN-GNN modules, which serve to extract the inner-structure of the sinogram. Subsequently, we revert the feature dimension of X_2' back to that of X_2 using a up-sampling operation. The details of the SSN-GNN and W-KNN-GNN modules will be described in the following sections.

SSN-GNN. To obtain graph data, Han et al. [5] reshape the size of X_2' into $N \times D$ form, resulting in a set of feature vectors $X_2' = \{x_1, x_2, ..., x_n\}$ that are viewed as a set of unordered nodes denoted as $\mathcal{V} = \{v_1, v_2, ..., v_n\}$. Here, $N = H \times W$, and $x_i \in \mathbb{R}^D$. The next step is then to determine the neighbor nodes for each node. Han et al. [5] calculate the characteristic distance between nodes in a graph and select the neighbors based on the K-nearest neighbor (KNN) approach. Overall, this approach determines the neighbor nodes according to the difference of the characteristic distances, allowing the central node to learn as much structural information as possible from the neighbor nodes.

However, the KNN method does not take into account the inner-structure of parallel-beam sinogram and hence the nodes that are significantly geometrically relevant but have less similarity to the central node may not be selected as neighbors. For example, as shown in Fig. 1(b), we take patch 1 as the central node, and patch 3 is a local neighbor node of the central node. The underlying sinogram exhibits simple structure without abundant texture feature. In this case, the feature discrepancy between the central node and patch 4 is smaller compared to that between the central node and patch 2. As a result, the KNN method considers the sinusoidal structure of the patch 4 to be more similar to central node and prefers to treat patch 4 as the neighbor node rather than patch 2. However, patch 2 is the conjugate node of patch 1, which shows strong similarity to patch 1 after the horizontal flip operation, and should be regarded as the neighbor of the central node.

To address the above issues, we propose a sinogram-structure-based self-defined neighbors (SSN) method resulting in the selection of more geometrically relevant neighbor nodes more effectively. The process for determining the neighboring nodes of each node according to the SSN method is illustrated in Fig. 4(a). Considering the mirror symmetry arising from the projection geometry of parallel-beam imaging, we first horizontally flip the bottom half of the feature map and then manually select the top, bottom, left, and right nodes as local neighbors of each central node, their (including the central node) corresponding conjugate nodes are typically selected as non-local neighbor nodes. In this way, we manually determine fixed $k = 9$ neighbor nodes $v_j \in \mathcal{N}(v_i)$ for the central node v_i. To handle the nodes located at the edges of the feature map, we employ zero padding to fill in the missing neighbors, which maintains the desired number of neighbors for all nodes. Following the graph construction process in ViG [5], we add an edge e_{ji} directed from v_j to v_i for all $v_j \in \mathcal{N}(v_i)$. By following this procedure for all central nodes, we obtain a graph $\mathcal{G} = (\mathcal{V}, \mathcal{E})$.

In Graph-level Processing, the feature representation of each central node is updated by using graph convolution to aggregate the features of its neighbors. To achieve this, we adopt the max-relative graph convolution (MRConv) for graph convolution [9]. Following the Graph-level Processing, we use the feed-forward network (FFN) module as proposed in ViG [5] to improve the feature transformation capability and alleviate the over-smoothing phenomenon. The FFN module comprises two fully-connected layers followed by the GeLU activation function. By incorporating the SSN-GNN, the network can better learn the representation ability of the sinogram local and non-local structures. For a more profound exploration of the sinogram's inner-structure, the feature map undergoes inverse operation of horizontal flip before entering W-KNN-GNN module.

W-KNN-GNN. A sinogram is typically composed of a series of overlapping sinusoidal strips, each characterized by specific amplitudes and phases. This distinct arrangement endows the sinusoidal strips with unique local topological features. To effectively utilize these topological features and enhance the quality of the restored sinograms, we propose the W-KNN-GNN module. As shown in Fig. 4(b), the input feature maps are initially partitioned into non-overlapping windows. In each of these windows, we calculate the characteristic distance between nodes and then find the K nearest neighbor nodes for each node using the KNN method. Following a similar implementation to SSN-GNN module, we construct separate subgraphs for each window using its center node and corresponding neighbor nodes. These subgraphs then go through Graph-level Processing and FFN processes to learn the sinogram structures unique to each window. Upon completion, all windows are merged back together by the inverse operation of initial window partitioning. It is worth noting that the proposed W-KNN method leverages a localized windowing mechanism for subgraph construction and graph convolution, which is more computationally efficient than the primitive vanilla VIG approach that performs these operations over the full window.

2.3 CNN Branch

The inductive bias property of CNN facilitates characterization of local spatial information during feature extraction process. Therefore, we integrate a CNN branch within the proposed GCUNet architecture. As illustrated in Fig. 3, the CNN branch consists of two consecutive convolutional blocks. Each block is composed of a convolutional layer, followed by a batch normalization layer and ReLU activation function. In this study, we set the filter size of the convolutional layer to 3×3 with a stride of 1. We further employ a zero-padding scheme to preserve spatial dimension between input and output.

3   Experimental Results

3.1   Dataset

In this study, we utilize the SIMIND [11] Monte Carlo program in combination with the XCAT [14] anthropomorphic computer phantom to simulate head and torso bone scan data. We simulate a virtual SIEMENS E.CAM gamma camera, which is equipped with a low energy high resolution (LEHR) parallel-beam collimator. The detector orbit is circular, covering 360°, and the radius of rotation is set to 25 cm. The SPECT projection data consists of 128 projection views in a 128-dimensional detector array, with each detector element of size 2.2 mm. In addition, we use an 18 % main energy window centered at 140 keV. The gamma photons within this energy window are considered as primary or first-order scattered photons.

To generate the dataset, ten different XCAT adult phantoms are utilized. Different photon histories are used to generate projection data of varying dose levels. Specifically, for each phantom, a scale of 10^7 photon histories per projection view is simulated to obtain the normal-dose projection data. Low dose projection data is simulated at three distinct levels-nn10,nn5, and nn1-corresponding to 1/100, 1/200, and 1/2000 of the photon histories of the normal dose, respectively. Note that more photon histories indicate higher doses simulated, resulting in less noisy data. Furthermore, to investigate the robustness of the proposed GCUNet, the data with noise levels of nn10 and nn5 are used for training, while the data with a noise level of nn1 is used for testing.

Without loss of generality, two of the ten phantoms are randomly selected to constitute the test set, while the rest form the training set. Following the extraction of 2D sinograms from the projection data, the sinograms obtained from the normal-dose projection data serve as labels, and the sinograms obtained from the low-dose projection data serve as the input to the network. The training and test sets contain 1803 and 386 pairs of input/label sinograms, respectively. To enlarge the training set and avoid overfitting, two data augmentation methods are utilized, including horizontal and vertical flips. Furthermore, we utilize the maximum likelihood expectation maximization (MLEM) [15] reconstruction algorithm to reconstruct these sinogram data to generate training data for competing methods operating in the image domain. Meanwhile, the reconstructed images of normal-dose sinograms serve as the reference images.

3.2 Implementation Details

All experiments are conducted on a 3090 GPU card using the PyTorch framework. The proposed network is trained with the Adam optimizer for a total of 300 epochs with a batch size of 16. The learning rate is held constant at 2×10^{-4} for the initial 200 epochs, then decreased to 1×10^{-4} for the final 100 epochs. In the GNN branch, the patch size of the GNN modules is 4, and the numbers of windows and neighbor nodes in W-KNN graph construction are 4 and 9, respectively. The L1 loss function is employed to measure the difference between the restored and normal-dose sinograms. To evaluate the performance of the proposed GCUNet network, we reconstruct the restored sinograms using the MLEM reconstruction algorithm, and then compare with five different state-of-the-art methods, including TV-PAPA [8], UNet [13], RED-CNN [2], AD-Net [18], and CTfomer [19]. TV-PAPA is a traditional iterative reconstruction method that utilizes the low-dose sinogram. Among the deep-learning-based reconstruction methods, UNet serves as the method for the projection domain, while the remaining methods serve as methods for the image domain. Both RED-CNN and ADNet are CNN-based methods, with ADNet enhancing its denoising capability by incorporating an attention-guided mechanism. CTformer is a state-of-the-art Transformer-based method for Low-Dose Computed Tomography (LDCT) denoising task. All methods are well-tuned to achieve the best performance on our dataset. To evaluate the performance of the proposed method, we utilize three image quality metrics: peak signal-to-noise ratio (PSNR), structural similarity index measure (SSIM) and root mean square error (RMSE). For a fair comparison, 5-fold cross-validation scheme is applied to all the competing deep learning-based methods.

3.3 Comparison with Other Methods

A representative chest image is selected to visually evaluate the performance of different reconstruction methods, as shown in Fig. 5. Although all competing methods suppress image noise and restore bone structures to various degrees, a notable performance gap exists between the traditional TV-PAPA algorithm and the deep learning-based methods. For a more detailed comparison, the second row of Fig. 5 shows zoomed-in images of a region of interest (ROI). It is observed that while UNet, RED-CNN, and AD-Net can eliminate most noise and recover the relatively complete bone structure, they exhibit inaccuracies in contrast recovery along the bone structure and suffer from structural deformation to different extents, as pointed out by the arrows in Fig. 5. Compared to the first three methods, the CTformer produces a clearer and more accurate image of the bone structure. However, slight structural deformation remains, and its performance in terms of contrast recovery along the bone structure is subpar. In contrast, the proposed GCUNet method outperforms all others in terms of noise suppression, contrast restoration and structure preservation.

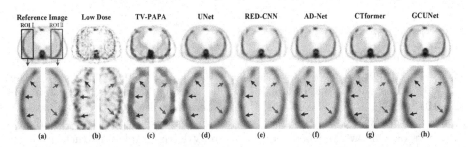

Fig. 5. The comparison of chest image in a male test phantom reconstructed by different methods. The first row is denoised images. The second row is zoomed regions in the red and blue boxes respectively. (a) Reference Image, (b) Low Dose, (c) TV-PAPA, (d) UNet, (e) RED-CNN, (f) AD-Net, (g) CTformer, (h) GCUNet(ours). (Color figure online)

Furthermore, we conduct a quantitative assessment of the performance across different methods, with results expressed as mean ± standard deviation (mean ± SD), as reported in Table 1. It can be observed that the proposed method outperforms other methods in all three image quality metrics, which is consistent with the results of the visual evaluation.

Table 1. Quantitative results (mean ± std) of different reconstruction methods.

Method	Metrics		
	PSNR↑	SSIM↑	NMSE↓
Low Dose	25.8121 ± 2.3487	0.8331 ± 0.0654	0.1226 ± 0.0254
TV-PAPA	27.7925 ± 3.6184	0.8812 ± 0.0572	0.1052 ± 0.0728
UNet	31.8918 ± 3.6454	0.9240 ± 0.0373	0.0366 ± 0.0170
RED-CNN	32.6533 ± 2.4186	0.9251 ± 0.0305	0.0257 ± 0.0066
AD-Net	32.9263 ± 2.3326	0.9313 ± 0.0206	0.0245 ± 0.0050
CTfomer	33.1775 ± 2.3897	0.9324 ± 0.0266	0.0233 ± 0.0075
GCUNet(Ours)	33.3571 ± 2.8828	0.9500 ± 0.0229	0.0227 ± 0.0081

3.4 Robustness of Proposed Method

Differences in individual human physiology and variations in imaging equipment used in different clinical settings might often result in acquired data with higher noise levels compared to those used during the training phase. This inconsistency presents a challenge to the robust performance of models that have been trained on specific datasets. A SPECT reconstruction method of clinical significance should exhibit high robustness to varying noise levels. Therefore, we evaluate the robustness of the proposed method and other competing methods under different noise levels. As shown in Fig. 6, a representative reconstructed images of the chest regions in the testing set are chosen to visually compare the performance of different methods. With increased noise levels in the test data, we can see that GCUNet outperforms other competing methods by effectively recovering the bone structure of reconstructed images and eliminating a significant portion of image noise and artifacts. This observation confirms GCUNet's robustness to noise, making it highly suitable for clinical applications.

Furthermore, we conduct quantitative evaluations of all methods with results listed in Table 2. Experimental results show that the proposed GCUNet outperforms the state-of-the-art methods across all metrics. This results are consistent with those of the visual evaluation and show that the proposed GCUNet might have strong robustness against noise.

3.5 Ablation Study

We conduct an ablation study to further verify the effectiveness of the proposed modules, with results in Table 3. We examine several model variants: 'Vanilla ViG-KNN' uses the original KNN-based graph construction of Vanilla ViG. 'GCUNet-CNN-Branch' keeps only the CNN branch of GCUNet. 'GCUNet-KNN' uses original KNN strategy for graph construction, while 'GCUNet-SSN' incorporates the SSN-GNN module. Lastly, 'GCUNet-SSN+W-KNN' integrates both SSN-GNN and W-KNN-GNN modules. Except for GCUNet-CNN-Branch,

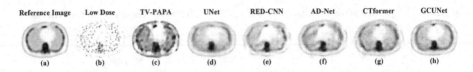

Fig. 6. Different methods compared the results of the robustness of chest images in a female test phantom. (a) Reference Image, (b) Low Dose, (c) TV-PAPA, (d) UNet, (e) RED-CNN, (f) AD-Net, (g) CTformer, (h) GCUNet(ours). (Color figure online)

Table 2. Quantitative results (mean ± std) of robustness performance of different reconstruction methods.

Method	Metrics		
	PSNR↑	SSIM↑	NMSE↓
Low Dose	16.1814 ± 2.3338	0.6962 ± 0.1095	1.1263 ± 0.2345
TV-PAPA	21.1565 ± 2.3290	0.7387 ± 0.0957	0.3610 ± 0.0862
UNet	26.8806 ± 2.4891	0.8615 ± 0.0610	0.0991 ± 0.0296
RED-CNN	26.6780 ± 2.3544	0.8476 ± 0.0634	0.1021 ± 0.0275
AD-Net	27.2632 ± 2.2695	0.8635 ± 0.0535	0.0884 ± 0.0026
CTfomer	27.4970 ± 2.1269	0.8719 ± 0.0601	0.0856 ± 0.0255
GCUNet(Ours)	28.0206 ± 2.6890	0.8821 ± 0.0516	0.0785 ± 0.0266

all models prefixed with 'GCUNet' are equipped with a global-local branch structure. Experimental results show that GCUNet-KNN improves PSNR by 0.54 dB and 0.51 dB over Vanilla ViG-KNN and GCUNet-CNN-Branch, respectively. When the SSN-GNN module replaces the KNN-strategy module, GCUNet-SSN enhances PSNR by 0.82 dB over GCUNet-KNN. Adding the W-KNN-GNN module to GCUNet-SSN results in an additional 0.11 dB improvement in PSNR. These results demonstrate the effectiveness of our dual-branch structure, SSN-GNN, and W-KNN-GNN modules.

Table 3. Ablation study of the proposed module.

Model	Metrics		
	PSNR↑	SSIM↑	NMSE↓
Vanilla ViG-KNN	31.8652 ± 2.3839	0.9208 ± 0.0232	0.0341 ± 0.0077
GCUNet-CNN-Branch	31.8918 ± 3.6454	0.9240 ± 0.0373	0.0366 ± 0.0170
GCUNet-KNN	32.4107 ± 3.4492	0.9394 ± 0.0297	0.0301 ± 0.0135
GCUNet-SSN	33.2395 ± 2.9444	0.9471 ± 0.0238	0.0236 ± 0.0087
GCUNet-SSN+W-KNN	33.3571 ± 2.8828	0.9500 ± 0.0229	0.0081 ± 0.0001

4 Conclusion

In this work, we propose a GCUNet network for low-dose SPECT image reconstruction. To enhance the quality of sinogram restoration, we propose two novel modules, SSN-GNN and W-KNN-GNN. Experimental results show GCUNet's superior performance on the XCAT dataset, outperforming several state-of-the-art methods. Furthermore, we assess GCUNet's robustness against noise, demonstrating its superior performance and potential for clinical applications. Finally, we conduct ablation studies to verify the effectiveness of each proposed module. In the future, we will focus on model compression to boost inference efficiency without compromising the quality of low-dose SPECT image reconstruction.

References

1. Brenner, D.J., Hall, E.J.: Computed tomography-an increasing source of radiation exposure. N. Engl. J. Med. **357**(22), 2277–2284 (2007)
2. Chen, H., et al.: Low-dose ct with a residual encoder-decoder convolutional neural network. IEEE Trans. Med. Imaging **36**(12), 2524–2535 (2017)
3. Dosovitskiy, A., et al.: An image is worth 16x16 words: transformers for image recognition at scale. arXiv preprint arXiv:2010.11929 (2020)
4. Hamilton, W., Ying, Z., Leskovec, J.: Inductive representation learning on large graphs. In: Advances in Neural Information Processing Systems 30 (2017)
5. Han, K., Wang, Y., Guo, J., Tang, Y., Wu, E.: Vision gnn: an image is worth graph of nodes. arXiv preprint arXiv:2206.00272 (2022)
6. Khalid, F., Javed, A., Ilyas, H., Irtaza, A., et al.: Dfgnn: an interpretable and generalized graph neural network for deepfakes detection. Expert Syst. Appl. **222**, 119843 (2023)
7. Kipf, T.N., Welling, M.: Semi-supervised classification with graph convolutional networks. arXiv preprint arXiv:1609.02907 (2016)
8. Krol, A., Li, S., Shen, L., Xu, Y.: Preconditioned alternating projection algorithms for maximum a posteriori ect reconstruction. Inverse Prob. **28**(11), 115005 (2012)
9. Li, G., Muller, M., Thabet, A., Ghanem, B.: Deepgcns: can gcns go as deep as cnns? In: Proceedings of the IEEE/CVF International Conference on Computer Vision, pp. 9267–9276 (2019)
10. Li, S., Ye, W., Li, F.: Lu-net: combining lstm and u-net for sinogram synthesis in sparse-view spect reconstruction. Math. Biosci. Eng. **19**(4), 4320–40 (2022)
11. Ljungberg, M., Strand, S.E., King, M.A.: Monte Carlo calculations in nuclear medicine: applications in diagnostic imaging. CRC Press (2012)
12. Luthra, A., Sulakhe, H., Mittal, T., Iyer, A., Yadav, S.: Eformer: edge enhancement based transformer for medical image denoising. arXiv preprint arXiv:2109.08044 (2021)
13. Ronneberger, O., Fischer, P., Brox, T.: U-Net: convolutional networks for biomedical image segmentation. In: Navab, N., Hornegger, J., Wells, W.M., Frangi, A.F. (eds.) MICCAI 2015. LNCS, vol. 9351, pp. 234–241. Springer, Cham (2015). https://doi.org/10.1007/978-3-319-24574-4_28
14. Segars, W.P., Sturgeon, G., Mendonca, S., Grimes, J., Tsui, B.M.: 4d xcat phantom for multimodality imaging research. Med. Phys. **37**(9), 4902–4915 (2010)
15. Shepp, L.A., Vardi, Y.: Maximum likelihood reconstruction for emission tomography. IEEE Trans. Med. Imaging **1**(2), 113–122 (1982)

16. Shi, P., Guo, X., Yang, Y., Ye, C., Ma, T.: Nextou: efficient topology-aware u-net for medical image segmentation. arXiv preprint arXiv:2305.15911 (2023)
17. Thrall, J.H., Ziessman, H.: Nuclear medicine: the requisites. Mosby-Year Book, Inc., p. 302 (1995)
18. Tian, C., Xu, Y., Li, Z., Zuo, W., Fei, L., Liu, H.: Attention-guided cnn for image denoising. Neural Netw. **124**, 117–129 (2020)
19. Wang, D., Fan, F., Wu, Z., Liu, R., Wang, F., Yu, H.: Ctformer: convolution-free token2token dilated vision transformer for low-dose ct denoising. Phys. Med. Biol. **68**(6), 065012 (2023)
20. Wells, R.G.: Dose reduction is good but it is image quality that matters. J. Nucl. Cardiol. **27**, 238–240 (2020)
21. Wu, W., Hu, D., Niu, C., Yu, H., Vardhanabhuti, V., Wang, G.: Drone: dual-domain residual-based optimization network for sparse-view ct reconstruction. IEEE Trans. Med. Imaging **40**(11), 3002–3014 (2021)
22. Yang, Q., et al.: Low-dose ct image denoising using a generative adversarial network with wasserstein distance and perceptual loss. IEEE Trans. Med. Imaging **37**(6), 1348–1357 (2018)
23. Zhang, Z., Yu, L., Liang, X., Zhao, W., Xing, L.: TransCT: dual-path transformer for low dose computed tomography. In: de Bruijne, M., et al. (eds.) MICCAI 2021. LNCS, vol. 12906, pp. 55–64. Springer, Cham (2021). https://doi.org/10.1007/978-3-030-87231-1_6
24. Zhou, B., Chen, X., Zhou, S.K., Duncan, J.S., Liu, C.: Dudodr-net: dual-domain data consistent recurrent network for simultaneous sparse view and metal artifact reduction in computed tomography. Med. Image Anal. **75**, 102289 (2022)

A Two-Stage Whole Body Bone SPECT Scan Image Inpainting Algorithm for Residual Urine Artifacts Based on Contextual Attention

Pingxiang Zhou[1], Gang He[1,2](✉), Zhengguo Chen[2], and Ling Zhao[2](✉)

[1] School of Computer Science and Technology, Southwest University of Science and Technology, Mianyang, China
ganghe@swust.edu.cn
[2] NHC Key Laboratory of Nuclear Technology Medical Transformation (Mianyang Central Hospital), Mianyang, China
zhaolingssaa@163.com

Abstract. Whole body bone SPECT scan is an important inspection method to detect early malignant tumor or evaluate bone metastases. As the specificity of the SPECT images is significantly affected by noise, metal artifacts, or residual urine, which may mislead the doctor's diagnosis. To address this issue, a two-stage whole body bone SPECT scan image inpainting algorithm for residual urine artifact based on contextual attention is proposed in this paper. In the first stage, TransUNet framework is utilized to segment the whole bone SPECT images according to the location of the residual urine. In the second stage, the artifact in the segmented image is recovered by a contextual attention network, which can effectively deal with residual urine artifact and improve the quality of the recovered image. Besides, an extended image dataset named as EX-BS-90K consisting more than 90k whole body bone SPECT scan images is presented in this paper and used to train the proposed two-stage inpainting model. The PSNR and SSIM are calculated as evaluation metrics of the proposed algorithm and the experimental results demonstrates that the proposed method can effectively inpaint the original bone structure in the artifact region, which can increase the specificity of the SPECT images. Meanwhile, the proposed method are compared with recent works and the comparison results demonstrates that our method provides better solution for whole body bone SPECT images inpainting problems. The code and dataset are available: https://github.com/Zhoupixiang/Two-stage-BS_inpainting.

Keywords: Whole body bone SPECT scan image · Image inpainting · Contextual Attention · Residual urine artifacts

1 Introduction

In whole body bone scanning, the radiopharmaceutical injected is distributed throughout the body via the circulatory system, with most of the drug being

Q. Liu et al. (Eds.): PRCV 2023, LNCS 14437, pp. 497–508, 2024.
https://doi.org/10.1007/978-981-99-8558-6_41

absorbed in the bones and then expelled from the body through the kidneys and urine. However, in clinical diagnosis, there can sometimes be an issue where residual urine causes the formation of a bright area on the bone SPECT scan images [6]. This bright area may obscure or confuse the images of the pelvis and lower abdominal bones, thereby impeding the detection of possible lesions in these areas. Consequently, this may result in misdiagnoses or missed diagnoses in patients. Until now, there has been no dedicated research addressing the problem of obstructions in bone scan images. Therefore, this paper proposes a two-stage whole-body bone SPECT scan image inpainting algorithm for residual urine artifacts based on contextual attention. This method is intended to resolve the problem of obstructions in bone scan images caused by residual urine and further enhance the quality of bone scan images.

Currently, common inpainting algorithms are divided into two categories: traditional image inpainting algorithms and image inpainting algorithms based on deep learning [18]. Previous image inpainting tasks have relied on traditional methods such as interpolation-based [1], diffusion-based [17,20], and patch-based [2,14] techniques. Most of the traditional methods are based on mathematical knowledge, using known information around missing regions in the image to derive complex formulas, and then iteratively update the image according to the formulas, so that the recover of missing regions slowly reaches relatively good results [22]. Generally, these methods can achieve good results on images with small damaged areas, simple structures, and repeated textures. However, since it predominantly relies on the inherent properties of a singular image, it fails to convey meaning and lacks the logical structure of the restored semantics. [11]. Aiming at the limitations of traditional methods, a method based on deep learning is proposed for image inpainting. Convolutional Neural Networks (CNN) and Generative Adversarial Networks (GAN) based on deep learning have achieved better results in generating recovered images that conform to semantic features. The CNN network is commonly used to transform the conditional image generation problem into a high-level recognition and low-level pixel synthesis problem and make an encoder-decoder network, and then train it together with the GAN network to generate reasonable content in highly structured images [9]. Pathak [15] et al. proposed a method combining the encoder-decoder network structure and GAN. The context encoder is used to learn image features and generate prediction maps for image missing areas. The GAN part judges the rationality of the prediction maps, but the recover results are false images with blurred edges. Yu [26] et al. proposed a two-stage feedforward generation network. The missing image first passes through a dilated convolution to obtain features as a convolution filter, matches the generated information with the known information, and passes through the scene attention channel and convolution respectively. The channels are integrated, and the results generated by them are combined to obtain the recover result. [16] Therefore, based on the design idea of Yu [26], this paper proposes a two-stage whole body bone SPECT scan image inpainting algorithm for residual urine artifacts. The main contributions of this paper are as follows.

1) BS-80K dataset has been expanded and renamed as EX-BS-90K.
2) The preprocessing of the bone SPECT scan image has been carried out, with an algorithm based on Weber's law proposed to enhance the bone SPECT scan images.
3) The location of the residual urine artifact is located using the TransUNet network. It is segmented into a 256×256 bone SPECT scan image, and the segmented image is input into the inpainting network based on the context attention mechanism for inpainting.
4) To facilitate the diagnosis at other positions, the cropped bone SPECT scan images are restored to a size of 256×1024 bone SPECT scan images after recovery.

The entire flow chart is shown in Fig. 1.

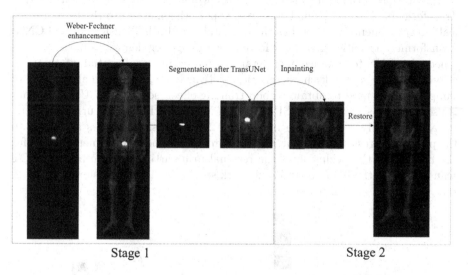

Fig. 1. Flow chart of bone SPECT scan images inpainting.

2 Methods

2.1 TransUNet

Existing medical images segmentation methods mainly rely on fully convolutional neural networks with U-shaped structures [25]. One of the most typical U-shaped structured networks, U-Net [25] consists of a symmetric encoder-decoder with skip connections. The network is designed to reduce the loss of spatial information caused by the downsampling process. Detail preservation is enhanced by skip connections. And it has achieved great success in many medical image segmentation tasks and occupies a dominant position. Following this idea, many variants of U-Net have emerged, such as Res-UNet [24], U-Net++ [27], for various medical image segmentation. In the aforementioned networks, due to the inherent limitations of convolution operations, it is difficult for pure CNN approaches to learn explicit global and long-range semantic information interactions. To overcome such limitations, existing research suggests establishing a self-attention mechanism based on CNN features. Transformers are powerful in modeling global context. But the pure Transformer method also has the problem of feature loss. Therefore, the method based on the combination of CNN architecture and Transformer module has received more and more attention. For example, literature [3] utilizes a self-attention mechanism to enhance the convolutional operator, concatenating a convolutional feature map emphasizing locality with a self-attention feature map capable of modeling the global. DETR [4] uses a traditional CNN skeleton network to learn the two-dimensional representation of the input image, and the encoder and decoder are composed of Transformers. Among them, TransUNet applies this method to the field of medical image segmentation and has the advantages of both Transformer and CNN. Transformer, as a strong encoder for medical image segmentation tasks, is combined with CNN to enhance finer details by recovering local spatial information. also due to the network characteristics of TransUNet, only a small number of samples can be used to obtain a good training effect, so the TransUNet network is used here to segment bone SPECT scan images with residual urine artifacts. It can accurately segment the image into 256×256 size pictures according to the position of residual urine artifacts. It provides a good experimental basis for the removal of the occlusion of the residual urine artifact in the bone SPECT scan images in the later stage. Its network structure diagram is shown in Fig. 2.

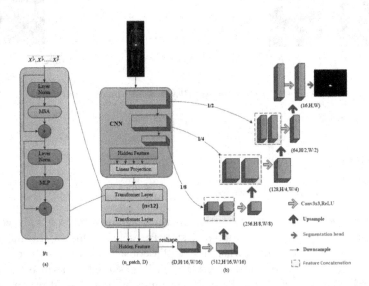

Fig. 2. Overview of the framework. (a) schematic of the Transformer layer; (b) architecture of the proposed TransUNet.

2.2 Image Inpainting with Contextual Attention

The generative image inpainting with contextual attention is divided into two stages. The first stage is a simple convolutional network. The loss value reconstruction loss is generated by continuously recovering the missing area, and the recovered result is relatively blurred [21]. The second stage is the training of the content-aware layer. The core idea is: to use the features of the known image patches as the convolution kernel to process the generated patches to refine the blurred recover result. It is designed and implemented in this way: Use the convolution method to match similar patches from the known image content, find the patch that is most similar to the area to be recovered by doing softmax on all channels, and then use the information in this area is deconvolved to reconstruct the patched area. Based on the above new deep generative model-based method, not only new image structures can be generated, but also the surrounding image features can be well used as a reference to make better predictions. For whole body bone SPECT scan images with residual urine artifacts, the location of residual urine artifacts may be in different positions according to individual differences, so it is necessary to use a content-aware network to recover bone SPECT scan images according to the structure of different human bodies. Its structure diagram is shown in Fig. 3.

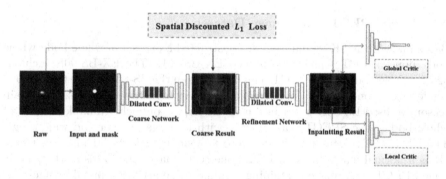

Fig. 3. Overview of the images inpainting with contextual attention framework.

2.3 Preprocessing of Bone SPECT Scan Images

The contrast of the bone SPECT scan images is too dark to see the bones clearly. To solve the problems, this paper proposed a bone SPECT scan images enhancement algorithm, which can improve the contrast between the background and the bone. The flow chart of the bone SPECT scan images enhancement algorithm is shown in Fig. 4. The enhanced bone SPECT scan images is shown in Fig. 5.

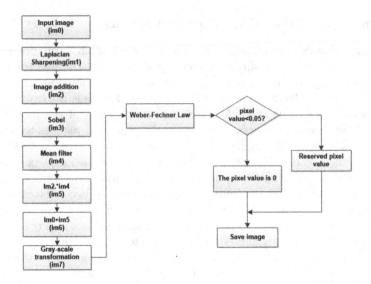

Fig. 4. Bone SPECT scan images enhancement flow chart.

3   Experiments

3.1   Bone SPECT Scan Image Dataset

Based on the BS-80K [8] dataset, we constructed a dataset of whole body whole body bone SPECT scan image named EX-BS-90K. The EX-BS-90K includes 82,544 whole body bone SPECT scan images from the BS-80K. More than 10,000 whole body bone SPECT scan image provided by other Hospitals. The main reason we used BS-80K is because it is the largest publicly available dataset of whole body bone SPECT scan images, with its large number and wide range. The EX-BS-90K dataset is divided into two small datasets D1 and D2, where D1 is a normal whole body SPECT bone scan image, and D2 is a whole body bone SPECT scan image containing residual urine artifacts, metal artifacts, or noise. D1 and D2 are used to train the proposed two-stage inpainting model.

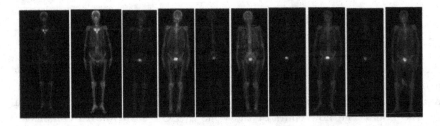

Fig. 5. Comparison of bone SPECT scan images before and after enhance.

3.2   Stage 1: Image Segmentation

To solve the problem that the bone information is not visible due to the occlusion of the residual urine artifacts, image inpainting is required to remove the occlusion of the residual urine artifacts to achieve the purpose of displaying the bones. For whole-body bone SPECT scan images, the size is 256×1024, to reduce the impact of noise in other areas on training, and it is impossible to accurately locate the place with residual urine artifact during recover. Due to the outstanding segmentation effect of TransUNet on medical images, TransUNet is used for image segmentation here, which can accurately locate the position and location of the residual urine artifacts in the bone SPECT scan images scope. The specific operation is as follows: First, according to the position of the residual urine artifacts, manually mark the whole body bone SPECT scan images to make a data set. Since the TransUNet network can achieve better results with a small amount of data, first manually mark the mask to make a training data set, and then train the TransUNet network, finally put the whole body bone scan picture into the trained model, get the corresponding mask picture, and finally get the segmented picture according to the mask picture, the effect is shown in Fig. 6.

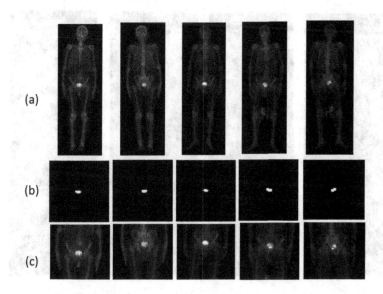

Fig. 6. Bone SPECT scan images segementation. (a) Enhanced bone SPECT scan images (b) TransUNet segmentation images (c) The segmented images.

3.3   Stage 2: Image Ipainting

For traditional inpainting methods, most of them are based on local pixel information. When the area to be repaired is across the important structure of the image, the inpainting result may lead to structural discontinuity. The residual urine artifacts in the bone scan image will appear near the pelvis, and the traditional inpainting method will destroy the structural information of the pelvis in the image or the effusion removal is not ideal. Deep learning-based image inpainting methods, such as contextual attention generation models, can better deal with these issues, such as restoring large areas of missing, maintaining structural continuity, creating new textures, and so on. The main process of repairing the whole bone SPECT scan image is as follows: the image size of 256×256 segemented by the TransUNet network is made into a data set, and then the data set is put into the trained inpainting network for recover, as shown in Fig. 7. The residual urine can be seen to be removed, restoring the obscured bone structure and allowing it to visually align with the surrounding area. At the same time, we used objective criteria such as PSNR and SSIM to evaluate the inpainting effect. In order not to affect the diagnosis of other parts of the bone scan image, the 256×256 size image obtained after image inpainting is restored to the original 256×1024 size image by using the image inpainting algorithm proposed in this paper.

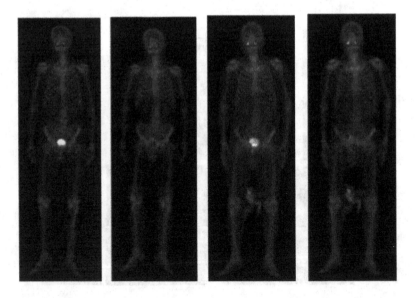

Fig. 7. Comparison of bone SPECT scan images before and after recover.

4   Results

4.1   Experimental Detail

Experimental hardware environment: Processor (CPU) is Intel Core i7-9700K @3.60 GHz octa-core, memory is 16 GB, The graphics card (GPU) is NVIDIA GeForce RTX 2080 Ti 22 GB experimental platform software environment is TensorFlow 1.1.5, CUDA 10.0. In the training process, Adam optimizer was used to optimize the parameters, and the learning rate was 0.0001, the first-order momentum was 0.5, and the second-order momentum was 0.999.

4.2   Evaluation Indicator

To compare recovered images with original images, PSNR (Peak Signal-to-Noise Ratio) [13] was introduced

$$PSNR = 10 \times \lg \left(\frac{MAX_I^2}{MSE} \right) \tag{1}$$

where : MAX_I^2 is the maximum value of the image pixel. The MSE (Mean Square Error) [10] is one of the indicators to measure image quality. Its expression is as follows:

$$MSE = \frac{1}{H \times W} \sum_{i=1}^{H} \sum_{j=1}^{W} (P_1(i-j) - P_2(i-j))^2 \tag{2}$$

where : P_1 represents the predicted value and P_2 represents the true value.

The SSIM (Structure Similarity Index Measure) [12] is another measure of image quality, which is usually used to measure the similarity of two digital images.

$$SSIM(x,y) = l(x,y) \cdot c(x,y) \cdot s(x,y) \tag{3}$$

where : $l(x,y)$,$c(x,y)$,$s(x,y)$ are calculated for brightness, contrast and similarity of structure, respectively. This paper selects the most commonly used PSNR and SSIM are used as objective evaluation indexes of the recover results of this method and the comparison method.

4.3   Evaluation

Subjective Evaluation. As can be seen from Fig. 8 and Fig. 9, the two traditional methods, Criminisi [5] and Fast Marching Method(FMM) [19], did not achieve the intended effect of residual urine artifacts removal. While Co-attentive Multi-task Convolutional Neural Network(CMCNN) [23] did remove part of the obstruction, an occlusion issue persists. Although GAN [7] has achieved better effusion removal effects, it also removes bone information in the process. The method proposed in this paper not only removes effusion occlusion but also retains bone information, and the surrounding noise is clearly removed.

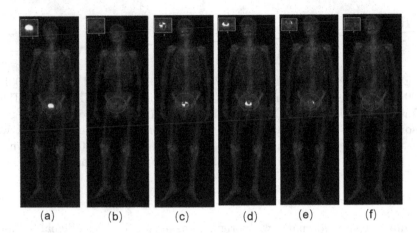

(a)   (b)   (c)   (d)   (e)   (f)

Fig. 8. The effect comparison of different algorithms. (a) Original image (b) Ours (c) Criminisi algorithm (d) FMM algorithm (e) GMCNN (f) GAN.

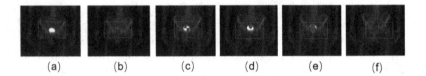

(a)   (b)   (c)   (d)   (e)   (f)

Fig. 9. The effect comparison of different algorithms. (a) Original image (b) Ours (c) Criminisi algorithm (d) FMM algorithm (e) GMCNN (f) GAN

Table 1. Objective Evaluation

Algorithms	Ours	Criminisi	FMM	GMCNN	GAN
PSNR	26.7	24.8	25.8	25.6	26.3
SSIM	0.87	0.83	0.76	0.84	0.88

Objective Evaluation. According to the comparison of PSNR and SSIM of these methods, the proposed algorithm has highest PSNR and SSIM values compared to latest works. As shown in Table 1.

5   Conclusion

To solve the problem of residual urine occlusion in SPECT images, a two-stage whole body bone SPECT scan image inpainting algorithm for residual urine artifacts based on contextual attention is proposed in this paper. In the first stage, the TransUNet network was used to segment the whole bone SPECT images based on the location of residual urine. In the second stage, the residual urine artifacts are removed and the image quality is improved through the context attention network. Comparative experiments were carried out in two aspects of subjective evaluation and objective evaluation, and the effectiveness of the proposed method was verified by comparing it with latest SPECT image inpainting methods. The recovered regions are clearer and the image quality is higher after inpainting by the proposed method. The proposed algorithm is beneficial to assist doctors to examine the pelvic area that cannot be seen due to residual urine occlusion and hence improve the diagnostic efficiency, which would provide a promising solution for residual urine artifacts in SPECT images.

Acknowledgement. This work was financially supported by Sichuan Science and Technology Program (No. 2020YFS0454), NHC Key Laboratory of Nuclear Technology Medical Transformation (MIANYANG CENTRAL HOSPITAL) (Grant No. 2021HYX024, No.2021HYX031).

References

1. Alsalamah, M., Amin, S.: Medical image inpainting with rbf interpolation technique. Inter. J. Adv. Comput. Sci. Appli. **7** (2016)
2. Alsaleh, S.M., Aviles-Rivero, A.I., Hahn, J.K.: Retouchimg: fusioning from-local-to-global context detection and graph data structures for fully-automatic specular reflection removal for endoscopic images. Comput. Med. Imaging Graph. (2019)
3. Bello, I., Zoph, B., Vaswani, A., Shlens, J., Le, Q.V.: Attention augmented convolutional networks. In: 2019 IEEE/CVF International Conference on Computer Vision (ICCV), pp. 3285–3294 (2019)
4. Carion, N., Massa, F., Synnaeve, G., Usunier, N., Kirillov, A., Zagoruyko, S.: End-to-end object detection with transformers. ArXiv abs/ arXiv: 2005.12872 (2020)
5. Depeng Ran, Z.W.: An improved criminisi algorithm for image inpainting, pp. 159–163 (2015)
6. Gao, JG, S.C.Y.W.Z.A.: 18f-psma-1007 analysis of influencing factors of nuclide retention in bladder by pet/ct imaging. Mod. oncology **30**(4504–4508) (2022)
7. Goodfellow, I.J., et al.: Generative adversarial nets (2014)
8. Huang, Z., et al.: Bs-80k: the first large open-access dataset of bone scan images. Comput. Biol. Med. **151 Pt A**, 106221 (2022)
9. Laube, P., Grunwald, M., Franz, M.O., Umlauf, G.: Image inpainting for high-resolution textures using CNN texture. Synthesis (2018). https://doi.org/10.2312/cgvc.20181212
10. Li, JQ., Zhou, J.: Learning optical image encryption scheme based on cyclic generative adversarial network. J. Jilin Univ. (Eng. Edn.) **51**(1060–1066) (2021). https://doi.org/10.13229/j.cnki.jdxbgxb20200521
11. Liu, HW., Y.J.L.B.B.X.: A two-stage cultural relic image restoration method based on multi-scale attention mechanism. Comput. Sci. **50**(334–341) (2023)
12. Liu, Z., Li, J., Di, X., Man, Z., Sheng, Y.: A novel multiband remote-sensing image encryption algorithm based on dual-channel key transmission model. Sec. Commun. Netw. (2021)
13. Lv, JW., Q.F.: Nighttime image defogging by combining light source segmentation and linear image depth estimation. Chin. Opt. Soc. **15**(34–44) (2022)
14. Milne, K., Sun, J., Zaal, E.A., Mowat, J., Agami, R.: A fragment-like approach to pycr1 inhibition. Bioorganic Med. Chem. Lett. **29**(18) (2019)

15. Pathak, D., Krähenbühl, P., Donahue, J., Darrell, T., Efros, A.A.: Context encoders: feature learning by inpainting. In: 2016 IEEE Conference on Computer Vision and Pattern Recognition (CVPR), pp. 2536–2544 (2016)

16. Peng, Y.F., G.L.W.G.: Second-order image restoration based on gated convolution and attention transfer. Liquid Crystal Display **38**(625–635) (2023)

17. Rodríguez, T., Spies, R.: A bayesian approach to image inpainting. Mecánica Computacional (2014)

18. Tao, YS., G.B.L.Q.Z.R.: Wheat image restoration model based on residual network and feature fusion. Trans. Agricult. Machinery **54**(318–327) (2023)

19. Telea, A.C.: An image inpainting technique based on the fast marching method. J. Graph. Tools **9**, 23–34 (2004)

20. Vlasánek, P.: Fuzzy image inpainting aimed to medical images (2018)

21. Wang, L.: Research on image restoration algorithm based on adversarial generation technology and application of biomedical image (2022)

22. Wang, Q.: Medical image inpainting with edge and structure priors. Measurement **185**, 110027–110027 (2021)

23. Wang, Y., Tao, X., Qi, X., Shen, X., Jia, J.: Image inpainting via generative multi-column convolutional neural networks. ArXiv abs/ arXiv: 1810.08771 (2018)

24. nan Xiao, X., Lian, S., Luo, Z., Li, S.: Weighted res-unet for high-quality retina vessel segmentation. 2018 9th International Conference on Information Technology in Medicine and Education (ITME), pp. 327–331 (2018)

25. Yin, X., Sun, L., Fu, Y., Lu, R., Zhang, Y.: U-net-based medical image segmentation. J. Healthcare Eng. **2022** (2022)

26. Yu, J., Lin, Z.L., Yang, J., Shen, X., Lu, X., Huang, T.S.: Generative image inpainting with contextual attention. In: 2018 IEEE/CVF Conference on Computer Vision and Pattern Recognition, pp. 5505–5514 (2018)

27. Zhou, Z., Rahman Siddiquee, M.M., Tajbakhsh, N., Liang, J.: UNet++: a nested u-net architecture for medical image segmentation. In: Stoyanov, D., et al. (eds.) DLMIA/ML-CDS -2018. LNCS, vol. 11045, pp. 3–11. Springer, Cham (2018). https://doi.org/10.1007/978-3-030-00889-5_1

Author Index

Printed in the United States
by Baker & Taylor Publisher Services